W9-BMA-114

SATISH K SHARMA
829 N MENARD AV
CHICAGO IL 60651

STRENGTH AND STRUCTURE
of
ENGINEERING MATERIALS

PRENTICE-HALL INTERNATIONAL, INC., *London*
PRENTICE-HALL OF AUSTRALIA, PTY, LTD., *Sydney*
PRENTICE-HALL OF CANADA, LTD., *Toronto*
PRENTICE-HALL OF INDIA (PRIVATE) LTD., *New Delhi*
PRENTICE-HALL OF JAPAN, INC., *Tokyo*

STRENGTH AND STRUCTURE of ENGINEERING MATERIALS

N. H. POLAKOWSKI

Professor of Metallurgical Engineering
Illinois Institute of Technology

E. J. RIPLING

Director of Research
Materials Research Laboratory, Inc.

PRENTICE-HALL, INC.

Englewood Cliffs, New Jersey

Current printing (last digit):
10 9 8 7 6 5

Library of Congress Catalog No. 64–21746
Printed in the United States of America
85179C

PREFACE

Except for the basic subjects of mathematics, physics, and chemistry, a course generally known as Strength of Materials is probably the most common in engineering curricula. There is universal interest in this subject because all branches of engineering require a broad understanding of the behavior of structural materials under load. Such breadth is not provided by the traditional textbooks, however, which are almost exclusively devoted to detailed numerical techniques applied to shafts, beams, and columns. More recently written texts, rather than correcting this deficiency in scope, have generally added a new level of sophistication to basically the same range of problems.

This book promotes an alternative approach to teaching a first course in strength of materials. It presents an introduction to the mechanics of deformation and fracture of solids, combined with a discussion of the physical meaning of each of the important mechanical properties. Concurrently, an attempt is made to correlate the bulk engineering properties with atomic scale mechanisms. The correlation is not always complete, however, since our knowledge of mechanical phenomenology and methods of describing its various aspects is ahead of an understanding of their origin.

The circumstance that not all aspects of engineering importance can as yet be satisfactorily interpreted in terms of atomic models does not detract from the importance of pointing out such relationships whenever possible. However, it often calls for simplification in order to show the idea in a tractable fashion. As an example, in the chapter concerned with macroscopic permanent deformation, the phenomenon of plastic flow is considered obvious and the major part of the discussion is concerned with the practical problem of how plastic yielding is affected by stress state. In the following chapter, dealing with micromechanics of plastic deformation, it is necessary to backtrack to simple structures and uniaxial stress conditions.

It is impractical to attempt to cover in a book of this type every conceivable facet of this very large field without sacrificing others or resorting to undesirable abridgements. As a criterion of what topics were to be included, the authors chose those that their personal experience indicated to be of most importance to all engineering students—both those for whom this text would serve as an introduction to their specialty and those for whom such a course is their only contact with the subject. Basic information on the structure of metals, ceramics, plastics, and glass is considered essential, and an entire chapter is devoted to this subject matter. Because dislocations in crystalline structures are the bridge between elastic and plastic deformation and are needed to explain a variety of macroscopic observations, they are discussed in some detail. A short section on rheological models is included: they are helpful in interpreting damping, creep, and certain properties of polymers. Similarly, residual stresses and their effects are thought to be of sufficient general importance to be included, as are the concepts of fracture mechanics. On the other hand, the equations of compatibility, membrane analogies, and all but the simplest stability problems, as well as the theory of metal forming and plastics conversion, are omitted since they are more appropriate in subsequent, specialized courses.

The text was specifically designed so that a reader with an average knowledge of calculus, physics, and some chemistry could readily understand it. This level was chosen to make the book useful to students in their third to sixth semester. In addition, it should be useful to practicing engineers who wish to supplement or up-date their knowledge in this field.

Many of the areas covered in the text are currently the subject of very active research. In these cases, general principles were emphasized rather than details, since the latter are changing so rapidly. Likewise, discussions of mechanical properties were devoted primarily to the interpretation of test results rather than describing procedures that can be readily found in various manuals and specifications.

N. H. P.

E. J. R.

CONTENTS

LIST OF PRINCIPAL SYMBOLS

a crystal lattice constant
A area
b Burger's vector
b scalar value of b
BHN Brinell hardness number
c lattice constant (h.c.p. crystals)
d diameter
D damping capacity
DPN Vickers hardness number
e elongation, per cent
E Young's modulus
f frequency (of vibration)
F force (load)
F_e critical Euler force
G shear modulus
$\mathscr{G}$ strain-energy-release-rate (fracture)
$\mathscr{G}_c(\mathscr{K}_c)$ fracture toughness
h height, thickness
ΔH activation energy
I moment of inertia
I_1, I_2, I_3 stress invariants
I.F. internal friction
J_1, J_2, J_3 strain invariants
k coefficient of thermal expansion
k, l, m directional cosines of a plane
k_y Petch's constant
K bulk modulus
K_1 theo. stress concentration factor
$\mathscr{K}$ stress intensity factor
$K_c, \mathscr{K}_c, (\mathscr{G}_c)$ fracture toughness
l length
m Meyer's index
M bending moment
MHN Meyer's hardness number
n strain-hardening exponent
N number (of)
p pressure
P load (force)
P_1, P_2, etc. creep parameters
q reduction of area, % fatigue notch sensitivity
q_n notch ductility
Q statical moment of surface
r radius
R gas content
R_B Rockwell "B" hardness
R_C Rockwell "C" hardness
S stress, stress amplitude
S_n notch strength
S_u tensile strength
t time
T temperature; torque
$\mathcal{T}$ transition time
u (v, w) displacement components
$\dot{u}$ $(\dot{v}, \dot{w})$ velocity components
V volume; shearing force
W (strain) energy
W_d strain energy of volume change
W_p modulus of resilience

W_s shear strain energy
W_a surface energy
y deflection (of beam)
Z section modulus (bending)
Z_p section modulus (torsion)

α, β, γ, etc. angles
ε normal (engineering) strain
$\dot{\varepsilon}$ strain rate
$\bar{\varepsilon}$ natural (log) strain
$\varepsilon_1, \varepsilon_2, \varepsilon_3$ principal strains
ε_a mean strain
ε_e effective strain
ε_v volume strain
$\in$ general strain
η shear viscosity
ϕ tensile viscosity

γ shear strain
$\dot{\gamma}$ shear strain rate
γ_Ω octahedral shear strain
λ Lame's constant
μ coefficient of friction
ν Poisson's ratio
ρ density
σ normal stress
σ_e effective stress
σ_f fracture stress
σ_h hydrostatic pressure
σ_F fatigue limit
σ_o yield stress (uniaxial)
σ_0' yield stress (constrained)
σ_{pl} proportional limit
σ_r radial stress
σ_θ hoop (tangential) stress
τ shear stress
τ_0 yield stress in shear

part I

I

INTRODUCTION

Each of the engineering materials with which we are familiar has attained its present importance because it possesses some unique property or combination of properties that makes it best suited for certain uses. Although the particular quality that is of paramount interest in selecting the material for any specific purpose depends on the use to which it is to be put, in the majority of cases a consideration of mechanical properties is essential. Frequently, satisfactory mechanical performance at a reasonable cost or sufficiently low weight is almost solely the basis for making a selection. Certainly this is the case in building a bridge or an aircraft frame. Even in those applications where mechanical considerations appear to be secondary, they generally cannot be neglected. For example, the high conductivity of copper makes it ideal for electric power transmission, yet, in outdoor use, the copper must be able to carry not only its own weight between supports, but, at times, also that of heavy ice layers as well.

The variation of mechanical properties found in different materials is unusually wide. Glass, metal, and rubber are representative of the range of this variation. The distinguishing features of these materials can be brought out clearly by comparing the amount that they extend as a continuously increasing pull is applied to a strand of each of these as shown schematically in Fig. 1.1. The macroscopic or phenomenological effect of loading these specimens is conveniently described by a diagram showing the load vs. change in length, with the latter conventionally taken as the independent variable, Fig. 1.2.

As the load is increased, starting from zero, all three materials gradually undergo an increase in length that is proportional to the applied load. The

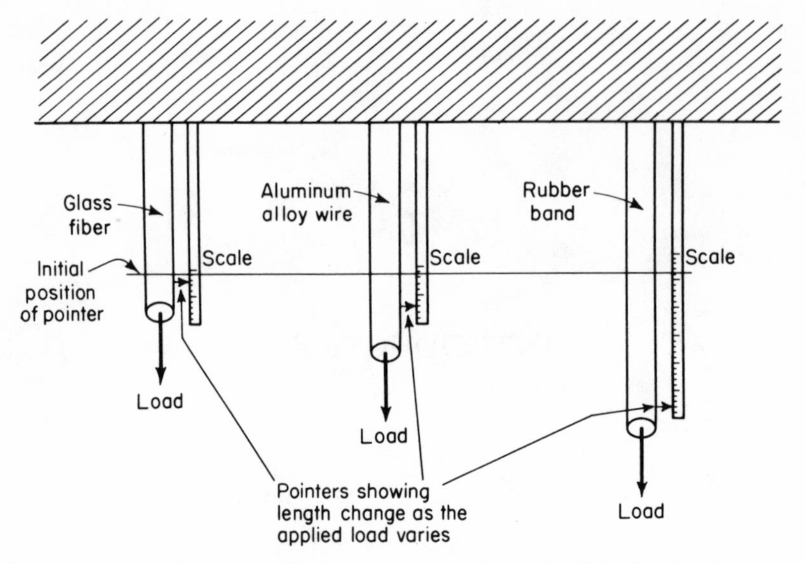

Fig. 1.1 Extension of glass fiber, aluminum alloy wire, and rubber band under load. (Schematic representation.)

length changes of the glass and aluminum observed during this period are of the order of a fraction of a per cent of the original length but of hundreds of per cent for the rubber. When the applied loads are removed, the specimens revert to their original dimensions. Deformations possessing this transient characteristic are termed elastic.

The basic difference between the glass and aluminum alloy is in the manner in which the elastic portions of the appropriate curves in Fig. 1.2 terminate. Glass ends its elastic extension by suddenly fracturing at point F. Aluminum, and most other metals, on the other hand, do not abruptly fracture at the end of their elastic range, but instead, starting at point A,

Fig. 1.2 Load vs. change in length diagram for the glass fiber, aluminum alloy wire and rubber band shown in Fig. 1.1.

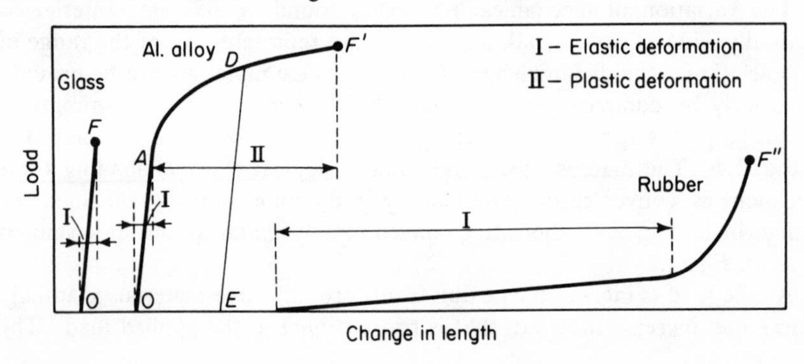

continue to deform at an increased rate. Deformation beyond point A is permanent (i.e., it does not vanish on release of load) and is termed plastic. Eventually increasing loads will also break the aluminum alloy wire, but by this time its length will have increased considerably.

The example above involved the action of a single load on a simply shaped body. In general, reversible transient dimensional changes, as well as permanent deformations, can be caused by any number of loads acting on bodies of arbitrary configurations.

Because the relationships between applied loads and elastic dimensional changes are basically simple, they were subjected rather early to successful mathematical analysis. As a result numerous problems of various degrees of complexity were solved, and general solution methods were devised that represent the cornerstone of the most important calculations in engineering design.

The application of the theory of elasticity to the design of structures and machines is mainly the province of civil and mechanical engineering. Yet, since the very act of measuring a material's properties, and certainly the interpretation of the data, requires techniques of the theory of elasticity, some elementary background on this subject is essential.

Another important application of the laws of elasticity is in evaluating internal forces that structural members at times contain even though no external loads are acting. These "locked in" or residual forces are, on occasion, purposely introduced to develop a desirable strengthening effect or they may arise spontaneously during manufacturing and assembly—in which case they distort the structure or even adversely affect its resistance to fracture. Such widely different items as cannon barrels and all-glass doors have their strength improved by residual forces purposely introduced during manufacture.

DEFORMATION OF CRYSTALLINE MATERIALS

The mechanism of elastic and plastic deformation of metals has been explained by the use of atomic models. Metals are composed of grains or crystals, each of which consists of atoms arranged in space according to simple, regular, three dimensional patterns. These patterns resemble a deep painter's scaffold, and are termed space lattices.

In Fig. 1.3a atoms, represented as spheres, are stacked so as to form a three dimensional space lattice which is commonly found in metals. The regularity of the lattice is apparent; the entire structure consists of plane layers of contacting spheres with many layers piled one on top of the other. The particular atom arrangement shown in Fig. 1.3a is representative of a so-called close-packed structure, meaning that each volume element contains the largest possible number of spheres. In order to achieve this, each of these spheres must not only touch all the neighbors in its own layer but also must

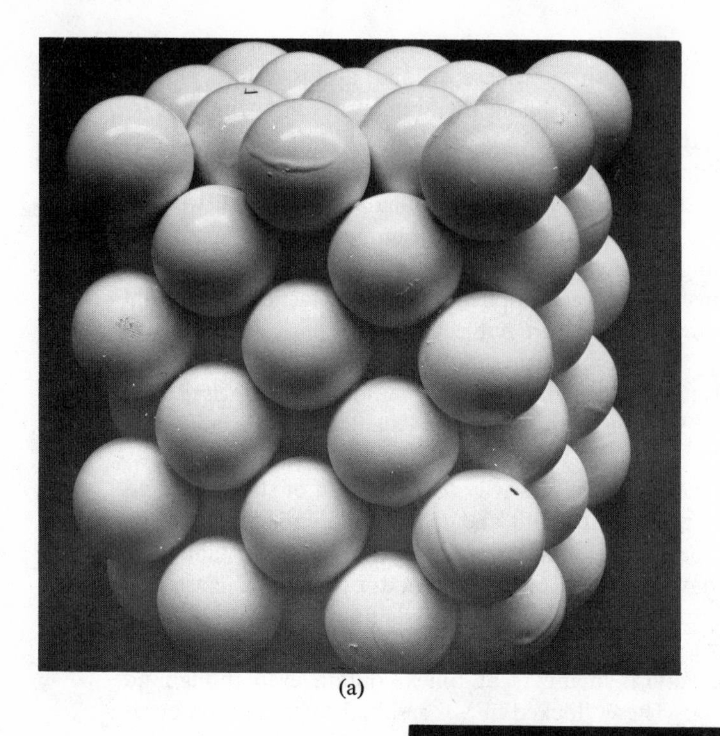

(a)

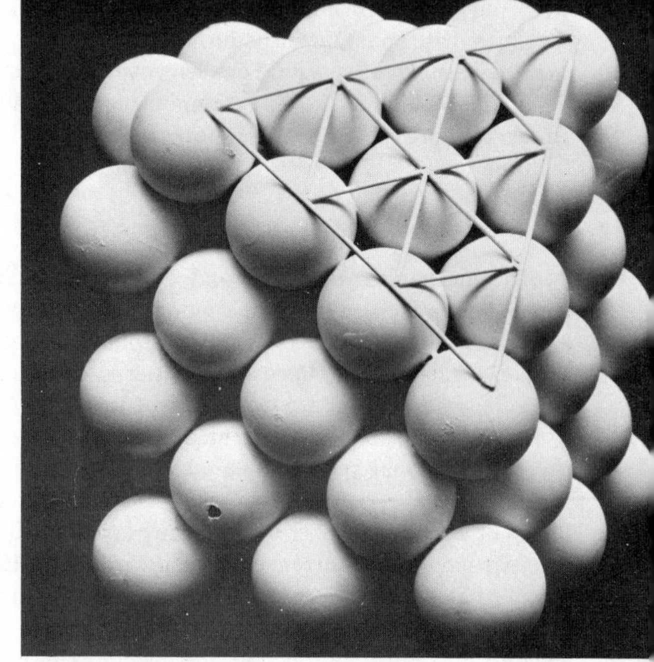

(b)

Fig. 1.3 (a) Three-dimensional space lattice formed by close packing of spheres; (b) Same as (a) with a corner pyramid of atoms removed to show close packing on a diagonal plane.

fit into the depressions formed between the spheres in the layers immediately above and below it as well. This close packing is not obvious from examining the external faces in Fig. 1.3a. However, it becomes apparent by cutting the stack with a plane that is equally inclined to all three edges as shown in Fig. 1.3b.

Deformation, from a submicroscopic viewpoint, consists of forced movement of atom layers with respect to their neighbors, resulting in an over-all distortion of the array. Because the distortions are relatively simple, the change in atom locations produced by external forces is not difficult to visualize.

When an increasing load is applied to a crystal, the atoms are forced to gradually move away from their equilibrium positions. Because of the interference of its neighbors, an individual atom generally cannot alter its position unless an entire layer of atoms moves along with it. Considering the cut plane of atoms in Fig. 1.3b, which is drawn schematically in Fig. 1.4a, one can see that these layers can either slide over each other or move apart. The successive location of atoms during sliding is depicted in Fig. 1.4, assuming the indicated direction of the applied force. The initial small load causes a uniform shift of each of the atoms to a new position as shown in Fig. 1.4b. The summation of all these small movements of the atoms produces the macroscopic elastic deformations between points O and A in Fig. 1.4e.

As long as the atom displacements are sufficiently small, the energy used to move the atoms is stored within the lattice so that the atoms spring back to their equilibrium positions on release of load. This is the mechanism that accounts for the disappearance of elastic strains upon removal of external forces as represented by loading and unloading along OA in Fig. 1.4e. In spite of the apparent identity of all points within the lattice, differences do exist that enable certain layers to slide easier than others. If one of these layers reaches a position such that its atoms can more easily occupy new equilibrium locations than their original position (Fig. 1.4c), they will spontaneously continue to the new location and remain there on release of load. Now one portion of the crystal is permanently displaced one atomic distance with respect to the other part (Fig. 1.4d). Repeating this process many times produces successive points along branch AB of the force-extension curve. This mechanism of deformation is termed slip. The elastic "tilt," shown in (b) and (c), persists in the lattice over AB, and the condition shown in (d) occurs only when the load is completely released at point C.

This description indicates that atom movement during slip occurs as "blocks" rather than as "sheets" of monoatomic thickness. Indeed, microscopic evidence of such block movement is obtained on polished single crystals in wire form. An originally smooth surface on the wire, after plastic deformation, takes on a stepped appearance. Not only the step thicknesses

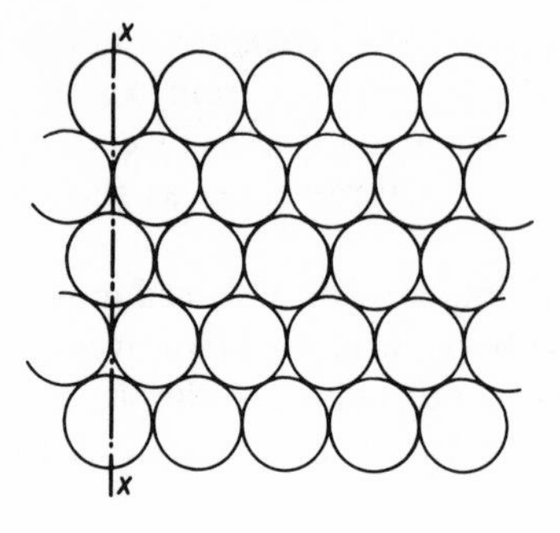

(a) Location of atoms in the absence of load

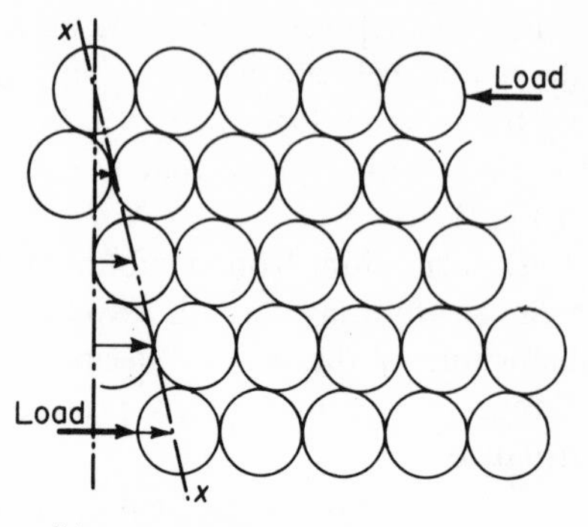

(b) Location of atoms when a small load is applied producing only elastic deformation

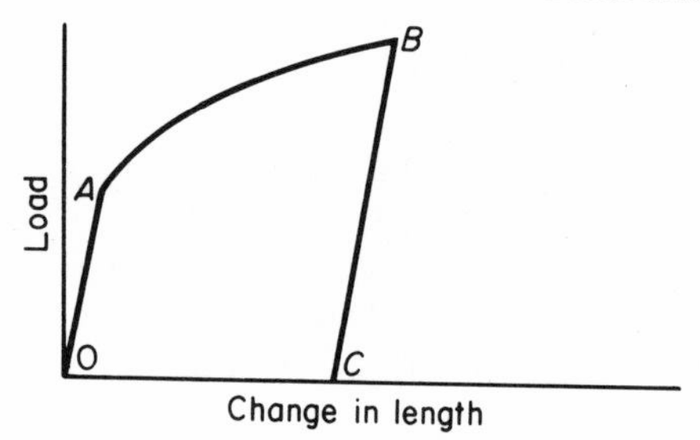

(e) Macroscopic load vs extension diagram resulting from atom movements in (b), (c) and (d)

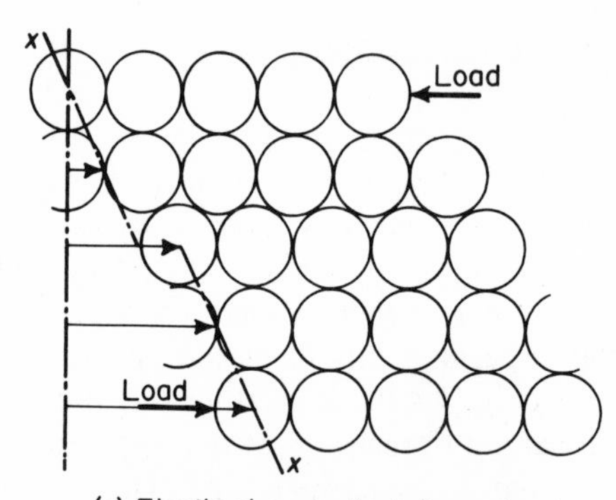

(c) Elastic plus plastic deformation. Notice that the third layer of atoms (from top) has shifted more than a half atomic diameter with respect to the second layer

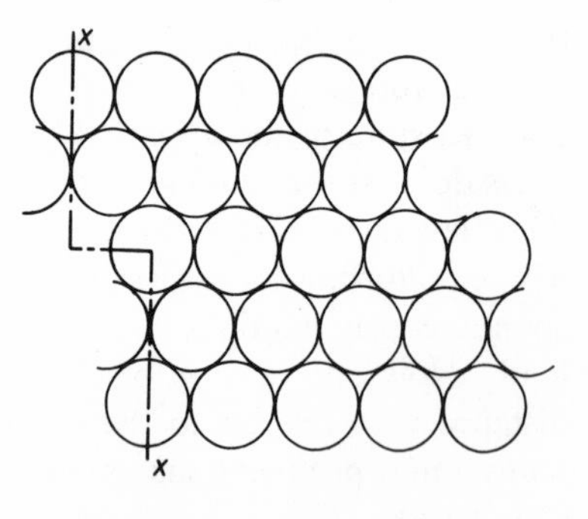

(d) On release of load, the third layer of atoms moved to a new location producing plastic flow while all others returned to their original positions

Fig. 1.4 Schematic representation of atom movements during loading. (Light arrows indicate displacements of layers of atoms under load.)

but also their lengths are many times greater than an atomic diameter, showing that slip does not occur over a single atomic distance (as in Fig. 1.4d), but instead packs of atom layers move as semirigid entities over a large number of atomic distances. Excepting for the over-all change of specimen shape, the lattice, after permanent deformation, is indistinguishable from the undeformed lattice.

It is possible to compute theoretically the force required to push a layer of atoms "over the hump" into the next equilibrium position, and hence to initiate plastic flow. The force holding the atoms together is readily obtained from elastic deformation measurements, and the distance through which the atoms must move to spring into new sites instead of returning to their original sites is also known. These data were first used in the 1920's to calculate the strength of crystals, but it was found that the calculated strength values based on block movements of atoms would exceed those actually measured by orders of magnitude. The source of the difficulty was easily traceable to the necessity of moving the entire block (considered as an essentially rigid unit) over the neighboring block and hence overcoming the interatomic binding forces between each pair of adjacent atoms all at once. In order to account for this very large discrepancy and to bring the calculated and actual strength figures into agreement, it was suggested that certain defects called dislocations are present in metal lattices. This particular type of crystal defect would allow plastic deformation to occur by a consecutive rather than by simultaneous atom movement. The underlying principle in this case is akin to the mechanism used by a worm to move across a sidewalk. When the worm travels, its body does not move as a rigid unit. Instead, while most of its body remains still, the head extends and becomes narrow. This narrow extended part of the worm moves along its length as a wave and, as it passes each segment of the worm, that part moves forward.

Not only can blocks of atoms slide with respect to each other over a large number of atomic distances to produce plastic deformation, but consecutive layers of atoms might shift in specific direction by a constant fraction of an interatomic distance as well. This type of deformation is known as twinning and operates as though one part of the crystal moved in a somewhat hingelike fashion with respect to the undeformed portion.

The twinned section of a crystal is a mirror image of its undeformed part, with the twinning plane serving as the reflecting surface as shown schematically in Fig. 1.5. Twinning produces more modest deformations than slip can.

The question of whether, under a system of external forces, a solid body will deform in an elastic or plastic manner depends on a variety of factors including the inherent properties and geometrical configuration of the material as well as the magnitude and orientation of the applied forces. Criteria were established to determine from basic data which of these will

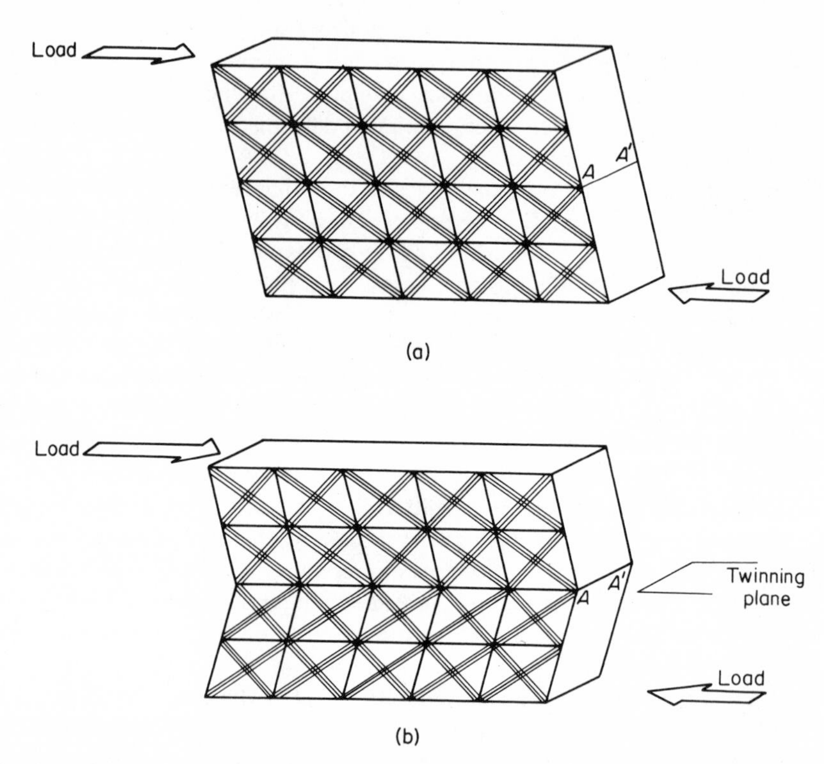

Fig. 1.5 Schematic diagram of lattice orientation change during twinning. (a) Before deformation; (b) After deformation (*A-A'* is edge of twinning plane).

occur, at least in metals. If the body deforms plastically, its deformation becomes the subject of the mathematical theory of plasticity which is an outgrowth of the theory of elasticity. In order to make the mathematics of plasticity problems solvable, metals must often be assumed to be perfectly homogeneous and isotropic (i.e., to have identical properties in all directions), an assumption which is not quite realistic in most cases. Despite these simplifications theoretical plasticity is a useful means for analyzing forming operations like drawing or rolling and for elucidating various perplexing features of the behavior of metals in service.

DEFORMATION OF RUBBERY MATERIALS

In the preceding discussion of metals, it was seen that atoms can move relative to their neighbors by only a very small distance if they are expected to return to their original positions when the external force is released. After deformations slightly beyond a certain maximum (which is still very small in absolute terms), the atoms will find it easier to occupy new sites. Hence the extension which metal crystals can undergo, while remaining elastic, is limited to one per cent or less. How, then, might one rationalize rubber elasticity where a rubber band may be stretched by several hundred per cent and on release of load return approximately to its initial dimensions? This spectacular difference between rubber and metals results because these two materials are held together by different types of interatomic or intermolecular forces. Unlike metals, two types of bonding forces are active in rubber and other polymers. Very strong interatomic forces bond the atoms into molecular units while much weaker van der Waals' forces join the molecules together to form the macroscopic substance, in this case a rubber band. When a force is applied to such substances, some of the intermolecular bonds are broken allowing the molecules to move about. This force is much too small to disjoin the individual molecules. In addition, the molecules in most polymers—and in rubber in particular—do not move as rigid bodies but are readily able to change their shapes under load.

Because of the complex configurations that polymer molecules may develop, the movement of individual atoms during deformation is less understood for these than for metals. Rubber is made up of long coiled molecules, a section of which might look as shown in Fig. 1.6a. A sufficiently long section of the molecule will curl up to such an extent that it probably would double back on itself.

The remarkable elasticity of rubber results from the high flexibility of its molecules. Each carbon atom which is joined to its neighboring carbon atom by a single bond has one degree of rotational freedom; i.e., these atoms can rotate with respect to their neighbors, with the only restrictions being that the center-to-center distance between adjacent atoms does not change and the angle between three atoms' centers is always about 109°. Consequently the molecule shown in Fig. 1.6a would not be expected to be planar; indeed the single bonds could assume any angle with the plane of the page so long as adjacent bonds lie on the surface of a cone with an apex angle of 109°. The most probable shape of a rubber molecule in the unstrained condition is a randomly curled up configuration such as shown schematically in Fig. 1.6b. It is this tendency of the molecules to take on a random shape that is a major factor in producing rubbery properties.

In addition to the great flexibility of the individual molecules and the weak intermolecular bonds, an interlocking of the molecules at a few points along their lengths is required so that they cannot slip apart under the applied load. The appearance of a group of molecules (of the type shown in Fig. 1.6b) locked together at a few points are shown schematically in Fig. 1.6c. On the application of a load to the rubber band, the very weak intermolecular forces are continuously broken and reformed while the kinked and curled up molecules are aligned in the direction of the applied load by rotation of

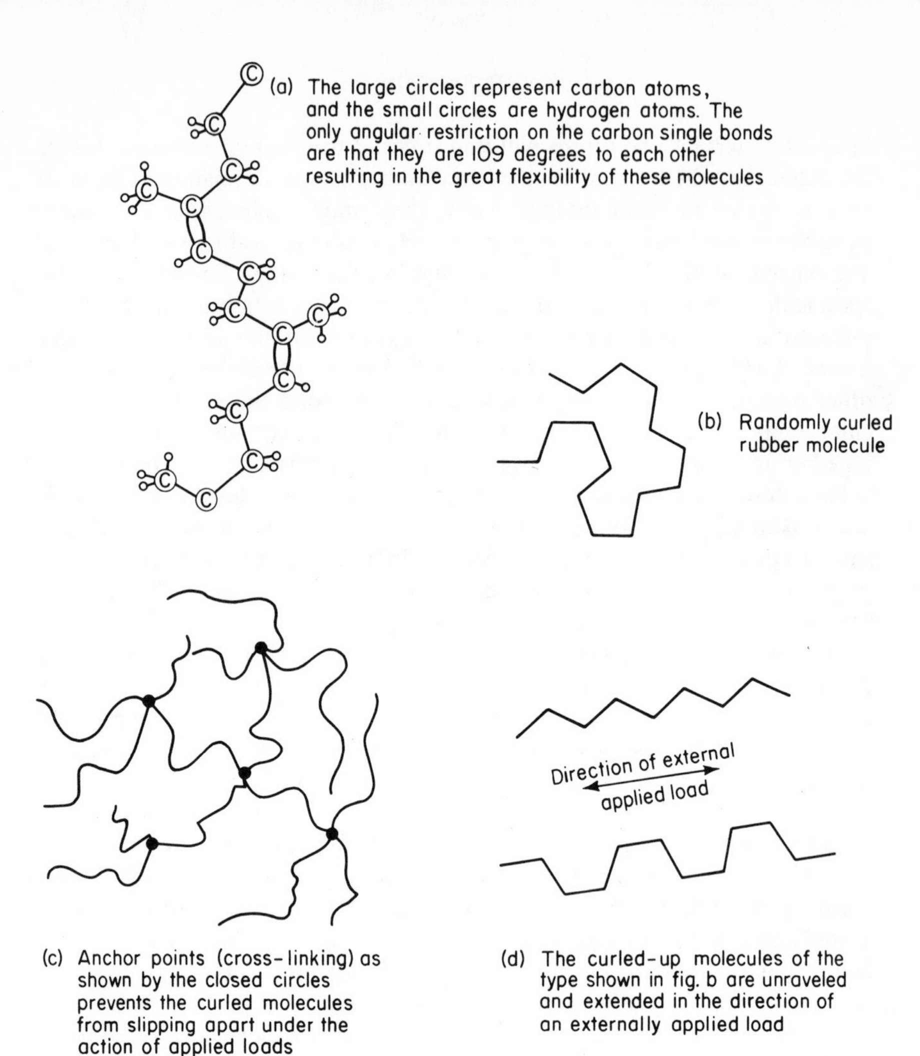

Fig. 1.6 Schematic diagrams of rubber molecules.

the single bonded carbon atoms. On stretching, the curled up molecule in Fig. 1.6b may assume the configuration shown, in Fig. 1.6d. Since the randomly curled up molecule is the equilibrium shape it recurls on releasing the load, accounting for the enormous elasticity of rubber.

DEFORMATION OF GLASS

The behavior of glass is distinctive because its atomic structure differs from that of either metals or rubber. Actually glass is a supercooled liquid so its deformation under load is similar to that of any other liquid; however this similarity becomes obvious only at sufficiently high temperatures. Liquids deform readily because many of the possible sites in the lattice are not occupied, leaving voids of an atomic size. These holes or vacancies account for the lower density of a liquid substance compared with that of the solid. Forces, even as small as the liquid's own weight, may cause its deformation by a continuous movement or diffusion of atoms into the holes. As the temperature of the liquid is increased, additional holes are formed in its space lattice (hence its density decreases). The thermal energy, present as vibration of the atoms, increases so that it becomes continuously easier for the atoms to move about by occupying the vacancies. This leaves behind new holes for additional diffusion, and so forth. The increased fluidity of liquids with increasing temperature is thus explained. Lower temperatures produce the opposite effect; i.e., fewer holes and slower motion of the atoms. At room temperature the mobility of glass atoms is so sluggish that the type of flow typically associated with liquids virtually ceases. Of course the atoms, even at room temperature, can move slightly and then snap back to their original sites much as metal atoms do when a force is applied and removed. It is not possible to apply forces large enough to produce atomic diffusion at ambient temperatures without causing the glass to break.

The ease with which glass breaks results from the presence of minute cracks even in macroscopically sound glass. These microcracks catastrophically grow and result in a brittle fracture from forces too small to move atoms from their initial sites into nearby holes within a reasonable period of time. They reduce the strength of glass to a fraction of its theoretically expected value. Experiments in which the formation of such cracks in thin glass filaments was temporarily suppressed resulted in strength values an order of magnitude higher than that found in ordinary glass fibers.

FRACTURE

Let us now consider the phenomenon of fracture in metals. In aluminum and most other metals, fracturing is a more complex phenomenon and occurs less spontaneously than in glass. The crack that eventually leads to failure frequently grows relatively slowly, and the appearance of the fracture surface varies for different metals and deformation conditions. For example, the separation surface formed in the aluminum wire of Fig. 1.1 will show, after fracture, either a pair of inclined planes or a "cup" and matching "cone" as in Fig. 1.7a and 1.7b. If the fracture is planar, it will be inclined at about 45° to the wire axis. Fractures that occur after appreciable permanent deformation are called ductile and are often found in metal forming operations like wire drawing or bending of bars and plates around too small a radius. Routine testing of ductile metals for strength also terminates normally in this type of failure.

On the other hand, many materials such as hardened steel and various cast alloys behave in a brittle fashion. They break with no or very little preceding plastic deformation and, when tested in the manner shown in

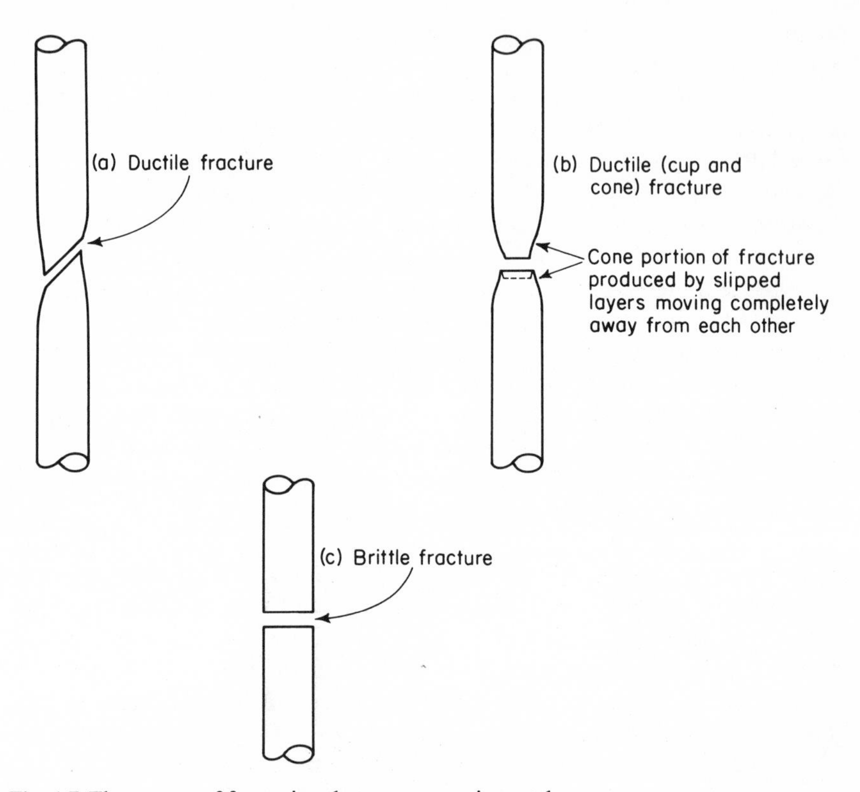

Fig. 1.7 Three types of fracturing that may occur in metals.

Fig. 1.1, they generally produce a separation surface normal to the specimen axis (Fig. 1.7c).

One of the most fascinating aspects of fracture is the fact that many common metals may break in either a brittle or ductile manner. The type of fracture depends on a variety of circumstances of which method of loading, temperature, and shape of the part are the most important. The problem of brittle fracturing in normally ductile materials has received much attention since World War II. Many ships constructed of mild steel plates during that period exhibited dangerous cracks with a brittle appearance. In some cases the entire vessel suddenly broke in two. This type of failure is not limited to ships, but it is of great concern in machines and equipment exposed to low temperatures, dynamic forces, and other adverse loading conditions.

Understanding of the motions of individual atoms or even layers of atoms during fracture is still incomplete. One reason for this is that fracturing can be the termination of many different types of atomic movements. For example, a ductile fracture arises when slipping that started on many planes eventually becomes concentrated on one, resulting in two divorced blocks.

An examination of the fracture surface would reveal that the metal within individual grains simply slipped apart. It is this type of atomic movement that results in the macroscopic fracture appearance shown in Figs. 1.7a and b.

Microscopic examination of a brittle fracture (Fig. 1.7c) generally shows an entirely different type of separation. The appearance of the individual grains suggests that this break results when the normal distance between layers of atoms exceeds some critical value. The interatomic bonding forces then could no longer balance the external forces so that fracturing (cleavage) follows with no evidence of slip.

Some structures fracture from excessive slipping alone while others show only the normal separation of layers of atoms. In still other cases, a combination of both fracture types is found.

Although it is difficult to rigorously rationalize metal fracturing on an atomic scale, much has been learned about the practical manifestations of fracturing on a microscopic scale. So far, in describing atomic layer separation, only fracturing within the metal crystal or grain was implied. Actually, certain types of fracture originate and propagate between grains rather than through the body of the grain. Failures occurring in turbine parts, or other equipment operating at elevated temperatures for long times, are generally of this type. Grain boundary separations also occur in steel under the combined action of applied forces and certain corrosive media. Intergranular fracturing can also result from the precipitation of brittle phases at the boundaries or as the result of certain heat treatments that form a semicontinuous network of a soft and weak phase around a hard matrix.

In the discussion of fracture, a metal's behavior under a continuously increasing load was considered. Some machine parts, and even complete large structures such as bridges or cranes, may fracture brittlely as a result of relatively small repeated loads. These failures are attributed as "fatigue" and they may result from forces that would be expected to produce only elastic deformation. Ideally, data on the behavior of a metal under continuously increasing loads should make it possible to predict the metal's resistance to the action of repetitive loading. This is not yet possible, however, with the present state of our knowledge. Although brittle fracturing produced by continuously increasing loads has some aspects that are similar to cracking by alternating loads, in some ways the two are different.

Understanding of the reaction of metals and other materials to forces is still not quite satisfactory in all respects so that many different types of mechanical tests are required. In general these tests simulate certain ideal simplified service conditions and are designed so as to be amenable to analysis. For example, compression testing and open die forging are quite similar, while the fatigue test simulates the service conditions of automotive axles or crankshafts. The application of the results to conditions different from the actual test requires intelligent interpretation through a combination of analytical and semiempirical methods.

part II

DEFINITIONS OF STRESS AND STRAIN

2

ANALYSIS OF STRESS

DEFINITION OF STRESS

In the Introduction, the qualitative relation between applied forces and deformation of three different materials in the form of slender rods was discussed. No mention was made of their size, and yet it is obvious that a given load, when applied to a thin rod, will cause a large amount of stretching and possibly fracturing, while this same load applied to a thick one would result in only a small extension. Apparently it is not the force alone that determines the amount of stretch or the incidence of fracture, but rather the amount of force acting on each unit of cross section. The term *stress* is used to define the ratio of load to the cross-sectional area over which it acts.

Although the stress is the same on each normal cross section of the rods discussed above, it generally will vary from point to point within a loaded body. For example, the force acting on a single ball in a ball bearing is distributed over a very small area at the point of contact between the ball and race so that the stress there is high. However, it diminishes rapidly on approaching the equatorial cross section. Hence, a specific stress value normally will only refer to a single point (or at times a group of points).

To define the stress at a specific point, a very small rectangular parallelepiped is constructed about it as shown in Fig. 2.1. The portion of the total applied load acting on this element is ΔF, and the area of face '*a*' is ΔA. The stress, S, acting at the point in the direction of the heavy arrow is thus defined as:

$$S = \lim_{A \to 0} \frac{\Delta F}{\Delta A} \tag{2.1}$$

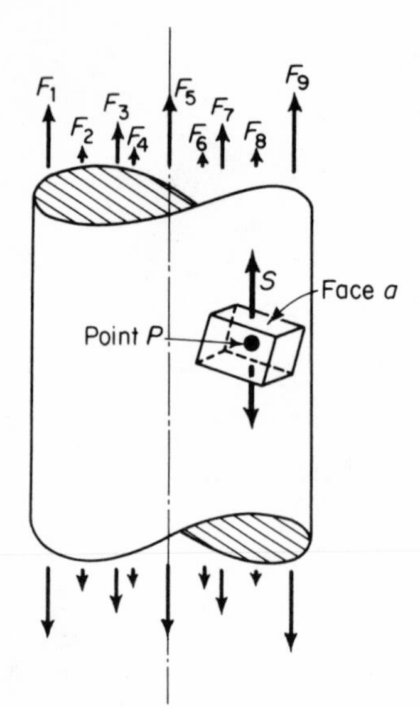

Fig. 2.1 Small rectangular parallelepiped drawn around point P in order to define stress, S.

COMPONENTS OF STRESS AT A POINT

Since the stress, S, generally is not perpendicular to the reference element face, it can be resolved into two components—one normal to the face on which S acts and the other tangential to this face—as indicated by the light arrows in Fig. 2.2a. The normal component is designated by a sigma (σ) followed by a subscript denoting its direction. To maintain a standard procedure for indicating directions, a Cartesian coordinate system is oriented so that its axes coincide with three mutually perpendicular edges of the elementary parallelepiped. Directions parallel with the element edges are given by referring to the coordinate axis through the appropriate edge, and various faces of the element are specified by the directions of their normals. For example, σ_z represents a *normal stress* on the 'z' face of our element. Normal stresses create tension if they are directed away from the faces as shown in Fig. 2.2, or they are compressive if directed toward the face. Tensile stresses are usually considered positive and compressive stresses, negative.

The tangential stress or *shear stress* is designated by a tau (τ). From the Pythagorean Theorem, as can be seen in Fig. 2.2a, the sum of the squares of the normal stress, σ, and the shear stress, τ, on any face is equal to the square of S, e.g.,

$$S^2 = \sigma_z^2 + \tau^2 \tag{2.2}$$

where σ_z = normal stress
and τ = shear stress on plane z.

The shear component of S will not usually be parallel with any of the axes of the chosen coordinate system. However, it will be convenient for subsequent numerical operations to refer to shear stresses in the direction of the coordinate axes so that τ is resolved into two components that are parallel with the reference directions as shown in Fig. 2.2b. These shear components are designated by a τ followed by two subscripts, the first indicating the face on which the shear stress acts and the second its direction. Consequently, τ_{zx} denotes a shear stress acting on the plane normal to the z axis and in the direction of the x axis. When the normal stress on a particular plane is positive, a positive shear stress on the same plane will point in a positive coordinate direction. If the normal stress is negative, a positive shear stress on the plane will then point in the negative direction.

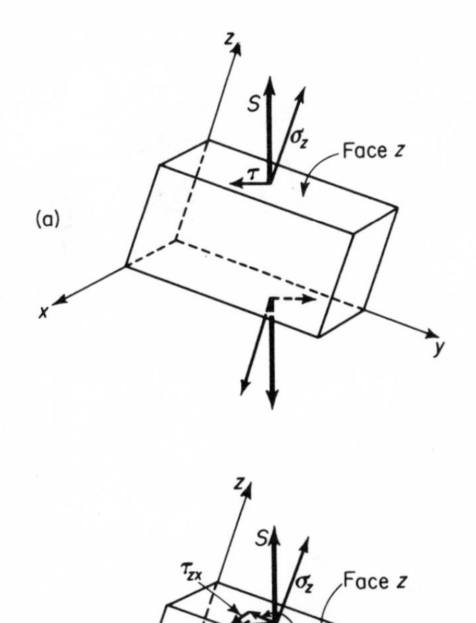

Fig. 2.2 (a) Stress, S, shown in Fig. 2.1 resolved into two components: σ_z normal to face, τ tangential to face. (b) Shear stress, τ, is further resolved into components, τ_{zx} and τ_{zy}, parallel with the coordinate axes.

In the general case, not two, but all six sides of the reference parallelepiped, are exposed to stresses of various magnitudes. Each of these stresses can be resolved into one normal and two shear components, resulting in a total of 18. Half of these are shown on the three near faces of the element in Fig. 2.3, whereas the remainder (omitted in the drawing) act on the far faces. Since the element is in static equilibrium within the body, the corresponding forces on opposite faces, spaced an infinitesimal apart, can be assumed to be equal in magnitude and opposite in direction. Moreover, the opposing faces are of equal area, so that the stresses on them are also equal. Hence, the number of independent stress components is reduced from 18 to 9—i.e., to three normal and six shear components—acting upon three neighboring and mutually perpendicular sides of our element.

The equality of stresses on opposite faces of the element will not prevent it from rotating and, therefore, represents a necessary but not a sufficient condition for equilibrium. The latter is secured only when in addition, the sum of the moments of the forces acting on the sides, with respect to any axis, is zero. If the moments about the z axis are considered, for example, and the parallelepiped in Fig. 2.3 is viewed from the top, τ_{yz} and τ_{xz} are directed

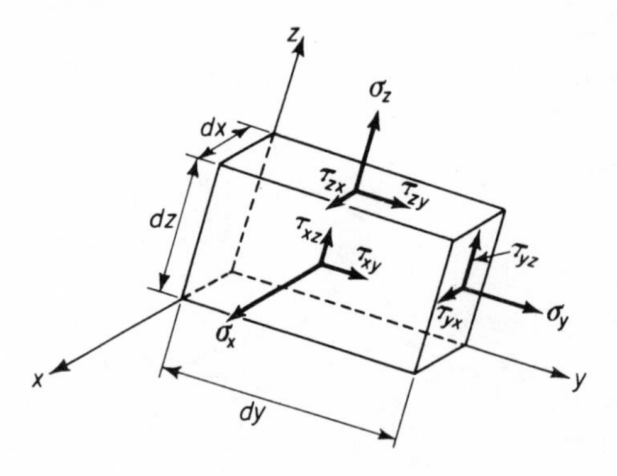

Fig. 2.3 Normal and shear components on near faces of reference element.

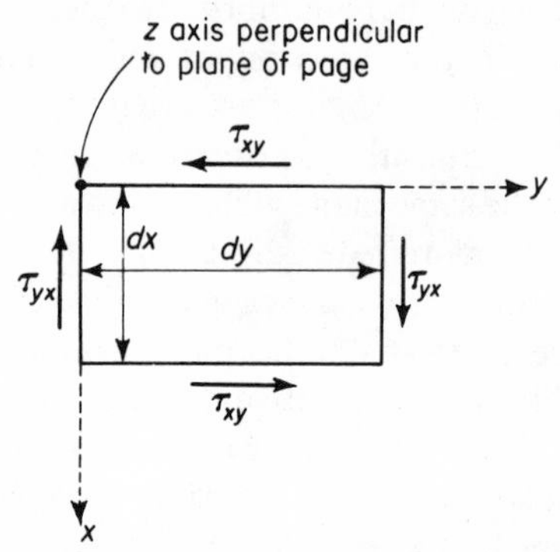

Fig. 2.4 Plan view of element in Fig. 2.3. Normal stresses are not shown since they cancel out.

parallel with the z axis and hence cannot contribute to such rotation. Likewise τ_{zy} and τ_{zx} are balanced by equal stresses on the "bottom" invisible z-surface and their moments cancel out. This applies also to σ_x and σ_y because each of these represents a pair of equal normal stresses on opposite sides of the element. Only the two shear components shown in Fig. 2.4 that do not cross the z axis can cause rotation.

The lengths of the sides of the parallelepiped are taken as dx, dy, and dz. The moment produced by stress τ_{xy} acting on area $(dy\, dz)$ at a distance dx is $\tau_{xy}(dy\, dz)\, dx$. Similarly, the moment of τ_{yx} about the z axis is $\tau_{yx}(dx\, dz)\, dy$. These two moments must act in opposite directions and be numerically equal for the element to be in equilibrium. Thus,

$$\tau_{xy}\, dy\, dz\, dx = \tau_{yx}\, dx\, dz\, dy$$

or

$$\tau_{xy} = \tau_{yx} \tag{2.3}$$

Similar equations are obtained by considering the equilibrium of the cube about the x and y axis, giving:

$$\tau_{yz} = \tau_{zy} \quad \text{and} \quad \tau_{zx} = \tau_{xz} \tag{2.3}$$

It is seen, therefore, that mutually perpendicular shear stresses on neighboring planes are equal. The complete stress system is then defined by only six rather than nine components:

$$\sigma_x, \sigma_y, \sigma_z, \tau_{xy}, \tau_{yz}, \tau_{zx}$$

HOMOGENEOUS STRESS

Before proceeding any further, an important restriction must be placed on the conditions under which forces on opposite sides of a parallelepiped can be equated. The total number of independent stresses acting on the parallelepiped have been reduced from 18 to 6 by assuming it to be in static equilibrium, and, on the basis of this assumption, normal forces on opposite faces as well as many of the shear stresses were found to be equal. Taking again the example of the loaded ball bearing and examining identically dimensioned parallelepipeds along the ball axis which connects the two contact points, it is apparent that the assumption that allowed us to equate forces is not valid. The stress on the test elements at the contact points is obviously greater than at the axis midpoint. In fact, normal stresses parallel with the ball axis on opposite faces of any volume element can never be equal. On the other hand, if the test element is so small that the length of its sides is given by infinitesimals, the stress may not change fast enough in going from one face to the next to introduce a significant error in the calculations even though such changes are disregarded. When the stresses change so slowly that they can be assumed equal on opposite faces of a small element, the stress at this point is said to be a *homogeneous stress*.

Some simply-shaped bodies are loaded so as to develop a true condition of homogeneous stress over large volumes. For instance, a long tie rod may be subjected to a uniform tension in which case the stress in any given direction is constant if the weight of the rod is negligible compared with the applied force. In this case, the stresses on opposite sides of even a large "test element" are equal, and the use of an infinitesimally small unit is not required. In the ball bearing, on the other hand, the error becomes increasingly large as the size of the elemental block increases.

Even in those cases where the stresses change extensively with location in the body, such as in the ball bearing example, the number of components required to describe the stress at any point is still six. This can be shown by establishing the equilibrium conditions without assuming stress homogeneity (see Problem 1). It may be added that this approach, albeit more complex, is more general and rigorous than the one used in the preceding section.

PRINCIPAL STRESSES

Summarizing what was stated above: The state of stress at any point within a body of arbitrary external shape, having a multitude of forces acting upon its surfaces is completely defined by the six components σ_x, σ_y, σ_z, τ_{xy}, τ_{yz}, τ_{zx}. Actually, these components apply to an elemental parallelepiped surrounding the point and oriented in a particular way within the body. If, however, this elemental block is tilted in various directions while still enveloping the same point, the stresses acting on its faces change.

It will now be our aim to determine these stress components as a function of the instantaneous position parameters. It will be seen that there is one particular position of the test element such that all the shear stress components vanish, and, therefore, three stresses instead of the usual six will suffice to completely define the stress condition in the surroundings of the point. This is indeed fortunate since it presents a simple method for describing stresses at a point regardless of the complexity of the body shape or external loading. These three stresses and their interrelation are of great importance in predicting whether a solid will deform elastically, plastically, or will fracture.

In order to determine how the magnitude of the stress components acting on the sides of our parallelepiped varies as the element is tilted, imagine the element as a small volume within a loaded body. Next, the body is cut by some arbitrary plane that runs through the parallelepiped. Since the cutting plane may have any direction so long as it passes through the volume element, in general it will cut off a corner of the element as shown in Fig. 2.5. The portion of the body that lies above the crosshatched cutting plane and that has been removed is shown dashed, while the remaining portion below the plane is shown by the solid outlines. The corner of the original parallele-

Fig. 2.5 Tetrahedron, $OABC$, formed by having corner of parallelepiped cut off by cross-hatched plane.

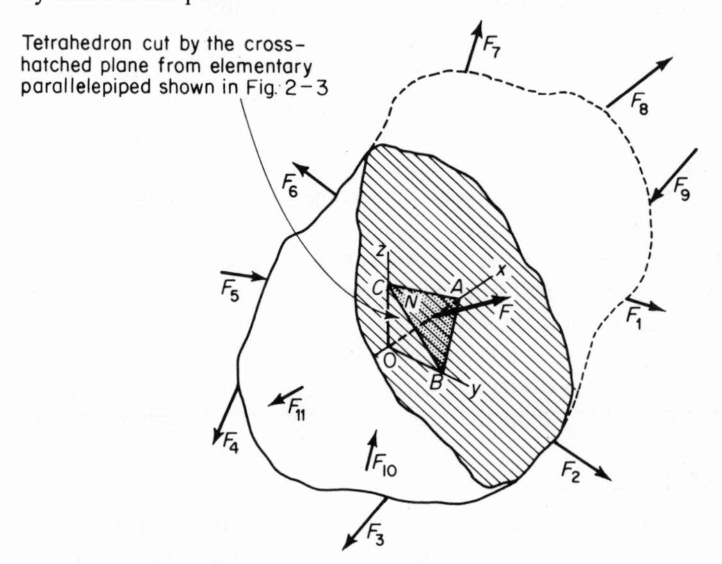

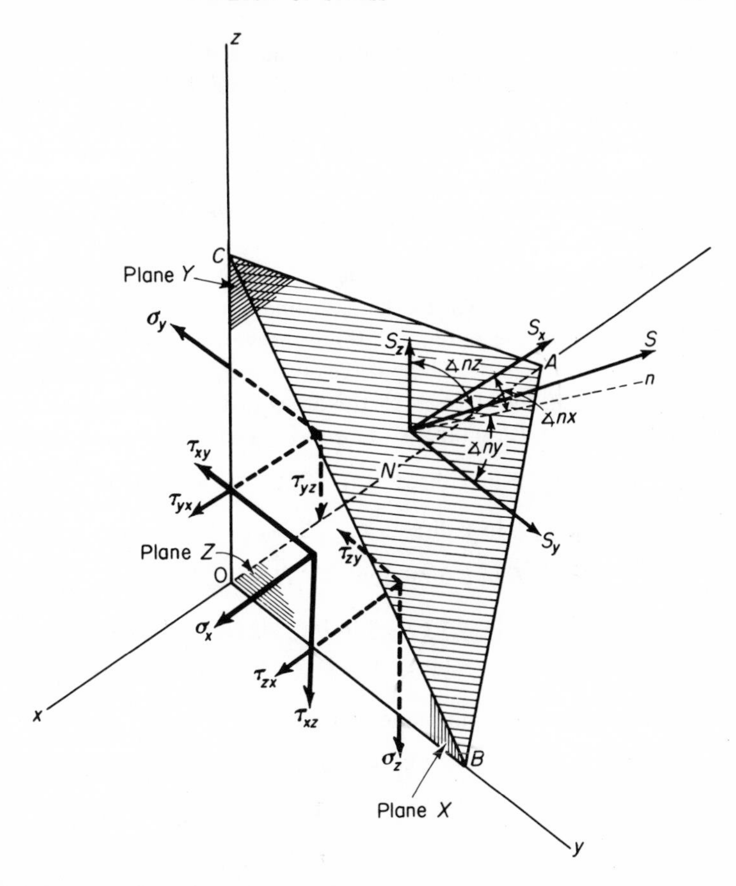

Fig. 2.6 Enlarged view of tetrahedron of Fig. 2.5 showing stresses acting on two of its sides. Stress, S, is not normal to plane N.

piped that is left after cutting is the tetrahedron $OABC$. The shaded side of the latter forms the triangle ABC that lies in the crosshatched plane of the cut, while the other three sides of the tetrahedron, i.e., triangles AOB, BOC, and COA, are the three coordinate planes of the original parallelepiped. For further convenience, triangle ABC is designated by N which is the base of the tetrahedron. We shall consider N to have an area of unity, and the force acting on this area to be equal to F. With $N = 1$, the areas of the other three sides of the tetrahedron become numerically equal to the direction cosines of the normals to these sides. This is shown in Fig. 2.6 where

$$\begin{aligned}
&\text{area of side } ABC = N = 1 \\
&\text{area of side } OBC = X = \cos \measuredangle (n, x) = k \\
&\text{area of side } OCA = Y = \cos \measuredangle (n, y) = l \\
&\text{area of side } OAB = Z = \cos \measuredangle (n, z) = m
\end{aligned} \tag{2.4}$$

Since force F in Fig. 2.5 acts on a unit area, it is identical with stress S acting on this same area, Fig. 2.6. Stress S generally will not be normal to N nor parallel to any of the coordinate axes. In order to be able to relate it to the already established directions, surfaces, etc., it is convenient to resolve it into components S_x, S_y, and S_z parallel with the coordinate axes. These three stress components on N can be converted to forces by multiplying them by the area of N.

Knowing now the magnitudes of all the component forces on plane N in the direction of each of the coordinate axes, one can compute the forces and stresses on the remaining three faces of $OABC$ by using the conditions of static equilibrium. These forces in the x, y, and z direction are obtained by multiplying the stresses by the areas of the sides on which they act. The areas, from Eq. (2.4), are X, Y, and Z. Hence,

$$\begin{aligned} &\textstyle\sum F \text{ in } x \text{ direction:} \quad S_xN = \sigma_x X + \tau_{yx} Y + \tau_{zx} Z \\ &\textstyle\sum F \text{ in } y \text{ direction:} \quad S_yN = \tau_{xy} X + \sigma_y Y + \tau_{zy} Z \\ &\textstyle\sum F \text{ in } z \text{ direction:} \quad S_zN = \tau_{xz} X + \tau_{yz} Y + \sigma_z Z \end{aligned} \qquad (2.5)$$

Dividing each of these by N, and bearing in mind that,

$$X = Nk \quad Y = Nl \quad \text{and} \quad Z = Nm \qquad \text{from (2.4)}$$

the resolved stresses on plane N are found to be

$$\begin{aligned} S_x &= \sigma_x k + \tau_{yx} l + \tau_{zx} m \\ S_y &= \tau_{xy} k + \sigma_y l + \tau_{zy} m \\ S_z &= \tau_{xz} k + \tau_{yz} l + \sigma_z m \end{aligned} \qquad (2.6)$$

When S_x, S_y, and S_z are projected on the direction normal to plane N, they produce components, S_xk, S_yl, and S_zm. The sum of these represents the total normal stress, σ_n, acting on N:

$$\sigma_n = S_xk + S_yl + S_zm \qquad (2.7)$$

Now substituting the values from (2.6) into (2.7):

$$\sigma_n = \sigma_x k^2 + \sigma_y l^2 + \sigma_z m^2 + 2\tau_{xy} kl + 2\tau_{yz} lm + 2\tau_{zx} mk \qquad (2.8)$$

In Eq. (2.8) the stresses appear as coefficients of the direction cosines k, l, and m. This equation can be converted into one in which the stresses are coefficients of the variables x, y, and z by introducing a vector, $\mathbf{r}$, such that the square of its length is inversely proportional to stress σ_n. The vector is in the direction of the normal, n, in Fig. 2.6. Hence $\sigma_n = \pm c/\mathbf{r}^2$, and if $\mathbf{r} = |\mathbf{r}|$ the coordinates of the end of the vector are $x = k\mathbf{r}$, $y = l\mathbf{r}$, and $z = m\mathbf{r}$. Substituting these values in Eq. (2.8) yields for constant C:

$$c = \sigma_x x^2 + \sigma_y y^2 + \sigma_z z^2 + 2\tau_{xy} xy + 2\tau_{yz} yz + 2\tau_{zx} zx \qquad (2.9)$$

This represents a surface of the second degree (*Cauchy's quadric*).

From analytical geometry it is known that the coefficients, σ and τ, of variables x, y, and z change as the coordinate reference axes are rotated, and at one position of rotation the coefficients that are the shear stress components can be shown to vanish. The directions of the coordinate axes corresponding to these positions are called *principal directions*, and the appropriate normal stresses are called *principal stresses*. The term *principal stress* is synonomous with *principal normal stress*. The planes on which these stresses act are the *principal planes*.

Since any stress system at a specified point, including the principal system, must describe the same quadric surface, the following rules can be stated:

1. A stress system at a point is completely defined by three principal stresses. The corresponding equation is the same as (2.9) but without the shear stress terms.

2. These principal stresses are always mutually perpendicular.

3. One of the principal stresses is the largest and another is the smallest normal stress acting at the point. It is customary to designate the largest principal stress as σ_1, the intermediate stress as σ_2, and the smallest as σ_3.

4. The principal stresses act on planes (the principal planes) on which the shear stresses are zero.

5. There is an infinite number of stress combinations that can be used to describe the stress state at a given point. If one of the coefficients in (2.9) is changed, then the other coefficients must undergo an adjustment if the equation is to represent the same state of stress.

For the loaded wire shown in Fig. 1.1, one of the principal directions must be the axis of the wire because this is the normal to a plane on which the shear stresses are zero. Because the other two principal directions must be perpendicular to the first, these directions must be those of the wire radii.

CALCULATION OF PRINCIPAL STRESSES

If the stresses at a point—i.e., σ_x, σ_y, σ_z, τ_{xy}, τ_{yz}, and τ_{zx}—are known, the principal stresses at the point can be calculated by taking advantage of the fact that the principal stresses are perpendicular to the surfaces on which they act. The calculation thus requires that one visualize S to be normal to the largest side of a tetrahedron; i.e., side N in Fig. 2.7. By definition N is then a principal plane. The components of S, in the direction of the x, y, and z coordinates, are:

$$S_x = Sk \quad S_y = Sl \quad S_z = Sm \qquad \text{(See Fig. 2.7)}$$

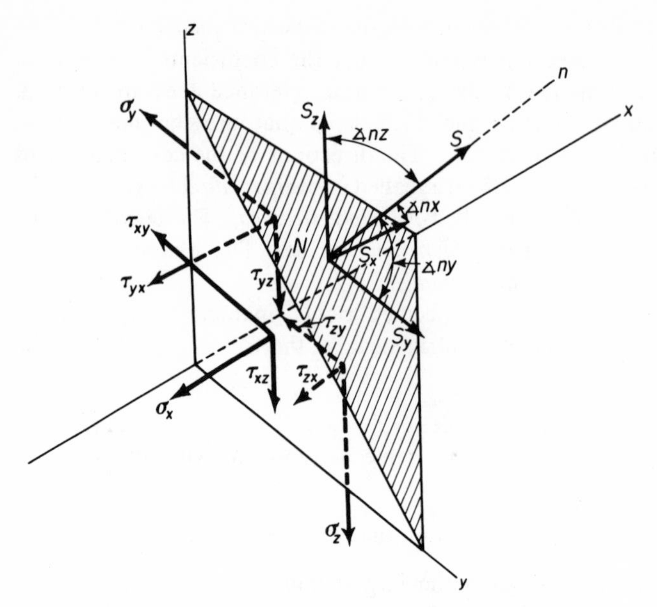

Fig. 2.7 Tetrahedron in which side N is the principal plane.

On substituting these in Eq. (2.6), we obtain:

$$\begin{aligned}(\sigma_x - S)k + \tau_{yx}l + \tau_{zx}m &= 0\\ \tau_{xy}k + (\sigma_y - S)l + \tau_{zy}m &= 0\\ \tau_{xz}k + \tau_{yz}l + (\sigma_z - S)m &= 0\end{aligned} \tag{2.10}$$

These three homogeneous linear equations in k, l, and m will yield solutions other than zero if their determinant is zero. Hence,

$$\begin{vmatrix}(\sigma_x - S) & \tau_{yx} & \tau_{zx}\\ \tau_{xy} & (\sigma_y - S) & \tau_{xy}\\ \tau_{xz} & \tau_{yz} & (\sigma_z - S)\end{vmatrix} = 0$$

Calculating this determinant leads to the following cubic equation in S:

$$S^3 - (\sigma_x + \sigma_y + \sigma_z)S^2 + (\sigma_x\sigma_y + \sigma_y\sigma_z + \sigma_x\sigma_z - \tau_{yz}^2 - \tau_{zx}^2 - \tau_{xy}^2)S \\ - (\sigma_x\sigma_y\sigma_z + 2\tau_{yz}\tau_{xz}\tau_{xy} - \sigma_x\tau_{yz}^2 - \sigma_y\tau_{xz}^2 - \sigma_z\tau_{xy}^2) = 0 \tag{2.11}$$

The three roots of (2.11) are S_1, S_2, and S_3. Since these are the principal stresses, we designate them as σ_1, σ_2, and σ_3. Using these roots successively in two of the (2.10) equations, along with the trigonometric identity $k^2 + l^2 + m^2 = 1$, will then give three sets of direction cosines for the principal planes.

It had been shown above that any number of combinations of normal and shear stress components could be used to describe the stress at a point. As the coordinates are transformed, yielding different values of σ_x, σ_y, σ_z, τ_{xy}, τ_{yz}, and τ_{zx}, certain relationships between the components must be maintained. These relationships are apparent from Eq. (2.11). Note that there are three combinations of the stress components that make up the coefficients of this equation. Since the value of these coefficients determine the three roots of S—i.e., the principal stresses—they evidently do not vary as the coordinate system changes. Hence,

$$\sigma_x + \sigma_y + \sigma_z = I_1$$

$$\sigma_x\sigma_y + \sigma_y\sigma_z + \sigma_x\sigma_z - \tau_{yz}^2 - \tau_{zx}^2 - \tau_{xy}^2 = I_2$$

and, (2.12)

$$\sigma_x\sigma_y\sigma_z + 2\tau_{yz}\tau_{xz}\tau_{xy} - \sigma_x\tau_{yz}^2 - \sigma_y\tau_{xz}^2 - \sigma_z\tau_{xy}^2 = I_3$$

where I_1, I_2, and I_3 are constants.

The three expressions in Eq. (2.12) are known as the three *stress invariants.* By far the most commonly used of these is the first which states that the sum of the normal stresses for any orientation of the coordinte system is equal to the sum of the normal stresses for any other orientation.

PRINCIPAL SHEAR STRESSES

Not only the normal stresses, but also the shear stresses acting on the element faces, change as the element is tilted, and the particular orientation of the reference element, at which one shear stress attains the highest possible value, will now be determined. The three shear stress components acting in the three mutually perpendicular faces, when the parallelepiped is oriented so as to produce this maximum shear component on one of its faces, are referred to as the *principal shear stresses.* These stress values, unlike the principal normal stresses, are seldom used in defining the stress state at a point. On the other hand, the principal shear stresses are very important in developing an understanding of the flow characteristics of bodies. It was pointed out that the principal normal stresses aided in predicting fracturing behaviors; the principal shear stresses serve a similar purpose in plastic flow.

The equations developed in the previous section can be extended to find expressions for the principal shear stresses. For this purpose stress relations in a triaxial system where the principal stresses coincide with the coordinate axes will be considered. Hence, Fig. 2.6 is altered so that no shear stresses are present on the coordinate planes as shown in Fig. 2.8 and, by selecting the coordinate axes to coincide with the principal directions, Eqs. (2.6) reduce to:

$$S_x = \sigma_1 k$$

$$S_y = \sigma_2 l \qquad (2.13)$$

$$S_z = \sigma_3 m$$

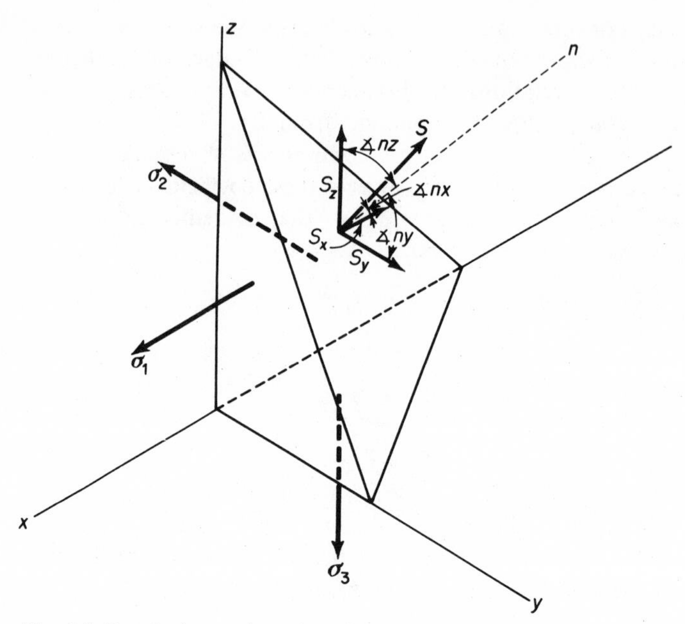

Fig. 2.8 Tetrahedron oriented so that the coordinate sides are principal planes.

and Eq. (2.8) becomes:

$$\sigma_n = \sigma_1 k^2 + \sigma_2 l^2 + \sigma_3 m^2 \tag{2.14}$$

Further, by considering the right triangles made by S_x, S_y, S_z, and S in Fig. 2.8:

$$S^2 = S_x^2 + S_y^2 + S_z^2 \tag{2.15}$$

Again from Pythagorean Theorem, Eq. (2.2):

$$\tau_n^2 = S^2 - \sigma_n^2$$

Substituting (2.13), (2.14), and (2.15) in (2.2b) yields:

$$\tau^2 = \sigma_1^2 k^2 + \sigma_2^2 l^2 + \sigma_3^2 m^2 - (\sigma_1 k^2 + \sigma_2 l^2 + \sigma_3 m^2)^2 \tag{2.16}$$

One of the direction cosines, for example, m, can now be eliminated from (2.16) by using the identity:

$$m^2 = 1 - k^2 - l^2$$

and Eq. (2.16) thus modified is now differentiated with respect to k and l. These two partial differentials are equated to zero thereby defining the positions of planes upon which τ reaches its extreme values (minimum or maximum). The two expressions are:

$$\frac{\partial}{\partial k} = k[(\sigma_1 - \sigma_3)k^2 + (\sigma_2 - \sigma_3)l^2 - \tfrac{1}{2}(\sigma_1 - \sigma_3)] = 0$$
$$\frac{\partial}{\partial l} = l[(\sigma_1 - \sigma_3)k^2 + (\sigma_2 - \sigma_3)l^2 - \tfrac{1}{2}(\sigma_2 - \sigma_3)] = 0 \qquad (2.17)$$

One of the solutions to these equations is found when:

$$k \text{ or } l = 0$$

For $k = 0$, the second equation of the (2.17) group yields:

$$l = \pm\sqrt{\tfrac{1}{2}}$$

For $l = 0$, the first of (2.17) yields:

$$k = \pm\sqrt{\tfrac{1}{2}}$$

If, now l and k are successively eliminated from (2.16) instead of m, two more identical pairs of numbers are obtained for m. All six values in addition to other obvious solutions ($k = l = 0$; $m = \pm 1$, etc.) are tabulated below:

TABLE 2.1
VALUES OF DIRECTION COSINES THAT GIVE MAXIMUM OR MINIMUM SHEAR STRESS

$k =$	0	0	± 1	0	$\pm\sqrt{\frac{1}{2}}$	$\pm\sqrt{\frac{1}{2}}$
$l =$	0	± 1	0	$\pm\sqrt{\frac{1}{2}}$	0	$\pm\sqrt{\frac{1}{2}}$
$m =$	± 1	0	0	$\pm\sqrt{\frac{1}{2}}$	$\pm\sqrt{\frac{1}{2}}$	0

For each of these columns of numbers, τ is a maximum or a minimum.

The first three columns represent the coordinate planes that are identical with the principal planes. On these planes τ is a minimum ($= 0$) [from (2.16)]. The last three columns denote planes passing through one of the coordinate axes and bisecting the angle between the other two. Hence each of these planes is inclined at 45° to two principal planes.

After substituting the extreme values of k, l, and m successively into formula (2.16), the following shear stresses result:

$$\tau_{12} = \pm\tfrac{1}{2}(\sigma_1 - \sigma_2)$$
$$\tau_{23} = \pm\tfrac{1}{2}(\sigma_2 - \sigma_3) \qquad (2.18)$$
$$\tau_{31} = \pm\tfrac{1}{2}(\sigma_3 - \sigma_1)$$

These are the principal shear stresses. Since σ_1 and σ_3 represent respectively the largest and smallest normal stresses appearing at the selected point, stress

$\tau_{31} = \tau_{13}$ and is also the absolutely largest shear stress acting within the element.

Normal stresses also act on the planes of the principal shear stresses in Eq. (2.18). They are obtained by substituting the values of k, l, and m in the last three columns of Table 2.1 successively into Eq. (2.14). When executed, this gives:

$$\frac{\sigma_1 + \sigma_2}{2}, \qquad \frac{\sigma_2 + \sigma_3}{2}, \qquad \frac{\sigma_3 + \sigma_1}{2} \tag{2.19}$$

Note that according to Eq. (2.18) all the shear stresses vanish (i.e., $\tau_{12} = \tau_{23} = \tau_{31} = 0$) when $\sigma_1 = \sigma_2 = \sigma_3$. This condition, known as a *hydrostatic state of stress*, is characterized by a total absence of shear stresses.

STRESS COMPONENTS AT A POINT AS THE COORDINATE AXES ARE ROTATED

In the previous two sections it was shown that, as the test element rotates, the stress components acting on its faces vary. At one particular orientation, where it becomes aligned with the principal directions, the shear components vanish while the normal stress on one face takes on a maximum value, and, on another face, a minimum value. At still another orientation we found that the maximum shear stresses were developed on certain faces. In many

Fig. 2.9 Relationship between original coordinate system and rotated system.

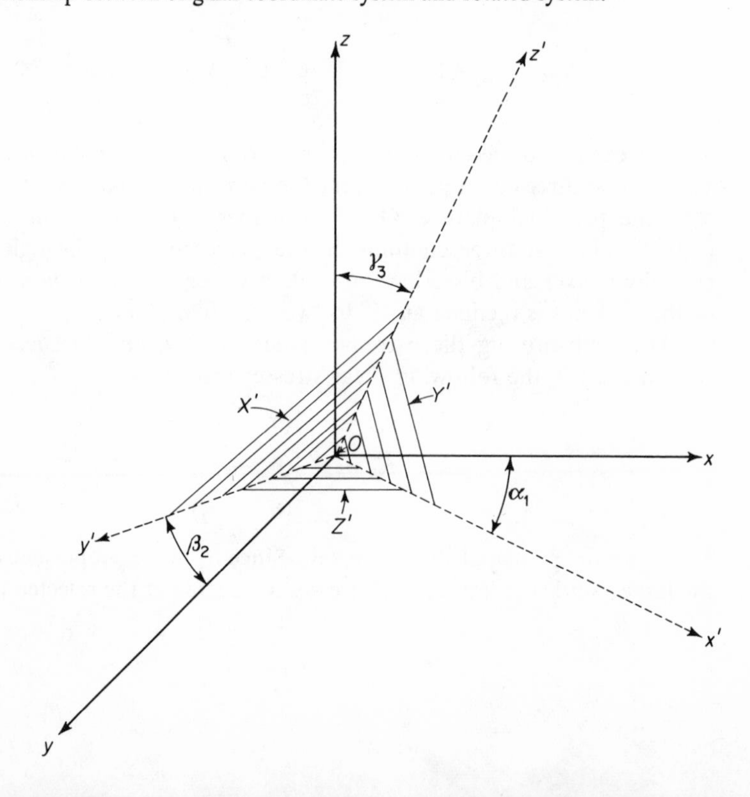

practical problems, however, it is necessary to know the stresses not only for these two special orientations but for any other rotation of the element as well. Consequently, expressions will now be developed for finding the six stress components when the test element is rotated (or as the coordinate system represented by its sides rotates).

Thus, knowing the values of σ_x, σ_y, σ_z, τ_{xy}, τ_{yz}, and τ_{zx} for the first coordinate system x, y, and z, the six stress components acting in new x', y', and z' coordinate directions will be computed. The relationship between the two Cartesian systems is shown in Fig. 2.9.

The position of the new system with respect to the old one will be completely determined by the direction cosines k_1, l_2, and m_3 of the three angles α_1, β_2, and γ_3 formed by x and x', y and y', z and z', respectively.

There are six other angles between the first and second systems ($x'Oy$, $x'Oz$, $y'Ox$, $y'Oz$, $z'Ox$, and $z'Oy$), the cosines of which can be determined from k_1, l_2, and m_3. All nine angles and their direction cosines are listed in Table 2.2.

TABLE 2.2

NOMENCLATURE OF ANGLES AND DIRECTION COSINES BETWEEN THE INITIAL AND THE ROTATED COORDINATE SYSTEM

Rotated Coordinate Axes \ *Initial Coordinate Axes*	x	y	z
x'	$\alpha_1 = \measuredangle(x'Ox)$ $\cos\alpha_1 = k_1$	$\beta_1 = \measuredangle(x'Oy)$ $\cos\beta_1 = l_1$	$\gamma_1 = \measuredangle(x'Oz)$ $\cos\gamma_1 = m_1$
y'	$\alpha_2 = \measuredangle(y'Ox)$ $\cos\alpha_2 = k_2$	$\beta_2 = \measuredangle(y'Oy)$ $\cos\beta_2 = l_2$	$\gamma_2 = \measuredangle(y'Oz)$ $\cos\gamma_2 = m_2$
z'	$\alpha_3 = \measuredangle(z'Ox)$ $\cos\alpha_3 = k_3$	$\beta_3 = \measuredangle(z'Oy)$ $\cos\beta_3 = l_3$	$\gamma_3 = \measuredangle(z'Oz)$ $\cos\gamma_3 = m_3$

Using the notation from Eq. (2.5), the new system will be described by stresses $\sigma_{x'}$, $\tau_{x'y'}$, and $\tau_{x'z'}$ on plane X'; $\sigma_{y'}$, $\tau_{y'z'}$, and $\tau_{y'x'}$ on Y'; and $\sigma_{z'}$, $\tau_{z'x'}$, and $\tau_{z'y'}$ on Z'.

Now we shall transpose the stresses from the x, y, z system to the x', y', z' coordinates by a simple but unavoidably lengthy procedure. The method consists of considering each of the new coordinate planes—i.e., X', Y', and Z'—as the plane inclined to the original coordinate system (as was done for plane N in Fig. 2.6). For example, let the X' plane (normal to x') be made equivalent to the inclined plane N in Fig. 2.6. The tetrahedron in this case consists of the coordinate planes X, Y, Z, and the base plane X'. The stresses acting on the coordinate sides are the known components σ_x, σ_y, σ_z, τ_{xy}, τ_{yz}, and τ_{zx}. A resultant stress, $S_{x'}$, acts on X' just as S acted on N in Fig. 2.6. The components of $S_{x'}$ are obtained with the aid of Eq. (2.6), using the proper direction cosines from Table 2.1. These components are found to be:

$$S_{x'x} = \sigma_x k_1 + \tau_{yx} l_1 + \tau_{zx} m_1$$

By an identical reasoning

$$S_{x'y} = \sigma_y l_1 + \tau_{zy} m_1 + \tau_{xy} k_1 \tag{a}$$

and

$$S_{x'z} = \sigma_z m_1 + \tau_{xz} k_1 + \tau_{yz} l_1$$

Note in Eqs. (a) that the stress components, $S_{x'x}$, $S_{x'y}$, and $S_{x'z}$, are designated in the customary fashion; i.e., the first subscript indicates the plane on which they act, and the second subscript gives direction.

The same procedure can be applied to the Y' plane by letting it be the base of the tetrahedron and by using the directional cosines k_2, l_2, and m_2 to find the $S_{y'}$ group of stresses. A similar procedure permits one to find the $S_{z'}$ stresses with the aid of k_3, l_3, and m_3.

Expressions are now available for all of the stress components acting on the X', Y', and Z' planes but these, instead of being directed in the x', y', and z' directions, are acting parallel with x, y, and z.

Recalling that the projection of a vector on an axis is equal to the sum of the projections of the components of the vector, we are now able to project $S_{x'x}$, $S_{x'y}$, $S_{x'z}$ on the new coordinates x', y', z'. Again with the aid of the directional cosines tabulated above, one obtains, by projecting the three $S_{x'}$ components on x':

$$\sigma_{x'} = S_{x'x} k_1 + S_{x'y} l_1 + S_{x'z} m_1$$

Their projections on y' and z' represent shear stresses with respect to plane X' and are:

$$\left.\begin{aligned} \tau_{x'y'} &= S_{x'x} k_2 + S_{x'y} l_2 + S_{x'z} m_2 \\ \tau_{x'z'} &= S_{x'x} k_3 + S_{x'y} l_3 + S_{x'z} m_3 \end{aligned}\right\} \tag{b}$$

Identical operations are now performed to project the $S_{y'}$ and $S_{z'}$ components from the old coordinates to the new ones. This results in two more sets of equations analogous (except for the subscripts on S) with those under (b) above. Altogether there are nine (b)-type equations. Expressions (a) are substituted into six of the (b) category so that $\tau_{x'y'}$, $\tau_{y'x'}$, etc., are not computed twice.

The end result will consist of six equations representing the new components in terms of the old ones:

$$\begin{aligned}
\sigma_{x'} &= \sigma_x k_1^2 + \sigma_y l_1^2 + \sigma_z m_1^2 + 2\tau_{xy} k_1 l_1 + 2\tau_{yz} l_1 m_1 + 2\tau_{zx} m_1 k_1 \\
\sigma_{y'} &= \sigma_x k_2^2 + \sigma_y l_2^2 + \sigma_z m_2^2 + 2\tau_{xy} k_2 l_2 + 2\tau_{yz} l_2 m_2 + 2\tau_{zx} m_2 k_2 \qquad (2.20a) \\
\sigma_{z'} &= \sigma_x k_3^2 + \sigma_y l_3^2 + \sigma_z m_3^2 + 2\tau_{xy} k_3 l_3 + 2\tau_{yz} l_3 m_3 + 2\tau_{zx} m_3 k_3
\end{aligned}$$

and

$$\begin{aligned}
\tau_{x'y'} = \tau_{y'x'} &= \sigma_x k_1 k_2 + \sigma_y l_1 l_2 + \sigma_z m_1 m_2 + \tau_{xy}(k_1 l_2 + k_2 l_1) \\
&\quad + \tau_{yz}(l_1 m_2 + l_2 m_1) + \tau_{zx}(m_1 k_2 + m_2 k_1) \\
\tau_{y'z'} = \tau_{z'y'} &= \sigma_x k_2 k_3 + \sigma_y l_2 l_3 + \sigma_z m_2 m_3 + \tau_{xy}(k_2 l_3 + k_3 l_2) \\
&\quad + \tau_{yz}(l_2 m_3 + l_3 m_2) + \tau_{zx}(m_2 k_3 + m_3 k_2) \qquad (2.20b) \\
\tau_{z'x'} = \tau_{x'z'} &= \sigma_x k_3 k_1 + \sigma_y l_3 l_1 + \sigma_z m_3 m_1 + \tau_{xy}(k_3 l_1 + k_1 l_3) \\
&\quad + \tau_{yz}(l_3 m_1 + l_1 m_3) + \tau_{zx}(m_3 k_1 + m_1 k_3)
\end{aligned}$$

BIAXIAL OR PLANE STATE OF STRESS

In many practically important problems, stresses acting in one direction may be so much smaller than those acting in the other two that the former can be disregarded. This is almost always the case with sheets or thin wall tubular bodies where the largest stress developed in the thickness direction is only a small fraction of those acting parallel with the surface. For example, the largest stress in the thickness direction of a boiler drum or compressed gas cylinder wall is the internal pressure, p. Yet the tangential stress exceeds p by the ratio of the vessel radius to its wall thickness. This ratio may vary from about 15 for a gas cylinder to about 150 for a beer can. Stress systems that are confined to surfaces (either plane or curved) are said to be biaxial or plane.

They can readily be analyzed with the aid of Eqs. (2.20a and b) which are simplified by the absence of stress components in one direction. Assuming this for example, to be the z-direction, $\sigma_z = \tau_{zx} = \tau_{zy} = 0$. Of the nine directional cosines (k_1 to m_3), only k_1, k_2, l_1, and l_2 do not vanish. The tetrahedron $OABC$ in Fig. 2.6 now degenerates into triangle OAB, Fig. 2.10, where the nomenclature is the same as in Fig. 2.9, with the normal direction, n, of Fig. 2.6 being identical with x'.

Equations (2.20) become:

$$\sigma_{x'} = \sigma_x k_1^2 + \sigma_y l_1^2 + 2\tau_{xy} k_1 l_1$$

$$\tau_{x'y'} = \tau_{y'x'} = \sigma_x k_1 k_2 + \sigma_y l_1 l_2 + \tau_{xy}(k_1 l_2 + k_2 l_1)$$

It is not necessary to present an expression for $\sigma_{y'}$ since it can be obtained readily from $\sigma_x + \sigma_y = \sigma_{x'} + \sigma_{y'} = I_1$.

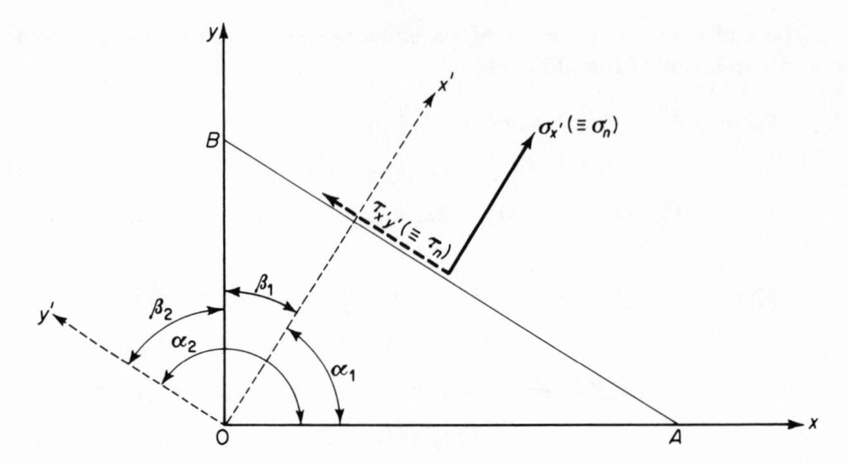

Fig. 2.10 Triangle into which tetrahedron of Fig. 2.6 degenerates when stresses in one (z) direction vanish.

From Fig. 2.10, $\cos \beta_1 = \sin \alpha_1 = -\cos \alpha_2$, and $\cos \beta_2 = \cos \alpha_1$, so that all directional cosines can be expressed in terms of α_1, resulting in:

$$\sigma_{x'} = \sigma_x \cos^2 \alpha_1 + \sigma_y \sin^2 \alpha_1 + 2\tau_{xy} \cos \alpha_1 \sin \alpha_1$$

$$_{x'y'} = -\sigma_x \cos \alpha_1 \sin \alpha_1 + \sigma_y \sin \alpha_1 \cos \alpha_1 + \tau_{xy}(\cos^2 \alpha_1 - \sin^2 \alpha_1)$$

Using the identities, $\sin^2 \alpha = (1 - \cos 2\alpha)/2$, $\cos^2 \alpha = (1 + \cos 2\alpha)/2$, and $2 \sin \alpha \cos \alpha = \sin 2\alpha$, one obtains:

$$\sigma_{x'} = \frac{\sigma_x + \sigma_y}{2} + \frac{\sigma_x - \sigma_y}{2} \cos 2\alpha_1 + \tau_{xy} \sin 2\alpha_1 \tag{2.21a}$$

and:

$$\tau_{x'y'} = -\frac{(\sigma_x - \sigma_y)}{2} \sin 2\alpha_1 + \tau_{xy} \cos 2\alpha_1 \tag{2.21b}$$

The principal stress σ, is computed by determining the angle α_1 in (2.21a) for which $\sigma_{x'}$ becomes an extreme, and substituting this value in (2.21a). Thus:

$$\frac{d\sigma_{x'}}{d\alpha_1} = -(\sigma_x - \sigma_y) \sin 2\alpha_1 + 2\tau_{xy} \cos 2\alpha_1 = 0$$

from which:

$$\tan 2\alpha_1 = \frac{2\tau_{xy}}{\sigma_x - \sigma_y} \tag{2.22}$$

After expressing $\sin 2\alpha_1$ and $\cos 2\alpha_1$ by the tangent, and substituting these into (2.21a), one obtains:

$$\sigma_{x'}\,(\text{principal}) = \sigma_1 = \frac{\sigma_x + \sigma_y}{2} + \sqrt{\left(\frac{\sigma_x - \sigma_y}{2}\right)^2 + \tau_{xy}^2} \qquad (2.23a)$$

The other principal stress results from:

$$\sigma_2 = (\sigma_x + \sigma_y) - \sigma_1 = \frac{\sigma_x + \sigma_y}{2} - \sqrt{\frac{\sigma_x - \sigma_y^2}{2} + \tau_{xy}^2} \qquad (2.23b)$$

Thus, knowing the values of σ_x, σ_y, and τ_{xy} for one coordinate system, the components in any other direction can be calculated.

Mohr's Stress Circle

Otto Mohr in 1882 suggested a geometrical construction, since known as *Mohr's Circle* or the *stress circle*, from which σ and τ can be readily obtained for any value of α if the normal and shear stresses acting on any pair of perpendicular planes are known (Fig. 2.11a).

Fig. 2.11 Test element and Mohr's Circle showing relationship of stress components acting on it as α_1 varies.

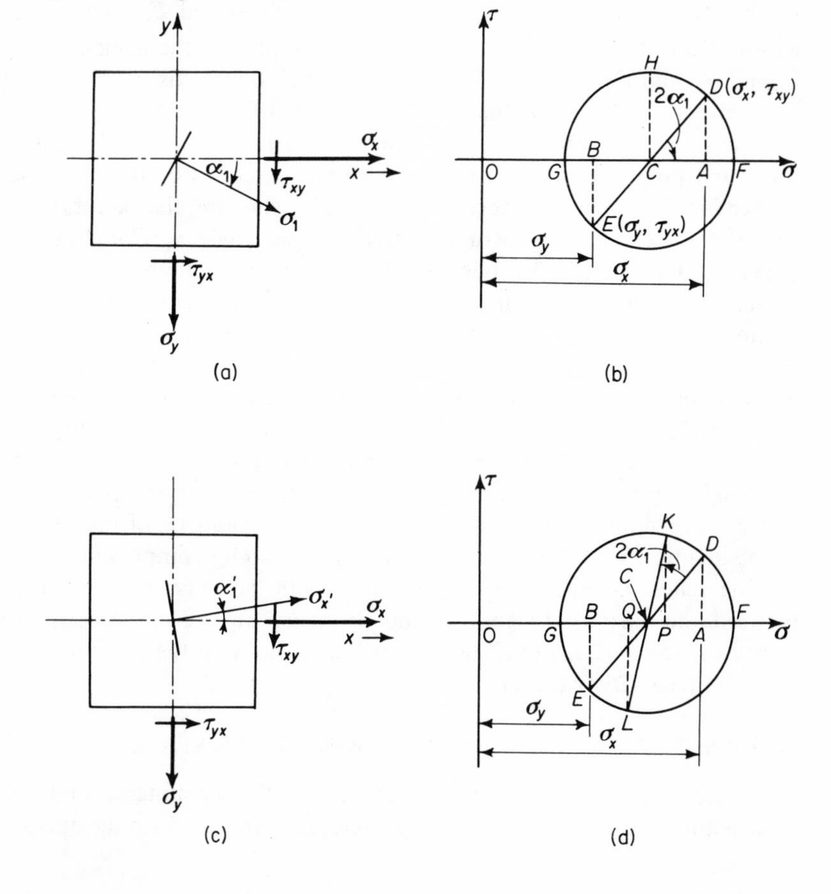

According to Mohr's method, the specified values of σ_x, σ_y, and τ_{xy} are used to plot a circle (Fig. 2.11b). The circle is drawn on Cartesian coordinates, with the abscissa representing normal stresses and the ordinate shear stresses. This circle has the property that the coordinates of points on its perimeter take on the value of the stress component at a point in the loaded body as the plane through this point is rotated through angle α_1. The method for drawing the circle is given below.

The normal stresses, $\sigma_x = \overline{OA}$ and $\sigma_y = \overline{OB}$, are plotted on the abscissa to obtain the proper length and direction with respect to O. The shear stresses τ_{xy} and τ_{yx} are laid out normal to the σ axis at A and B. The convention for the sign of τ used to construct a stress circle differs from the one on p. 21. A shear stress acting on the side of the elementary rectangle (which replaces the parallelepiped in plane stress) is considered positive if it tends to rotate the rectangle clockwise and vice versa. The end points $D(\sigma_x, \tau_{xy})$ and $E(\sigma_y, \tau_{yx})$ are now joined by a straight line which intersects the σ axis at C, midway between A and B.

A circle, centered at C, and passing through D and E, intersects the σ axis at F and G. The distances, $\overline{OF}$ and $\overline{OG}$ represent the principal stresses σ_1 and σ_2, and the circle radius, CH, normal to the σ axis, is the maximum shear stress. Angle $2\alpha_1$ is the same as computed from (2.22).

The stress components for any other orientation of the test element about this same point can also be read off Mohr's circle. For this purpose the element in Fig. 2.11a is reproduced in 2.11c, showing the orientation of a sought normal stress $\sigma_{x'}$ at a counterclockwise angle α_1' from the known stress σ_x. The circle in 2.11b is also redrawn in 2.11d. Point D on the circle locates σ_x; hence, $2\alpha_1$ is measured counterclockwise from CD, point K is plotted, and the new diameter KL drawn. The direction was measured counterclockwise in Fig. 2.11d because α_1 is counterclockwise in going from σ_x to $\sigma_{x'}$ in 2.11c. In 2.11b, on the other hand, the principal stress σ_1 was $2\alpha_1$ in a clockwise direction from σ_x, thus accounting for the direction α_1 in Fig. 2.11a. The distance between the two end points of this diameter and the σ axis represents the shear stresses, $\tau_{x'y'}$ and $\tau_{y'x'}$, on the planes rotated α_1' from the initial orientation. The projections, P and Q, of the end points of the diameter give the new $\sigma_{x'} = OP$ and $\sigma_{y'} = OQ$ components.

The reader should have no difficulty with proving that the relations between the various dimensions and distances on the circle construction satisfy equations (2.21) to (2.23). He will also find that the principal stresses thus obtained satisfy (2.11).

MOHR'S CIRCLES FOR SOME COMMON STRESS STATES

Certain types of biaxial stresses are frequently encountered in a variety of structural applications, in testing materials, and in forming operations.

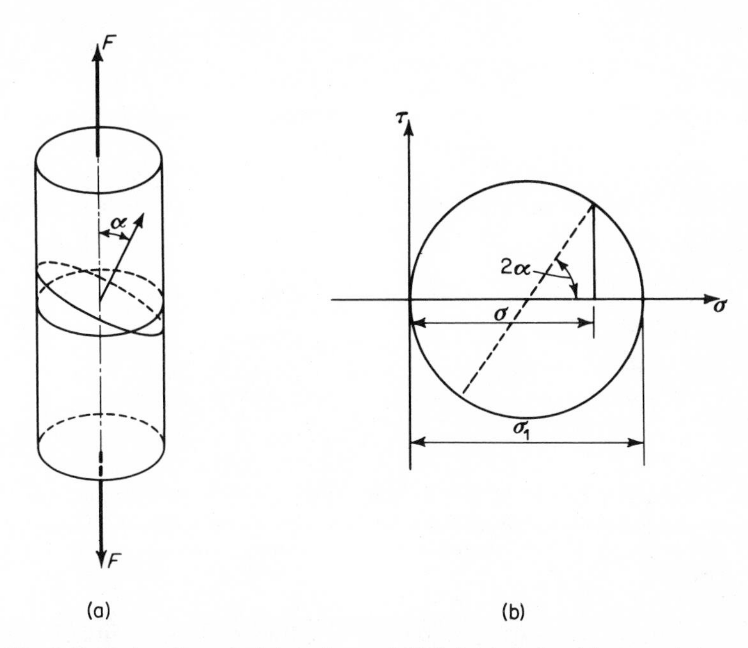

Fig. 2.12 Rod under uniaxial tension and Mohr's circle for this stress state.

These stress systems, or stress states, are defined by the ratios of their principal stresses. Mohr's circles for three of the most common stress states are shown below:

(a) *Uniaxial Stress State*: When a rod or wire carries a single applied force that coincides with the wire axis, only one principal stress, σ_1, acts at each point while $\sigma_2 = \sigma_3 = 0$, as shown in Fig. 2.12a. By following the procedure given above for drawing Mohr's circle, a circle tangent to the τ axis is obtained for this case. The circle lies in the first and fourth quadrant if the applied load is tension or in the second and third if it is compression (Fig. 2.12b). The maximum shear stress is equal to the radius of the circle, or $\sigma_1/2$, and lies at 45° to the direction of the applied load. This explains the occurrence of "Lüders lines" at aproximately 45° to the specimen axis during tension testing of steel (discussed in Chapter 10).

(b) *Simple Shear* (*Torsion*): If a thin-walled tube is twisted at its ends, as indicated by the arrows in Fig. 2.13a, a small element in the body has shear stresses, τ_{xy}, set up on its faces *a*. According to Eq. (2.3), shear stresses also act on perpendicular faces of the reference element. The latter are equal to τ_{xy} but oppositely directed, so that stress $\tau_{yx} = -\tau_{xy}$ acts on face *b*. No normal forces are applied to the tube so that no normal

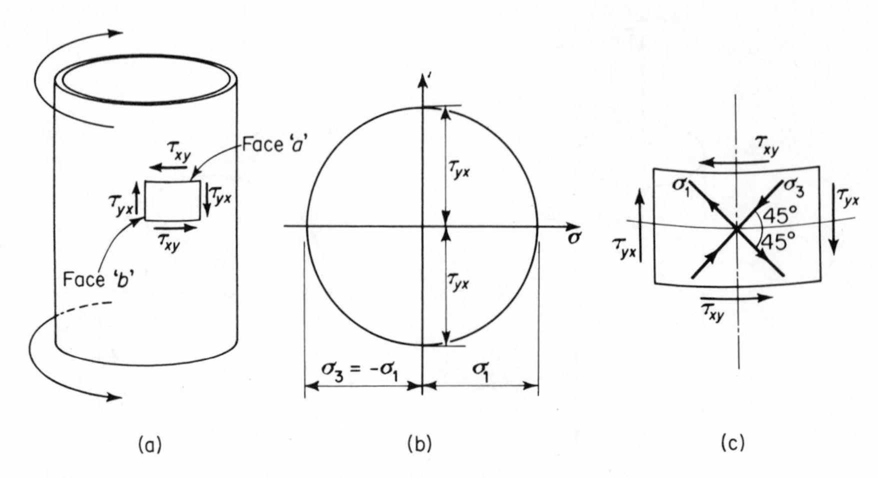

Fig. 2.13 (a) Tube under simple shear; and (b) Mohr's circle for this stress state; (c) Enlarged view of element showing relation between applied shear stress and resultant equal but oppositely directed normal stresses at 45° to the shear stresses.

stress occurs on faces *a* or *b*. On drawing Mohr's circle by the procedure given above, the ends of one diameter of the circle are given by the point $\sigma = 0$, $\tau = \tau_{xy}$ and the point $\sigma = 0$, $\tau = \tau_{yx} = -\tau_{xy}$. Hence, the circle is symmetrical about the origin (Fig. 2.13b). Observe that simple shear, such as in torsion, creates a stress condition in which the maximum shear stress and the two principal stresses σ_1, and σ_2, under 45° are of the same absolute magnitude, with σ_1 being tension and σ_3 compression (Fig. 2.13c).

(c) *Biaxial Stress with Tension and Compression on Perpendicular Faces*: In many metal forming operations the process of deformation subjects the material to compression in one direction and to tension in a perpendicular direction. For example, in metal rolling with front tension (which is the means generally used for making flat products such as strips and foils) the rolls apply compression to the metal while the front pull applies tension at right angles to the former. Mohr's circle for this stress state is similar to that shown above for torsion. The difference is that the center of the circle need not be at origin of the coordinate system. Indeed, torsion, or the application of equal tensile and compressive components on perpendicular planes, might be considered a special case of the type of biaxial stress described in this paragraph.

PLANE STRESS PROBLEM IN POLAR COORDINATES

Problems involving stresses in tubes, rings, or discs are usually analyzed and solved more easily by the use of polar rather than Cartesian coordinates.

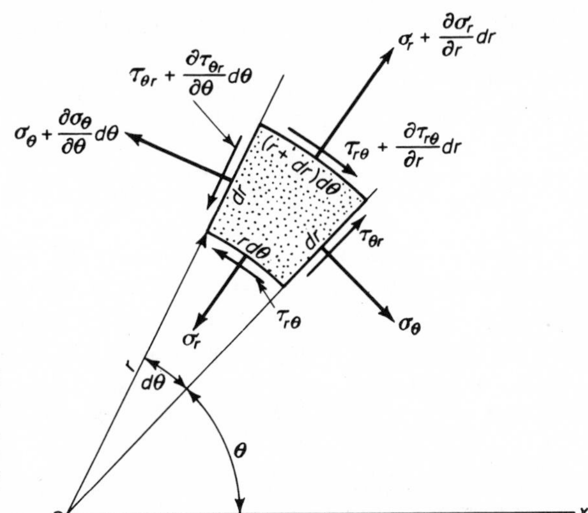

Fig. 2.14 Stresses acting on elementary segment in plane polar coordinates. Symbols inside the segment denote lengths of appropriate sides.

The position of an element and its dimensions are then expressed in terms of radius r and angle θ or their infinitesimal increments, dr and $d\theta$, respectively (Fig. 2.14).

The normal stresses acting on faces normal to a radius have a subscript r and are called *radial stresses*; those on planes normal to constant r circles receive subscript θ and are termed *tangential, circumferential,* or *hoop stresses.* The shear stresses on the same planes will carry subscripts $r\theta$ and θr.

Angle θ is constant along any radius so that, for any fixed θ, σ_r and $\tau_{r\theta}$ vary with r only. If r is held constant but θ is variable, hoop stress σ_θ and the radial shear component $\tau_{\theta r}$ are functions of θ. For instance, the radial increment of σ_r over a small distance dr is $(\partial\sigma_r/\partial r)\,dr$; the increment of shear stress on a radial plane $(= \tau_{\theta r})$ is $(\partial\tau_{\theta r}/\partial\theta)\,d\theta$, etc. As usual $\tau_{r\theta} = \tau_{\theta r}$. Assuming that θ increases counterclockwise, the stresses on the four bounding faces of the element are as illustrated in Fig. 2.14.

Taking the thickness of the element in the z direction as unity, the forces acting on each face are products of stress and length. Thus, on the cylindrical face nearer to 0 in Fig. 2.14, the force is $\sigma_r r\,d\theta$; on the opposite face, $[\sigma_r + (\partial\sigma_r/\partial r)\,dr]\,(r + dr)\,d\theta$, etc.

Noting that $d\theta$ is small so that $d\theta \approx \sin(d\theta) \approx \tan(d\theta)$, projection of all the elementary forces in the radial direction results in the following equation after simplification and omission of the higher powers and products of $d\theta$ and dr:

$$\frac{\partial\sigma_r}{\partial r} + \frac{1}{r}\frac{\partial\tau_{r\theta}}{\partial\theta} + \frac{\sigma_r - \sigma_\theta}{r} = 0 \tag{2.24}$$

Occasionally a body force R may act, say, in the radial direction. In such case the full value of R (referring to a unit of volume) is added to the left-hand side of (2.24). Such will be the case with the centrifugal force in a rotating cylinder or disc.

Similarly, by projecting all the elementary forces in the tangential direction, a second equation of equilibrium is obtained:

$$\frac{1}{r}\frac{\partial \sigma_\theta}{\partial \theta} + \frac{\partial \tau_{r\theta}}{\partial r} + \frac{2\tau_{r\theta}}{r} = 0 \tag{2.25}$$

Especially important is the axisymmetrical case when the radial and tangential stresses are principal. The axial direction is then principal by definition. In this case $\tau_{r\theta} = 0$ so that all the terms in (2.25) vanish. In the presence of a radially acting body stress, R, Eq. (2.24) becomes:

$$\frac{\partial \sigma_r}{\partial r} + \frac{\sigma_r - \sigma_\theta}{r} + R = 0 \tag{2.26}$$

Equation (2.26) is basic for calculating the stresses in thick-walled tubular bodies like high-pressure hydraulic cylinders, gun barrels, or turbine rotors. Using the relations between stresses and deformations (later developed in Chapters 3 and 5), the principal stresses, σ_r and σ_θ, in a thick-walled tube subjected to an internal pressure p_a and external pressure p_b can be calculated. The resulting equations developed in 1852 by Lamé are:

$$\sigma_r = \frac{p_a a^2 - p_b b^2}{b^2 - a^2} - \frac{(p_a - p_b)a^2 b^2}{(b^2 - a^2)r^2} \tag{2.27a}$$

$$\sigma_\theta = \frac{p_a a^2 - p_b b^2}{b^2 - a^2} + \frac{(p_a - p_b)a^2 b^2}{(b^2 - a^2)r^2} \tag{2.27b}$$

where a and b are the internal and external radii, respectively. To calculate σ_θ or σ_r at the inside or outside surface, one must put $r = a$ or $r = b$. The above expressions are greatly simplified when p_a (or p_b) is zero.

PROBLEMS

1. Prove that:

   a. Even though the stress changes with distance from a reference point, mutually perpendicular shear stresses on adjacent planes are equal ($\tau_{xy} = \tau_{yx}$, etc.)

   b. The equilibrium of a weightless, stationary body requires the following *equations of equilibrium* to be satisfied:

$$\frac{\partial \sigma_x}{\partial x} + \frac{\partial \tau_{xy}}{\partial y} + \frac{\partial \tau_{xz}}{\partial z} = 0$$

$$\frac{\partial \sigma_y}{\partial y} + \frac{\partial \tau_{yz}}{\partial z} + \frac{\partial \tau_{yx}}{\partial x} = 0$$

$$\frac{\partial \sigma_z}{\partial z} + \frac{\partial \tau_{zx}}{\partial x} + \frac{\partial \tau_{zy}}{\partial y} = 0$$

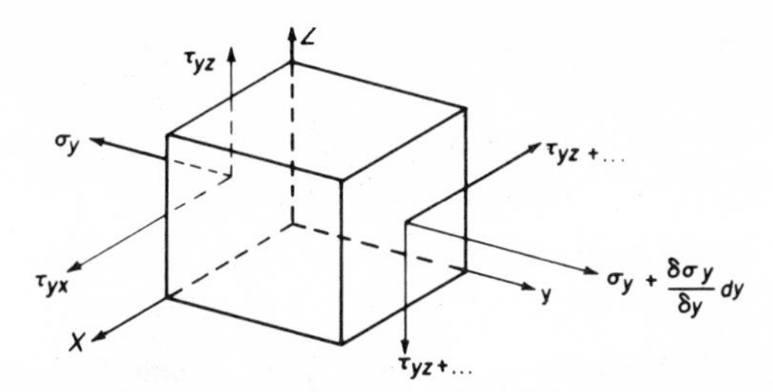

Use the stress and stress increment notation indicated (by example) on the above infinitesimal element.

2. a. Determine the values and directions of principal stresses at a point acted upon by the following components:

$$\begin{aligned} \sigma_x &= 2{,}000 & \tau_{xy} &= 250 \\ \sigma_y &= 1{,}000 & \tau_{yz} &= 650 \\ \sigma_z &= -500 & \tau_{zx} &= 200 \end{aligned}$$

b. Check the numbers obtained by elementary trigonometrical relations and the stress invariants I_1 and I_2.

3. Find the principal stresses and directions in a thin-walled tube to which a shear stress, $\tau = 12{,}000$ psi, is applied through end couples simultaneously with an axial pull, $\sigma = 8{,}000$ psi.

4. Specify a system of principal stresses σ_1, σ_2, σ_3 resulting in principal shear stresses: $\tau_{12} = 3{,}000$, $\tau_{23} = 4{,}000$, and $\tau_{31} = 1{,}500$ psi.

5. Stresses at a point defined by $r = 10$ and $\theta = \pi/6$ are: $\sigma_r = 2{,}000$, $\sigma_\theta = 4{,}000$ and $\tau_{r\theta} = 800$ psi. Determine analytically σ_x, σ_y, τ_{xy} as well as σ_1 and σ_2 and their directions. Assume plane stress conditions with angle θ measured counterclockwise from the positive x-coordinate.

6. Solve Problem 5 using Mohr's circle.

7. With the aid of (2.27), plot the ratio of σ_θ at the inside and outside surfaces of a tube as a function of the b/a ratio, starting from $b/a = 10$ and decreasing asymptotically toward $b/a = 1$.

8. A thin-walled cylindrical vessel with hemispherical ends is subjected to an internal pressure p. Compute the three principal stresses and shear stresses τ_{12}, τ_{23}, and τ_{31} in the cylindrical wall and at the pole of the hemisphere. $I.D. = d = 36$ inches, wall thickness $t = 0.375$ inches. Ignore all resulting data that are less than 5% of the maximum stress component (e.g., radial stress caused by internal pressure).

REFERENCES FOR FURTHER READING

(R2.1) Crandall, S. H., and N. C. Dahl, (eds.) *An Introduction to the Mechanics of Solids*. New York: McGraw-Hill Book Company, 1959.

(R2.2) Frocht, M. M., *Photoelasticity*, Vol. 1 New York and London: John Wiley & Sons, Inc., 1951.

(R2.3) Hoffman, O. and G. Sachs, *Introduction to the Theory of Plasticity for Engineers*. New York: McGraw-Hill Book Company, 1953.

(R2.4) Nadai, A., *Theory of Flow and Fracture of Solids*, Vol. 1 New York: McGraw-Hill Book Company, 1950.

3

ANALYSIS OF STRAIN

DISPLACEMENT

When force is applied to a body such as a piece of rubber attached to a rigid surface (Fig. 3.1), the points on its periphery as well as those inside the body move to new positions. In this chapter these point movements shall be examined since it is by these that strain is defined. If it is assumed that force F is gradually increased, the shape of the body changes as shown by the outlines 1, 2, 3, etc. An arbitrarily chosen internal point, initially at I, will move to positions II, III, etc., during deformation. The vector connecting the two positions of the point is its *displacement*. Naturally, as the point moves, the magnitude of the displacement increases. Generally, the simultaneous displacement of any other point would differ from that of the chosen point.

If, now, the origin of a coordinate system were attached to the rigid support, the successive positions of the moving point can be described as $(x_{\mathrm{I}}, y_{\mathrm{I}}, z_{\mathrm{I}})$, $(x_{\mathrm{II}}, y_{\mathrm{II}}, z_{\mathrm{II}})$, $(x_{\mathrm{III}}, y_{\mathrm{III}}, z_{\mathrm{III}})$, etc. By means of the reference system it is also possible to describe the displacement in terms of its components by projecting the displacement vector onto the three coordinate axes. These projections, or *displacement components* in the x, y, and z directions, are denoted by u, v, and w, respectively. If the displacement were from point I to point II, its components would be:

$$u_{\mathrm{I}} = x_{\mathrm{II}} - x_{\mathrm{I}} \qquad v_{\mathrm{I}} = y_{\mathrm{II}} - y_{\mathrm{I}} \qquad w_{\mathrm{I}} = z_{\mathrm{II}} - z_{\mathrm{I}}$$

Had the body not been attached, and if u, v, and w were constant at all points, the body would not have been deformed but instead would have moved

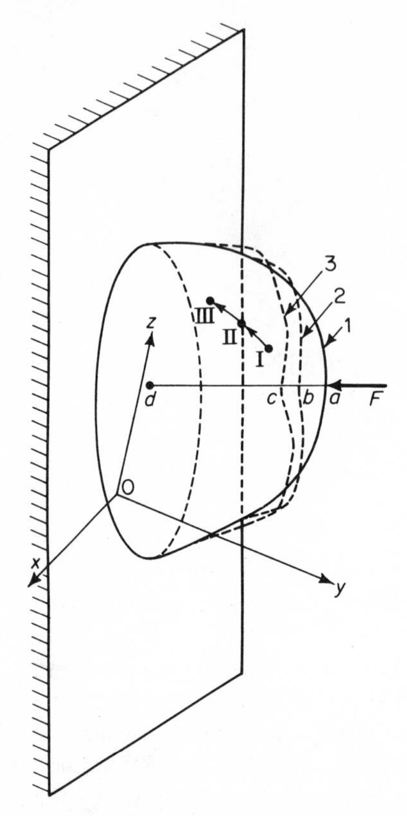

Fig. 3.1 Movement of a point (I) within a body as it is progressively deformed by an external force, F.

as a rigid unit. Motion of this type is known as *translation*. Another type of rigid unit motion that might be anticipated is *rotation* around a momentarily fixed axis.

DEFINITION OF STRAIN AND COMPONENTS OF STRAIN

Strain is defined as deformation over a unit length. As the strain created by force F, for example, one could consider the displacement ac divided by the initial thickness of the rubber body in the direction of F. Therefore, the strain would be ac/ad. One could also have subdivided the distance ad by, say, ten equally spaced points and then defined strain for each of these sections separately. For this purpose, the difference between the displacements of the beginning and end points of each section, i.e., its *deformation* must be divided by the initial spacing of these points. Experience shows that the strain values thus obtained will differ, decreasing from a maximum value in the sections adjacent to F to a minimum at the rigid wall. Obviously strain ac/cd obtained in the first place has some intermediate value between these extremes. As the length of the sections are made progressively smaller and eventually become of infinitesimal magnitude, one arrives at the notion of *strain at a point*.

Naturally either average strain or strain at a point can be measured not only in the direction of F but in any other direction as well. For numerical calculation it is convenient to resolve a strain in an arbitrary direction into six *strain components* in the directions of the coordinate axes. The latter are chosen to be of specific interest for the problem at hand.

The practical determination of strain components at a point begins with constructing at point I, a parallelepiped with edges parallel to the coordinate axes and with infinitesimal edge lengths dx, dy, and dz, as shown in light outline in Fig. 3.2a. During deformation this parallelepiped moves to new positions II and III, just as did the point in Fig. 3.1. The components of

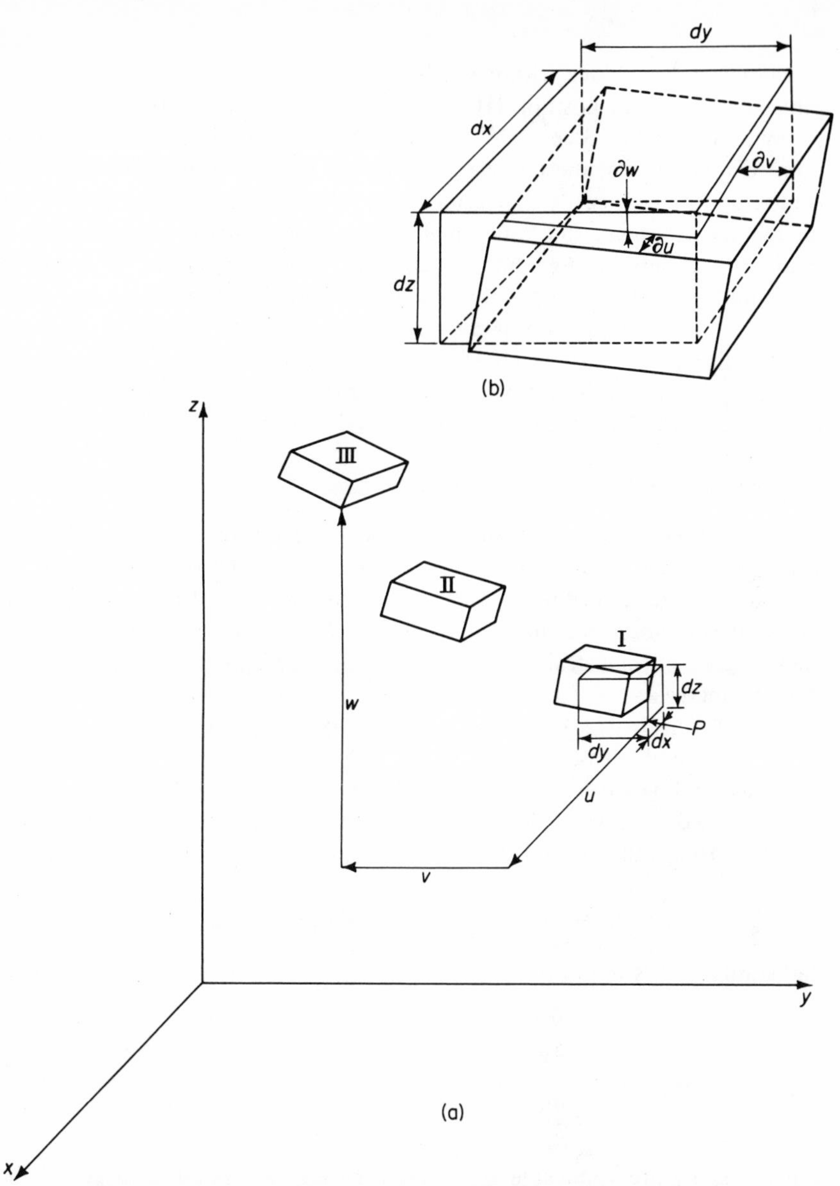

Fig. 3.2 (a) Change in the location, orientation, and shape of a rectangular parallelepiped constructed at point *I* of Fig. 3.1 as the body is distorted. (b) Enlarged view of our element before deformation (light outline) and after deformation (heavy outline).

displacement of the front bottom right corner, point *P*, will be u, v, and w. Another point P', an infinitesimal distance from *P*, will undergo a displacement of $u + du$, $v + dv$, and $w + dw$. Because these increments du, dv, and dw are small but generally not equal to zero, the element is not only moving as a rigid block within the body but also continuously changes its dimensions; and at position III its shape may be as shown by the enlarged heavy outline in Fig. 3.2b. Now, regardless of the complexity of the point to point displacements, the test element can do nothing more than translate, rotate, and/or change its dimensions. In other words, deformation of the bulk body can cause the test element to undergo a linear translation as a rigid unit, a rotation around a fixed axis, and a change of dimensions. In defining strains, we are only concerned with the latter. Consequently, the two types of rigid movement must be eliminated. This shall be done graphically (Fig. 3.2b) by constructing the parallelepiped at position III with the dimensions it had before deformation so the undeformed and deformed elements have a common corner and a common base plane.

NORMAL STRAIN

Now, by comparing the dimensions of the test element before and after straining, we can define the various strain terms. The edge extensions or contractions are termed *normal strains* and are designated by an ε (epsilon) followed by a subscript indicating the direction in which the strain is measured. Hence ε_x is a normal strain in the direction of the x axis. If the length of the element increases as a result of the deformation, the strain is positive or tensile; if it decreases, the strain is negative or compressive.

If the initial element length is dx before deformation (Fig. 3.2b), it becomes $dx + (\partial u/\partial x)\,dx$ after deformation. The partial differential $\partial u/\partial x$ must be used because u is not only a function of x but also a function of y and z. Hence the normal strain:

$$\varepsilon_x = \left[\left(dx + \frac{\partial u}{\partial x}\,dx\right) - dx\right]\Big/ dx = \frac{\partial u}{\partial x}$$

and similarly it is found that:

$$\varepsilon_y = \frac{\partial v}{\partial y},$$

and (3.2)

$$\varepsilon_z = \frac{\partial w}{\partial z}$$

Equations 3.2 are applicable where deformations are small or large. If the strains are small, the reference lengths ∂x, ∂y, and ∂z can be considered constant during the deformation so that, in integrating Eqs. (3.2), the denominators are treated as constants. Before integrating, however, it is convenient to change the symbols to those commonly used in engineering practice. Thus, Eqs. (3.2) can be rewritten as:

$$\varepsilon = \frac{dl}{l} \tag{3.2a}$$

where dl is the amount by which l changes in the direction in which it is measured. Assuming l to be constant and equal to the initial length l_o, integration of (3.2a) between l_o and the instantaneous length l_f yields:

$$\varepsilon = \frac{l_f - l_o}{l_o} = \frac{l_f}{l_o} - 1 \tag{3.3}$$

Multiplying (3.3) by 100 gives ε in per cent which is denoted by e and is the common strain term (per cent elongation) encountered in engineering problems. It is usually called *conventional* or *engineering strain.*

Occasionally it is more convenient to refer length changes to the variable instantaneous length rather than to the constant initial dimensions. Hence the denominators in Eqs. (3.2) or (3.2a) are represented by the variable l, and, on integrating between two arbitrary lengths, one obtains:

$$\bar{\varepsilon} = \int_{l_1}^{l_2} \frac{dl}{l} = ln \frac{l_2}{l_1} \tag{3.4}$$

This value for which the symbol $\bar{\varepsilon}$ is used is called *natural, logarithmic,* or *true* strain; and when l is the instantaneous length, and l_o the original, length:

$$\bar{\varepsilon}_o = ln\left(\frac{l}{l_o}\right) = ln(1 + \varepsilon) \tag{3.4a}$$

The logarithmic strain has a number of important advantages in developing numerical relations between stresses and strains in problems involving large deformations. These will become apparent to the reader on a number of occasions in this text.

SHEAR STRAIN

In addition to length changes, the angles between the sides of the parallelepiped may be distorted by a sliding of parallel layers, much as one might do with a deck of cards. The amount of shift undergone by two parallel planes that are a unit distance apart is termed *shear strain.* An equivalent, and sometimes more convenient, definition of shear strains is the cotangent of the angle between two planes that were perpendicular before deformation. Shear strains are designated by γ (gamma) and two subscripts giving the direction of the two sides of the right angle that is being distorted. Hence, γ_{xy} indicates a deformation that changes the right angle between the x and y direction to α. If α is less than $\pi/2$, the strain is said to be positive; if it is greater than $\pi/2$, the strain is negative.

In defining shear strain one assumes that strains are small compared to unity so that all the approximations made with small angles are applicable to the definition; specifically, the sum of the tangents of two small angles nearly equals the tangent of the sum of the angles. Hence, in Fig. 3.2c,

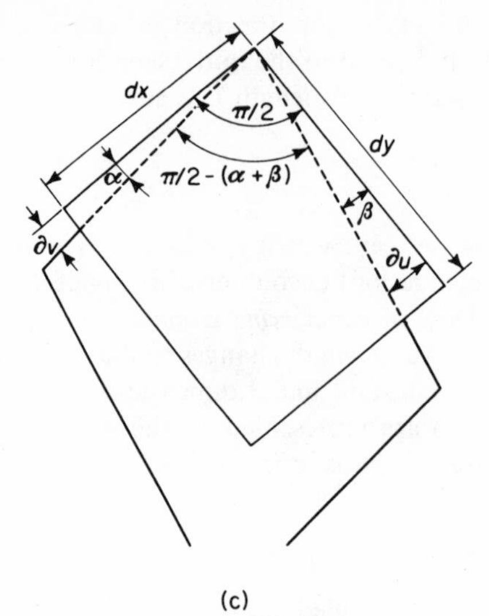

Fig. 3.2 (c) Bottom view of Fig. 3.2b.

$\tan \alpha + \tan \beta \approx \tan (\alpha + \beta)$. Since shear strains are equal to the cotangent of an angle formed by two initially perpendicular lines:

$$\gamma_{xy} = \cot \left[\frac{\pi}{2} - (\alpha + \beta)\right] = \tan (\alpha + \beta)$$

$$\approx \tan \alpha + \tan \beta = \frac{\partial v}{\partial x} + \frac{\partial u}{\partial y}$$

Following the same procedure with other faces of our reference element, results in all three equations:

$$\gamma_{xy} = \frac{\partial v}{\partial x} + \frac{\partial u}{\partial y}$$

$$\gamma_{yz} = \frac{\partial w}{\partial y} + \frac{\partial v}{\partial z} \tag{3.5}$$

$$\gamma_{zx} = \frac{\partial u}{\partial z} + \frac{\partial w}{\partial x}$$

These definitions of shear strains are only valid for small deformations because of the assumptions introduced above. Shear strain terms have also been developed for defining large deformations by avoiding the small angle assumptions. These, however, are used for special purposes only.

One characteristic of shear strains not obvious in Fig. 3.2 is that shear strains on perpendicular planes are equal. This becomes apparent by drawing a square with sides h (Fig. 3.3). After shearing the square, it becomes

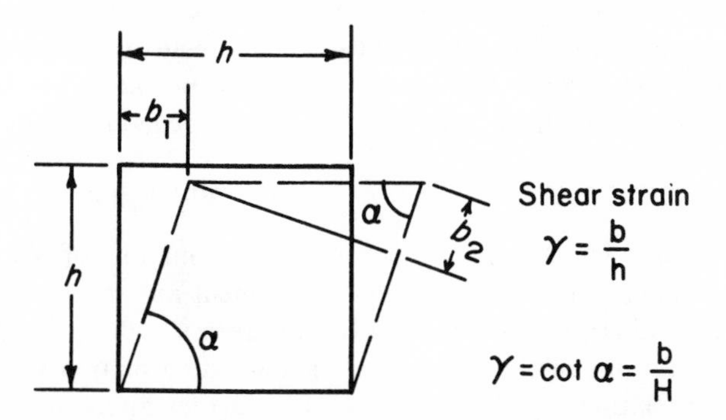

Fig. 3.3 Shear strain on two pairs of prependicular planes. The planes are h units apart.

a parallelogram with corners α and $\pi - \alpha$. The shear strain undergone by the horizontal sides is b_1/h, and by the vertical sides, b_2/h. Since $b_1 \approx b_2$ ($\approx h \cot \alpha$), the shear strain on perpendicular planes are equal. This figure also shows that the shear strain, γ, is identical with the cotangent of the angle between initially perpendicular planes; i.e., $\gamma = \cot \alpha$.

The application of the above definitions of strain to a practical problem would involve elementary mathematical methods so long as u, v, and w are proportional to x, y, and z, respectively. Deformations of this type are called *homogeneous* and are characterized by the fact that, after deformation, all points that initially lie on a single plane remain coplanar, and initially parallel planes remain parallel. Examples of homogeneous and *nonhomogeneous* deformation are shown in Fig. 3.4.

The two mutually perpendicular sets of parallel lines scribed on the face of bar (a) are altered on straining; as shown in (b), the straight lines are distorted in the vicinity of the hole although they remain straight and

Fig. 3.4 Flat bar with hole in center: (a) Before and; (b) After straining.

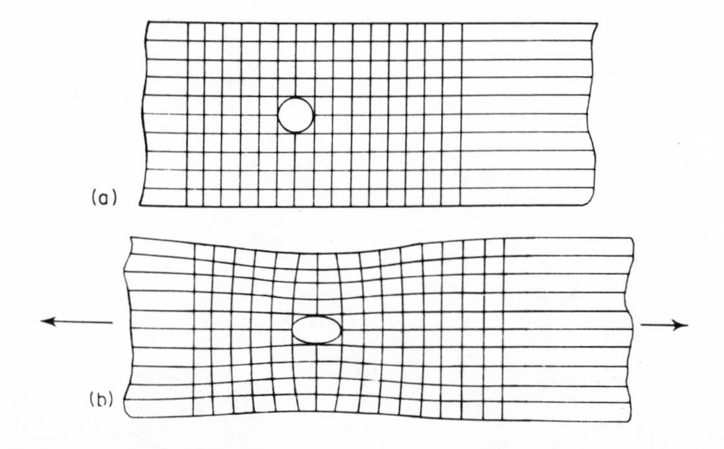

parallel at locations remote from it. The deformation is thus nonhomogeneous near the hole but homogeneous in the rest of the bar. Homogeneous deformation is characterized by a constancy of strain along any direction in the body.

Absence of large-scale homogeneity does not invalidate the assumption, identical with the treatment of stress, of a homogeneous strain at a point.

COMPONENTS OF STRAINS AT A POINT: PRINCIPAL STRAINS

Since strains are determined from the deformation of the edges and angles of a small test element, tilting this element in various directions will cause these strains to alter continually. Because it is frequently helpful to know the directions of the largest and smallest deformation within a body, and also the relationships between strains in various directions, the effect of the elements orientation on the magnitudes of the strain components will now be determined. The appropriate equations are very much like those found for stress, leading to the conclusion that deformations can be completely described by either the six components ε_x, ε_y, ε_z, γ_{xy}, γ_{yz}, γ_{zx}, or by three *principal strains.*

Let $MN = ds$ be the infinitesimal distance between two adjacent points prior to deformation, as shown in Fig. 3.5. This changes after deformation to M_1N_1, which length will be called ds_1. The displacement components of point M in going to M_1 are u, v, and w, while those of point N are u', v', and w'. The directional cosines of MN are:

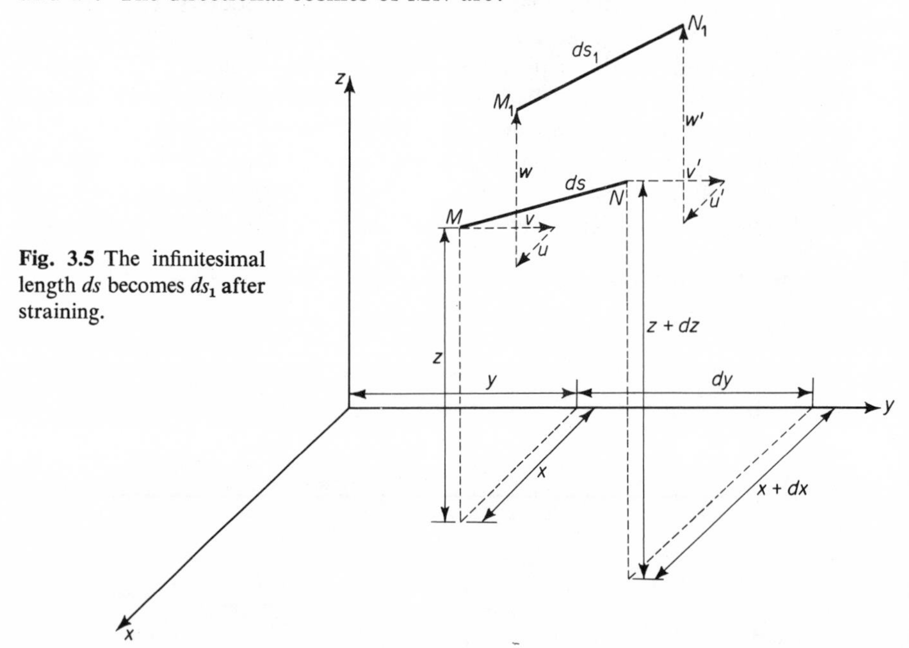

Fig. 3.5 The infinitesimal length ds becomes ds_1 after straining.

$$k = \frac{dx}{ds}; \qquad l = \frac{dy}{ds}; \qquad \text{and} \qquad m = \frac{dz}{ds} \tag{a}$$

also;

$$ds_1^2 = (dx + u' - u)^2 + (dy + v' - v)^2 + (dz + w' - w)^2 \tag{b}$$

and

$$ds_1 = ds(1 + \varepsilon) \tag{c}$$

where ε, by definition, is the normal strain in ds produced by the deformation as given above. Carrying out (b), one obtains:

$$ds_1^2 = dx^2 + dy^2 + dz^2 + 2\,dx(u' - u) + 2\,dy(v' - v) + 2\,dz(w' - w) + (u' - u)^2 + (v' - v)^2 + (w' - w)^2 \tag{d}$$

On the other hand, squaring (c) yields:

$$ds_1^2 = ds^2(1 + 2\varepsilon + \varepsilon^2) \tag{e}$$

The terms, $(u' - u)^2$, etc., are very small compared with dx, etc., and can be disregarded since it is assumed that the deformations are small. Remembering that $ds^2 = dx^2 + dy^2 + dz^2$, and subtracting this quantity from each side of (d) yields:

$$ds_1^2 - ds^2 = 2\,dx(u' - u) + 2\,dy(v' - v) + 2\,dz(w' - w) \tag{f}$$

From (e) one obtains:

$$ds_1^2 - ds^2 = 2\,ds^2\varepsilon \tag{g}$$

since ε^2 is very small for small strains. Equating (f) and (g) gives:

$$ds^2\varepsilon = dx(u' - u) + dy(v' - v) + dz(w' - w) \tag{h}$$

or:

$$\varepsilon = \frac{dx}{ds}\frac{(u' - u)}{ds} + \frac{dy}{ds}\frac{(v' - v)}{ds} + \frac{dz}{ds}\frac{(w' - w)}{ds} \tag{i}$$

Using (a):

$$\varepsilon = k\frac{u' - u}{l} + l\frac{v' - v}{l} + m\frac{w' - w}{l} \tag{j}$$

The quantities $(u' - u)$, $(v' - v)$, and $(w' - w)$ are the projected length changes of the distance ds in the coordinate directions. The length of u' can be found by adding onto u the product of the rate at which u changes in each of the three coordinate directions times the length of the element in the appropriate direction.

Hence:

$$u' = u + \frac{\partial u}{\partial x}dx + \frac{\partial u}{\partial y}dy + \frac{\partial u}{\partial z}dz$$

Similarly, v' and w' can also be found:

$$v' = v + \frac{\partial v}{\partial x} dx + \frac{\partial v}{\partial y} dy + \frac{\partial v}{\partial z} dz \tag{k}$$

$$w' = w + \frac{\partial w}{\partial x} dx + \frac{\partial w}{\partial y} dy + \frac{\partial w}{\partial z} dz$$

Rearranging equation (k) and substituting in (j) gives:

$$\varepsilon = k\left(\frac{\partial u}{\partial x}\frac{dx}{l} + \frac{\partial u}{\partial y}\frac{dy}{l} + \frac{\partial u}{\partial z}\frac{dz}{l}\right) + l\left(\frac{\partial v}{\partial x}\frac{dx}{l} + \frac{\partial v}{\partial y}\frac{dy}{l} + \frac{\partial v}{\partial z}\frac{dz}{l}\right) + m\left(\frac{\partial w}{\partial x}\frac{dx}{l} + \frac{\partial w}{\partial y}\frac{dy}{l} + \frac{\partial w}{\partial z}\frac{dz}{l}\right)$$

which, after regrouping and substituting from equation (a), becomes:

$$\varepsilon = \frac{\partial u}{\partial x} k^2 + \frac{\partial v}{\partial y} l^2 + \frac{\partial w}{\partial z} m^2 + \left(\frac{\partial u}{\partial y} + \frac{\partial v}{\partial x}\right) kl + \left(\frac{\partial v}{\partial z} + \frac{\partial w}{\partial y}\right) lm + \left(\frac{\partial w}{\partial x} + \frac{\partial u}{\partial z}\right) mk$$

Using (3.2) and (3.5) this finally leads to:

$$\varepsilon = \varepsilon_x k^2 + \varepsilon_y l^2 + \varepsilon_z m^2 + \gamma_{xy} kl + \gamma_{yz} lm + \gamma_{zx} mk \tag{3.6}$$

Expression (3.6) is analogous with that obtained for the resultant stress in Eq. (2.8). Hence, Eq. (3.6) can be reduced to the form of a *strain quadric* analogous to the stress quadric defined by Eq. (2.9). Accordingly, ε_1, ε_2, and ε_3 are the *principal strains*; the directions in which they act are the *principal axes of strain*; and the corresponding planes, the *principal planes of strain*. All the characteristics of principal stresses, which had previously been developed for the stress quadric, are applicable to principal strains.

The normal and shear strain components at a point are also related by a cubic equation of the type developed for stress [Eq. (2.11)]. The corresponding strain expression is:

$$\epsilon^3 - (\varepsilon_x + \varepsilon_y + \varepsilon_z)\epsilon^2 + [\varepsilon_x\varepsilon_y + \varepsilon_y\varepsilon_z + \varepsilon_z\varepsilon_x - \tfrac{1}{4}(\gamma_{xy}^2 + \gamma_{yz}^2 + \gamma_{zx}^2)]\epsilon - \varepsilon_x\varepsilon_y\varepsilon_z - \tfrac{1}{4}(\gamma_{xy}\gamma_{yz}\gamma_{zx} - \varepsilon_x\gamma_{yz}^2 - \varepsilon_y\gamma_{zx}^2 - \varepsilon_z\gamma_{xy}^2) = 0 \tag{3.7}$$

The symbol ϵ has the same function here as S had in Eq. (2.11). Again, the coefficients of ϵ are constant for any orientation of our test element. These three *strain invariants* are:

$$J_1 = \varepsilon_x + \varepsilon_y + \varepsilon_z$$

$$J_2 = \varepsilon_x\varepsilon_y + \varepsilon_y\varepsilon_z + \varepsilon_z\varepsilon_x - \tfrac{1}{4}(\gamma_{xy}^2 + \gamma_{yz}^2 + \gamma_{zx}^2)$$

and:

$$J_3 = \varepsilon_x\varepsilon_y\varepsilon_z + \tfrac{1}{4}(\gamma_{xy}\gamma_{yz}\gamma_{zx} - \varepsilon_x\gamma_{yz}^2 - \varepsilon_y\gamma_{zx}^2 - \varepsilon_z\gamma_{xy}^2) \qquad (3.8)$$

Note that (3.6) through (3.8) are identical to the appropriate expressions for the stress, except for the normal strain and one-half the shear strains replacing the normal and shear stresses.

STRAIN COMPONENTS AT A POINT AS THE COORDINATE AXES ARE ROTATED

In Chapter 2 equations were developed for the six stress components of a coordinate system, inclined in some arbitrary fashion to the original system on which these six values were known. A similar procedure makes it possible to derive equations for the strain components after the system is rotated as long as the components before rotation are known. If, again, the angles between the two systems are α_1, β_2, and γ_3 (Fig. 2.9), the angular relations and direction cosines between each of the new and original axes are given by Table 2.2. The equations equivalent to (2.20a) and (2.20b) are:

$$\begin{aligned}
\varepsilon_{x'} &= \varepsilon_x k_1^2 + \varepsilon_y l_1^2 + \varepsilon_z m_1^2 + \gamma_{xy}k_1l_1 + \gamma_{yz}l_1m_1 + \gamma_{zx}m_1k_1 \\
\varepsilon_{y'} &= \varepsilon_x k_2^2 + \varepsilon_y l_2^2 + \varepsilon_z m_2^2 + \gamma_{xy}k_2l_2 + \gamma_{yz}l_2m_2 + \gamma_{zx}m_2k_2 \qquad (3.9a) \\
\varepsilon_{z'} &= \varepsilon_x k_3^2 + \varepsilon_y l_3^2 + \varepsilon_z m_3^2 + \gamma_{xy}k_3l_3 + \gamma_{yz}l_3m_3 + \gamma_{zx}m_3k_3
\end{aligned}$$

and

$$\begin{aligned}
\gamma_{x'y'} = \gamma_{y'x'} &= 2\varepsilon_x k_1k_2 + 2\varepsilon_y l_1l_2 + 2\varepsilon_z m_1m_2 + \\
&\quad \gamma_{xy}(k_1l_2 + k_2l_1) + \gamma_{yz}(l_1m_2 + l_2m_1) + \gamma_{zx}(m_1k_2 + m_2k_1) \\
\gamma_{y'z'} = \gamma_{z'y'} &= 2\varepsilon_x k_2k_3 + 2\varepsilon_y l_2l_3 + 2\varepsilon_z m_2m_3 + \\
&\quad \gamma_{xy}(k_2l_3 + k_3l_2) + \gamma_{yz}(l_2m_3 + l_3m_2) + \gamma_{zx}(m_2k_3 + m_3k_2) \qquad (3.9b) \\
\gamma_{z'x'} = \gamma_{x'z'} &= 2\varepsilon_x k_3k_1 + 2\varepsilon_y l_3l_1 + 2\varepsilon_z m_3m_1 + \\
&\quad \gamma_{xy}(k_3l_1 + k_1l_3) + \gamma_{yz}(l_3m_1 + l_1m_3) + \gamma_{zx}(m_3k_1 + m_1k_3)
\end{aligned}$$

Observe that (2.20) becomes (3.9) when σ and τ are replaced by ε and $\gamma/2$ with the proper subscripts.

VOLUME STRAIN (DILATATION)

Volume strain or dilatation is defined as the volume change per unit of volume of a deformed body. This quantity is readily defined in terms of principal strains.

Consider a rectangular parallelepiped, whose sides are the principal planes of strain, with an initial volume V and edge lengths a, b, c (Fig. 3.6). After straining, its volume becomes V' and the edge lengths a', b', and c'.

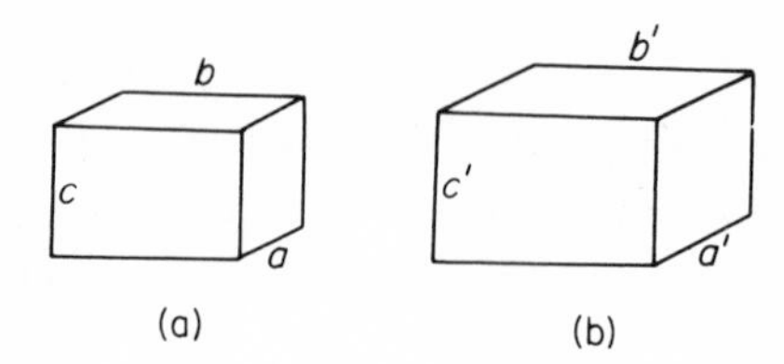

Fig. 3.6 Dimensions of rectangular parallelepiped with initial volume V and edge lengths a, b, and c and final volume V' with edge lengths a', b', and c'.

Then:

$$a' = a(1 + \varepsilon_1)$$
$$b' = b(1 + \varepsilon_2)$$
$$c' = c(1 + \varepsilon_3)$$

and

$$V' = a'b'c' = a(1 + \varepsilon_1)b(1 + \varepsilon_2)c(1 + \varepsilon_3)$$
$$= abc(1 + \varepsilon_1 + \varepsilon_2 + \varepsilon_3 + \varepsilon_1\varepsilon_2 + \varepsilon_1\varepsilon_3 + \varepsilon_2\varepsilon_3 + \varepsilon_1\varepsilon_2\varepsilon_3)$$

When linear strains are small, their products can be neglected so that:

$$\varepsilon_V = \frac{\Delta V}{V} = \frac{V' - V}{V} \approx \frac{abc(1 + \varepsilon_1 + \varepsilon_2 + \varepsilon_3) - abc}{abc} = \varepsilon_1 + \varepsilon_2 + \varepsilon_3 \quad (3.10a)$$

According to Eq. (3.8), the sum of the normal strains in three perpendicular directions is invariant so that:

$$\varepsilon_V = \varepsilon_x + \varepsilon_y + \varepsilon_z = 3\varepsilon_a = J_1 \quad (3.10b)$$

where ε_a is the mean strain.

If the strains are large, their products may exceed their sums and cannot be disregarded. This becomes important in forming problems and rubber elasticity. Logarithmic strains (3.4) are used in these cases (see Chapter 6).

Since the volume strain is equal to the first invariant, it follows that shear strains, at least small ones, have no effect on volume. This can be proved directly by distorting a rectangular parallelepiped into a rhombohedron by small strains γ_{xy}, γ_{yz}, and γ_{zx} and comparing the volumes before and after straining. The volume strain caused by the shear distortion will be of the order of γ_{avg}^2 which is a negligible value.

PLANE STRAIN STATE

As was the case with stress, two dimensional or plane strain is of special interest. It will be recalled that a plane stress state is developed in loading bodies in which one dimension is very small compared with the other two; e.g., in bulging a diaphragm. Plane strain states, on the other hand, occur in problems in which the body has one long dimension compared with the other two—such as in a tunnel, thin-walled tube, or wide plate. Because in many two dimensional strain problems the strain components on one coordinate system will be known while their value on some other rotated system is needed, expressions for ε and γ as the reference system rotates will now be developed.

In Chapter 2, the equations for plane stress were derived by using Eq. (2.20a) and (2.20b) and allowing all the stresses in one direction (the z direction) to vanish. A similar technique is applied for plane strain by letting $\varepsilon_z = \gamma_{yz} = \gamma_{zx} = 0$. Again, on going from a three to a two dimensional case, only four direction cosines (k_1, k_2, l_1, l_2) out of the total of nine do not vanish. This simplifies Eqs. (3.9) to:

$$\begin{aligned} \varepsilon_{x'} &= \varepsilon_x k_1^2 + \varepsilon_y l_1^2 + \gamma_{xy} k_1 l_1 \\ \varepsilon_{y'} &= \varepsilon_x k_2^2 + \varepsilon_y l_2^2 + \gamma_{xy} k_2 l_2 \end{aligned} \tag{3.11a}$$

and

$$\gamma_{x'y'} = \gamma_{y'x'} = 2\varepsilon_x k_1 k_2 + 2\varepsilon_y l_1 l_2 + \gamma_{xy}(k_1 l_2 + k_2 l_1) \tag{3.11b}$$

The trigonometric identities, used to derive Eqs. (2.21) with the aid of Fig. 2.11, enable (3.11) to be rewritten in a form structurally similar to (2.21):

$$\begin{aligned} \varepsilon_{x'} &= \frac{\varepsilon_x + \varepsilon_y}{2} + \frac{\varepsilon_x - \varepsilon_y}{2} \cos 2\alpha_1 + \frac{\gamma_{xy}}{2} \sin 2\alpha_1 \\ \varepsilon_{y'} &= \frac{\varepsilon_x + \varepsilon_y}{2} - \frac{\varepsilon_x - \varepsilon_y}{2} \cos 2\alpha_1 - \frac{\gamma_{xy}}{2} \sin 2\alpha_1 \end{aligned} \tag{3.12a}$$

and

$$\gamma_{x'y'} = \gamma_{y'x'} = -(\varepsilon_x - \varepsilon_y) \sin 2\alpha_1 + \gamma_{xy} \cos 2\alpha_1 \tag{3.12b}$$

These equations, like the plane stress equations, define a circle—*Mohr's circle of strain*. In constructing the figure, a technique identical with that for stress circles is used. The details of the construction are left to the reader.

BIAXIAL STRAIN MEASUREMENT WITH STRAIN ROSETTES. RESISTANCE STRAIN GAUGES

It will be shown now that, on knowing the normal strains in any three directions at a point on a surface, strain components in any other directions can be calculated from them. Such procedure is often used in practice since normal strains can be measured quite easily. In this manner it is possible to determine the magnitude and directions of the principal stresses or any other stresses of interest acting at a particular location.

For this purpose the first equation of (3.11) will be rewritten by inserting the appropriate functions of angle α in Fig. 2.10 (suffix "1" is omitted as redundant). With $k_1 = \cos \alpha$, and $l_1 = \sin \alpha$:

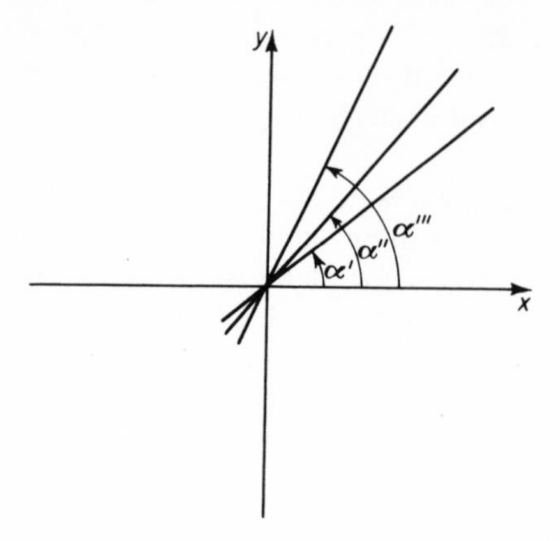

Fig. 3.7 Directions in which strain is measured are defined by angles α', α'', and α'''.

$$\varepsilon_\alpha(=\varepsilon_{x'}) = \varepsilon_x \cos^2 \alpha + \varepsilon_y \sin^2 \alpha + \gamma_{xy} \sin \alpha \cos \alpha \tag{a}$$

If strain measurements are made in three coplanar directions inclined to an arbitrarily selected x direction at angles α', α'', and α''' (Fig. 3.7), three simultaneous equations of the (a) type can be written:

$$\begin{aligned}
\varepsilon_{\alpha'} &= \varepsilon_x \cos^2 \alpha' + \varepsilon_y \sin^2 \alpha' + \gamma_{xy} \sin \alpha \cos \alpha' \\
\varepsilon_{\alpha''} &= \varepsilon_x \cos^2 \alpha'' + \varepsilon_y \sin^2 \alpha'' + \gamma_{xy} \sin \alpha'' \cos \alpha'' \\
\varepsilon_{\alpha'''} &= \varepsilon_x \cos^2 \alpha''' + \varepsilon_y \sin^2 \alpha''' + \gamma_{xy} \sin \alpha''' \cos \alpha'''
\end{aligned} \tag{3.13}$$

The above set can be readily solved for ε_x, ε_y, and γ_{xy} from which the principal (or any other) strain components are found analytically or geometrically (Mohr's circle). The directions of the principal strains are read

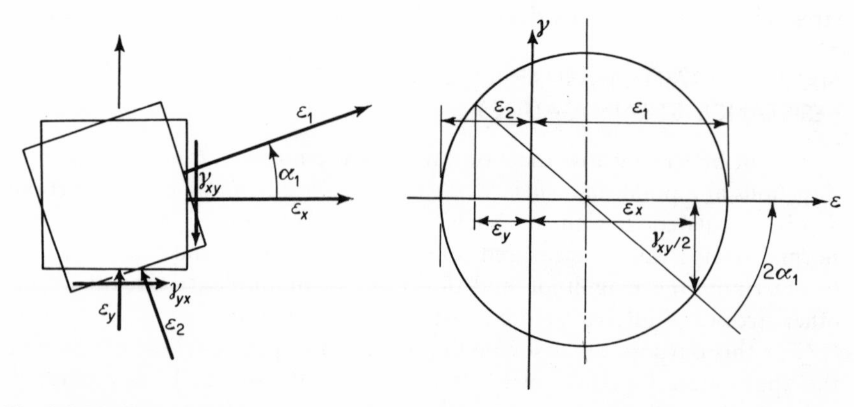

Fig. 3.8 Finding principal strains ε_1 and ε_2 from known ε_x, ε_y, and γ_{xy}.

off the circle (Fig. 3.8), or the angle between ε_1 (ε_2) and ε_x (or ε_y) is calculated from:

$$\tan 2\alpha_1 = \frac{\gamma_{xy}}{\varepsilon_x - \varepsilon_y} \tag{3.14}$$

an equation which has two roots and is analogous to (2.22).

In practical applications the directions in which the three strains $\varepsilon_{\alpha'}$, $\varepsilon_{\alpha''}$, and $\varepsilon_{\alpha'''}$ are measured are selected so as to simplify the calculations involved in using (3.13). Most frequently, the three angles α', α'', and α''' increase in 45° or 60° steps so that:

$$\alpha'' = \alpha' + 45 \text{ (or } 60°\text{), and } \alpha''' = \alpha' + 90 \text{ (or } 120°\text{)}$$

A set of strain measuring devices (gauges) arranged in the 45° pattern is called a *rectangular rosette* (Fig. 3.9a). A 60° or equilateral arrangement follows the scheme in diagram (b) but is often executed as a *delta rosette,* according to (c), which is more convenient and compact.

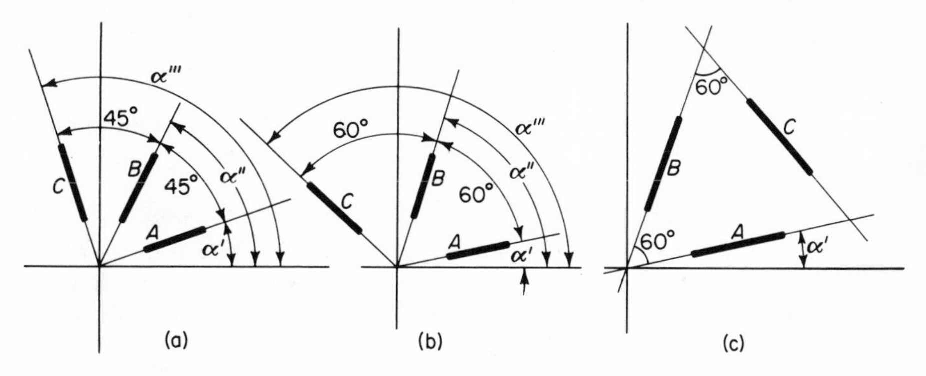

Fig. 3.9 Strain rosettes: (a) Rectangular; (b) Equiangular; (c) Delta. *A, B, C*-gauges.

Strains in existing structures and equipment are commonly measured with the aid of *resistance strain gauges.* Such gauges consist of a length of very thin resistance wire sandwiched between strips of paper or other flexible insulating material. To obtain a sufficiently short gauge, the wire is run to and fro many times (Fig. 3.10).

The gauge is bonded with a suitable adhesive onto the surface on which the strain is measured. After drying, the gauge, and the wires inside it, will extend and contract together with the elements to which it is bonded. The electrical resistance of the wire will change in proportion to its extension, and the strain is equal to the observed change of resistance multiplied by a constant factor.

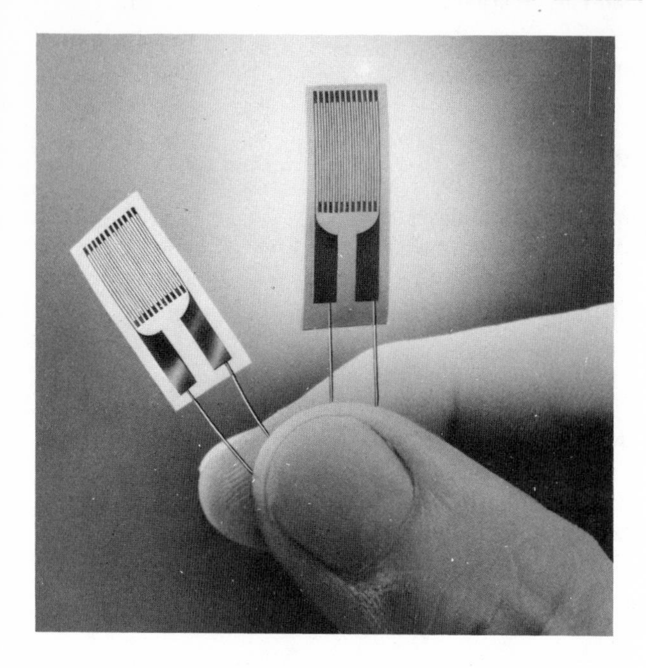

Fig. 3.10 Photograph of resistance strain gauges. (Baldwin-Lima-Hamilton Corp., Price List Booklet 4310-62.)

Strain measurements based on this principle have been developed into a highly versatile technique which is dealt with in many special monographs (R3.1).

Since, more often than not, one is more interested in stresses than in strains at a particular location in a structural element, these stresses can be determined from strain measurements as long as some fundamental relation between the two exists in a particular material or group of materials. This relation will be discussed in Chapter 5.

STRESS AND STRAIN TERMINOLOGY

As previously pointed out, the equations developed for stress and strain are identical if the corresponding stress and strain terms shown in Table 3.1 are substituted one for the other.

TABLE 3.1

CORRESPONDING STRESS AND STRAIN TERMS

Stress	σ_x	σ_y	σ_z	τ_{xy}	τ_{yz}	τ_{zx}
Strain	ε_x	ε_y	ε_z	$\dfrac{\gamma_{xy}}{2}$	$\dfrac{\gamma_{yz}}{2}$	$\dfrac{\gamma_{zx}}{2}$

The symbols used for normal and shear stresses and strains in this and the preceding chapters are thought, by the authors, to be the most commonly used in engineering practices. They are by no means unique, however, so that some of the other designations that may be encountered by the reader are shown in Table 3.2. Unfortunately, many other symbols are used for stress and strain especially in describing the properties of high polymers.

TABLE 3.2

ALTERNATE NOTATION FOR STRESS AND STRAIN

Stress				*Strain*	
I	II	III	IV	I	II
σ_x	σ_{xx}	X_x	S_{xx}	ε_x	e_{xx}
σ_y	σ_{yy}	Y_y	S_{yy}	ε_y	e_{yy}
σ_z	σ_{zz}	Z_z	S_{zz}	ε_z	e_{zz}
τ_{xy}	σ_{xy}	X_y	S_{xy}	γ_{xy}	e_{xy}
τ_{yz}	σ_{yz}	Y_z	S_{yz}	γ_{yz}	e_{yz}
τ_{zx}	σ_{zx}	Z_x	S_{zx}	γ_{zx}	e_{zy}

PROBLEMS

1. A bar of plasticine of uniform cross section is extended until its length is tripled. Plot the conventional vs. natural strains as the length increases from its initial to final value. Does the plot suggest any advantage of one type of plotting compared with the other?

2. A wire is reduced by passing it through seven successive dies, where the cross section is reduced by 35% in each pass. Plot the relative length change ($= l/l_o$) in terms of conventional and natural strains and discuss the results.

3. Assume that the volume of a cube did not change when the length of one of its sides is doubled ($\varepsilon_1 = 1$). Calculate $\varepsilon_2 = \varepsilon_3$ using (3.10b) and a more accurate method.

4. Draw Mohr's circle represented by Eqs. (3.12).

5. Construct Mohr's circle for the plane strain condition:

$$\varepsilon_x = 0.011, \qquad \varepsilon_y = 0.004, \qquad \gamma_{xy} = 0.0065.$$

6. Three resistance strain gauges are affixed to a small area of an airplane wing. Gauge No. 1 is parallel to the wing length while 2 and 3 are at ±120° to the first one. The values read when the wing is loaded are 1,000, 720, and 600 micro-inches/inch for gauges 1, 2, and 3, respectively. What are the values and directions of the principal strains?

REFERENCES FOR FURTHER READING

See References at end of Chapter 2 and the following:

(R3.1) Perry, C. C. and H. R. Lissner, *Strain Gage Primer*. New York: McGraw-Hill Book Company, 1962.

part III

RHEOLOGY

4

CLASSIFICATION AND STRUCTURE OF ENGINEERING MATERIALS

Although stress and strain can be described without a knowledge of the nature of specific materials, different materials respond differently to applied loads as indicated in the Introduction. It therefore is necessary to be concerned with specific substances, or at least categories of substances.

As a result of the progress made in solid state science since the 1920's, it is now possible, at least in a general way, to correlate the bulk properties of engineering materials with their atomic and microscopic structural features.

There are two facets of importance on the atomic level which affect a substance's reaction to stress. These are the forces binding the atoms to each other and the arrangement of atoms within the materials. Studies of these features are not independent of each other but, since each is investigated by different means, they complement each other.

If one were to try to explain all material properties, mechanical as well as others—such as density, electrical resistivity, etc.—on the basis of atomic models, it would be found that properties are classified as either *structurally insensitive* or *structurally sensitive*. The former designates properties that depend only on average atomic characteristics. For example, in calculating a metal's density, only the weight of an individual atom and the ideal arrangement of atoms that the aggregate approaches must be known. Slight irregularities extending over very small volumes of the body can be neglected. So far as mechanical properties are concerned, only elastic behaviors can be considered as structurally insensitive. Thus, if one knows the binding energies as a function of distance between atoms in the aluminum wire referred to in the Introduction, he should be able to calculate the slope of the elastic portion of the curve in Fig. 1.2. On the other hand, the maximum load that the wire could withstand, or the amount by which it could stretch before breaking are structurally sensitive. That is, they depend not only on average binding energies of the atoms and the over-all atomic structure but also on localized structural irregularities. One of these is a line defect in the arrangement of atoms and is known as a dislocation. These were mentioned in the Introduction. Dislocations, or any imperfections, however, only alter the behavior of perfect aggregates, and, therefore, to understand them we must first study the ideal atom arrangements that real materials approach. Hence, in this chapter perfect atomic structures are discussed. The important imperfections will be considered later.

FORCES BETWEEN ATOMS

Molecular bonding and cohesion in solids result because the energy of the atoms is less in the aggregate state than it is when the atoms are separated by large distances. This situation could equally well be described in terms of interatomic forces, but usually energy is used for consistency with quantum mechanics. The interatomic force between two atoms is quantitatively equal to the slope of the energy-distance curve. At the equilibrium position the interaction energy curve goes through a minimum so that the attractive and repulsive forces between adjacent atoms are equal.

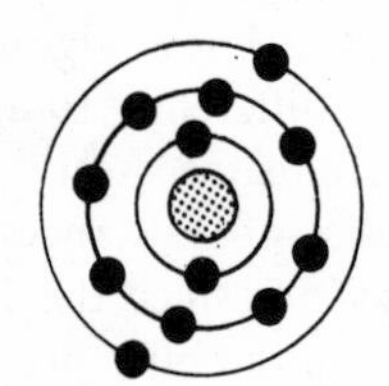

Fig. 4.1 Shell model of magnesium.

An isolated atom consists of a nucleus surrounded by clouds of electrons grouped in shells, Fig. 4.1. The nucleus, and the inner, filled electron shells or levels constitute the *ion core*, while the *valence electrons* are in the outer shell farthest from the nucleus. Addition or removal of an electron from the outer shell causes the atom to become a negative or positive *ion*.

Atoms become bonded to each other and take up equilibrium positions because of the changes in potential energy of interaction that occurs as they move toward each other. This energy is conventionally taken as zero when the atoms are infinitely far apart (Fig. 4.2). It does not change significantly as the atoms approach each other until they are within a few atomic spacings, at which time the attractive force between the atoms causes the total potential energy to drop continuously to negative values. When the atom cores begin to overlap, however, the force repelling the two atoms rises rapidly as the interatomic distance decreases still further; hence, work must now be

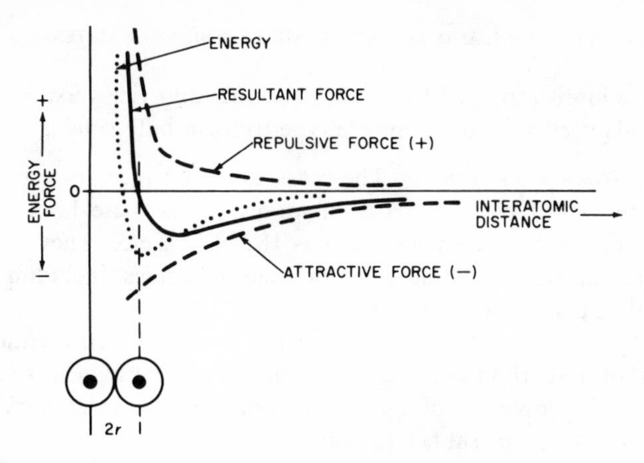

Fig. 4.2 Variation of potential energy and forces with distance between two atoms.

supplied if the atoms are to be brought still closer together so that the total interaction energy increases. The equilibrium position of the atoms is found where the attractive and repulsive forces are equal, and the potential energy is at a minimum as schematically shown in Fig. 4.2.

All of the various interatomic bonds can be classified as *primary*, strong bonds, or weaker, *secondary bonds*. The primary ones include (a) *ionic*, *electrovalent*, or *heteropolar bonds* between oppositely charged ions; (b) *covalent* or *homopolar bonds* between neutral atoms and (c) *metallic bonds*. The secondary bonds are *van der Waals*' and *hydrogen bonds*. All these really represent extremes; in many materials the bonds may be intermediate between the basic types. In some cases, the same type of force may be effective in all directions in an aggregate, while in others different types of forces act in different directions.

Primary Bonds

Ionic Bonds. Probably the best understood type of binding is an ionic coupling of an electropositive and an electronegative ion, for example, sodium and chlorine. The binding force in this case is the Coulomb attraction of oppositely charged ions. The charged ions result from the movement of the single electron in the sodium outer shell to the unfilled position in the chlorine outer shell leaving a positively charged sodium ion and a negatively charged chlorine ion. This is the type of bond found in inorganic salts such as the alkali halides, in magnesium oxide, etc. The ions in these salts act like rigid spheres maintaining, to a first approximation, a definite radius that remains almost unchanged in passing from one compound to another. The single atoms do not pair-up, one with another, but instead each positive ion is attracted to all negative ones and vice-versa so that each ion surrounds itself with the opposite type to form bulk solids.

Covalent or Homopolar Bonds. These are a second primary type in which electrons are shared between pairs of atoms. It is these bonds that join atoms into molecules in elements such as H_2, N_2, O_2 etc. They are also the binding force in diamonds and most organic molecules, including the long ones typical of plastics and rubber.

A variety of covalent bonds are possible, even between identical atoms such as carbon-to-carbon as shown in Table 4.1. Because of this variability, the mechanical properties of covalently bonded molecules vary between such extremes as that of rubber to diamond.

TABLE 4.1*

DISSOCIATION ENERGIES OF THE VARIOUS CARBON-CARBON BONDS

Bond	*Bond energy* (10^5 *dynes/cm.*)	*Distance between atom centers* (*A*)
C≡C	15.6	1.20
C=C	9.8	1.33
C—C	7.6	1.40
C—C	4.4	1.56

* From (R4.6).

One of the most important features of these bonds is the rotational freedom of single C—C, C—O, or C—N bonds. This freedom is eliminated by double bonding. Actually, the ability for free rotation even in single bonds is not without restrictions. For example, both polyethylene and polyhexafluoroethane (Teflon) are composed of molecules whose backbone consists of C—C type bonds. In soft polyethylene, the carbon atom valence electrons that are not shared with other carbon atoms to form the long molecule, are saturated by hydrogen; thus

```
   H  H  H  H
   |  |  |  |
 —C—C—C—C—
   |  |  |  |
   H  H  H  H
```

while in Teflon they are saturated with fluorine:

```
   F  F  F  F
   |  |  |  |
 —C—C—C—C—
   |  |  |  |
   F  F  F  F
```

The hydrogen substituents in polyethylene, of course, repel each other so that the potential energy of the molecule varies as the carbon atoms rotate since this rotation changes the relative positions of the hydrogen atoms attached to adjacent carbon atoms. The energy of the molecule is minimized when the hydrogen atoms are as far apart as possible. Hence at room temperature some stiffness is developed in the molecule because the hydrogen atoms want to stay apart. As the temperature is raised the stiffness decreases because the additional thermal agitation introduced into the molecule is sufficient to allow more rotation.

Because of the large fluorine atoms, the Teflon molecule develops even more repulsive interaction between the substituents hence resulting in still more hindrance to free rotation. Consequently at a temperature where polyethylene is quite soft, Teflon remains reasonably hard.

The mechanical behavior of solids consisting of covalently bonded macromolecules depends to a large degree on the regularity of the molecular shape. The more uniform it is, the greater the opportunity for close packing of the molecules. For example, polystyrene has bulky phenyl side groups:

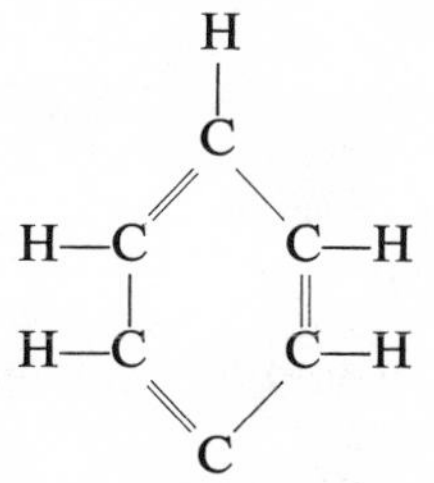

```
          H
          |
          C
        //  \
    H—C      C—H
      |      ||
    H—C      C—H
        \\  /
          C
```

which are normally attached at random positions along the molecule.

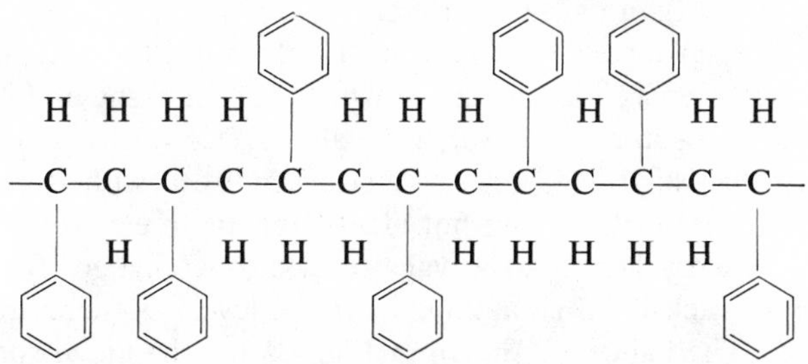

```
H  H  H  H  Ph  H  H  H  Ph  H  Ph  H  H
|  |  |  |  |   |  |  |  |   |  |   |  |
—C—C—C—C—C—C—C—C—C—C—C—C—C—
 |  |  |  |  |  |  |  |  |  |  |  |  |
 Ph H  Ph H  H  H  Ph H  H  H  H  H  Ph
```

Structures with randomly placed side groups such as these are termed *atactic*. They are not as strong as the same materials with a *syndiotactic* arrangement, wherein a regular pattern is formed by the side groups alternating across the molecule, or when all side groups are on the same side or *isotactic*. In the latter two conditions, close packing is facilitated and, as shown in the sections that follow, the closer packing explains the strength increase of the polymer.

In addition to randomly or regularly placed side groups, molecules will pack more or less densely depending on the shape of the molecule chain. The latter is influenced by the angular relation that can occur in certain polymers when two single bonds are on opposite ends of a double bond. For example, the structural formula —C—C═C—C—C—C═C— can represent polymers that have very different properties because they take on one of two different configurations, both of which are stable. Rubber, with this formula has a *cis* structure shown at (a) while far stiffer gutta-percha has the *trans* structure shown at (b)

```
              —C         C—C'         C"—C
(a)             \       /    \       /    \
                 C═C          C═C          C═C

(b)   C                  C—C'                          C
       \                /    \                        /
        C═C        C═C        C═C          C═C
           \      /              \        /
            C—C                   C"—C
```

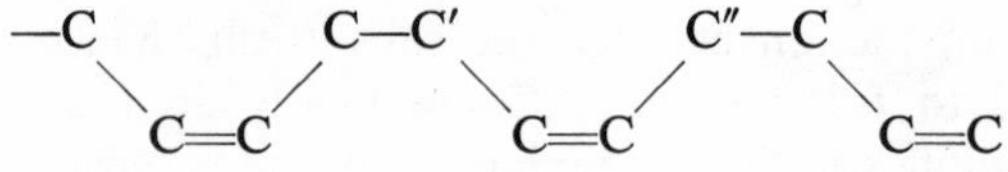

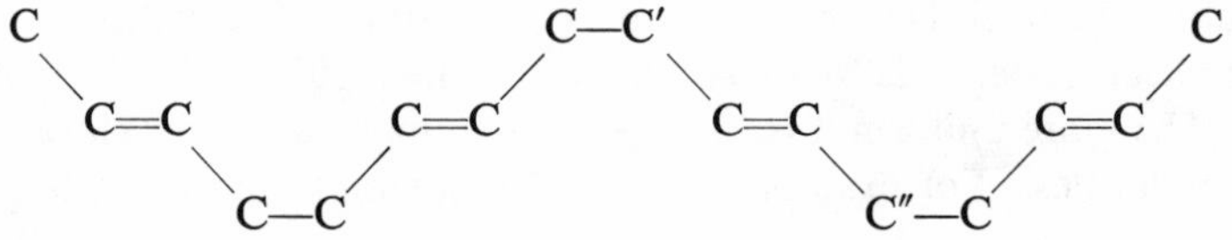

Although carbon atoms C′ and C″ are the same distance apart if one travels along the molecule in either case, the actual distance between these atoms is less in the *cis* than in the *trans* configuration so that the former molecule tends to coil more than the latter.

Metallic bonds. These are similar to covalent ones in that they occur between identical or similar atoms. Metals differ from nonmetals, however, in that their valence electrons are held with much weaker forces and hence are relatively easy to separate from the ion cores. This ease of separation leads to a sharing rather than pairing of electrons.

So long as each metal atom is an isolated entity, its electrons exist in discrete allowable energy levels, as shown in the far right of Fig. 4.3. As the atoms move together, however, the valence electrons begin to overlap and the electrons which had previously been identified with a specific atom are shared with many atoms and not identifiable as to electron state. Now the allowable energy level of the valence electrons changes from discrete levels to bands, each band containing as many levels as there are atoms in the aggregate. The bands, as shown in Fig. 4.3 for the metal sodium, continue to broaden and eventually overlap as the atoms are brought closer and closer together so that the valence electrons are no longer confined to moving within a small space.

Van der Waals' or Secondary Valence Bonds; Hydrogen Bonds

Van der Waals' or secondary valence bonds are generic terms denoting a number of different types of relatively weak bonds which frequently occur

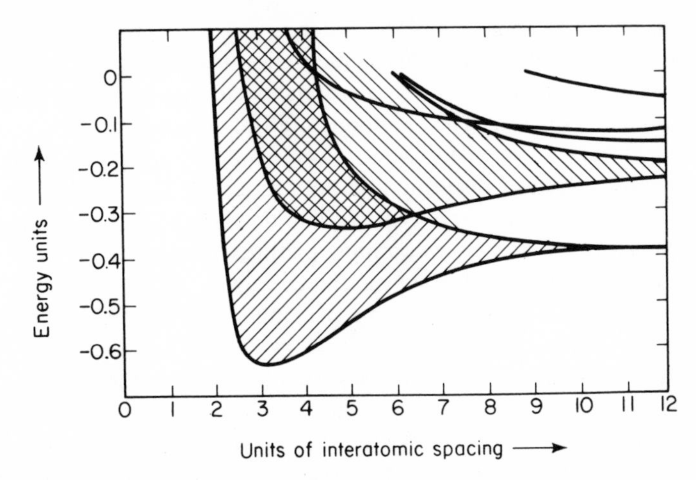

Fig. 4.3 Behavior of the electronic energy levels of sodium atoms as these are brought together to form the arrangement in which they occur in the metal. (After Slater, J. C., *Rev. Modern Physics*, 6, 209, 1934).

together. These are the forces that cause rare gases to solidify at low temperatures, and are of particular interest because they hold together long covalently bonded molecules to form solid polymeric materials such as polyethylene. Van der Waals' forces exist between atoms whose primary valences have been saturated. In spite of this saturation, each atom is still a dynamic entity, and its variously charged portions may interact with other atoms.

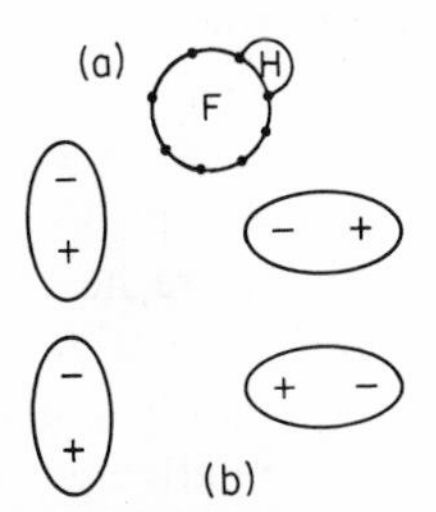

Fig. 4.4(a) Polarization in an HF molecule (schematic), (b) Ellipses represent dipolar molecules in positions of mutual attraction.

One component of Van der Waals' bond results from the *orientation effect* of rigid dipoles. For example, the hydrogen fluoride molecule has the configuration shown schematically in Fig. 4.4a. Note that the fluorine ion is more completely surrounded by negative charge than the hydrogen ion. As a consequence the centers of the positive charge and negative charge do not coincide so that polar molecules can be represented by ellipses of the type shown in Fig. 4.4b, and it is apparent that certain orientations of these dipoles are preferred. This is one cause of Van der Waals' force. In addition to this permanent dipole effect, dipolar molecules will induce a moment in adjacent molecules giving rise to an additional attractive force (*induction effect*).

Both of these attractions act only between dipolar molecules. They do not explain how non-polar molecules such as H_2, O_2 or the rare gases might be made to condense. The latter is attributed to the third source of weak binding, the *dispersion effect* which arises because at any instant each molecule has its electrons and nucleus in some position which gives it an instantaneous dipolar moment. The resultant of all these moments is zero. However, the temporary moment in each molecule induces a dipole in other molecules in phase with itself resulting in a net attraction.

Generally, all three types of van der Waals' forces occur simultaneously. This leads to a complex bonding picture since each type is affected differently by temperature and other parameters.

Stronger lateral cohesion of long molecules is obtained by *hydrogen bonds* or *hydrogen bridges* found in certain aggregates of molecules containing hydrogen atoms in lateral positions to the long dimension of the molecules. For example, if the hydrogen atom is initially attached to an oxygen atom in one molecule, it will at the same time be attracted to the unshared electrons in another oxygen or similar atom such as a carbon or nitrogen in its immediate vicinity. Hence, it serves as a joint between two molecules or molecular segments. Hydrogen bonds account for the strong force that holds the molecules of nylon together.

ELEMENTS OF CRYSTALLOGRAPHY

Having discussed the forces that bind atoms into aggregates, let us now consider the manner in which the atoms arrange themselves to form crystalline substances. The fact that metals consist of crystals was mentioned in the Introduction. This is also true for most other solids since less energy is required to form a regular atomic packing than an irregular packing. Even when these crystalline solids, such as metallic potassium, are heated above their melting point, the atoms composing the liquid do not take completely random positions. Especially at temperatures only slightly above the melting temperature, clusters of atoms continue to occupy relative positions that are the same as in the solids. Of course, since the atoms in liquids are in continuous motion, an orderly pattern of any particular group of atoms is transient. *Short range order* results; i.e., the pattern exists only over a small volume and only produces periodicity with its near neighbors. At this instant, an identical pattern with this same orientation may not be found in any other portion of the liquid. The atoms in crystalline solids, on the other hand, not only form a pattern with near neighbors, but this identical configuration extends for long distances developing *long range order*. Basically, the characteristic that uniquely defines crystalline solids is that they possess long range order. Indeed, in solid state science, it is necessary to distinguish between two types of solids: those that exhibit long range order and those that do not. Such a distinction is helpful because noncrystalline, or *amorphous*, "solids" pass from a semirigid state to a liquid state continuously; i.e., on heating they gradually soften until they flow like any other liquid. On the other hand, crystalline materials abruptly transform from solid to liquid with discontinuous changes in many of their properties, for example, density. In this text any material that sustains a shear stress for an appreciable time is regarded as a solid. For example, glass displays elastic reactions to loads (see Chapter 1) even though it is amorphous. At times, however, it will be necessary to differentiate amorphous "solids" from real or crystalline solids.

In some instances an entire body may be composed of atoms arranged in a single periodic pattern or array, in which case the material is referred to as a *single crystal*. Single crystals are frequently prepared from both metals and inorganic nonmetals. Single crystals not only are of great scientific interest but are sometimes of commercial interest as well; e.g., in transistors or in sonar equipment.

Structural crystalline materials consist of *grains* generally of microscopic size, which have identical arrays while adjacent grains may have different arrays or, more commonly, the same array in a different orientation. Each of these grains is separated from its neighbors by grain boundaries. The size of the grains and their orientations depend on their thermal and mechanical history. Since grain size, shape, and orientation greatly influence the mechanical and physical properties of metals, control of grain growth is an important problem of physical metallurgy.

Space or Bravais Lattices and Unit Cells

Atomic arrays or crystals are described in terms of their atom arrangement and the different type of symmetry displayed by the resulting structures. The symmetry considerations are of great importance to crystallographers since these are used in the experimental techniques that make it possible to establish crystal structures. For the present purpose, only the characteristics of atom arrangements are of interest since these are important in explaining macromechanical properties.

The manner in which points can be located in space to form periodic arrays will now be considered. Each point can subsequently be used to represent an individual atom, as is the case for many pure elements, or the site of a pattern-unit or cluster of atoms called a *basis*, as is sometimes the case in crystalline polymers or inorganic solid compounds. The atoms generally will not be stationary at these points but will oscillate about them. In some materials the pattern-unit or basis may only represent an area or volume in which two or more different types of atoms are rotating.

Certain characteristics are apparent in the three dimensional array of points formed by the intersection of the parallel lines shown in Fig. 4.5a. In particular, this configuration has the property that the arrangement of points about any one point is identical with the arrangement around any other internal point. This property can be stated mathematically by means of three vectors such as **a**, **b**, and **c** in Fig. 4.5a. Indeed, one could travel

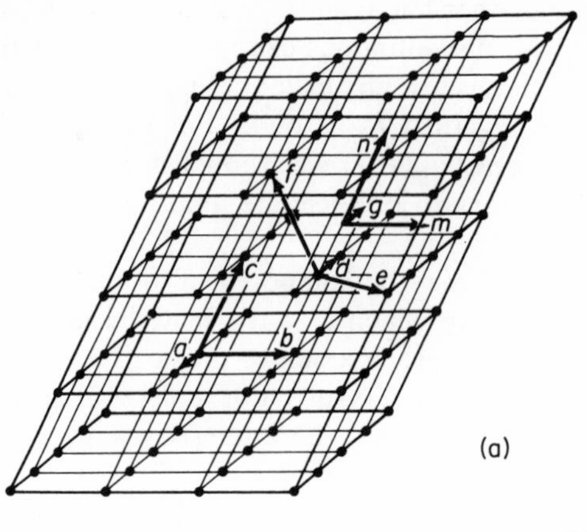

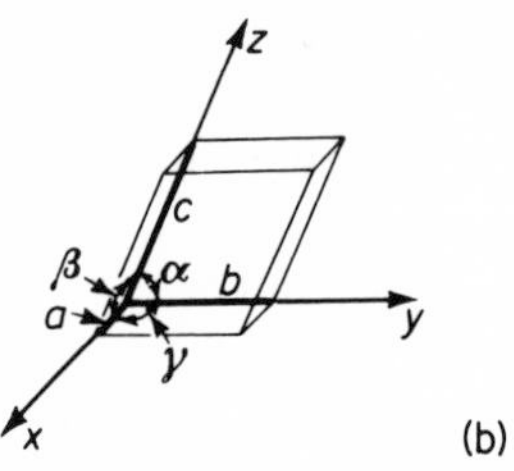

Fig. 4.5 (a) An array of periodically repeating points in space forming a space lattice. The lattice could be defined by the vectors **a**, **b**, and **c**, or **d**, **e**, and **f** or **g**, **m**, and **n**. (b) The unit cell with edge lengths a, b, and c and corner angles α, β, and γ will form the space lattice in (a) by a series of translations.

from one point, **r**, to any other point, **r**′, by moving some integral number of vector lengths in the direction of **a**, **b**, and **c**. Hence point **r**′ is located by

$$\mathbf{r}' = \mathbf{r} + n_1\mathbf{a} + n_2\mathbf{b} + n_3\mathbf{c} \tag{4.1}$$

where n_1, n_2, and n_3 are arbitrary integers.

A continuous repetition of translations using various integers for n_1, n_2, and n_3 will develop the complete three dimensional network (shown in Fig. 4.5a) an indefinite distance in any direction. The points formed by the intersection of the lines of this network form a *space lattice*, or a *Bravais lattice*, which, according to Eq. (4.1), is characterized by the fact that every point within it has identical surroundings. Bravais has shown that all possible periodic arrangements of points in space can be described by only fourteen space lattices in three dimensions. These lattices are specified by using vectors—like **a**, **b**, and **c** in Eq. (4.1)—as various systems of coordinate axes. All fourteen space lattices require just seven different coordinate systems. The angles between the axes are specified by the crystal system. The lengths of the vectors **a**, **b**, and **c** are the *lattice constants*.

The coordinate axis system that would be used to define the space lattice in Fig. 4.5a is that shown in Fig. 4.5b with angles α, β, and γ between the axes, and unit lengths a, b, and c in the x, y, and z directions, respectively. A complete listing of all the crystallographic systems, including the relative axis lengths and the angles between axes, with some examples in each system, are listed in Table 4.2. These are the basis for the seven crystal systems used to classify minerals. The fourteen space lattices using these seven crystal systems are shown in Fig. 4.6. The common engineering materials fall within just a few of them.

Fig. 4.6 The fourteen Bravais lattices: (1) triclinic, simple; (2) monoclinic, simple; (3) monoclinic, base centered; (4) orthorhombic, simple; (5) orthorhombic, base centered; (6) orthorhombic body centered (b.c.) ; (7) orthorhombic face centered (f.c.) ; (8) hexagonal; (9) rhombohedral; (10) tetragonal, simple; (11) tetragonal, b.c.; (12) cubic, simple; (13) b.c.cubic ; (14) f.c cubic.

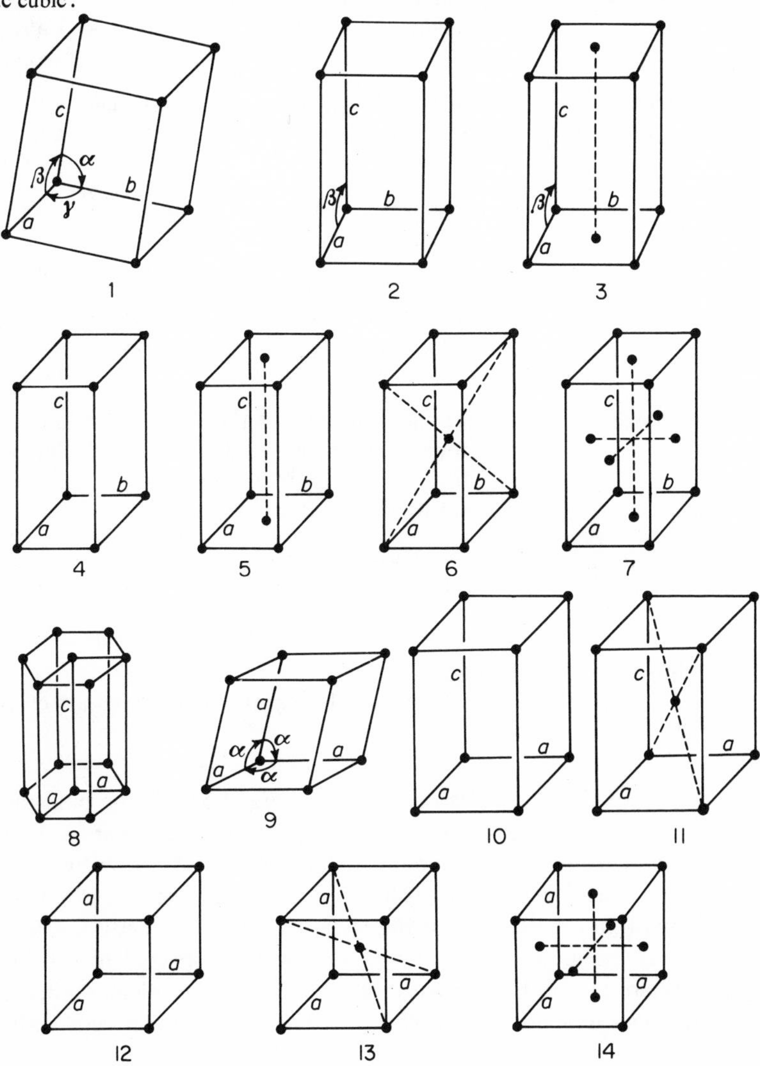

TABLE 4.2

THE CRYSTAL SYSTEMS*

(In this table $\neq$ means "not necessarily equal to, and generally different from")

System	*Axes and Interaxial Angles*	*Examples*
Triclinic	Three axes not at right angles, of any lengths $a \neq b \neq c \qquad \alpha \neq \beta \neq \gamma \neq 90°$	K_2CrO_7
Monoclinic	Three axes, one pair not at right angles, of any lengths $a \neq b \neq c \qquad \alpha = \gamma = 90° \neq \beta$	β-S $CaSO_4 \cdot 2H_2O$ (gypsum)
Orthorhombic (rhombic)	Three axes at right angles; all unequal $a \neq b \neq c \qquad \alpha = \beta = \gamma = 90°$	α-S Ga Fe_3C (cementite)
Tetragonal	Three axes at right angles; two equal $a = b \neq c \qquad \alpha = \beta = \gamma = 90°$	β-Sn (white) TiO_2
Cubic	Three axis at right angles; all equal $a = b = c \qquad \alpha = \beta = \gamma = 90°$	Cu, Ag, Au Fe NaCl
Hexagonal	Three axes coplanar at 120°, all equal; fourth axis at right angles to these $a_1 = a_2 = a_3 \neq c \quad \alpha = \beta = 90°$ (or $a_1 = b \neq c$)) $\quad \gamma = 120°$	Zn, Cd NiAs
Rhombohedral† (trigonal)	Three axes equally inclined, not at right angles; $a = b = c \qquad \alpha = \beta = \gamma \neq 90°$	As, Sb, Bi Calcite

* After C. S. Barrett, (R4.2).

† The rhombohedral system can also be regarded as a subdivision of the hexagonal system.

Any space lattice, such as shown in Fig. 4.5a, can be built up of a series of blocks or prisms whose edges are given by the vectors **a**, **b**, and **c**. The basic unit from which the complete lattice can be built by simple translations is known as the *unit cell.* The choice of the unit cell is not unique, and although, in the example of Fig. 4.5a, the prism with edges **a**, **b**, and **c** could be used (as shown in Fig. 4.5b), one is not restricted to this choice. This lattice could also have been developed from a completely different unit cell, with edges **d**, **e**, and **f** or **g**, **m**, and **n** as well. The unit cells with edges **a**, **b**, and **c** or **e**, **f**, and **g** have a pattern-unit (basis) at each corner, and, since each of the eight corners is shared by eight unit cells, each cell has a volume equal to one pattern-unit. The unit cell with edges **g**, **m**, and **n** has a basis at the center of each of its vertical edges with each basis being shared with four cells, again giving one pattern-unit per cell. It is frequently convenient to select a *primitive cell* as the unit cell; i.e., a prism with the shortest possible sides and with unit patterns at its corners. It will be found in the following sections, however, that mechanical properties are frequently better understood if the atomic structure is described by a *compound unit cell* which contains more than one basis per cell.

Miller and Miller–Bravais Indices

The intersection points within any space lattice can be considered to lie on a set of parallel planes. Again, as is apparent from Fig. 4.5a, a series of planes parallel with the base of the figure will contain all the lattice points. This set of parallel planes is not the only one containing all these points. Planes parallel with the figure sides, or those lying diagonally within the figure, could have served as well. In dealing with the deformation of materials, it is important to be able to describe the orientation of these planes with respect to the axes of the unit cell without having to consider the position of the plane in space. Indeed, the experimental methods used by crystallographers do not distinguish individual planes in a crystal, only sets of parallel planes. Planes are universally designated by means of Miller indices, which are related to the intercepts of these planes with the crystallographic axes. To find the Miller indices of plane MNO in Fig. 4.7 lay off the vectors **a**, **b**, and **c** on the crystallographic axes x, y, and z. The plane has intercepts at $p\mathbf{a}$, $q\mathbf{b}$, and $r\mathbf{c}$ on the three axes. The Miller indices, designated

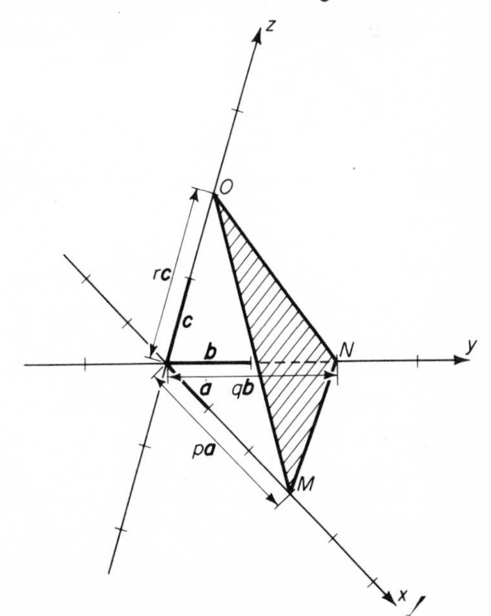

Fig. 4.7 Method used for specifying Miller intercepts of plane MNO.

as h, k, and l, of this plane are obtained by setting the reciprocals of p, q, and r proportional to h, k, and l. Thus:

$$h:k:l = \frac{1}{p} : \frac{1}{q} : \frac{1}{r} \tag{4.2}$$

It is customary to reduce h, k, and l to the lowest set of integral numbers that preserves the ratios of Eq. (4.2). Hence the parallel planes with intercepts 6, 6, and 3 or 2, 2, and 1 would both be written as 1, 1, and 2.

By enclosing the Miller indices in parentheses, (hkl), one signifies a single plane or a set of parallel planes. If one or more of the intercepts are negative, the corresponding index is also negative, and the negative sign is placed above the appropriate index, $(1\bar{1}2)$. Enclosing the indices with a double set of parenthesis ((hkl)) or with single braces $\{hkl\}$ indicates a *form*, or those planes which are equivalent in a crystal even though their Miller indices are different. In the cubic system all sides of the cube are equivalent so that ((100)) includes the planes (100), (010), (001), $(\bar{1}00)$, $(0\bar{1}0)$ and $(00\bar{1})$.

Sometimes it is helpful to think of the Miller indices as the number of parallel planes that must be crossed in going from one lattice point to the next along each of the crystallographic axes. This is readily seen in the example of the (123) planes in Fig. 4.8. In this figure, if we were to draw the (011) planes, or the (111) planes, these lower index planes would be more densely packed than those with the higher indices (123). Since deformation occurs along dense packed planes, one need seldom be concerned with indices higher than 4 or 5. The Miller indices of some of the most densely populated planes in the cubic system are shown in Fig. 4.9.

Most industrially important metals crystallize in either the cubic or hexagonal systems. The advantage of Miller indices in the cubic system,

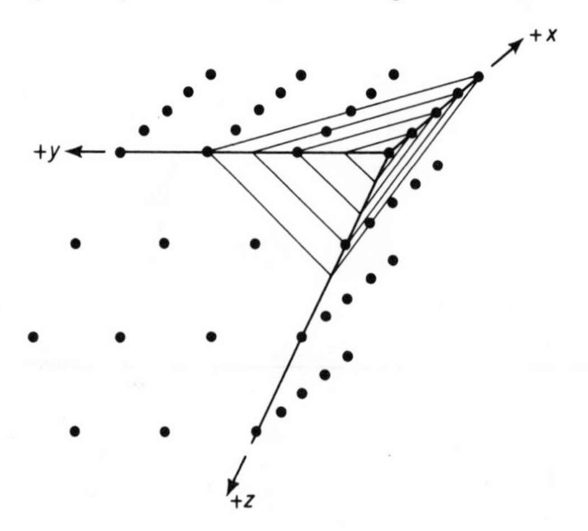

Fig. 4.8 Intercepts of 1, $\frac{1}{2}$, and $\frac{1}{3}$ give Miller indices of (123).

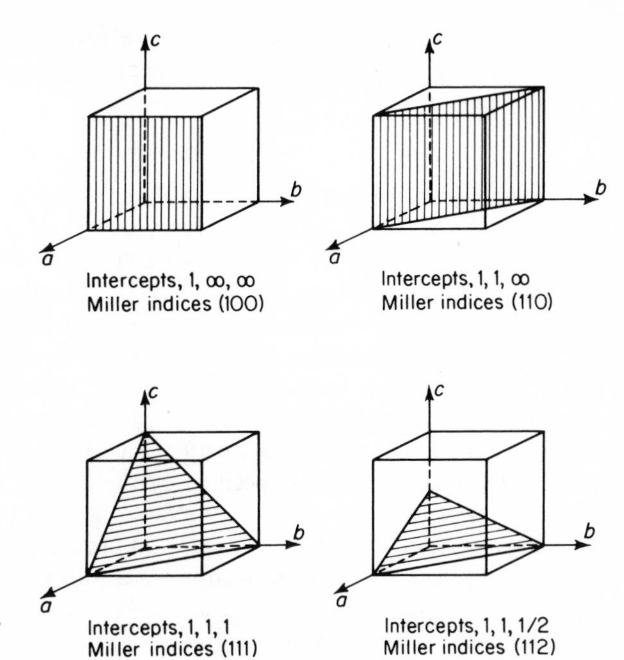

Fig. 4.9 Close packed planes in the cubic structure.

wherein equivalent planes have similar indices, is not true for the hexagonal system. For example, the sides of the hexagonal prism in Fig. 4.10, although equivalent, have Miller indices of (100), (010), ($\bar{1}$10), etc. To avoid this difficulty, *Miller–Bravais indices* are used for identifying planes in the hexagonal systems. Instead of three crystallographic axes, this type of indices uses four, of which three lie in one plane, 120 degrees apart, while the fourth is perpendicular to this plane (Fig. 4.10).

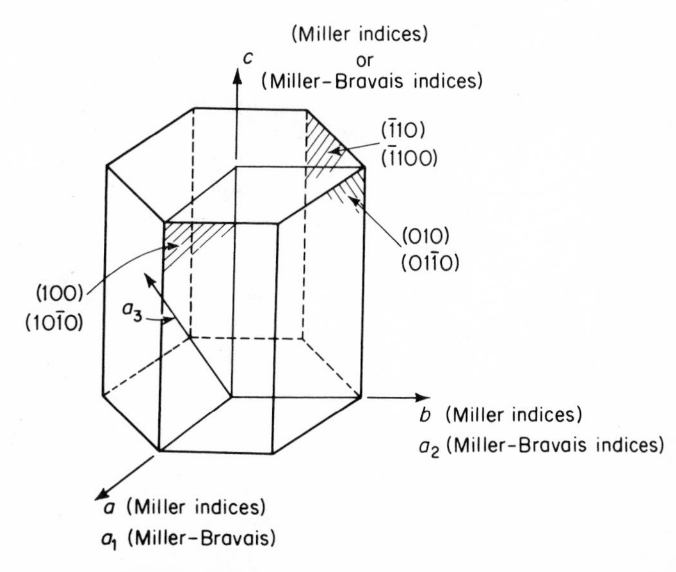

Fig. 4.10 Comparison of the Miller and Miller-Bravais Indices for the Hexagonal system. a, b, c—Miller coordinate axes. a_1, a_2, a_3, c—Miller-Bravais coordinate axes. Arrows indicate positive directions.

Again the reciprocal intercepts of the planes are used for indices and after reducing these to the smallest integers, they are written (*hkil*). The third index is related to the other two by:

$$i = -(h + k) \tag{4.3}$$

so that it is often omitted and replaced with a dot, $(hk \cdot l)$. In the Miller–Bravais system equivalent planes have similar indices. The sides of the prism in Fig. 4.10 are $(10\bar{1}0)$, $(01\bar{1}0)$, $(\bar{1}100)$ or (10.0), (01.0), $(\bar{1}1.0)$, etc.

Indices of a direction are given as successive translations along the **x**, **y**, and **z** axes through which one must move from the origin to locate the end of a vector pointing in the sought direction. Note that reciprocals are not used to indicate directions. For example, if the vector in Fig. 4.11a shows the wanted direction, its end is reached by moving from the origin distances $u\mathbf{a}$ along the **x** axis, $v\mathbf{b}$ parallel with the **y** axis, and $w\mathbf{c}$ parallel with the **z** axis. The direction is then specified by the indices u, v, and w, with these integers selected to have no common factor greater than unity. To indicate a single direction, the indices are enclosed in a square bracket, [*uvw*]. A full set of equivalent directions (directions of a form) are indicated by doubling the brackets, [[*uvw*]], or by carets, $\langle uvw \rangle$. Double brackets will be used throughout this text since they are easier to remember than the carets. The indices of a direction are seen to be proportional to the direction cosines of a vector in the specified direction.

Fig. 4.11a Method used to define Miller Indices of a direction.
Fig. 4.11b Example of it used in hexagonal system (*OA* is $[\bar{1}010]$ direction and *OB* is $[11\bar{2}0]$ direction).

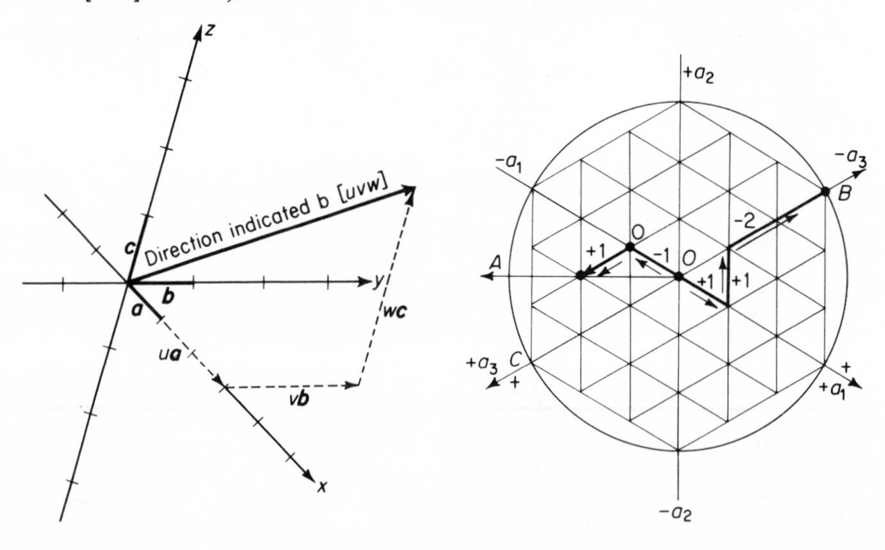

It is apparent, for example, by referring to Fig. 4.8, that the smaller the values of the directional indices, the less is the distance between points.

To designate directions in the hexagonal lattice, symbols are used which are formally identical with the Miller–Bravais four element indices. The procedure here is such that, starting from the origin of the hexagonal coordinates, one proceeds first along a_1, then a_2, and finally a_3 directions to a point located along the direction to be described. This path is so chosen as to satisfy Eq. (4.3). With reference to Fig. 4.11b, the direction A which bisects the angle between coordinates a_3 and $-a_1$ is described as $[\bar{1}010]$. Similarly, the direction of $-a_3$, indicated as B, is $[11\bar{2}0]$. Although the latter could be just as well denoted as $[00\bar{1}0]$ or $[1100]$, neither of these satisfies $h + k = -i$.

Directions that are not parallel with the (0001) planes simply receive a fourth index different from zero.

Geometrical Relations in Crystal Lattices

Many known formulae of analytical geometry can be expressed in terms of lattice dimensions and Miller indices of planes and directions. Such relations are very useful for determining the relative positions of various planes and lines within the lattices. Without reproducing the proofs which can be found in any textbook on crystallography or X-ray analysis (e.g. R4.2), a few that are particularly useful are quoted below. We denote planes by (hkl) and directions by $[uvw]$.

(a) For a direction to be normal to a plane in a cubic lattice, the following must be satisfied:

$$h = u; \qquad k = v; \qquad l = w \tag{4.4}$$

(b) For a direction to be parallel to a certain plane in a cubic lattice:

$$hu + kv + lw = 0 \tag{4.5}$$

(c) The angle δ between two directions—$u_1v_1w_1$ and $u_2v_2w_2$—in a cubic lattice is:

$$\cos\delta = \frac{u_1u_2 + v_1v_2 + w_1w_2}{\sqrt{u_1^2 + v_1^2 + w_1^2}\sqrt{u_2^2 + v_2^2 + w_2^2}} \tag{4.6}$$

(d) The interplanar distance d in a cubic lattice with a constant a is:

$$d = \frac{a}{\sqrt{h^2 + k^2 + l^2}} \tag{4.7}$$

and the hexagonal system:

$$d = \frac{a}{\sqrt{\frac{4}{3}(h^2 + hk + k^2) + (a/c)^2l^2}} \tag{4.8}$$

ANISOTROPIC AND ISOTROPIC MATERIALS

It was just explained that the density of points along a line in a crystal varies as the direction of the line changes. Therefore, it is not surprising to find that the mechanical and physical properties of single crystals vary greatly as the direction of measurement is changed. For example, the coefficient of thermal expansion of zinc, which crystallizes in the hexagonal system, is more than four times greater along the hexagonal axis than it is perpendicular to this direction. Even more startling than this is oriented graphite which has a thermal conductivity 100 times greater along close packed planes than in a direction normal to the planes. Materials whose properties vary as a function of direction are termed *anisotropic*. Directional variability of properties resulting from the crystalline nature of materials is termed *crystallographic anisotropy* to distinguish it from other types of anisotropy. In most structural crystalline materials the grain size is very small compared to the size of the body. If the grains are randomly oriented, the anisotropy found in the individual grains is averaged out so that the material has virtually identical properties in all directions. Any material whose properties are independent of direction is said to be *isotropic*. Isotropic materials are generally striven for in structural applications although in heavily deformed metals, e.g., sheets, isotropy is hard to obtain. In some nonmechanical applications, like the use of silicon steels in transformers, a high degree of crystallographic anisotropy is sought in order to attain superior magnetic properties. One notable exception to the desire for isotropic structural bodies is in polymer fibers such as nylon threads; in this case, by aligning as many molecules as possible along the fiber axis, the strength of the fiber is greatly increased. This alignment is termed *orientation*. Oriented fibers may be amorphous or partly crystallized. The crystalline nature of polymers is discussed later.

STEREOGRAPHIC PROJECTIONS

The orientation of a particular grain or multitude of grains in an anisotropic sheet can be related to the geometrical features of the body of which they are a part. *Stereographic projections* are generally used for this purpose. This method of showing the orientations of planes utilizes the fact that the same angular relationships exist between planes as between their normals. To develop a stereographic projection, one first imagines the crystal to be very small and at the center of a *reference sphere* (Fig. 4.12a). The normal to each of the crystal planes can now be projected onto the surface of the sphere, where it appears as a point. Each point is a *pole* of the plane with which it is associated, and all the points thus projected on a sphere form a *pole figure*. The stereographic projection is a map of the reference sphere. The map is formed by imagining the reference sphere to be transparent so that a light source placed on its surface (as in Fig. 4.12b) will project the poles as shadows onto the projection plane; for example, pole P appears on the plane as point P'. The two dimensional stereographic projection has the same relation to the reference sphere as a wall map of the world has to the globe. Stereographic projections can be made on a stereographic net, which is a map of a ruled globe with the north and south poles on a vertical axis and in which the equator appears as a horizontal axis. The projection might also be made on a polar net, which is the map formed by viewing a ruled globe from the north or south pole, so that the globe axis is in the center and the equator forms the peripheral circle.

When a stereographic projection is made so that a crystal plane with low indices is parallel with the projection plane, it is known as a *standard projection*. A projection of this type is shown in Fig. 4.13. The (001) plane of the crystal is parallel with the projection plane so that the pole of this plane is on the center of the circle. Note that the angular relations on the projections are identical with those in the crystal; for example, the angle between the (111) and the (101) planes is 45° (compare with Fig. 4.12a). Point P in Fig. 4.13 shows the relation of the axis of a single crystal specimen to the low indices planes (001), ($\bar{1}$11), and (011). Since all cubic crystals would have the same standard projection if the (001) pole is at the projection center, the position of point P in the heavily outlined triangle completely identifies its orientation.

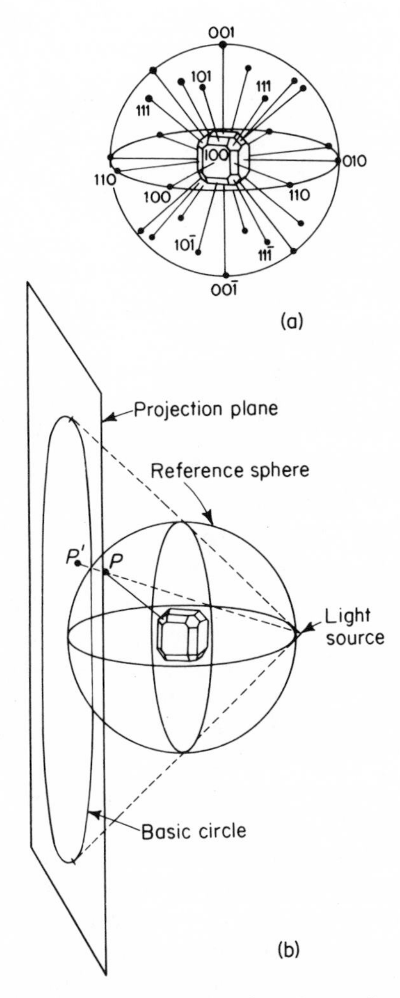

Fig. 4.12 (a) The low indice planes of a cubic crystal shown as poles on a reference circle. (From C. W. Bunn, *Chemical Crystallography*, Oxford-Clarendon Press, p. 30, 1946.) (b) Stereographic projection of (111) plane on the projection plane showing the relationship between the poles on the reference circle and the stereographic projection.

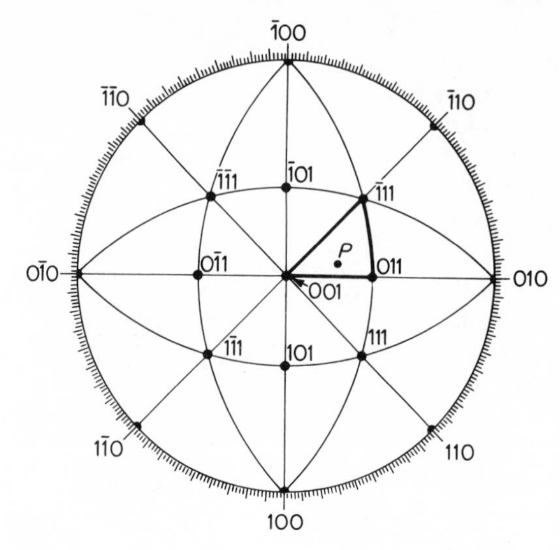

Fig. 4.13 Standard projection of cubic crystal. Unit stereographic triangle which contains specimen axis *P*, is outlined. From (R4.2)

STRUCTURE OF SOME COMMON ENGINEERING MATERIALS

Metals

In discussing metal bonds it was stated that the attractive forces between the atoms in these materials is caused by the interaction between the positively charged ion core and the electron gas. This type of bonding results in an equal attraction in all directions, leading to relatively simple atomic structures with close packing and large *coordination numbers*. The latter specify the number of nearest neighbors that each atom has.

The three systems that can be used to describe almost all commercial metals are represented by simply shaped compound unit cells in which each basis point represents a single atom or ion. The most commonly occurring cell is face-centered cubic (abbreviated, f.c.c.), shown by the single lines in Fig. 4.14. The elements Al, Cu, Ni, and dilute alloys of these metals crystallize in this system as does austenitic stainless steel. The primitive cell, with edges $\mathbf{a}_1$, $\mathbf{a}_2$, and $\mathbf{a}_3$ is also shown in this diagram. Although the primitive cell has edges of equal length, the angles between them are not 90 degrees.

Another commonly occurring unit cell is body-centered cubic (b.c.c.)—characteristic of steel, Cr, Mo, and W. The compound cell (single lines) and primitive cell (double lines) representing this structure are shown in Fig. 4.15. Again, the primitive cell has edges of equal length but these are not mutually perpendicular. It is apparent in examining these two structures why the right angled compound cell with the simplicity of a cube is preferred over the primitive cells.

The third system in which metals commonly crystallize is the hexagonal close packed (abbreviated h.c.p. or c.p.h.). In this case the primitive cell,

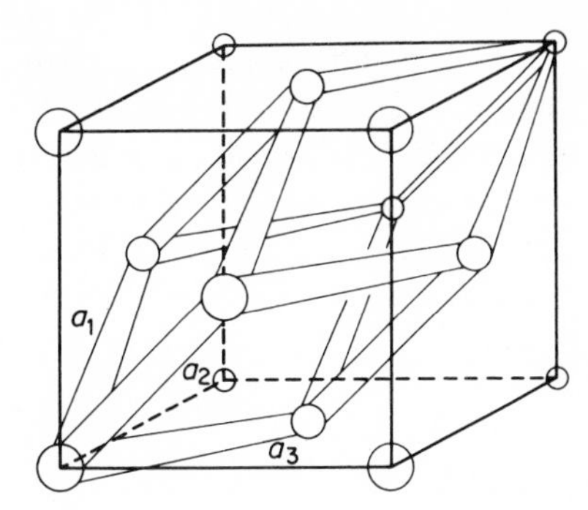

Fig. 4.14 Face centered cubic structure shown as single lines. Primitive cell for this system, shown as double lines, lies completely within the compound cell.

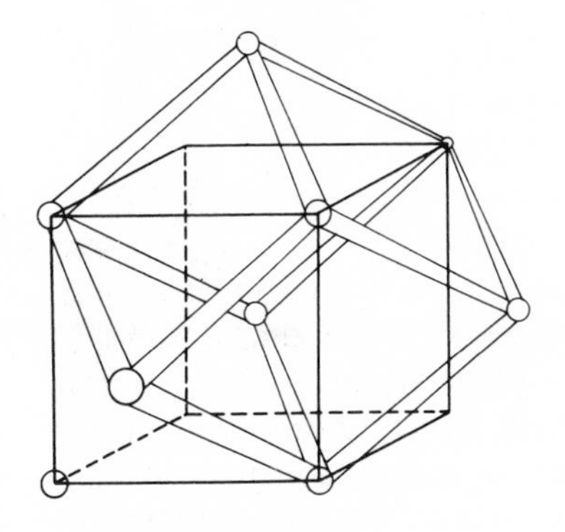

Fig. 4.15 Body centered cubic structure shown as single lines. Primitive cell for the same system shown as double lines. The primitive cell uses the body centered atoms of the compound cells lying in front of the one shown, the one to its right, and the one above it.

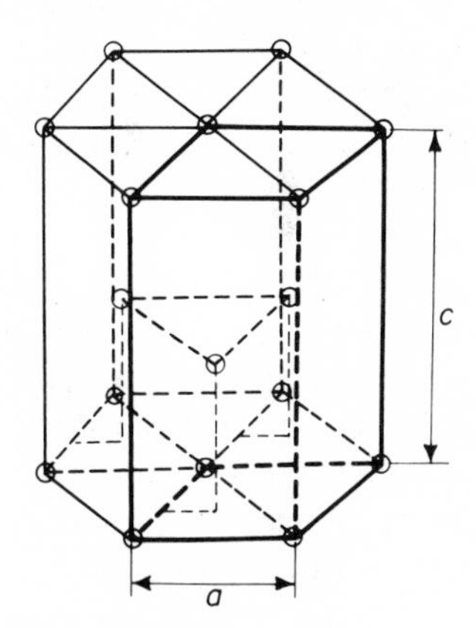

Fig. 4.16 Hexagonal structure shown in the form of the generally used hexagon and as the primitive cell (heavy outline). The closest possible packing results when the ratio of $c/a = 1.6330$.

shown by heavy outlines in Fig. 4.16, is occasionally used, although the structure generally is described by a hexagonal prism with a volume of three primitive cells. Unlike the cubic system which is completely defined by one lattice constant, a, the cube edge length, the hexagonal system requires two lattice constants; i.e., the distance between atom centers on the basal plane *a* and the prism height *c*. If the ratio of *c/a* is 1.6330, the atoms are packed as closely as one can put identically sized spheres. If the *c/a* ratio is greater than this, the atoms on the (0002) planes are not in contact with the atoms on the (0001) plane. If *c/a* is less than 1.6330, the atoms on the basal planes are not in contact. Some metals crystallizing in the hexagonal system and their *c/a* ratios are shown in Table 4.3.

TABLE 4.3

RATIO OF *c/a* FOR SOME HEXAGONAL METALS

Metal	*c/a ratio*
Be	1.5682
Mg	1.6235
h.c.p. (theoretical)	1.6330
Zn	1.8563
Cd	1.8859

The face-centered cubic and hexagonal lattices represent the densest packing of atoms possible and each has a coordination number of twelve. In the b.c.c. structure this number is eight while in the simple cubic it would be six. The f.c.c. and h.c.p. structures are much more alike than the unit cells in Figs. 4.14 and 4.16 suggest. Both structures are built up by stacking close packed planes. The (0001) plane in the hexagonal system and the (111) plane in the face-centered cubic system represent a packing of the same size spheres, in which each sphere is in contact with its six coplanar neighbors (Fig. 4.17a). The next layer, in which each sphere fits into the depressions formed in the first layer, is also identical for the two systems. In the third layer, however, a choice must be made because the depressions in the second layer occur at positions directly above the spheres in the first layer and at 60 degrees to these positions. In Fig. 4.17a the spheres in the third layer are over those in the first, leading to the hexagonal system with a stacking sequence of *ABABAB* . . . . If the spheres had been placed in the other group of depressions, and then the fourth layer made identical with the first, the stacking would have been *ABCABC* . . . , which results in a face-centered cubic structure. The side views of spheres with these stackings are shown in Fig. 4.17b.

Crystal systems are sometimes described in terms of the method of atom stacking rather than by means of unit cells. This is particularly helpful in the f.c.c. and h.c.p. systems when one is dealing with lattice defects or imperfections. In this method of description a close packed layer—the (111) in f.c.c. or (0001) in h.c.p.—is represented by a triangle, $\triangle$, and this symbol is written either as shown or inverted, ∇, according to the manner in which adjacent layers are stacked. To show how these symbols are used, let us again build a space lattice by stacking spheres. For this purpose the circles

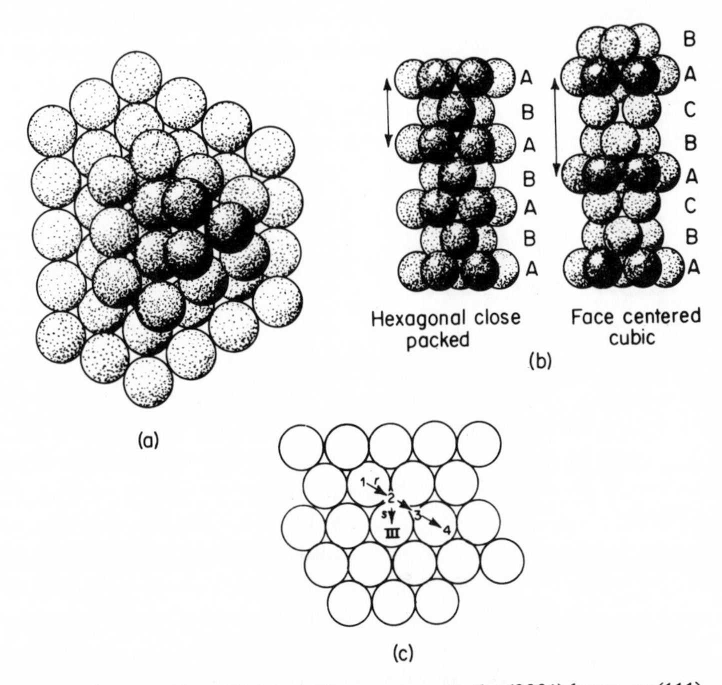

Fig. 4.17 (a) Close packing of bottom plane represents the (0001) h.c.p. or (111) f.c.c. The second layer in the two structures is also identical. In the hexagonal structure shown at (a) the third layer atoms lie above those in the first layer. In the f.c.c. the fourth layer lies above the first; (b) A side view of the ABAB . . . packing of the hexagonal system (left), and the ABCABC . . . packing of the f.c.c. system (right); (c) f.c.c. structure is formed by translating each stacked layer by vector ***r***, h.c.p. layers are stacked by translation ***r***, ***s***, ***r***, ***s***, etc. (a) and (b) from (R4.7).

in Fig. 4.17c represent the packing of one layer. To indicate the location of the next higher layer above this one, the sphere centers are moved from 1 to 2; i.e., a translation of r is effected and the symbol $\triangle$ is used for the operation. The next layer has its sphere centers moved from 2 to 3, again the translation is r and the symbol, $\triangle$. Each additional layer will also be displaced by r from the one below it so that the system is designated by $\triangle\triangle\triangle\triangle\triangle$ . . . . Observe that the centers of the fourth layer of atoms are directly above those of the first layer, giving a stacking of *ABCABC* . . . ; that is, a f.c.c. structure.

To build an h.c.p. structure, the second layer is again shifted with respect to the first by vector r, the symbol being $\triangle$. The third layer, however, requires a translation s, hence the symbol ∇. The fourth layer, since it is above the first, is located by r, hence $\triangle$, so that the system is given as $\triangle\nabla\triangle\nabla$ . . . .

Ionically Bonded Materials

Ionically bonded materials—which include most ceramics—crystallize in a wide variety of systems varying from relatively simple to quite complex.

Two factors are important in establishing crystal structures of these materials; first, the valences establish the ratio of the two atoms that make up the ionic crystal, and second, the ratio of the radii of the two ions in a two-component system determines their packing. The latter establishes the coordination number which, in turn, determines the crystal system.

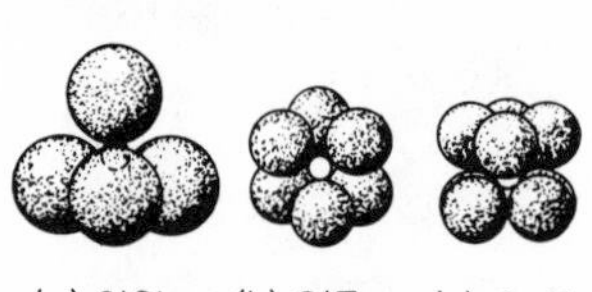

Fig. 4.18 Geometric explanation of the different co-ordination numbers (according to A. Magnus).

Ions behave like rigid spheres, and the stable crystal structure will be the one in which the smaller spheres exactly fill the voids formed in the packing of the larger ones. Magnus presented the schematic drawings (in Fig. 4.18) to demonstrate that the space formed by four contacting chlorine atoms with a radius of 1.80 Å left a void just large enough for a silicon atom with a radius of 0.4 Å. Similarly, one silicon ion fills the void formed by six fluorine ions, and one osmium atom that between eight fluorine atoms.

The crystal structure of ionic salts is frequently described by analogy with some other well known structure. Two commonly occurring structures for crystals with the stoichiometric formula of A^+X^- are the NaCl and CsCl structure (Fig. 4.19). The former consists of two interpenetrating f.c.c. cells. The corner of one cell is located at point $\frac{1}{2}$, 0, 0 and the corner of the other cell at 0, 0, 0. Many of the common oxides such as MgO, CaO, and BaO crystallize in this system, as do the alkali metal halides.

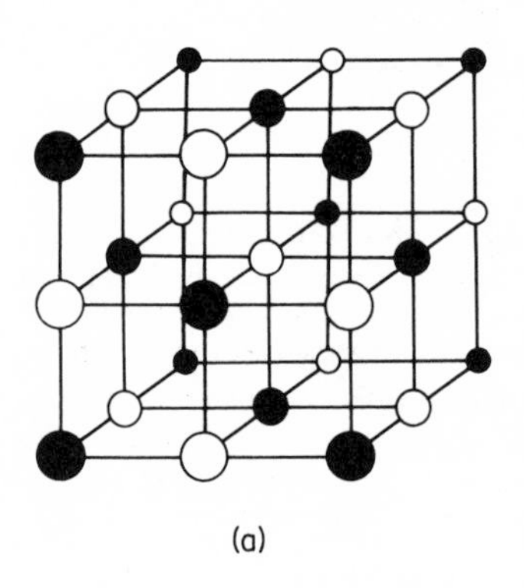

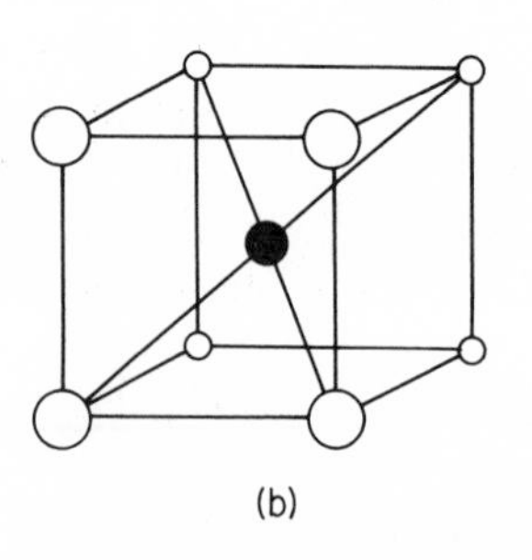

Fig. 4.19 (a) Sodium chloride structure; (b) Cesium chloride structure.

The CsCl structure is body-centered cubic with one type of atom taking the cube center position while the corner positions are occupied by the other component (Fig. 4.19b).

Covalent Bonded Materials

Atoms that are bound together by covalent forces in all three directions are found in a number of engineering materials. One example is diamond, the crystal structure of which is shown in Fig. 4.20. Although it is difficult in this diagram to recognize the unit cell for the diamond lattice, it is apparent that each ion has a coordinate number of four and that the four neighboring ions are at the corner points of a tetrahedron. The distance between atom centers is 1.54 angstroms, which is close to the 1.56 of aliphatic bonds (see Table 4.1). The great strength of diamonds results because these bonds exist in all directions throughout the entire body. A unit cell of diamond is shown in Fig. 4.21. This consists of a f.c.c. lattice, with each lattice point representing two atoms—one at (0, 0, 0) and the other at ($\frac{1}{4}$, $\frac{1}{4}$, $\frac{1}{4}$) from the first.

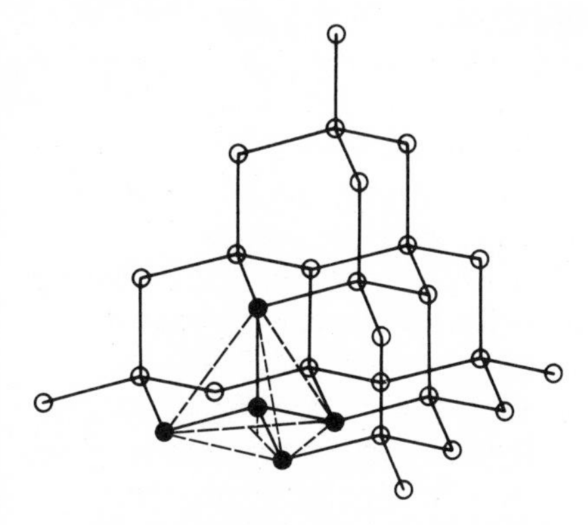

Fig. 4.20 Diamond structure showing that each atom has four near neighbors located at the corner of a tetrahedron.

The three dimensional cross-linked polymers are another example of atoms covalently bonded in all three directions. Unlike diamond, however, these materials are amorphous. Polymers that are included in this class are the thermosetting resins such as Bakelite and vulcanized rubber. In these materials the macromolecules that are characteristic of all polymers have become so cross-linked and intertwined that the complete body can be considered as one supermacromolecule. Since all the bonding in these cases is covalent, the body does not soften on heating. Instead, when sufficient heat is supplied to disrupt the primary bonds, the material decomposes.

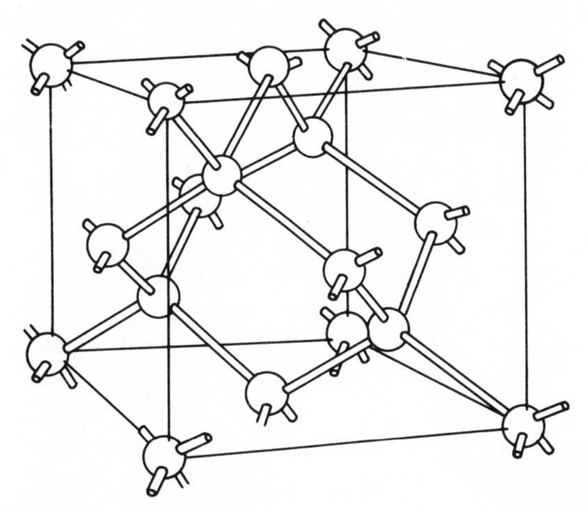

Fig. 4.21 Unit cell for the diamond structure.

Aggregates with Intermediate Types of Binding Forces and with Different Types in Different Directions

The atoms making up many important engineering materials may be bound by forces that are intermediate to the types described above. All sorts of variously bound structures may be found in nature. For example, there are many exceptions to the rule previously stated that the systems in which compounds crystallize depend on the ratio of their ion radii. The cause for this appears to be that the atoms in these cases are "incompletely ionized" so that each atom exhibits ionic bonds in some directions and covalent or metallic bonds in others.

Semiconductors are another group of materials that do not display any one specific type of bond. These are held together by covalent linkage at low temperatures and metallic bonds at high temperatures. Therefore, they act in some instances as an insulator and in others as a conductor.

Another interesting example of aggregates formed by different types of forces in different directions is graphite, the space lattice of which is shown in Fig. 4.22. The distance between atom centers in the close packed plane is 1.42 angstroms so that this bond is much like those found in aromatic organic compounds like benzene. The distance between the densely packed layers is 3.41 Å indicating that these are only held together with Van der Waals' forces. This weak bonding in one direction allows one layer to slide over the other, which accounts for the excellent lubricating properties of graphite.

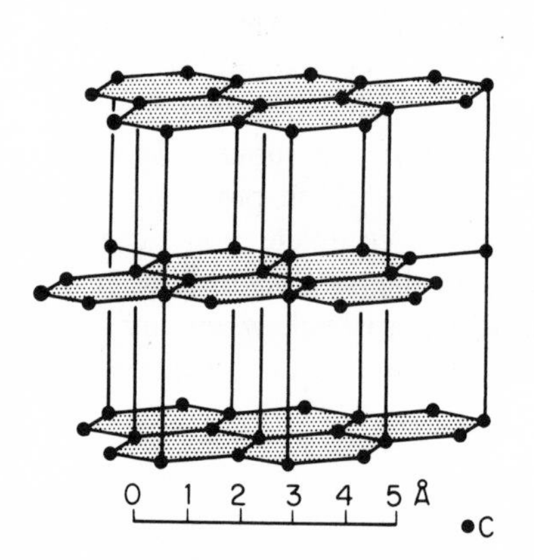

Fig. 4.22 Layered structure of graphite. (From (R4.6) p. 195.)

The most important constructional materials with mixed bonding are the linear polymers. These include the thermoplastics and soft rubber which are always bound within the molecule by covalent bonds, while the intermolecular bonds are Van der Waals' forces or hydrogen bonds.

Because of the variability allowable when two types of bonds join the atoms within bodies, linear polymers have been developed with a much wider range of properties than those of three dimensional cross-linked polymers. Linear polymers can vary from soft and rubbery to hard and brittle.

In commenting on the relative rigidity of these polymers, H. Mark lists five important factors. The first of these is the degree of overlapping of the molecules. This is proportional to the molecular weight and is expressed as degree of polymerization (DP) (the number of monomers that have been linked together to form the macromolecule of average length). There is a characteristic relationship between strength and DP for all linear polymers, as shown schematically in Fig. 4.23. The strength increases with degree of polymerization up to some saturation level.

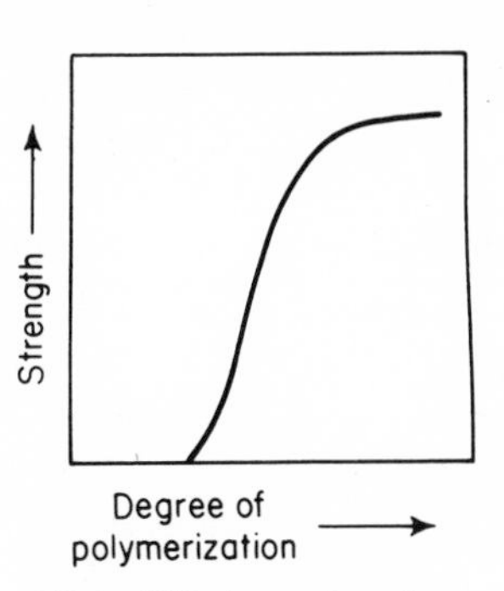

Fig. 4.23 Strength of a typical polymer as a function of degree of polymerization.

The second factor he lists is *axial order*, or *orientation*, which establishes the density and type of intermolecular bonds obtained. In solution, or in the molten state, polymer molecules are randomly coiled. If the solution is coagulated, or the molten mass frozen, this random coiling is maintained. Even though there is short range order in this state, long range ordering is lacking so that the lateral attachment of molecules, either by van der Waals' forces or by hydrogen bonds, is at a minimum. The molecules, however, can be "parallelized" along certain directions by deformation; for example, by stretching a fiber. This type of orientation was mentioned previously. A schematic drawing of the molecules in an amorphous, a crystalline, and an oriented crystalline polymer is shown in Fig. 4.24. After the molecules are made to lie alongside of each other, more lateral attachment and hence increased strength becomes possible.

A third factor, the *flexibility of the macromolecules*, generally involves interferences with free rotation about a single bond. This was discussed on p. 68. Obviously, the more rigid the molecule the less flexible the polymer.

Fig. 4.24 Schematic representation of (a) amorphous polymer; (b) A crystalline polymer; and (c) An oriented crystalline polymer. From T. Alfrey, Jr., *Mechanical Behavior of High Polymers*, Interscience, Publishers, Inc., p. 344, (1948).

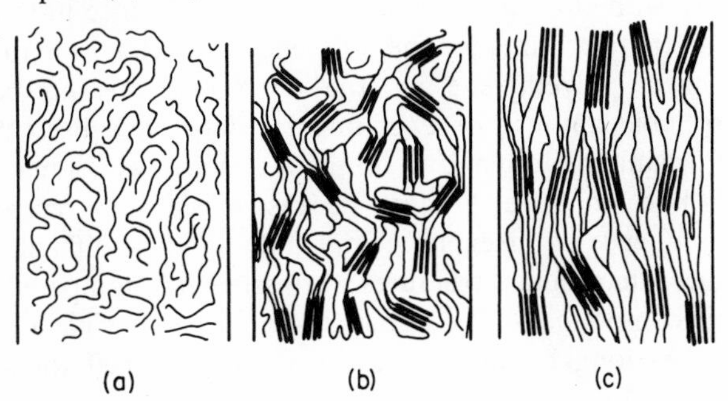

A fourth consideration is the type of *intermolecular forces* that are involved (as discussed under "Secondary Bonds"). It was pointed out that the strongest lateral bonds are hydrogen bridges.

Finally, there is the important factor of *stereoregularity* in lateral bonding. Many of the contributants to this factor were considered in the section on "Covalent Bonds." One additional factor not previously mentioned is the degree to which side branching affects stereoregularity. For example, high pressure polyethylene may form side branches as shown here:

```
                             |
                           H—C—H
                           H—C—H
                           H—C—H
                           H—C—H
                           H—C—H
                             |
 H   H   H   H   H   H   H   |   H   H   H   H   H
 |   |   |   |   |   |   |   |   |   |   |   |   |
—C———C———C———C———C———C———C———C———C———C———C———C———C—
 |   |   |   |   |   |   |   |   |   |   |   |   |
 H   H   H   H   H   |   H   H   H   H   H   H   H
                     |
                   H—C—H
                   H—C—H
                   H—C—H
                   H—C—H
                   H—C—H
                     |
                     H
```

These interfere with close packing, and it is the difference in the amount of side branching that accounts for the difference between reasonably rigid low-pressure or linear polyethylene and the softer high pressure polyethylene.

An appreciable amount of experimental data have been collected on the shape of organic molecules and on the manner in which macromolecules pack to form linear polymer bodies. Because it is a more recent science, as well as a consequence of the experimental and analytical difficulties involved, only a few macromolecular structures were studied as compared with metals and inorganic compounds. The structural formulas that have been used by organic chemists long before techniques were developed for studying the shape of these molecules are surprisingly representative of the actual molecule appearance. Mark and Tobolsky, for example, have compared the experimentally derived shape of the nickel phthalocyanine molecule obtained by Robertson with the structural formula of phthalocyanine (Fig. 4.25). The series of concentric circles in the center of the experimentally derived shape represents the nickel atom. Since none of these circles overlap neighboring atoms, the metal atom must be ionically bonded to the remainder of the molecule. The rest of the molecule is covalently bonded. One important difference between structural formulas, and, especially, the molecule shape of linear polymers, is that the structural formula does not show valence angles or the rotational freedom of single bonds.

Techniques, used for many years to determine the crystal structure of metals and salts, have been applied to establish the crystal structure of some polymers. This is a difficult task, however, because the structures of polymers are complex since each basis point in the unit cell does not represent a single ion but, in general, a section of a molecule. The first organic high

Fig. 4.25 (a) Structural formula of phthalocyanine. (b) Electron density map of nickel phthalocyanine. Note the similarity of the actual atom placement in the molecule with the structural formula. The center series of concentric circles is the nickel atom which is not shown in the formula. (From (R4.6) p. 23.)

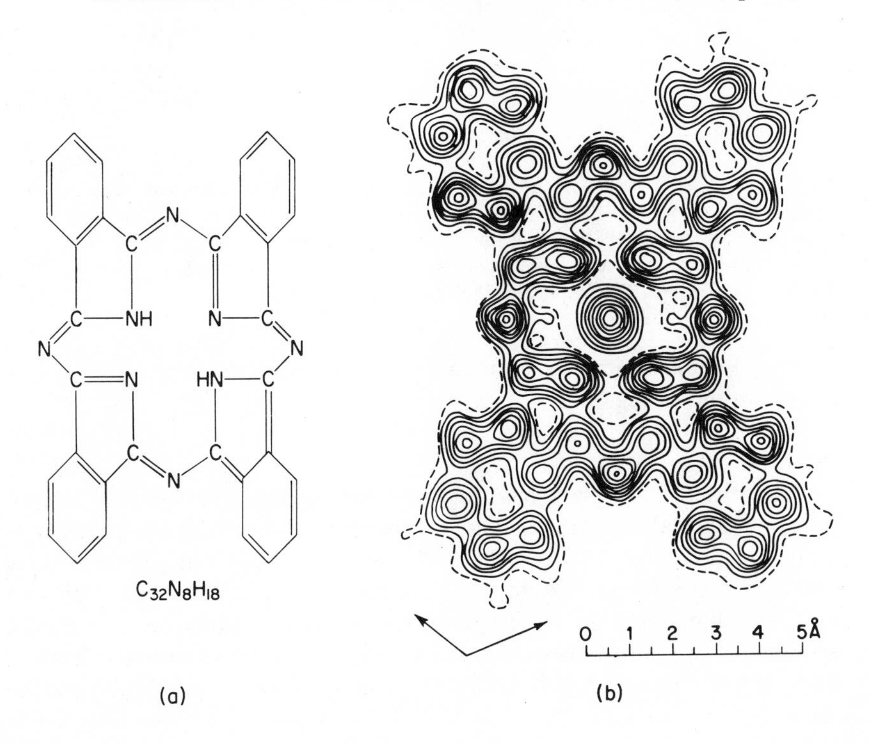

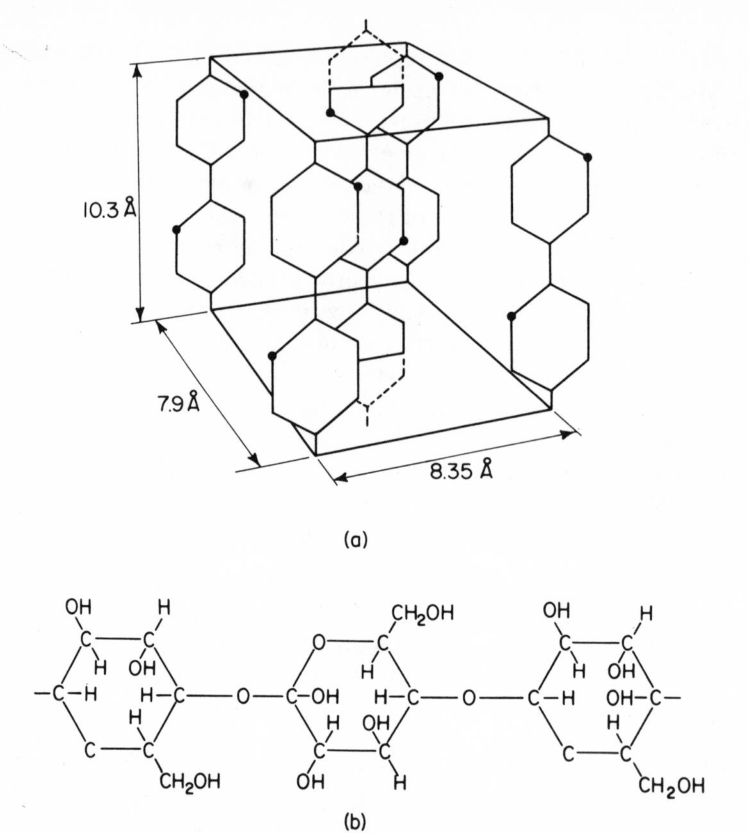

Fig. 4.26 Unit cell of cellulose. Each of the rings in (a) represents one of the rings in the structural formula of the polymer as shown in (b). The dot in each of the rings in the unit cell represents the position of the oxygen atom in that particular ring. The oxygen atoms between rings are not shown in the cell. (From R. D. Preston, *Scientific American, 197*, No. 3, 197 1957.)

polymer whose crystal structure was determined is cellulose, with a unit cell according to Fig. 4.26.

Glass

The structure of glass has not been unequivocally established as yet. Experimental data suggest that it is bonded by forces that are covalent in most directions but ionic in others. It has short range but not long range order, as is typical of a liquid (a supercooled liquid in this case). Whether the available data should be interpreted to indicate a real amorphous material, extremely fine grained crystalline material, or some other structure has not been agreed upon, however. The short range structure of glass has been suggested by Zachariasen to consist of oxygen atoms placed tetrahedrally about silicon atoms (Fig. 4.27a). Each of these oxygen atoms joins two tetrahedra together, and these tetrahedra are joined at at least three of their apices. The randomness of his proposed structure, which is required for an amorphous material, results from allowing the bond angle at which the oxygen joins the tetrahedra to vary. In a crystalline material this angle would always be 180°; a two dimensional schematic drawing of the structure of glass is compared with that of a crystal (Fig. 4.27b). Although the silica tetrahedron is the basis for the common glasses, an additional large cation is usually also present, resulting in a general formula for glass of:

$$A_mB_nO$$

where A = Large cation with a small charge, such as sodium (Na^+)
B = Large trivalent or tetravalent ion, such as silicon (Si^{++++}) or boron (B^{+++})
O = oxygen

The B and O atoms are covalently bonded to form the continuous chain. The oxygen atoms that do not serve as tetrahedron bridges are in terminal positions where they form ionic bonds with the alkali (Na) ions. The A atoms can migrate through the glass without affecting the chain network, and it is this freedom that allows the glass to become viscous without breaking down at elevated temperatures.

The mechanical properties of glass are controlled to a large extent by the manner in which it is corroded by water. Indeed, when glass fractures, it is generally the result of a combined action of stress and corrosion (see Chapter 10). Because of its importance in the mechanics of fracture, one of the proposed mechanisms for corrosion as suggested by Charles* is reviewed

* See (R4.3)

Fig. 4.27 (a) Silica Tetrahedron. (b) Two dimensional representation of the difference between the structure of a crystal (left) and a glass (right) according to Zachariasen.

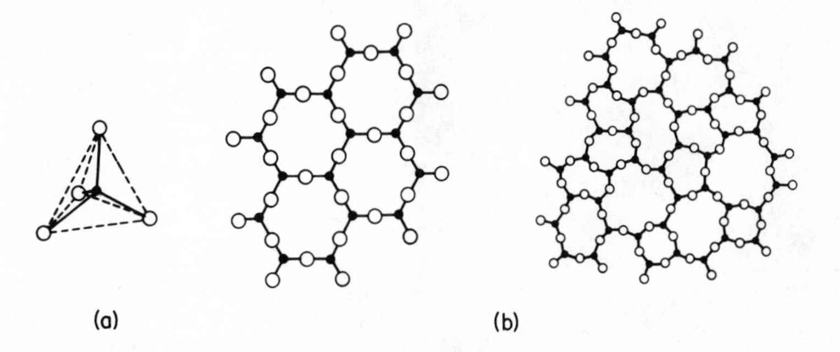

here. These reactions apply to a soda-lime glass. Glasses that consist only of the network structure (fused silica) are more stable in the presence of water vapor.

The first and most important reaction for the corrosion consists of the migration of a sodium ion. The broken sodium-oxygen bond allows the remaining oxygen atom to dissociate a water molecule, forming a hydride and hydroxyl ion:

$$\left[-\overset{|}{\underset{|}{\text{Si}}}-\text{O}-\boxed{\overset{\uparrow}{\text{Na}}} \right] + \text{H}_2\text{O} \rightarrow -\overset{|}{\underset{|}{\text{Si}}}\text{OH} + \text{OH}^-$$

The free hydroxyl ion combines with silicon:

$$\left[-\overset{|}{\underset{|}{\text{Si}}}-\text{O}-\overset{|}{\underset{|}{\text{Si}}}- \right] + \text{OH}^- \rightarrow -\overset{|}{\underset{|}{\text{Si}}}\text{OH} + \overset{|}{\underset{|}{\text{Si}}}\text{O}^-$$

The SiO^- dissociates another water molecule so that the reaction can continue:

$$\left[-\overset{|}{\underset{|}{\text{Si}}}-\text{O}^- \right] + \text{H}_2\text{O} \rightarrow -\overset{|}{\underset{|}{\text{Si}}}\text{OH} + \text{OH}^-$$

In the course of these three reactions, Na^+ ions are released. This increases the pH of the corrosion layer making the dissolution autocatalytic. Slowing down the rate at which the pH increases does not allow the accelerated dissolution, explaining why water vapor is more damaging than a liquid phase and why the addition of an acid causes less dissolution.

This very severe damage caused by water vapor is apparent in Fig. 4.28.

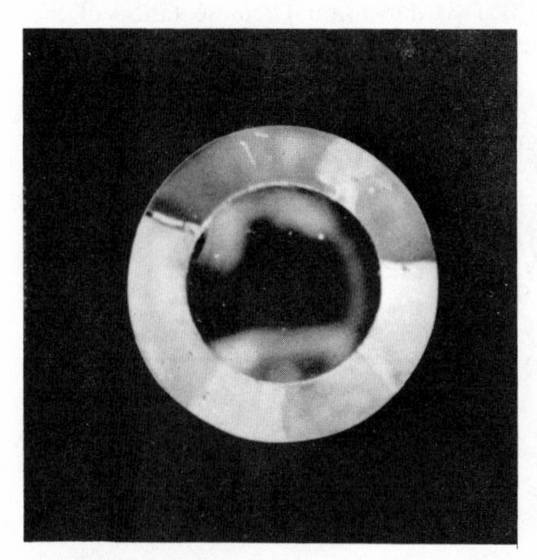

Fig. 4.28 Severe stress corrosion of a soda-lime glass rod (0.1 in. diameter) subjected to saturated steam (250°C) for 5 hours. (From (R. 4.3).)

PROBLEMS

1. Find the handbook values of structurally sensitive and structurally insensitive properties of copper and of polyethylene.

2. The equation for the potential curve in Fig. 4.2 (curve 2) can be written as:

$$U = -\frac{a}{r^n} + \frac{b}{r^m}$$

where r = distance between atom centers and a, b, n, m are constants depending on the type of bond.
This is simplified to.

$$U = \frac{Ae^2}{r} + \frac{B}{r^n}$$

where A and B are constants and e is an electrical charge for a univalent alkali halide. Using this equation, plot U vs. r and find the equilibrium radius, r_e.

3. Calculate the relative density of f.c.c. and b.c.c. irons.

4. Calculate the distance between atom centers in a single crystal of zinc. Zinc crystallizes in the h.c.p. system with c = 4.945Å and a = 2.664 Å.

5. What is the radius of the largest sphere that can be placed at the $\frac{1}{2}, \frac{1}{2}, 0$ position in a b.c.c. structure with a = 2 Å.

6. Show that in the cubic system:

a—[110] is perpendicular to (101)
b—[111] is perpendicular to (111)
c—[112] is perpendicular to (112)

Is this true for noncubic systems?

7. What is the densest direction in the following planes of a f.c.c. ((100)), ((110)), and ((111))?

8. Silver crystallizes in the f.c.c. cubic system with a = 4.077 Å. What is the radius of the silver atom?

9. Derive Equation 4.7.

10. Find the density of cesium and chlorine from a handbook. CsCl forms a b.c.c. structure with alternating layers of Cs and Cl. Calculate the density of the compound.

11. Describe the properties of graphite in terms of its interatomic bonds.

12. Why is nylon fiber stretched as one operation in its production?

13. Glass is sometimes classified as an inorganic polymer. Why?

REFERENCES FOR FURTHER READING

(R4.1) Azaroff, L. V., *Introduction to Solids.* New York: McGraw-Hill Book Company, 1960.

(R4.2) Barrett, C. S., *Structure of Metals* (2nd ed.). New York: McGraw-Hill Book Company, 1952.

(R4.3) Charles, R. J., "The Strength of Silicate Glasses and Some Crystalline Oxides," in Averbach, Felbeck, Hahn, and Thomas (eds.) *Fracture.* New York: Technology Press and John Wiley & Sons, Inc., 1959.

(R4.4) Dekker, A. J., *Solid State Physics* Englewood Cliffs, N.J.: Prentice-Hall, Inc., 1957.

(R4.5) Kittel, C., *Introduction to Solid State Physics* (2nd ed.) New York: John Wiley & Sons, Inc., 1956.

(R4.6) Mark, H. and A. V. Tobolsky, *Physical Chemistry of High Polymeric Systems* (2nd ed.) New York: Interscience Publishers, Inc., 1950.

(R4.7) Pauling, L., *Nature of the Chemical Bond.* Ithaca, New York: Cornell University Press, 1945.

(R4.8) Seitz, F., *The Physics of Metals.* New York: McGraw-Hill Book Company, 1943.

5

ELASTICITY

In Chapters 2 and 3 stresses and strains in solids were discussed and analyzed independently of each other, the former being treated as a problem of statics and the latter as one of geometry. In reality these two are closely interrelated since strains are created in a body when stress is applied to it. The specific function that describes the relationship between a particular applied stress and the resulting strain depends on the nature of the substance from which the body is made. One type of reaction to stress is a temporary or transient deformation termed *elastic*. This was discussed in general terms in the Introduction, and it is the relationship between stress and strain within this elastic range that is the subject of this Chapter. The elastic properties are assumed to be isotropic except in one section of the Chapter devoted to elastic anisotropy.

HOOKE'S LAW AND THE ELASTIC CONSTANTS E, G, AND ν

The unique feature of truly elastic behavior is that strain is a linear function of stress and, when one is zero, so is the other. Hence stress and strain are related simply by a proportionality factor. Nevertheless, since there are two basically different types of stresses and strains, their respective proportionality constants also differ. The relationship between normal and shear stresses and their appropriate strains is given by Hooke's Law:

$$\sigma = E\varepsilon \tag{5.1a}$$

and

$$\tau = G\gamma \tag{5.1b}$$

TABLE 5.1

YOUNG'S MODULUS AND POISSON'S RATIO OF A VARIETY OF MATERIALS

Material	*Test Temp. (°F.)*	*E (psi)*	ν
	+1000	38.8×10^6	
Beryllium*	RT	41.7×10^6	0.05
	−300	42.7×10^6	
	+1000	56.1×10^6	
Tungsten*	RT	59.3×10^6	0.30
	−300	60.4×10^6	
Wrought iron	+1000	25.6×10^6	
and steel*	RT	30.8×10^6	0.28
	−300	32.2×10^6	
Cast iron (gray iron)	RT	$11\text{–}23 \times 10^6$	0.17
Aluminum	RT	$10\text{–}11 \times 10^6$	0.33
Concrete	RT	$3.5\text{–}5 \times 10^6$	0.19
Quartz fiber	RT	7.4×10^6	
Brick	RT	$1.5\text{–}2.5 \times 10^6$	
Limestone	RT	8.4×10^6	
Rubber (soft)	RT	$0.15\text{–}2 \times 10^3$	0.49
Wood	RT	$0.9\text{–}2.2 \times 10^6$	
Nylon	RT	$0.18\text{–}0.45 \times 10^6$	
Epoxy	RT	$0.4\text{–}20 \times 10^6$	
Glass	RT	$9\text{–}10 \times 10^6$	0.25
Pyroceram	RT	$12.5\text{–}17.3 \times 10^6$	0.25

* After W. Köster, from N.B.S. Circular 520, 1952, Supt. of Documents, U.S. Govt. Printing Office, Washington, D.C.

The linear relation between stress and strain for small strains has also been established from thermodynamic principles†. Since the discussions in this chapter are limited to small deformations, conventional strain symbols [see Eq. (3.3)] are used. The constant E is the *elastic* (or *Young's*) *modulus* whereas G is the *shear modulus* or *modulus of rigidity*. Their dimensions are the same as those of stress (psi, Kg/cm^2, etc.) and their values vary within extremely wide limits depending on the particular material and its temperature (Table 5.1).

When an elastic rod is stretched, not only does it lengthen in accordance with Hooke's law but it contracts laterally as well. The lateral strain has

† See, for example, I. N. Sneddon and D. S. Berry, *Encyclopedia of Physics*, ed. by S. Flügge (Berlin: Springer-Verlag, 1958), VI, 15–17.

been found by experience to be a constant fraction, ν, of the longitudinal extension. Like E and G (but to a lesser extent) this fraction, which is called *Poisson's Ratio,* depends on the specific material (Table 5.1). For metals, ν is generally taken as 0.3. If the strain in the direction of the applied force is ε, the strain normal to it is $-\nu\varepsilon$. When a single stress σ_x acts, the normal strains in terms of stress are found by Hooke's law as:

$$\varepsilon_x = \frac{\sigma_x}{E}, \qquad \varepsilon_y = \varepsilon_z = -\frac{\nu\sigma_x}{E}$$

The negative sign merely indicates that the longitudinal and transverse strains have opposite signs since extension in the one direction is accompanied by contractions in the other two.

The deformation of a cube of unit dimensions is shown in Fig. 5.1. When this cube is subjected simultaneously to σ_x, σ_y, and σ_z, the resulting strains are calculated by superposition with the aid of the following table:

TABLE 5.2

STRAINS PRODUCED IN A UNIT CUBE BY STRESSES σ_x, σ_y, AND σ_z

Stress	*Strain in the x Direction*	*Strain in the y Direction*	*Strain in the z Direction*
σ_x	σ_x/E	$-\nu\sigma_x/E$	$-\nu\sigma_x/E$
σ_y	$-\nu\sigma_y/E$	σ_y/E	$-\nu\sigma_y/E$
σ_z	$-\nu\sigma_z/E$	$-\nu\sigma_z/E$	σ_z/E

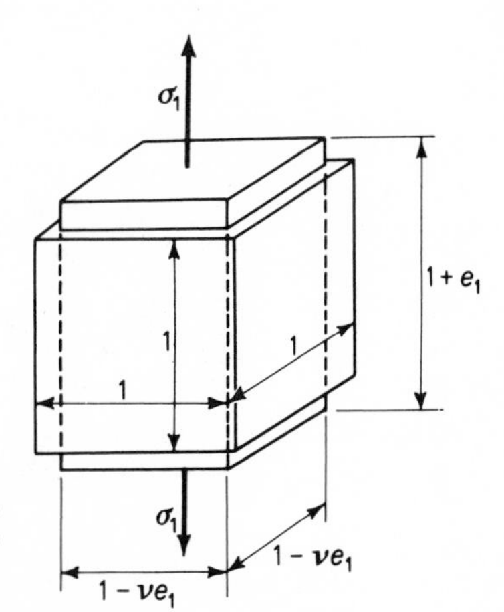

Fig. 5.1 Elastic deformation of unit cube under uniaxial stress σ_1.

By adding the terms in each vertical strain column, the resultant total strains are found to be

$$\varepsilon_x = \frac{1}{E}[\sigma_x - \nu(\sigma_y + \sigma_z)]$$

$$\varepsilon_y = \frac{1}{E}[\sigma_y - \nu(\sigma_z + \sigma_x)] \qquad (5.2a)$$

$$\varepsilon_z = \frac{1}{E}[\sigma_z - \nu(\sigma_x + \sigma_y)]$$

Equation (5.2a) is the *generalized form of Hooke's law* for isotropic materials.

If the normal strain in one direction, say, along the z-axis is zero, the above group is simplified because when:

$$\varepsilon_z = \frac{1}{E}[\sigma_z - \nu(\sigma_x + \sigma_y)] = 0$$

then:

$$\sigma_z = \nu(\sigma_x + \sigma_y) \qquad (5.2b)$$

Such a condition of *plane strain*, with $\varepsilon_z = 0$, can be intentionally induced on an element. More commonly, however, it occurs spontaneously by a *lateral restraint* or *constraint* exerted upon the highly stressed area by its neighboring relatively unstressed portion. For instance, the sharp edge of a knife is prevented from expanding in the length direction, even when cutting a hard material, because of the constraint exerted upon it by the thick and lightly stressed main section.

When σ_z from (5.2b) is substituted in the first two equations of (5.2a), the nonzero strains become:

$$\varepsilon_x = \frac{1}{E}[(1 - \nu^2)\sigma_x - \nu(1 + \nu)\sigma_y]$$

$$\varepsilon_y = \frac{1}{E}[(1 - \nu^2)\sigma_y - \nu(1 + \nu)\sigma_x] \qquad (5.2c)$$

The shear stresses τ_{xy}, τ_{yz}, and τ_{zx}, acting simultaneously with σ_x, σ_y, and σ_z, will produce [from (5.1b)] the following shear strains:

$$\gamma_{xy} = \frac{1}{G}\tau_{xy}$$

$$\gamma_{yz} = \frac{1}{G}\tau_{yz} \qquad (5.3)$$

$$\gamma_{zx} = \frac{1}{G}\tau_{zx}$$

Note that a shear stress, unlike a normal stress on one plane, does not affect the strain on perpendicular planes.

THE VOLUME STRAIN (DILATATION) EXPRESSED BY STRESSES

The expression for volume strain when the individual linear strains are small has previously been shown by (3.10b) to be:

$$\varepsilon_V = \varepsilon_x + \varepsilon_y + \varepsilon_z$$

Using (5.2a):

$$\varepsilon_V = \frac{1}{E}(1 - 2\nu)(\sigma_x + \sigma_y + \sigma_z) \tag{5.4a}$$

or:

$$\varepsilon_V = \frac{1}{E}(1 - 2\nu)3\sigma_a = \frac{1}{E}(1 - 2\nu)I_1 \tag{5.4b}$$

where $\sigma_a = \frac{1}{3}(\sigma_x + \sigma_y + \sigma_z)$ is the *mean normal stress.*

An alternative form for the same relation is:

$$\sigma_a = K\varepsilon_V = 3K\varepsilon_a \tag{5.4c}$$

where $K = E/3(1 - 2\nu)$ is the *modulus of volume expansion* or *bulk modulus,* and ε_a is the mean normal strain.

RELATION BETWEEN THE ELASTIC CONSTANTS *E* AND *G*. LAMÉ'S CONSTANT AND ALTERNATIVE FORM OF HOOKE'S LAW

To find the relationship between E and G, one writes expression (2.18):

$$\tau_{xy(\max)} = \tau_{12} = (\sigma_1 - \sigma_2)/2 \tag{a}$$

and by analogy with this equation and Table 3.1:

$$\gamma_{xy(\max)} = \gamma_{12} = \varepsilon_1 - \varepsilon_2 \tag{b}$$

By rewriting the first two equations of (5.2a) to refer to principal stresses and strains, and then subtracting the second from the first, one obtains:

$$\varepsilon_1 - \varepsilon_2 = \frac{1}{E}[\sigma_1 - \sigma_2 + \nu\sigma_1 - \nu\sigma_2] = \frac{1 + \nu}{E}(\sigma_1 - \sigma_2) \tag{c}$$

From (a) and (b), with $\gamma = \tau/G$:

$$\varepsilon_1 - \varepsilon_2 = \gamma_{12} = \frac{\tau_{12}}{G} = \frac{\sigma_1 - \sigma_2}{2G} \tag{d}$$

Substituting the final element of (d) for $\varepsilon_1 - \varepsilon_2$ in (c), one finally has:

$$\frac{1}{2G} = \frac{1 + \nu}{E} \quad \text{or} \quad G = \frac{E}{2(1 + \nu)} \tag{5.5}$$

In most metals $\nu \approx \frac{1}{3}$ to $\frac{1}{4}$, so that the shear modulus is $0.37E$ to $0.40E$.

By combining E, G, K, and ν, a variety of elastic constants have been derived. One of the most commonly used of these is *Lamé's constant*, λ, where:

$$\lambda = \frac{\nu E}{(1 + \nu)(1 - 2\nu)} \tag{5.6}$$

Hooke's generalized law can also be expressed in the form $\sigma = f(\varepsilon_x, \varepsilon_y, \varepsilon_z)$ by solving (5.2a) for σ_x, σ_y, and σ_z. The resulting equations are cumbersome but they can be considerably abridged with the aid of Lamé's constant resulting in:

$$\left.\begin{aligned} \sigma_x &= 3\lambda\varepsilon_a + 2G\varepsilon_x \\ \sigma_y &= 3\lambda\varepsilon_a + 2G\varepsilon_y \\ \sigma_z &= 3\lambda\varepsilon_a + 2G\varepsilon_z \end{aligned}\right\} \tag{5.7}$$

ELASTIC STRAIN ENERGY

When an increasing force is applied to an elastic body, the area or point upon which it acts is displaced as a result of the deformation of the material. Hence, a certain amount of work is performed by the applied force. This work is stored in the deformed material in the form of *elastic strain energy*.

If the external force is now gradually reduced, the deformed body will gradually return to its original shape. The area upon which the force acts will now move in the direction opposite to the first displacement, thereby converting its elastic strain energy into external work.

In determining the strain energy stored in an elastic material deformed by one or more external forces, the principles in Chapters 2 and 3 are used. The strain energy dW in the volume dV of an elementary parallelepiped with sides dx, dy, dz ($dV = dx\,dy\,dz$) (Fig. 5.2) is computed. Since the various stresses are assumed to be uniformly distributed over the appropriate element faces, the deformation is substantially homogeneous, and the relation between stress and strain or force and strain is linear.

Consequently, the work of deformation performed by a single elementary force P will be $Ps/2$, where s is the displacement. The constant $\frac{1}{2}$ appears because the force increases linearly from zero on initial contact and reaches the value of P after it has moved past distance s. In the general case (Fig. 5.2) the normal forces are $\sigma_x\,dy\,dz$, $\sigma_y\,dz\,dx$, $\sigma_z\,dx\,dy$, and the shear forces $\tau_{xy}\,dy\,dz$, $\tau_{yz}\,dz\,dx$, $\tau_{zx}\,dx\,dy$. The linear dimensional changes are $\varepsilon_x\,dx$, $\varepsilon_y\,dy$, and $\varepsilon_z\,dz$ in the x, y, and z directions, respectively. The shears or tangential displacements are $\gamma_{xy}\,dx$, $\gamma_{yz}\,dy$, and $\gamma_{zx}\,dz$. The deformation energy of the parallelepiped is, therefore:

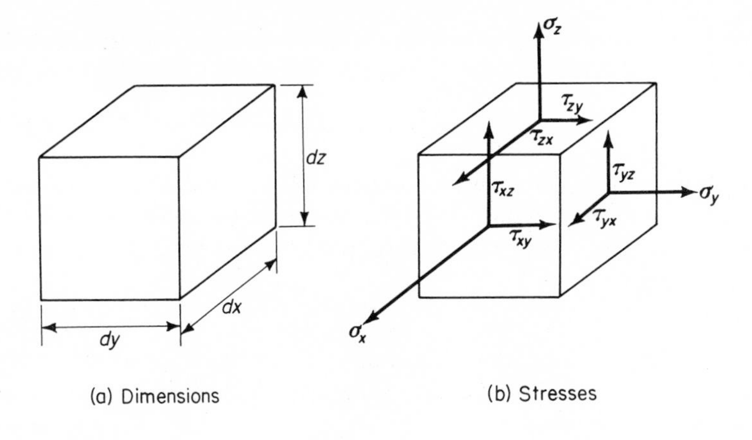

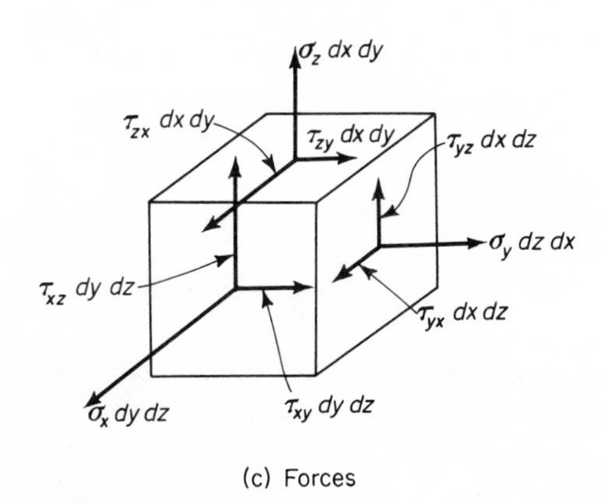

(c) Forces

Fig. 5.2 Dimensions, stresses and forces in elementary parallelepiped for strain energy calculation.

$$\begin{aligned} dW &= \tfrac{1}{2}\sigma_x\, dy\, dz \cdot \varepsilon_x\, dx + \tfrac{1}{2}\sigma_y\, dz\, dx \cdot \varepsilon_y\, dy \\ &\quad + \tfrac{1}{2}\sigma_z\, dx\, dy \cdot \varepsilon_z\, dz + \tfrac{1}{2}\tau_{xy}\, dy\, dz \cdot \gamma_{xy}\, dx \\ &\quad + \tfrac{1}{2}\tau_{yz}\, dz\, dx \cdot \gamma_{yz}\, dy + \tfrac{1}{2}\tau_{zx}\, dx\, dy \cdot \gamma_{zx}\, dz \\ &= \tfrac{1}{2}(\sigma_x\varepsilon_x + \sigma_y\varepsilon_y + \sigma_z\varepsilon_z + \tau_{xy}\gamma_{xy} + \tau_{yz}\gamma_{yz} + \tau_{zx}\gamma_{zx})\, dx\, dy\, dz \end{aligned}$$

or

$$dW = \tfrac{1}{2}(\sigma_x\varepsilon_x + \cdots + \tau_{xy}\gamma_{xy} + \cdots)\, dV$$

Since dW/dV represents the strain energy per unit volume or *specific strain energy* W_o

$$W_o = \frac{1}{2}(\sigma_x\varepsilon_x + \sigma_y\varepsilon_y + \sigma_z\varepsilon_z + \tau_{xy}\gamma_{xy} + \tau_{yz}\gamma_{yz} + \tau_{zx}\gamma_{zx}) \tag{5.8}$$

In a body of finite dimensions, the total strain energy is represented by the integral

$$W = \iiint W_o \, dx \, dy \, dz \tag{5.9a}$$

which can be computed when the various stresses and strains are known functions of x, y, and z. Only when the state of stress is homogeneous over the entire body (see Chapter 2 for definition) can the total strain energy be directly obtained as:

$$W = W_o V \tag{5.9b}$$

Equation (5.8) can be transformed so as to eliminate either the strain or the stress elements by utilizing Hooke's laws (5.1, 5.2). Suitable substitutions will then yield

$$W_o = \frac{1}{2E}(\sigma_x^2 + \sigma_y^2 + \sigma_z^2) - \frac{\nu}{E}(\sigma_x\sigma_y + \sigma_y\sigma_z + \sigma_z\sigma_x) + \frac{1}{2G}(\tau_{xy}^2 + \tau_{yz}^2 + \tau_{zx}^2) \tag{5.10a}$$

in terms of stress, and

$$W_o = \frac{9}{2}\lambda\varepsilon_a^2 + G(\varepsilon_x^2 + \varepsilon_y^2 + \varepsilon_z^2) + \frac{1}{2}G(\lambda_{xy}^2 + \lambda_{yz}^2 + \lambda_{zx}^2) \tag{5.10b}$$

in terms of strains. It can be shown that differentiation of W_o with respect to any strain component will produce the stress in the appropriate direction so that:

$$\frac{\partial W_0}{\partial \varepsilon_x} = \sigma_x, \text{ etc.} \tag{5.11}$$

Expressions (5.8) through (5.10) are quite cumbersome but in many problems they can be considerably simplified. For instance, if the stress system is plane, terms containing σ_z, τ_{yz}, and τ_{zx} in (5.10a) vanish. Likewise, plane strain will result in a reduced Eq. (5.10b), in view of $\varepsilon_z = \gamma_{yz} = \gamma_{zx} = 0$.

Quite often the stress conditions at a point are specified in terms of principal stresses. This eliminates all the shear components, with (5.10a), for example, becoming:

$$W_o = \frac{1}{2E}(\sigma_1^2 + \sigma_2^2 + \sigma_3^2) - \frac{\nu}{E}(\sigma_1\sigma_2 + \sigma_2\sigma_3 + \sigma_3\sigma_1) \tag{5.12}$$

STRAIN ENERGY OF VOLUME CHANGE (DILATATIONAL ENERGY)

Only the normal strain components are responsible for changes of volume in elastically deformed bodies (see p. 103). Since the sum of all normal stresses is constant in the vicinity of a particular point (the first stress invariant I_1), the volume change caused by an arbitrary stress system is equivalent to that due to a mean normal stress, $\sigma_a = \frac{1}{3}(\sigma_x + \sigma_y + \sigma_z)$, acting in three mutually perpendicular directions. It follows from (5.4) that such a stress system will produce identical linear strains

$$\varepsilon_a = \frac{(1 - 2\nu)\sigma_a}{E} \tag{5.13}$$

in all three coordinate directions. The strain energy of this "averaged" deformation is

$$\begin{aligned} W_d &= 3(\tfrac{1}{2}\sigma_a \varepsilon_a) \\ &= 3\left(\frac{1 - 2\nu}{2E}\right)\sigma_a^2 \\ &= \frac{1 - 2\nu}{6E}(\sigma_x + \sigma_y + \sigma_z)^2 \end{aligned} \tag{5.14}$$

the subscript, d, in W_d standing for dilatation.

Assuming that the volume element is a cube, one can see that the mean stress σ_a acting in three mutually perpendicular directions transforms it into another cube that may be larger or smaller than the original one, depending on the sign of σ_a (or ε_a). A condition in which the three principal stresses are actually equal is said to represent a *hydrostatic state of stress*. Such condition can be readily produced by submerging a body into a liquid that is subsequently pressurized.

Since the three principal σ components usually differ, they transform the cube into a parallelepiped or rectangular prism, with its longest side in the direction of the algebraically largest stress. The volume of the prism will be equal to that of a cube dilated by the action of three equal components, σ_a. Likewise, the strain energy, W_d, required to produce the observed volume change will not depend on the final shape. However, the total strain energy W_o used to effect the combined changes of volume and shape will be larger than W_d as is obvious from a comparison of (5.10a) and (5.14).

Related to the notion of hydrostatic state of stress is the *hydrostatic stress (or strain) component*, which can be readily separated from the actual principal stresses (strains) whenever all of them are of the same sign. For instance, a stress system with

$$\sigma_1 = -25{,}000$$

$$\sigma_2 = -35{,}000$$

$$\sigma_3 = -60{,}000$$

is equivalent to one where a hydrostatic stress (pressure), $\sigma_h = -25{,}000$, is superimposed upon

$$\sigma_1' = 0$$
$$\sigma_2' = -10{,}000$$
$$\sigma_3' = -35{,}000$$

so that

$$\sigma_1 = \sigma_1' + \sigma_h = \sigma_h$$
$$\sigma_2 = \sigma_2' + \sigma_h$$
$$\sigma_3 = \sigma_3' + \sigma_h.$$

The hydrostatic stress component is identical with the numerically smallest principal (compressive) stress. It does not affect the principal shear stresses because $\sigma_1 - \sigma_2 = (\sigma_1' + \sigma_h) - (\sigma_2' + \sigma_h) = \sigma_1' - \sigma_2' = \tau_{12}$.

The same manipulation can be performed on an all-tensile principal stress system by regarding the smallest tension ($= \sigma_3$) as σ_h. Since the term "hydrostatic" implies pressure rather than tension, the term *balanced triaxial tension* better describes σ_h in these conditions. However, *hydrostatic tension*, while awkward, has the advantage of brevity.

STRAIN ENERGY OF DISTORTION (SHEAR STRAIN ENERGY)

If the dilatational energy term (5.14) is subtracted from the specific strain energy (5.10), the difference must represent the portion of energy attributable to shear stresses alone. Shear deformations distort the shape of the body by altering the angles between intersecting planes, and this also requires an energy expenditure. Consequently, the difference, $W_o - W_d = W_s$, is called *distortional* or *shear strain energy*.

By utilizing the identity

$$\sigma_x\sigma_y + \sigma_y\sigma_z + \sigma_z\sigma_x = (\sigma_x^2 + \sigma_y^2 + \sigma_z^2) - \tfrac{1}{2}[(\sigma_x - \sigma_y)^2 + (\sigma_y - \sigma_z)^2 + (\sigma_z - \sigma_x)^2]$$

the distortional energy can be expressed in the form

$$W_s = W_o - W_d = \frac{1 + \nu}{6E}[(\sigma_x - \sigma_y)^2 + (\sigma_y - \sigma_z)^2 + (\sigma_z - \sigma_x)^2] + \frac{1}{2G}(\tau_{xy}^2 + \tau_{yz}^2 + \tau_{zx}^2) \quad (5.15a)$$

or, with $E = 2(1 + \nu)G$

$$W_s = \frac{1}{12G}[(\sigma_x - \sigma_y)^2 + (\sigma_y - \sigma_z)^2 + (\sigma_z - \sigma_x)^2 + 6(\tau_{xy}^2 + \tau_{yz}^2 + \tau_{zx}^2)] \quad (5.15b)$$

If the principal stresses are specified, formula (5.15a) becomes

$$W_s = \frac{1+\nu}{6E}[(\sigma_1 - \sigma_2)^2 + (\sigma_2 - \sigma_3)^2 + (\sigma_3 - \sigma_1)^2] \qquad (5.15c)$$

and equivalent formulae can be readily developed with the aid of (5.2a) to express W_s by strains.

In uniaxial tension, $\sigma_1 \neq 0$, $\sigma_2 = \sigma_3 = 0$, so that $W_o = \sigma_1^2/2E$

and
$$W_d = \sigma_1^2(1 - 2\nu)/6E$$
$$W_s = \sigma_1^2(1 + \nu)/3E. \qquad (5.16)$$

In simple torsion $\sigma_1 = -\sigma_3$ and $\sigma_2 = 0$; $W_o = \sigma_1^2(1 + \nu)/E$ and $W_d = 0$; $W_s = W_o$ representing a case of pure distortion with no dilatation.

When $\sigma_1 = \sigma_2 = \sigma_3$, then $W_o = W_d = 3(1 - 2\nu)\sigma_1^2/2E$ and $W_s = 0$. We have thus a case of "hydrostatic" volume change without an attendant change of shape (distortion).

To better visualize the physical meaning of the two "components" of the total strain energy, consider at a point A within a large body a small arbitrarily oriented cube with stresses σ_x, σ_y, σ_z, τ_{xy}, τ_{yz}, τ_{zx} acting on its sides in the usual way. In order to follow the shape changes associated with each type of elastic energy, assume first that the τ stresses are absent and the cube is loaded in all three normal directions with a mean stress $\sigma_a = (\sigma_x + \sigma_y + \sigma_z)/3$ which may, for example, be tensile. As a consequence of this loading, cube I in Fig. 5.3a will expand uniformly into another larger cube II.

Fig. 5.3 Changes of volume V and shape of elementary elastic cube by: (a) three equal normal stresses, σ_a; (b) Three different normal stresses $(\sigma_x + \sigma_y + \sigma_z)/3 = \sigma_a$; (c) Same as (b) with shear stresses added.

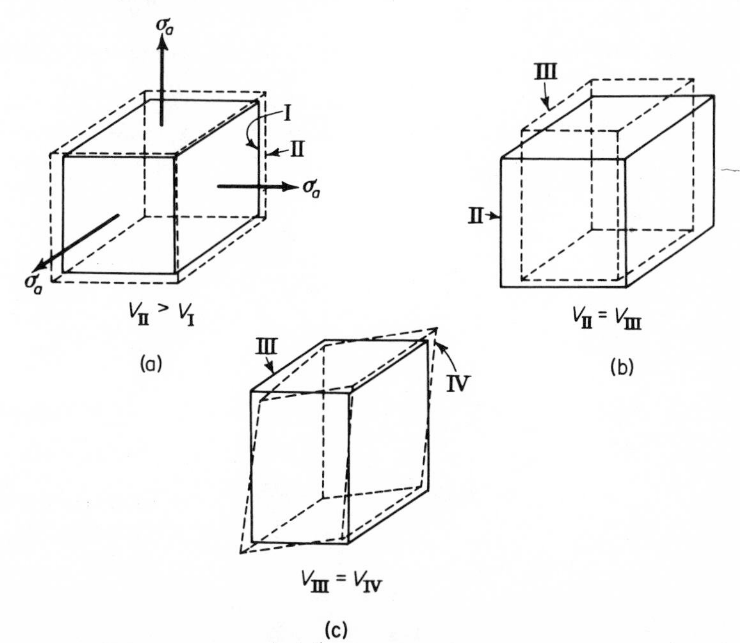

The individual σ_a components are now gradually altered, keeping their sum constant until each of these takes on its initial value; i.e., σ_x, σ_y, or σ_z. As the stresses change, the expanded cube II gradually elongates in the direction of the largest normal stress and contracts in one or both of the other directions, while maintaining a constant volume. As a result, a rectangular parallelepiped III (Fig. 5.3b) will be obtained. Note that work is being consumed to effect this transformation which involves solely a change of shape.

The only elements that were ignored so far are the shearing stresses, and, when these are applied to the appropriate sides, the rectangular block III in 5.3b distorts into a rhombohedron (Fig. 5.3c). Again, no volume change is involved.

Had we initially selected the "test cube" so that its edges coincide with the principal directions 1, 2, and 3, the resulting deformation would have transformed it into a rectangular parallelepiped. In this case the right angles between the sides would have remained unchanged. Note, however, that the shape of the "principally oriented" cube would have suffered more linear distortion than that associated with the II → III transition in Fig. 5.3b (for the same σ_a). This is because σ_1 is always larger, and σ_3 smaller, than any of the three stresses σ_x, σ_y, or σ_z. Nevertheless, it can be readily proved that as long as this cube and the one in Fig. 5.3a refer to the same point in the body, the distortional (or shear) energies W_s will be exactly the same in both cases. The fraction of the total strain energy required in the "random" cube to distort the angles between the sides is used in the "principal" cube to produce the extra difference between the linear strains in different directions.

STRAIN ENERGY EXPRESSED BY OCTAHEDRAL STRESSES

The two parts of the specific energy W_0—viz., W_d and W_s—can also be expressed as a function of a single stress (or stress component) acting on any

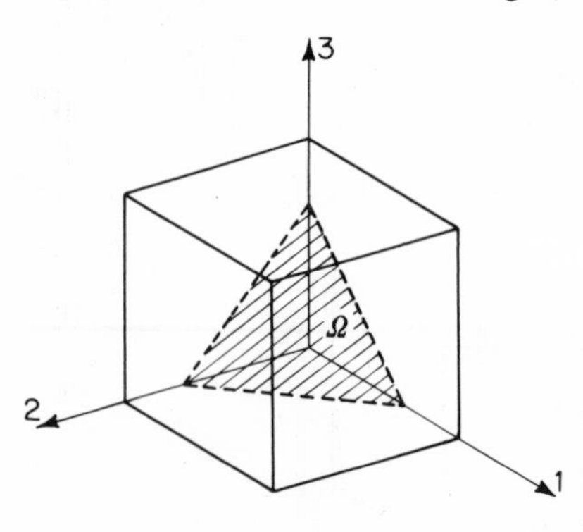

Fig. 5.4 Octahedral plane Ω in a principal cube.

one of the planes equally inclined to the three principal directions. These are the *octahedral* planes and one of them, marked Ω, is shown in Fig. 5.4.

Crystallographically, the octahedral planes represent the ((111)) family, and, if all eight of them are drawn through the eight corners of a cube, a regular octahedron is obtained—with the bounding planes (111), $(\bar{1}11)$, $(\bar{1}\bar{1}1)$, and $(1\bar{1}1)$ on the top and $(11\bar{1})$, $(\bar{1}1\bar{1})$, $(\bar{1}\bar{1}\bar{1})$, and $(1\bar{1}\bar{1})$ on the bottom (Fig. 5.5).

All three angles, ω, between the normal to plane Ω and the coordinate axes are equal, and, since $k^2 + l^2 + m^2 = 3\cos^2\omega = 1$, we obtain, as the direction cosines of the octahedral plane, $\cos\omega = 1/\sqrt{3} = 0.577$, and $\omega = 54° \, 44'$.

Observing that the octahedral plane Ω cuts off a tetrahedron from the elemental cube, the stress components on its surface can be found with the aid of equations (2.13) through (2.15) and (2.2). From (2.14), with $k = 1/\sqrt{3}$, the stress normal to the octahedral plane is:

$$\sigma_\Omega = \tfrac{1}{3}(\sigma_1 + \sigma_2 + \sigma_3) \tag{5.17a}$$

Thus the *octahedral normal stress* σ_Ω is equal to the mean normal stress $\sigma_a = I_1/3$.

The *octahedral shear stress* τ_Ω is obtained from:

$$\tau_\Omega^2 = S^2 - \sigma_\Omega^2 \qquad \text{(see Eq. 2.2).}$$

Bearing in mind that $S^2 = S_x^2 + S_y^2 + S_z^2$, and using (2.6) for S_x, S_y, and S_z

$$\tau_\Omega = \tfrac{1}{3}\sqrt{(\sigma_x - \sigma_y)^2 + (\sigma_y - \sigma_z)^2 + (\sigma_z - \sigma_x)^2 + 6(\tau_{xy}^2 + \tau_{yz}^2 + \tau_{zx}^2)} \tag{5.17b}$$

or, in terms of principal stresses

$$\tau_\Omega = \tfrac{1}{3}\sqrt{(\sigma_1 - \sigma_2)^2 + (\sigma_2 - \sigma_3)^2 + (\sigma_3 - \sigma_1)^2} \tag{5.17c}$$

The relation between σ_Ω and the dilatational strain energy W_d (5.14) is evident because $\sigma_\Omega = \sigma_a$. Likewise, by comparing (5.15b) with (5.18), it is seen that the shear strain energy can be written in the form:

$$W_s = \frac{3}{4G}\tau_\Omega^2 = \frac{3(1+\nu)}{2E}\tau_\Omega^2$$

The total deformation energy is thus a simple function of the octahedral normal and shear stresses.

Similar operations lead to the notion of *octahedral normal strain* and *octahedral shear strain*. Without carrying out the calculations, the respective terms are given below:

$$\varepsilon_\Omega = \tfrac{1}{3}(\varepsilon_1 + \varepsilon_2 + \varepsilon_3) = \tfrac{1}{3}(\varepsilon_x + \varepsilon_y + \varepsilon_z) \tag{5.18a}$$

$$\gamma_\Omega = \tfrac{2}{3}[(\varepsilon_x - \varepsilon_y)^2 + (\varepsilon_y - \varepsilon_z)^2 + (\varepsilon_z - \varepsilon_x)^2 + \tfrac{3}{2}(\gamma_{xy}^2 + \gamma_{yz}^2 + \gamma_{zx}^2)]^{\frac{1}{2}} \quad (5.18b)$$

or in terms of principal strains:

$$\gamma_\Omega = \tfrac{2}{3}\sqrt{[(\varepsilon_1 - \varepsilon_2)^2 + (\varepsilon_2 - \varepsilon_3)^2 + (\varepsilon_3 - \varepsilon_1)^2]} \quad (5.18c)$$

They are related to dilatation and distortion energy in a manner similar to that in which σ_Ω and τ_Ω were related to these quantities (except for the coefficients).

For uniaxial stress, the three principal strains are:

$$\varepsilon_1,\ \varepsilon_2 = \varepsilon_3 = -\nu\varepsilon_1,$$

This corresponds to an octahedral shear strain:

$$\gamma_\Omega = \frac{2}{3}\sqrt{2\varepsilon_1^2(1+\nu)^2} = \frac{2\sqrt{2}(1+\nu)}{3}\varepsilon_1 \quad (5.19)$$

SIGNIFICANT (EFFECTIVE) STRESS AND STRAIN

Two other terms are often used to define the "average" stress-strain relations at a point. They have the dimensions of stress and strain and are called the *significant*, *effective*, or *generalized* stress

$$\sigma_e = \frac{3}{\sqrt{2}}\tau_\Omega \quad (5.20a)$$

and the *significant*, *effective*, or *generalized* strain

$$\varepsilon_e = \frac{3}{2\sqrt{2}(1+\nu)}\gamma_\Omega \quad (5.20b)$$

Because these two quantities have no actual physical meaning, it is not obvious that they are related by Hooke's law. However, this can be proven by first writing Hooke's law for octahedral shear stress and strain:

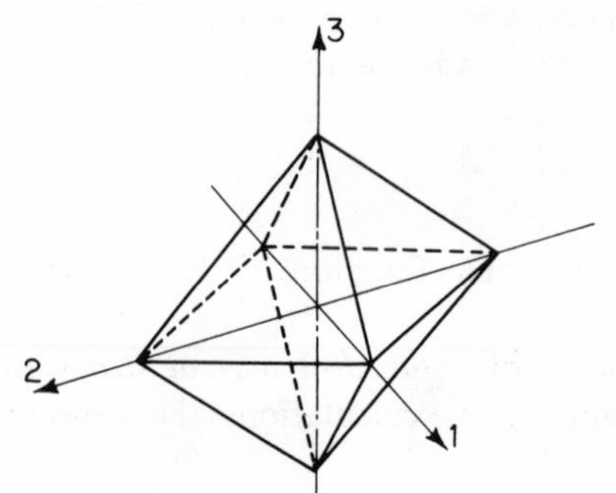

Fig. 5.5 Regular octahedron formed by 8 octahedral planes passing through corners of a principal cube.

$$\tau_\Omega = G\gamma_\Omega \tag{5.21}$$

The relation between the significant units is now obtained by expressing τ_Ω and γ_Ω in terms of σ_e and ε_e, using (5.20). The shear modulus G is replaced with E and ν, with the aid of (5.5). These substitutions lead to

$$\frac{\sqrt{2}}{3}\sigma_e = \frac{E}{2(1+\nu)} \times \frac{2\sqrt{2}(1+\nu)}{3}\varepsilon_e$$

and

$$\sigma_e = E\varepsilon_e \tag{5.22}$$

THERMAL STRESSES AND SHRINK FITS

Elastic stresses and strains are usually thought to result from forces applied externally to a body. They are also developed, however, in the absence of external forces when a member is simply heated or cooled. This results because all materials expand or contract when their temperature is changed. If the temperature of the body could gradually change from T_o to T_n in such a fashion that at any instant the temperature were uniform throughout its mass—i.e., the heating (or cooling) were isothermal, no stresses would be produced by the temperature change. However, it requires some finite period of time for the heat to pass from the surface to the interior. Hence, at any instant various elements of the body are at different temperatures, so that each element wants to take on a different size, depending on its instantaneous temperature. Such independent localized distortions are not possible if each element is to remain compatible with its neighbors, and, therefore, stresses are developed within the body.

To clarify the mechanism by which this takes place, consider a warm metal cylinder at temperature T_o, with a coefficient of thermal expansion k, suddenly plunged into a refrigerated bath of temperature T_b. Immediately after immersion its surface layers cool faster than its core so that they exert a pressure upon the interior while shrinking. The core resists the compression since it wants to retain a size dictated by its own higher temperature. As a consequence, the cold outer shell is effectively expanded relative to the size it would have assumed if its contraction were unhindered. A radial compressive stress (pressure) is thus developed, both in the cold case and in the relatively warm core. Using the nomenclature shown in Fig. 5.6a and b, the tangential stresses are compressive in the core but tensile in the outer "bandage" (Fig. 5.6c). Stresses also arise in the axial direction and they are tensile in the contracting crust but compressive farther inside. If the cylinder were immersed in a hot bath, all these stresses would have their signs reversed.

As the cooled layer thickens with time, the stresses in all three directions change in magnitude. We can readily follow these changes in the axial direction since at all times the tension forces F_p in the periphery are balanced

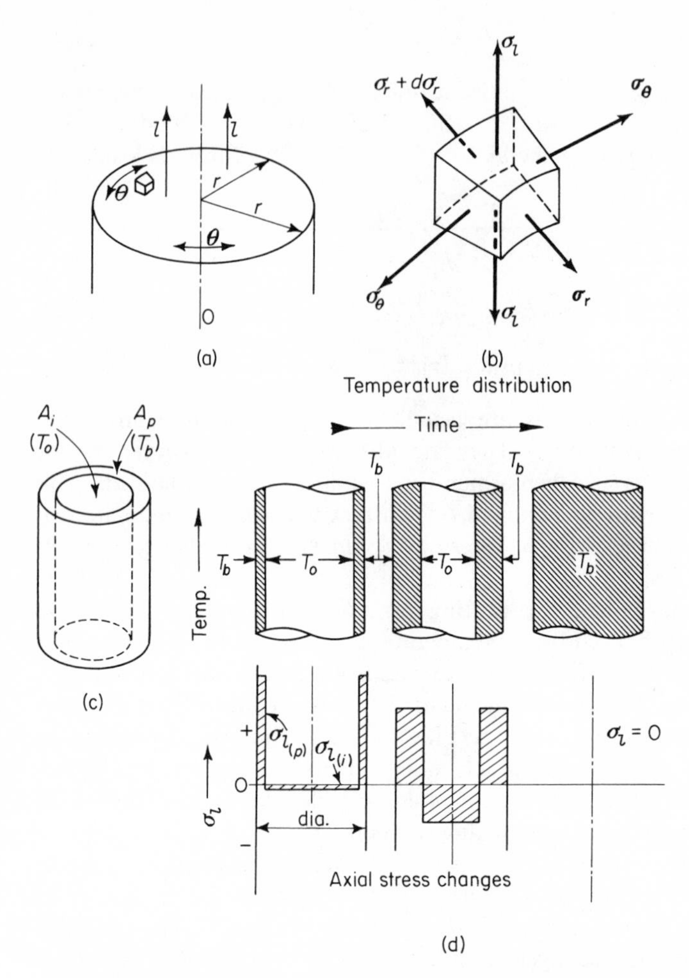

Fig. 5.6 Development of thermal stresses in a quenched cylinder. (a) and (b) show stress notations in polar coordinates (l-longitudinal, r-radial, θ-tangential); (c) Thermal conditions of quenched cylinder (schematic); (d) Change of temperature (top) and of axial stress σ_l (bottom) after short, intermediate and long times of immersion.

by the compressive forces F_i inside. For the simplified temperature distribution of Fig. 5.6d, a state of equilibrium requires that $A_i\sigma_{l(i)}\ (= F_i) = A_p\sigma_{l(p)}$ $(= F_p)$ where A stands for area.

Assuming that the stresses in the cylinder never exceed the elastic range, the stress in the outer shell must be highest immediately after it reached the bath temperature T_b. At this moment $A_p \ll A_i$, and, therefore, $\sigma_{l(i)} \ll \sigma_{l(p)}$ in order to keep $F_i = F_p$. Hence, $\sigma_{l(i)} \approx 0$, and $\varepsilon_i = E\sigma_{l(i)} \approx 0$. The core is thus almost undeformed, and the cold surface layer, A_p, which, if it were free to contract, would have assumed a length $1 + k(T_o - T_b)$ per inch, retains its original size. This can be interpreted as a forced extension applied after thermal contraction to restore the original unit length. The elastic strain required for this is obtained from

$$1 + k(T_b - T_o) + \varepsilon = 1$$

or

$$\varepsilon = -k(T_b - T_o) = k(T_o - T_b)$$

In this case, ε is tension because $T_o - T_b > 0$. The maximum stress in the skin is thus:

$$\sigma_l = E\varepsilon = Ek(T_o - T_b). \tag{5.23}$$

As the cold front penetrates deeper, the ratio A_p/A_i increases, and the stress developed in the "skin" decreases and eventually vanishes when the temperature of the element is uniform.

The actual temperature distribution shows a gradient and, therefore, the thermally induced stresses change from negative to positive gradually rather than abruptly.

Transient thermal stresses almost always arise during heating or cooling through a surface. Important exceptions are heating by electric resistance and nuclear fission, where the entire cross section is heated simultaneously.

Thermal stresses can be detrimental and, if of sufficient intensity, can even cause fracture. Cracking of glassware when hot water is poured into it is an example. On the other hand, transient thermal stresses can be put to use. For instance, in bimetallic switching elements two strips of dissimilar metals with different thermal expansion coefficients are welded together and they bend one way or the other, depending on temperature (Fig. 5.7).

Thermal dilatation is often utilized in assembling elements that are semi-permanently bonded after the temperature returns to equilibrium. The wheel in Fig. 5.8 is expected to fit on the axle so tight that the two can be regarded as monolithic for most practical purposes. This can be achieved only when $D_w < D_a$ by several thousandths of an inch. The wheel can then be forced onto the axle by a suitable press or it can be heated until D_w expands so as to clear D_a. The wheel is then simply dropped into position and allowed to cool.

Gun barrels and dies for extruding metals are "armored" in that way.

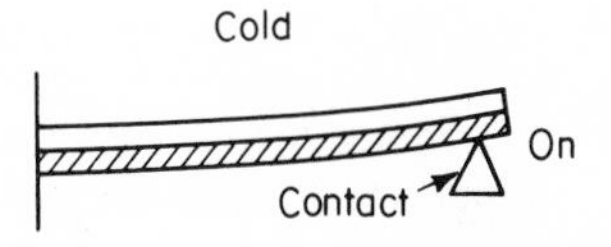

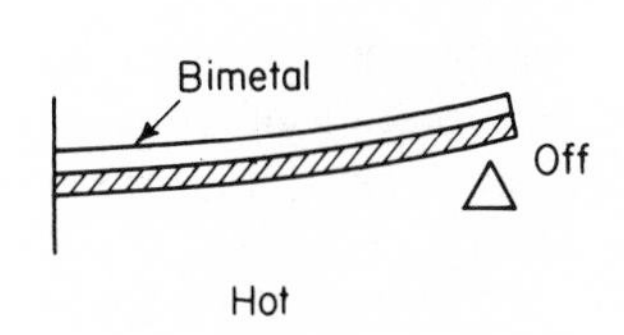

Fig. 5.7 Principle of thermostatic switch in electric circuit.

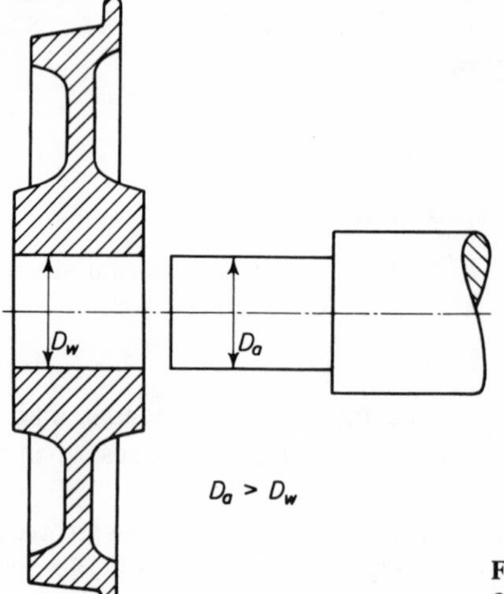

Fig. 5.8 Example of press fit (wheel on axle).

Thermal stresses are computed by using the principles discussed above although the formulae are more complicated for thick-walled cylinders (2.27). Since the stresses that finally develop in the assembly are caused by dimensional mismatch and not by the temperature condition, *shrink fit stresses* cannot be properly regarded as a form of thermal stress.

Other types of internal stresses known as *residual stresses* are of considerable importance. They are frequently found in parts that have undergone various phase transformations, mechanical or thermal treatment. These are discussed in some detail in Chapter 17.

ELASTICITY OF ANISOTROPIC MATERIALS

Elastic stresses and strains were seen to be related, even in the three dimensional case, by just two constants, E and ν in Eq. (5.3), as long as we were limited to isotropic bodies. Anisotropic materials such as wood, leather, sedimentary rock, or single crystals, and even, under certain conditions, polycrystalline metals exhibit, on the other hand, a far more complex relationship between stress and strain. Consider, for example, the elastic deformation of wood, a familiar anisotropic construction material. Imagine a block cut out of a tree so that its x direction is parallel with the trunk axis while y is tangential and z radial to the trunk. If a tensile load were applied parallel with the y axis of the block (tangential to the tree trunk), it would not contract by the same amount in the x direction (along the grain) as in the z direction (across the grain). Indeed, in order just to describe the three normal strains produced by σ_y, requires five constants: E_1, E_2, E_3, ν_{12}, and ν_{32}:

$$\varepsilon_x = \frac{1}{E_1}(\nu_{12}\sigma_y)$$

$$\varepsilon_y = \frac{1}{E_2}\sigma_y \tag{5.24}$$

$$\varepsilon_z = \frac{1}{E_3}(\nu_{32}\sigma_y)$$

To define completely all of the elastic stress–strain relationships for the block in the general case where three normal stresses and three shear stresses are acting would require nine independent constants. The more anisotropic the material (i.e., the lower the symmetry of its internal structure), the larger the number of constants required to relate its stress and strain components.

Crystallographers have long been concerned with the elastic properties of bodies with varying degrees of anisotropy because single crystals vary from reasonably symmetrical (e.g., cubic) to extremely anisotropic (e.g., triclinic). Rather than writing Hooke's law in a different fashion for each of the crystallographic systems, it is used in its most general form, Eq. (5.25).

$$\begin{aligned}
\sigma_x &= c_{11}\varepsilon_x + c_{12}\varepsilon_y + c_{13}\varepsilon_z + c_{14}\gamma_{xy} + c_{15}\gamma_{yz} + c_{16}\gamma_{zx} \\
\sigma_y &= c_{21}\varepsilon_x + c_{22}\varepsilon_y + c_{23}\varepsilon_z + c_{24}\gamma_{xy} + c_{25}\gamma_{yz} + c_{26}\gamma_{zx} \\
\sigma_z &= c_{31}\varepsilon_x + c_{32}\varepsilon_y + c_{33}\varepsilon_z + c_{34}\gamma_{xy} + c_{35}\gamma_{yz} + c_{36}\gamma_{zx} \\
\tau_{xy} &= c_{41}\varepsilon_x + c_{42}\varepsilon_y + c_{43}\varepsilon_z + c_{44}\gamma_{xy} + c_{45}\gamma_{yz} + c_{46}\gamma_{zx} \\
\tau_{yz} &= c_{51}\varepsilon_x + c_{52}\varepsilon_y + c_{53}\varepsilon_z + c_{54}\gamma_{xy} + c_{55}\gamma_{yz} + c_{56}\gamma_{zx} \\
\tau_{zx} &= c_{61}\varepsilon_x + c_{62}\varepsilon_y + c_{63}\varepsilon_z + c_{64}\gamma_{xy} + c_{65}\gamma_{yz} + c_{66}\gamma_{zx}
\end{aligned} \tag{5.25}$$

Depending on the symmetry of the crystal structure, some of the constants are found to be equal to each other while still other constants are equal to zero.

Rather than write the stress as a function of strain, as is done in (5.25), the strain components could have been given in terms of stress. In either case 36 moduli are involved, but for any crystal, even in the most anisotropic material, $c_{12} = c_{21}$, $c_{13} = c_{31}$, etc., so that at most 21 independent constants remain (R5.4). A single triclinic crystal which has the greatest possible degree of anisotropy requires all 21, while other crystals, as well as most anisotropic bodies, require less. Rhombic single crystals, just like wood, require nine; hexagonal close-packed crystals, five; and cubic, three. For polycrystalline solids in which the grains are randomly oriented, as well as for amorphous substances

$$c_{11} = c_{22} = c_{33} = E$$

$$c_{12} = c_{21} = c_{13} = c_{31} = c_{23} = c_{32} = -\frac{E}{\nu}$$

$$c_{44} = c_{55} = c_{66} = \frac{E}{2(1 + \nu)} = G$$

while all other constants are zero.

The elastic properties of single crystals have been studied in great detail. For example, Young's modulus of α iron is represented as a function of the testing direction by the model in Fig. 5.9.

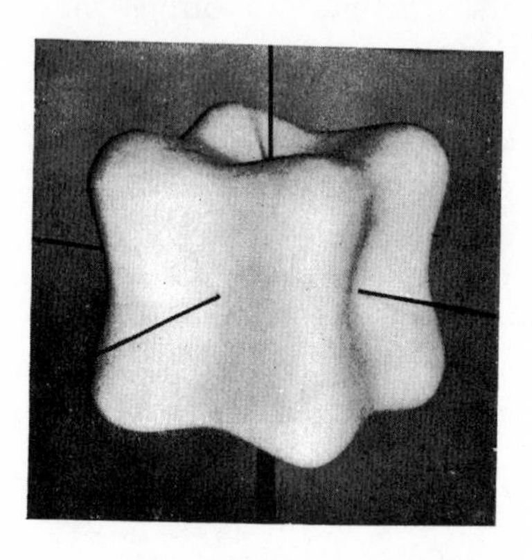

Fig. 5.9 Young's modulus of a single crystal of iron as a function of testing direction. After E. Schmid and W. Boas.

In defining the elastic stress–strain relationships of anisotropic materials other than single crystals, it is convenient to describe them by analogy with the crystallographic system having the same symmetry.

HOOKEAN ELASTICITY VS. RUBBER ELASTICITY

The model picture of elastic deformations, as it was discussed up to this point, involves relatively small stretching or compressing of the interatomic bonds between adjacent atoms. There is no movement of the atoms from one site to some other new position of equilibrium. As explained in the Introduction, the latter results in plastic deformation. Because modest distortion of the atomic bonds yields a time independent linear relationship between stress and strain, it is referred to as *Hookean elasticity* to differentiate it from *rubber elasticity*. The latter, as discussed previously, involved uncoiling and recoiling of flexible molecules so that atoms do change neighbors, not along the molecule backbone, but between adjacent molecular segments. It is interesting to compare these two types of elastic deformation from a thermodynamic viewpoint since this leads to equations which define some basic differences between them.

The first law of thermodynamics states that the change in internal energy of a body, dU, is equal to the sum of the heat it adsorbs, dQ, and the work done on it, dW

$$dU = dQ + dW \tag{5.26}$$

The second law defines the entropy change, dS, in any reversible process by the relation

$$T\,dS = dQ \tag{5.27}$$

where T is the absolute temperature. Combining (5.26) and (5.27) gives:

$$dU = T\,dS + dW \tag{5.28}$$

Now, if a simple system such as a cylindrical bar with length l is loaded in tension by force F (Fig. 5.10), its length increases by dl, and the work done is

$$dW = F\,dl. \tag{5.29}$$

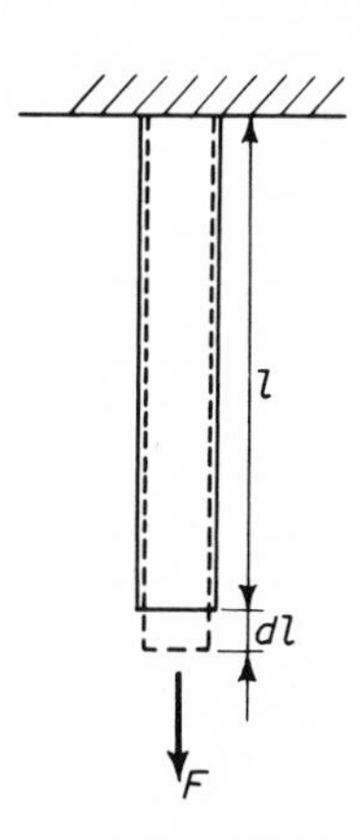

Fig. 5.10 Extension of a bar under force F.

With this, (5.28) becomes:

$$dU = T\,dS + F\,dl \tag{5.30}$$

If the rod is maintained at a constant temperature, (5.30) can be solved for the applied force,

$$F = \left(\frac{\partial U}{\partial l}\right)_T - T\left(\frac{\partial S}{\partial l}\right)_T \tag{5.31}$$

showing that the applied force involves an energy term and an entropy term. It had been suggested that the first term, $(\partial U/\partial l)_T$, is the potential energy stored as strain of the interatomic bonds; therefore, this is the strain energy discussed in the preceding sections. The significance of $(\partial S/\partial l)_T$ lies in the concept of entropy as a measure of thermodynamic probability. If the entropy decreases during stretching, the system becomes more ordered. This was suggested as the mechanism for rubber elasticity in the Introduction. It can be shown (by classical thermodynamics*) that:

$$\left(\frac{\partial S}{\partial l}\right)_T = -\left(\frac{\partial F}{\partial T}\right)_l \tag{5.32}$$

Hence, (5.31) can be rewritten as:

$$\left(\frac{\partial U}{\partial l}\right)_T = F - T\left(\frac{\partial F}{\partial T}\right)_l \tag{5.33}$$

Note from Table 5.1 that Young's modulus decreases with increasing temperature. Consequently, if a steel bar is extended and held to this new length while its temperature is raised, the force on the bar decreases. Rubber behaves in the exact opposite fashion. If a rubber band is stretched and held at this new length while it is heated, the force to keep it extended increases. Indeed this force, at large extensions, is proportional to the

* See, for example, L. A. G. Treloar, *The Physics of Rubber Elasticity* (New York: Oxford University Press, 1949).

absolute temperature, and at $T = 0$, $F = 0$, so that the energy contribution to the force is zero. Since the force is proportional to absolute temperature, the temperature of rubbery materials increases with adiabatic stretching. (The reader can convince himself of this by holding a rubber band to his lips and quickly stretching it. On abrupt extension, it gets warm and, on contracting, it gets cold.) This is opposite to what is found in metals which undergo adiabatic volume expansion and become cooler during stretching while suffering a rise of temperature in compression (*thermoelastic effect*). It might be mentioned that, because of these temperature changes, stresses and strains are not quite single-valued functions of each other. This is one cause of what is referred to as *anelastic behavior*, which is discussed in more detail in Chapter 8.

PROBLEMS INVOLVING NONHOMOGENEOUS STRESS (STRAIN) DISTRIBUTIONS. "STRENGTH OF MATERIALS" VS. "THEORY OF ELASTICITY"

Considerations of stress and strain so far were limited essentially to a single point in a body. An analysis at just one point, however, will completely describe the behavior of a finite body if the stress and strain conditions were homogeneous. Such conditions are encountered in tie rods or shanks of long bolts. Long vertical elements, like a deep well pump plunger or arch bridge suspenders, approximate them except for the modifying effect of their own weight. This, however, can be easily taken into account.

Most structures are not subjected to the same stress and strain at all points within their interior. Even a simple axial force in a member of variable section, as is the case, for example, with a tapered column, will produce lower stresses in the larger cross sections than in a smaller one. In this particular example, the stresses along the column height can be assumed to be equal to the applied load divided by the cross-sectional area at the section of interest, as long as the taper is slight. In many other practical cases it is also possible to make certain simplifying but demonstrably inocuous assumptions regarding the stress distribution over a section, especially when dealing with rod–like elements. As a consequence, one can determine the stresses and associated strains with relative ease for a variety of loading conditions. The mathematical aids required for this purpose are simple and rarely go beyond elementary calculus. The discipline dealing with the quantitative analysis of stress–strain relations in these simple members is known as "Strength of Materials." Several of its applications are included in Part III.

The methods of Strength of Materials become increasingly inaccurate and eventually fail completely when one leaves the basic form of a rod and attempts to analyze stresses in plates, shells, and, especially, in massive objects or at contact points between bodies (e.g., in rollers and races of roller bearings). Strength of Materials is also inadequate for calculating local

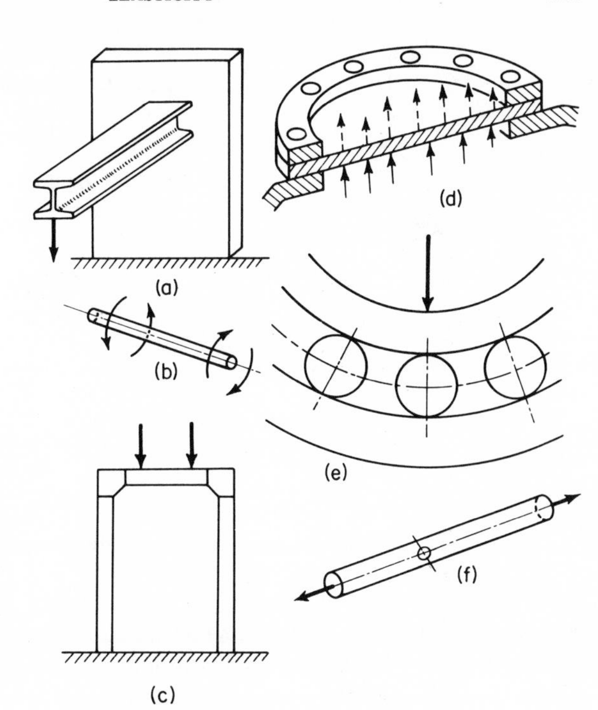

Fig. 5.11 Methods of Strength of Materials satisfactory for analysis of (a-c); Theory of Elasticity required for (d-f). (a) *I*-beam, (b) transmission shaft, (c) portal frame (d) boiler manhole closure, (e) roller bearing, (f) tension bar with transverse hole.

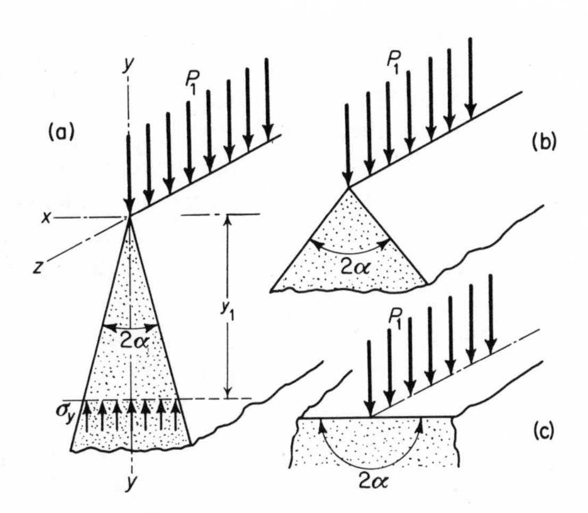

Fig. 5.12 Wedge with varying included angle 2α under uniform linear load P_1. Stress, σ_y, distributed according to Strength of Materials.

stresses created by such disturbances in continuity as holes, fillets, or notches (Fig. 5.11).

The Theory of Elasticity steps in where the methods of the former discipline are inadequate. Its treatment is rigorous, and the only assumptions it uses are those concerning the physical nature of the material such as continuity and isotropy. The required mathematical tools are more sophisticated.

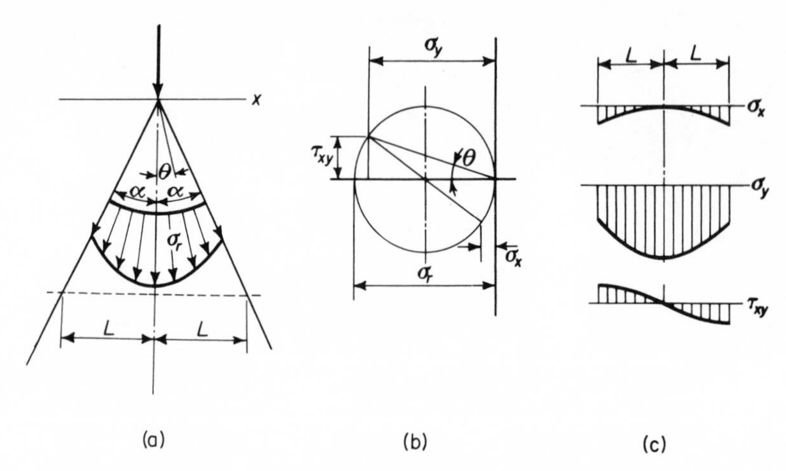

Fig. 5.13 Stress distribution of wedge in Fig. 5.12 in (a) polar; (b) Mohr's circle for this stress system; (c) Cartesian coordinates.

Ordinary or linear elasticity theory is based on a linear relation between stress and strain (Hooke's law) and is, therefore, confined to problems involving small deformations. The nonlinear theory deals with conditions where Hooke's law does not apply because of the nature of the material (rubber) or because the strains are so large that (3.10) is seriously in error.

In order to give the reader a more direct "feel" of the limitations inherent in the elementary approach without going into the mathematical details, let us consider the stresses in the wedge loaded as shown in Fig. 5.12a. This is akin to the axially loaded tapered column already mentioned. The rigorous treatment of this case is the simpler of the two due to the two dimensional character of the problem. A uniformly distributed load P_1 per unit of length acts along the tip of a wedge which is very long in the z–direction. The surface area normal to P_1 at a distance y_1 from the tip is $A_1 = 2y_1 \tan \alpha$ per unit length.

According to the elementary concepts of Strength of Materials, x and y are principal directions, and the pressure is uniformly distributed over A_1. Thus:

$$\sigma_y = \frac{P_1}{A_1} = \frac{P_1}{2y_1 \tan \alpha} \tag{5.31}$$

From this it follows that $\sigma_y \to 0$ when $\alpha \to \pi/2$, independent of y (the possibility of y_1 ever becoming zero is excluded). Putting this in words, the pressure in the direction of the applied force decreases gradually as the wedge becomes blunt (Fig. 5.12b) and it should become zero when the wedge turns into a semi–infinite mass (Fig. 5.12c).

This is clearly not true. For example, strains caused by a narrow punch or roller pressed onto a thick slab can be detected by direct measurement even at a considerable depth beneath the contact area. Moreover, their magnitude agrees closely with those predicted by the expressions developed by the methods of the theory of elasticity.

In terms of stress, the exact solution to the axially loaded wedge in polar coordinates (Fig. 5.13a) is:

$$\sigma_r = \frac{P_1 \cos \theta}{r(\alpha + \frac{1}{2} \sin 2\alpha)}$$
$$\sigma_\theta = \tau_{r\theta} = \tau_{\theta r} = 0 \tag{5.32a}$$

The radial direction is, therefore, a principal one. With the aid of Mohr's circle (Fig. 5.13b), we find the Cartesian stress components to be:

$$\sigma_x = \sigma_r \sin^2 \theta$$
$$\sigma_y = \sigma_r \cos^2 \theta \tag{5.32b}$$
$$\tau_{xy} = \tfrac{1}{2}\sigma_r \sin 2\theta$$

The appropriate stress distributions are drawn in Fig. 5.13c. These results show that the elementary "solution" (5.31) is applicable only to sharp wedges when α is small and $\sin^2 \alpha$ is negligible. The correct solution (5.32), on the other hand, is valid for any α, and, for a semi–infinite solid ($\alpha = \pi/2$), it becomes

$$\sigma_r = -\frac{2P_1 \cos \theta}{\pi r} \tag{5.33}$$

with a maximum for $\theta = 0$

COMPATIBILITY OF STRESSES AND STRAINS

These discrepancies indicate that, in the above example, the elementary approach used in Strength of Materials contains an inherent error, which should be readily traceable. To illustrate this point, consider the triangular extremity ABC (Fig. 5.14) of a thin layer contained between two planes parallel to x-z (Fig. 5.12). According to the assumption in (5.31), the top surface of the layer carries a uniformly distributed stress, σ_y', while the bottom surface a slightly lower stress, σ_y''. Considering the equilibrium of

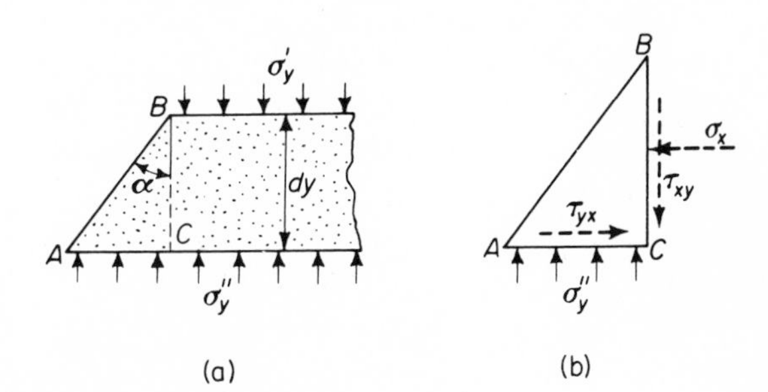

Fig. 5.14 Proof that x and y in Fig. 5.12 are generally not principal directions.

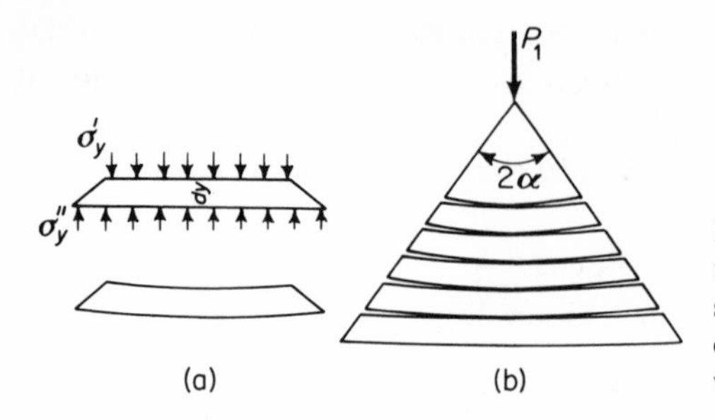

Fig. 5.15 Assumption of uniform stress σ_y over the section (a) results in incompatibility of strains within the wedge (b).

ABC (Fig. 5.14b), one finds that the resultant of σ_y'' along *AC* is not balanced by an equivalent force directed downward. Surface *AB* is completely stress free, and force $\sigma_y'' \times AC$ can be countered only by a shear stress τ_{xy} along *BC* such that $\tau_{xy} \times BC = \sigma_y'' \times AC$ (Fig. 5.14b). By definition, there must be also a shear stress, $\tau_{yx} = \tau_{xy}$, along *AC*. This, in turn calls for a balancing σ_x component directed to the left. Hence, contrary to what was initially assumed neither *y* nor *x* are principal directions.

What would be the practical outcome of an attempt to impose an internal stress system such as defined by (5.31)? This could be done by cutting up the entire wedge into thin slices parallel to *x-z* and then loading each of them by equal and uniformly distributed stresses, σ_y' and σ_y'', on the upper and lower faces (Fig. 5.15a). Since their lengths differ, each lamina bends somewhat. But when they are stacked to reproduce the shape of the wedge under load, the successive laminae do not match exactly, and voids will develop between them (Fig. 5.15b).

Such a condition is described as *incompatibility* of strains and, because of Hooke's law, it also results in an incompatibility of stress. This simply means that, once the geometry of the body in question and the mode of external loading are established, one cannot assume arbitrarily a plausible internal stress distribution. Likewise, a wrongly assumed (or rather guessed) displacement function may call for stresses that cannot be produced by the applied force due to its actual value or manner of application. This is so because stresses are uniquely related to strains through Hooke's law, while strains are functions of displacements defined by Equations (3.5). Thus there is no freedom of choice with regard to any of these three magnitudes.

In the theory of elasticity, so–called *compatibility* or *continuity equations* were derived. They represent a particular type of relation between the various stresses or strains specified in functional form. If the continuity equations are satisfied, the assumed or computed stress-strain distribution is not only correct but also the only possible.

More examples of inaccurate assumptions in the strength-of-materials methods can be easily found. However, in many cases, extreme accuracy is not needed and shortcuts are acceptable. For this reason, both disciplines coexist and are widely used in their respective fields.

PHOTOELASTICITY

In Chapter 3, the use of resistance-wire strain gauges for measuring strains at a point was explained, and these strains can readily be converted into elastic stresses by Hooke's law. Another method for measuring the stresses directly takes advantage of the optical anisotropy displayed by certain transparent solids when under load.

Some crystalline materials such as calcite or mica are optically anisotropic even in the absence of loads. When light is passed through them, it is divided into two beams vibrating at right angles to each other.* Each of these travels through the crystal at a different velocity; i.e., the index of refraction encountered by the beams is different. If the light that is transmitted through such *double refractive* or *birefringent materials* is initially plane polarized, the intensity of the emitted beam depends on the orientation of the plane of polarization and the relative velocities of the two beams.

Mica possesses two optical axes that are in a plane parallel to its cleavage plane and at right angles to each other. If a plane polarized beam is normal to a sheet of mica, and parallel with either optical axes of the mica, the light passes through the sheet without changing its character; that is, it

* In this discussion, light is considered to have the properties of waves rather than particles.

Fig. 5.16 Schematic diagram of optical principle of photoelasticity.

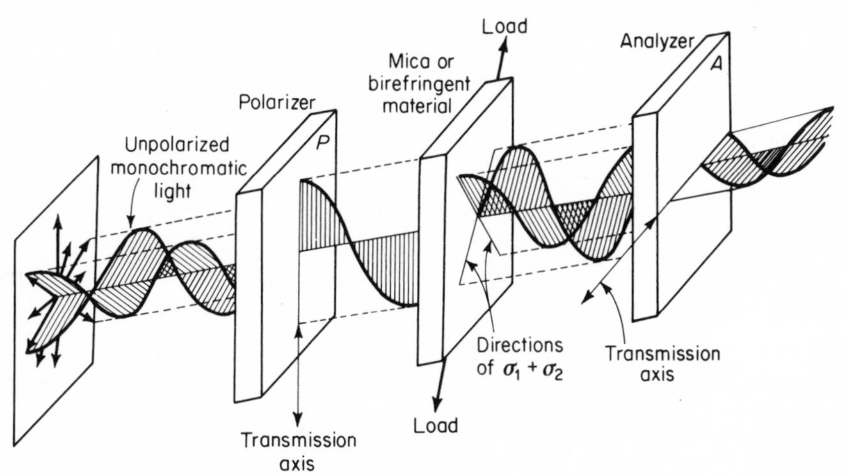

emerges as a single beam. If, on the other hand, the direction of vibration of the polarized beam is at, say, 45° to the optical axes, the beam divides into two beams, each at 45° to the incident one. This condition is shown schematically in Fig. 5.16. The element used for plane polarizing the beam from the source is referred to as the *polarizer*, *P*. If a second piece of polarizing material is placed in the path of the beam beyond the birefringent subject, it is termed the *analyzer*, *A*. In the case in Fig. 5.16, the polarizer and analyzer are placed so that their transmission axes are at right angles to each other. Hence, if the optical axis of the mica is aligned with the transmission axis of the polarizer, the single beam emerging from the mica sheet is blocked by the analyzer, resulting in complete extinction. As the mica is rotated, the intensity of the beam emerging from the analyzer varies, obtaining its maximum value at 45°.

For a given orientation of the polarizer, mica, and analyzer, where complete extinction does not occur, the intensity of the emerging beam depends on the phase relation of the two beams leaving the birefringent material. This, in turn, is established by the relative indices of refraction of the mica along its optical planes and the material thickness. Since the indices, μ_1 and μ_2, are defined as the ratio of the velocities of light along the particular axes with its velocity in air, c, the relative retardation of the two waves, δ, is equal to

$$\delta = \frac{d}{c}(\mu_1 - \mu_2) \tag{a}$$

where d = thickness.

Hence, the phase difference in radians of the two beams transmitted through the birefringent crystal is

$$\begin{aligned} \omega\delta &= \frac{\omega d}{c}(\mu_1 - \mu_2) \qquad \text{(b)} \\ &= \frac{2\pi d}{\lambda}(\mu_1 - \mu_2) \end{aligned}$$

where ω = angular rotation
and λ = wave length or period = $2\pi\, c/\omega$;
or, if n = retardation in wave lengths:

$$\omega\delta = 2\pi n = \frac{2\pi d}{\lambda}(\mu_1 - \mu_2) \tag{c}$$

or:

$$n = \frac{d}{\lambda}(\mu_1 - \mu_2) \tag{5.34}$$

Brewster discovered in 1816 that a transparent and optically isotropic material, such as glass (and, as found later, other transparent materials, including epoxy, bakelite, or cellophane), becomes *temporarily birefringent* when subjected to stress, and $(\mu_1 - \mu_2)$ is proportional to $(\sigma_1 - \sigma_2)$. Hence, (5.34) can be written as

$$n = C\,d(\sigma_1 - \sigma_2) \tag{5.35}$$

where C is a constant for any temporarily birefringent material known as the *stress-optic coefficient*.

Consider now the intensity of light emerging from the analyzer if the mica in Fig. 5.16 is replaced with a stressed transparent material, and the transmission planes of P and A are parallel.

Each element of the stressed body behaves like the mica example discussed above, and at each point the difference in μ_1 and μ_2 is proportional to $(\sigma_1 - \sigma_2)$. If a monochromatic light is used as the source, and the birefringent material is of uniform thickness, an observer looking at the image emerging from the analyzer will see a pattern of light and dark lines if $(\sigma_1 - \sigma_2)$ varies throughout the body.

The intensity of light transmitted through the analyzer (see Prob. 6, p. 131) is

$$I = I_1 \sin^2 \theta \sin^2 \frac{\omega\delta}{2} \tag{5.36}$$

where I_1 = intensity of incident light
θ = angle between plane of polarization and nearer principal stress.

Obviously $I = 0$ when either θ or $\omega\delta = 0$. The latter occurs for each value of $\omega\delta = 2m\pi$ (where $m = 0, 1, 2, \ldots$). The first of these conditions—viz., $\theta = 0$—occurs at each spot on the image where the principal stress axis of the photoelastic model is parallel with the transmission plane of P and A. These loci of points of constant inclination of the principal stresses are termed *isoclinic lines* or *isoclinics*. The locus of points where $\sin^2(\omega\delta/2) = 0$ are called *fringes*. It is because the intensity due to the latter changes cyclically as stress is increased that stress analysis by photoelastic procedures is possible. A schematic diagram of intensity vs. stress is shown in Fig. 5.17.

A *polariscope* consists of a light source, polarizer, analyzer, and a variety of auxiliary equipment. If, in a particular polariscope, monochromatic light is used, with either parallel or perpendicular P and A, it would not be

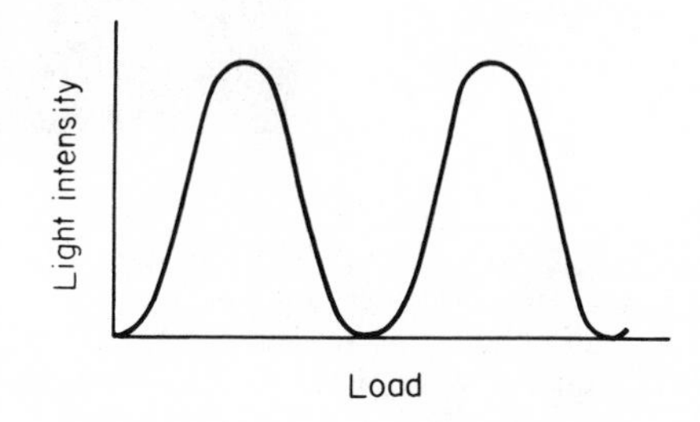

Fig. 5.17 Variation of light intensity at a point as load is continuously increased.

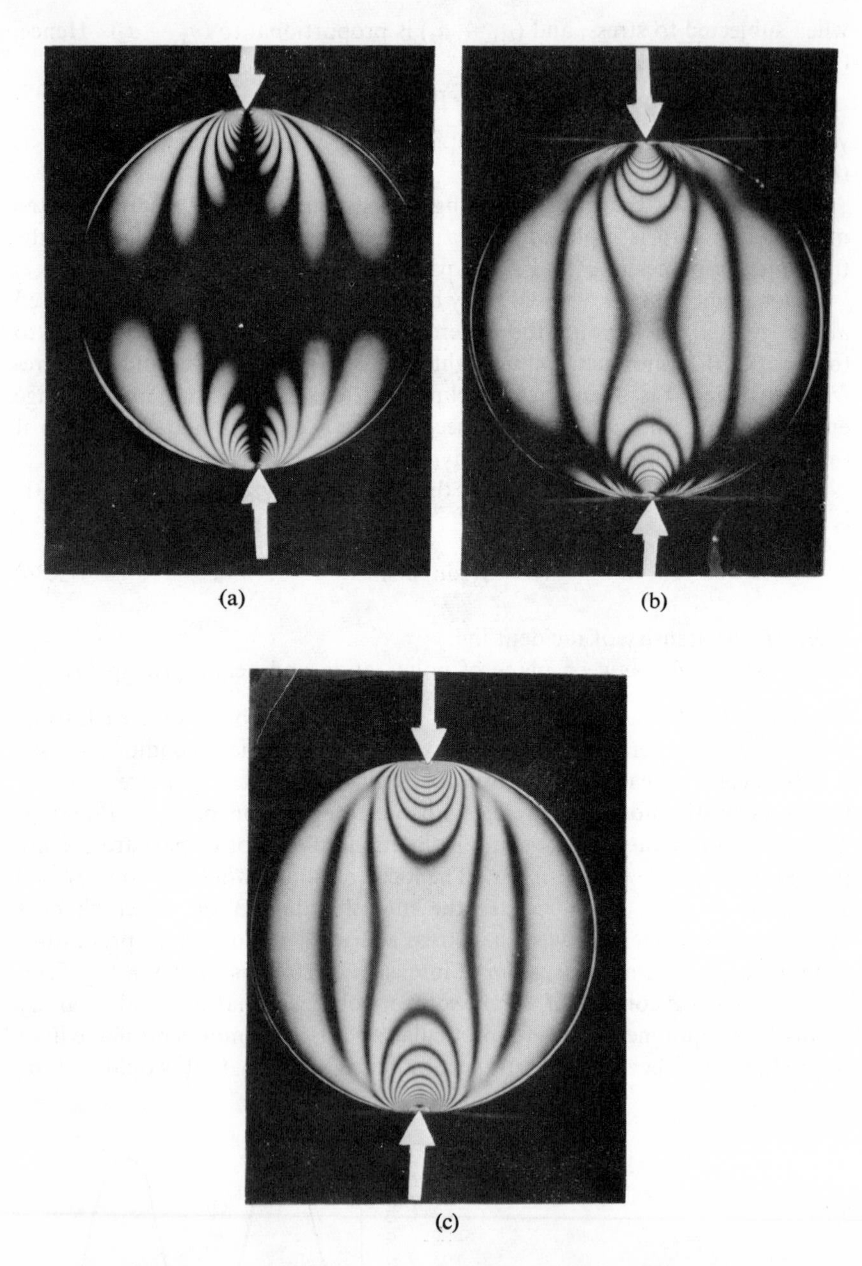

Fig. 5.18 Stress pattern on a circular disk subjected to concentrated edge loading (a) and (b) fringes and isoclinics superimposed with no quarter-wave plate. Plane of polarization vertical in (a); and 45° in (b); (c) isoclinics eliminated by using quarter-wave plate.

possible to distinguish isoclinics from fringes in the birefringent material. By rotating the polarizer and analyzer together, however, θ is changed while $\omega\delta$ remains constant so that the isoclinics continuously move while the fringes do not, making separation of the two possible, Fig. 5.18a and b. Even in this case, however, it is difficult to distinguish between the isoclinics and fringes. If more than one color of light is used as the source, each of the component colors (i.e., each light wave frequency) will be extinguished at different values of $\omega\delta$. If, for example, red and blue were used, at those points where $\omega\delta = 0$ for red a blue dot would appear, and vice versa. Hence a white light source develops fringes that vary from yellow through red to blue and green so that the fringes or *isochromatic lines* are distinguished from the black isoclinics. The sharpest color change occurs in going from the reds to the blue or greens, and this fringe is referred to as the *tint of passage*.

The isoclinics are completely avoided, even with monochromatic light, by using *circularly polarized light*. Such a condition is developed by placing a *quarter-wave plate* between the polarizer and stressed member. Any permanent birefringent material (e.g., mica) can be used for this purpose. The transmission axis of the mica is placed at 45° to the axis of the polarizer, and the thickness of the mica is selected so that the relative retardation of the two components is exactly a quarter-wave length; i.e., 90°. Because of this dependence of thickness on wavelength, only monochromatic light can be circularly polarized. With two perpendicular waves of equal intensity incident on the transparent plate being analyzed, θ in Eq. (5.36) is never zero so that isoclinics cannot occur (Fig. 5.18c). By placing a second quarter-wave plate, again oriented at 45°, between the model and analyzer, the beam incident on the analyzer is again plane polarized. If the quarter-wave plates are parallel, this final beam vibrates in a plane perpendicular to the initial polarized beam; if they are crossed, the two beams vibrate in the same plane. Hence, rotating the quarter-wave plates with respect to each other is equivalent to rotating the analyzer with respect to the polarizer.

USE OF PHOTOELASTIC DATA IN STRESS ANALYSIS

Equation (5.35) can be written as

$$\frac{\sigma_1 - \sigma_2}{2} = \frac{n}{2Cd} \tag{5.37a}$$

or, by using (2.18),

$$\tau_{\max} \times Fn \tag{5.37b}$$

where $F =$ the model fringe value $= \dfrac{1}{2Cd}$

The value of F at $d = 1$ is called the *material fringe value*, f, and its value is easily obtained by applying known loads to a uniform prismatic section of the birefringent material.

To use (5.37) in stress analysis, a model of the member of interest is made out of a birefringent material such as bakelite or epoxy, and a calibration is made to find the material fringe value. Since most photoelastic studies are two dimensional, the models are generally cut out of transparent plates of uniform thickness of the photoelastic material. The model is then loaded and a record of the isochromatic fringe pattern vs. load is obtained. The pattern of isoclinics is not always needed but, if it is, it is obtained by using white light, no quarter-wave plate, and by keeping the polarizer and analyzer crossed at all times. As P and A are rotated together, a record of the isoclinics vs. θ is obtained. The highest possible load is used in this step since the isoclinics become narrower and sharper as the stresses are increased. Having the isoclinic pattern, one is able to draw the *stress trajectories*. The latter are a family of curves whose tangent and normal at any point are principal stress directions (Fig. 5.19).

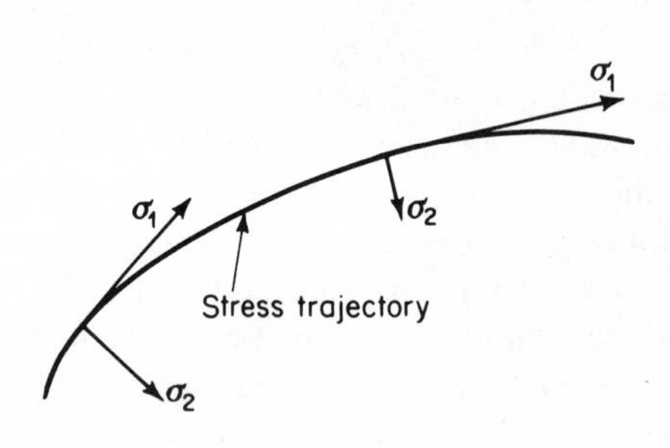

Fig. 5.19 Stress trajectory.

The shear stress can be read directly off the model by counting the fringes, starting at any free surface, and using (5.37b). Obtaining the value for the individual principal stresses is somewhat more complex, and a variety of procedures have been developed for evaluating them. The most obvious is a direct application of Hooke's laws. Since the analysis is two dimensional, Eq. (5.2a) can be written as:

$$\varepsilon_z = -\frac{\nu}{E}(\sigma_1 + \sigma_2) \tag{5.38}$$

A direct measurement of strains in the thickness direction yields $(\sigma_1 + \sigma_2)$, and the photoelastic method gives $(\sigma_1 - \sigma_2)$, so that each principal stress can be calculated. Accurate measurement of ε_z is difficult, however, and this limits the use of this method. Other methods for finding the values of the principal stresses are too cumbersome to be described here but are discussed in a number of other texts [see, for example, (R5.2)].

Although photoelastic stress analysis is commonly limited to the use of plane models, three dimensional stress distribution problems can also be solved, and the method can be applied to opaque structures. Three dimensional analysis takes advantage of the *frozen stress method*. In this process, the model is loaded and then annealed while the deformation is maintained. The birefringency is retained in the model after unloading so that it can be sliced into plane sections each of which is analyzed as a plane problem. To avoid the need for polishing the cut surfaces, the sections are examined in a fluid whose index of refraction is close to that of the model materials.

To examine structures directly, the photoelastic material is cemented to the unloaded structure. After the body is loaded, the coating fringes are analyzed in a manner similar to that used in the two dimensional model. In this case, the polarized light passes through the overlaid birefringent material and is reflected from the opaque surface back through the layer for analysis.

PROBLEMS

1. Three steel rods of 1 sq. in. cross-section are suspended on hinges as shown. Compute the displacement of point A under force $F = 40{,}000$ lbs.

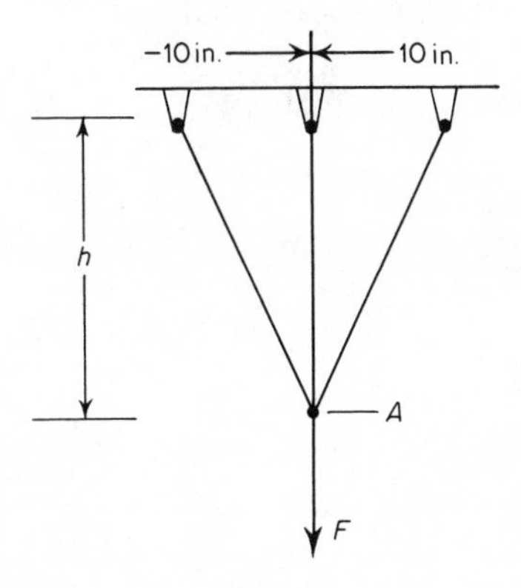

2. Carry out the same calculation, assuming that the two side rods are of bronze ($E = 16 \times 10^6$ psi).

3. A solid 5 in. diam. glass cylinder is exposed to a uniform radial compressive stress of 150,000 psi. What stresses develop at the cylinder's axis?

4. The cylinder in (2) has a 0.5 in. diam. axial hole. What are the stresses at the hole surface?

5. A hollow carbon steel cylinder of 4 in. O.D., 2 in. I.D., and 20 in. long was plugged with a snugly fitting 304 stainless steel bolt tightened lightly with a nut against the face of the cylinder. The operation was performed at 70°F. The assembly was first cooled to −100°F. and then heated to 200°F. Calculate the stresses in the bolt and cylinder wall at each temperature, using handbook data for dilatation.

6. Derive Eq. (5.36) [*Hint:* The vector representing the plane polarized beam leaving the polarizer is $V_p = a \sin \phi$. The vectors leaving the stressed member are found next, remembering these are out of phase. After finding the amplitude of the vector leaving the analyzer, note that the intensity is proportional to the square of the amplitude.]

REFERENCES FOR FURTHER READING

(R5.1) Crandall, S. H. and N. C. Dahl, *An Introduction to the Mechanics of Solids*. New York: McGraw-Hill Book Company, 1959.

(R5.2) Frocht, M., *Photoelasticity*. New York: John Wiley & Sons, Inc., 1951.

(R5.3) Hoffmann, O. and G. Sachs, *Introduction to the Theory of Plasticity for Engineers*. New York: McGraw-Hill Book Company, 1953.

(R5.4) Huntington, H. B., *The Elastic Constants of Crystals*. New York: Academic Press, 1958.

(R5.5) Timoshenko, S. and J. N. Goodier, *Theory of Elasticity*. New York: McGraw-Hill Book Company 1951.

6

PLASTIC DEFORMATION

The elastic range of solids is limited by the occurrence of fracture in some cases and by permanent deformation in others. The ability to undergo permanent—i.e., plastic—deformations rather than fracture is an extremely valuable property of crystalline materials, especially of metals. It enables a part of a machine or structure to "yield" or "give" instead of breaking when instantly overloaded by a stress beyond the elastic range. This prevents catastrophic accidents and is the reason that metals are so outstanding as construction materials. Furthermore, plastic deformation on an extensive scale represents the most potent and versatile way of converting crude metals into useful primary, semi-finished shapes like wire, bars, beams, plates, then into innumerable secondary, finished products from bolts and nails to gas bottles or aircraft wing sections. Equipment is normally designed to avoid permanent deformation in service. Conversely, in metal forming, a maximum amount of plastic flow with a minimum effort is strived for. Calculations of whether a body will behave elastically or plastically must be based on a knowledge of easily measurable basic properties. The material property used for these calculations is the *yield strength* (or *yield stress*) obtained from a tension test on a rod of uniform cross section. It is denoted by σ_o and defines the stress at which deformation changes from elastic to plastic (Fig. 6.1).

Beyond σ_o or σ_{o-1} the material changes its dimensions permanently, and the stress required to continue deformation increases along *flow curve* b (or b_1). This increase is due to *strain-hardening* or *work-hardening* and

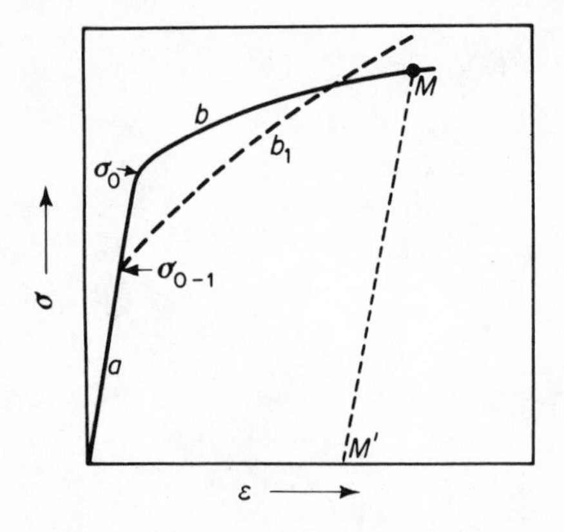

Fig. 6.1 Stress-strain diagrams of two ductile metals: *a*-elastic deformation; b, b_1-plastic deformation with different strain-hardening rates. σ_0, σ_{0-1}-yield strength.

its rate depends on the particular metal. For instance, curve b_1 characterizes a higher work-hardening rate than is the case in b although its initial yield stress σ_{o-1} is lower than in the former case. Since work-hardening is a permanent property only at temperatures less than about 35 to 40% of the absolute melting temperature, it is said to result from *cold-working* or *cold-deformation*. It will be seen later in this chapter that the effect of cold work can be removed through *recrystallization* by heating to a sufficiently high temperature.

Work-hardening is often beneficial because it raises the yield stress. If plastic deformation is interrupted at M, for example (Fig. 6.1), the new elastic range is M'–M, and a stress higher than the original σ_o can be applied to the material without causing it to deform permanently.

Assuming that the yield strength is known, we shall now see how this is used for design purposes. According to Chapter 2, each element of a stressed body may be acted upon by as many as six stress components, and it is necessary to relate these to σ_o in such a way as to determine whether plastic yielding will or will not occur. The reader may guess that over the years there were many attempts to develop such relationships, and the resulting theories or hypotheses were usually called *failure*, *yield*, or *strength theories* or *criteria*. We shall designate them as *yield criteria* in this part of the text. The other terms were introduced because, historically, studies of this problem were done by engineers who considered plastic yielding a failure or strength limitation in design. Of the number of empirical relationships that have been suggested, only the two discussed below remain in common use.

YIELD CRITERIA

Maximum Shear Stress Criterion

Tresca, around 1865, proposed that plastic deformation begins when the shear stress anywhere in the body reaches the value it has in simple tension or compression at the yield point.* This is known as the *Maximum Shear Stress Criterion*. If the tensile yield stress is σ_o, the compressive yield stress = $-\sigma_o$, the corresponding shear stress at 45° to the tension axis is $\tau_o = \sigma_o/2$ (see p. 39). Yielding will not occur, according to the Tresca criterion, as long as

$$\begin{aligned} -\tau_o < \tau_{12} < \tau_o \\ -\tau_o < \tau_{23} < \tau_o \\ -\tau_o < \tau_{31} < \tau_o \end{aligned} \tag{6.1a}$$

where τ_{12}, τ_{23}, and τ_{31} are the principal shearing stresses.

With the aid of (2.18) these three terms can be expressed in terms of normal stresses as:

$$\begin{aligned} -\sigma_o < \sigma_1 - \sigma_2 < \sigma_o \\ -\sigma_o < \sigma_2 - \sigma_3 < \sigma_o \\ -\sigma_o < \sigma_3 - \sigma_1 < \sigma_o \end{aligned} \tag{6.1b}$$

Because of the convention $\sigma_1 > \sigma_2 > \sigma_3$, only the third inequity in (6.1b) need be satisfied. Hence plastic flow occurs when the difference between the largest and smallest principal stress reaches $\pm\sigma_o$. To avoid a negative difference between two positive stresses, the order of σ_3 and σ_1 is reversed. The maximum shear stress criterion is then:

$$-\sigma_o < \sigma_1 - \sigma_3 < \sigma_o \tag{6.2}$$

The various stress combinations which may cause the material to behave either elastically or plastically can be conveniently shown on a three dimensional diagram. When this is done by plotting the various planes (six in all) defined by (6.1b), one finds that they delineate an infinitely long straight hexagonal prism coaxial with the [111] line passing through the origin (Fig. 6.2). For this plot, the three coordinate axes represent the principal stresses, and σ_1 is not

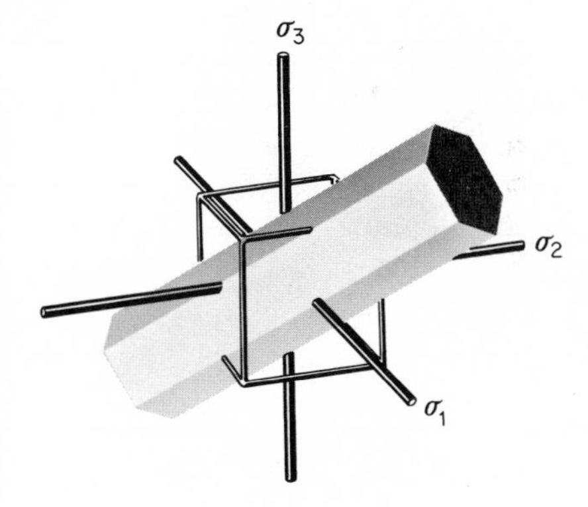

Fig. 6.2 The yield surface according to the maximum shear stress (Tresca) criterion.

* The tensile and compressive yield strengths σ_o are assumed to be of identical absolute magnitude in the yield criteria discussed in this chapter. Materials with different strengths in tension and compression are listed in Table 11.1.

necessarily the largest nor σ_3 the smallest of the three. Combinations of stress components that lie within this *yield surface* produce only elastic distortion. Those outside it cause plastic flow as well. Because the prism extends to infinity, it is possible to have combinations of very large normal stresses that still produce an elastic deformation only.

The position of the prism relative to the coordinates can be more readily visualized by considering its trace on any one coordinate plane; e.g., the $\sigma_1 - \sigma_2$ plane. Putting $\sigma_3 = 0$, the group (6.1b) is transformed into

$$-\sigma_o < \sigma_1 - \sigma_2 < \sigma_o$$

$$-\sigma_o < \sigma_2 < \sigma_o \tag{6.3}$$

$$-\sigma_o < \sigma_1 < \sigma_o$$

which delineates the shaded surface formed by the intersection of six lines in Fig. 6.3. The traces on the $\sigma_1 - \sigma_3$ and $\sigma_2 - \sigma_3$ surfaces are analogous with those in Fig. 6.3.

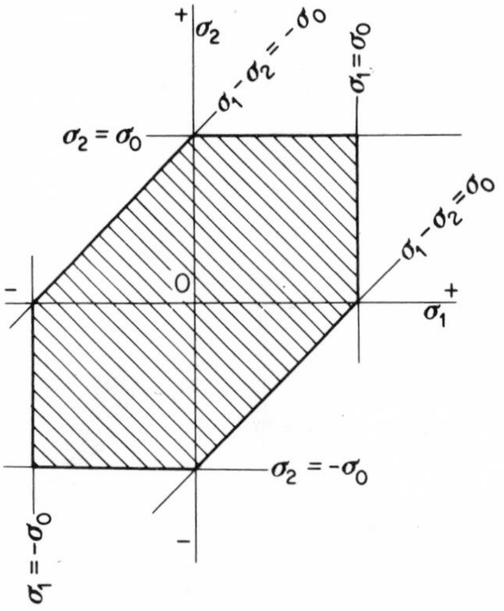

Since the shear stress is independent of the hydrostatic pressure and it is known from experience that yielding is not influenced by hydrostatic pressure, the theory is consistent in this regard.

The predictive value of the Tresca theory is generally satisfactory despite its simplicity. The main objection raised to it is that relation (6.2) ignores the possible effect of the intermediate stress σ_2. While this is true, it is demonstrated below that the error introduced in this manner only underestimates the elastic strength by about 15% at the most.

Fig. 6.3 The maximum shear stress yield boundary for plane stress.

The Distortion Energy Criterion

M. T. Huber in 1904, and later R. v. Mises and H. Hencky, suggested that plastic deformation begins when the elastic distortion energy induced in the material by any combination of stresses reaches a certain critical value. The latter is equal to the elastic distortion energy (W_s) that is stored in a unit volume of a body uniaxially loaded to its yield strength. This quantity, termed W_{so}, is obtained from Eq. (5.16) by putting $\sigma_1 = \sigma_o$, $\sigma_2 = \sigma_3 = o$, resulting in:

$$W_{so} = \frac{1 + \nu}{3E} \sigma_o^2 = \frac{\sigma_o^2}{6G} \tag{6.4}$$

When all three principal stresses are acting, the condition of no-yield requires that the distortional energy W_s from (5.15c) be less than W_{so}. Hence:

$$(\sigma_1 - \sigma_2)^2 + (\sigma_2 - \sigma_3)^2 + (\sigma_3 - \sigma_1)^2 < 2\sigma_o^2 \tag{6.5a}$$

When the stress system is specified in terms of nonprincipal components, (5.15a) is used for W_s, and

$$(\sigma_x - \sigma_y)^2 + (\sigma_y - \sigma_z)^2 + (\sigma_z - \sigma_x)^2 + 6(\tau_{xy}^2 + \tau_{yz}^2 + \tau_{zx}^2) < 2\sigma_o^2 \tag{6.5b}$$

The yield surface represented by these expressions is the periphery of a cylinder coaxial with the [111] direction passing through the origin of the coordinates (Fig. 6.4). This cylinder is circumscribed on the hexagonal prism of Fig. 6.2, indicating that the quantitative discrepancies between Tresca's and Huber's theories are modest.

For $\sigma_2 = \sigma_1$ or $\sigma_2 = \sigma_3$, expression (6.5a) becomes identical with (6.2). Thus σ_2 is only a modifying element. Its maximum effect is obtained when the left-hand side of (6.5a) is differentiated with respect to σ_2 and the result equated to zero. Hence

$$\frac{\partial}{\partial \sigma_2} = 2\sigma_2 - \sigma_3 - \sigma_1 = 0$$

and

$$\sigma_2 = \frac{(\sigma_1 + \sigma_3)}{2} \tag{6.6}$$

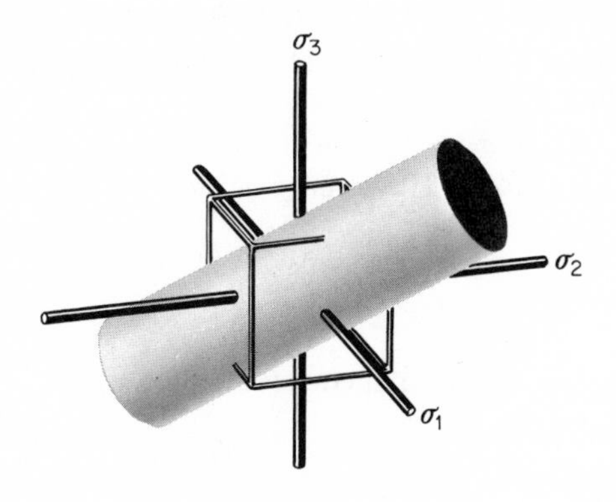

Fig. 6.4 The yield surface according to the shear strain energy (Huber-Mises) criterion.

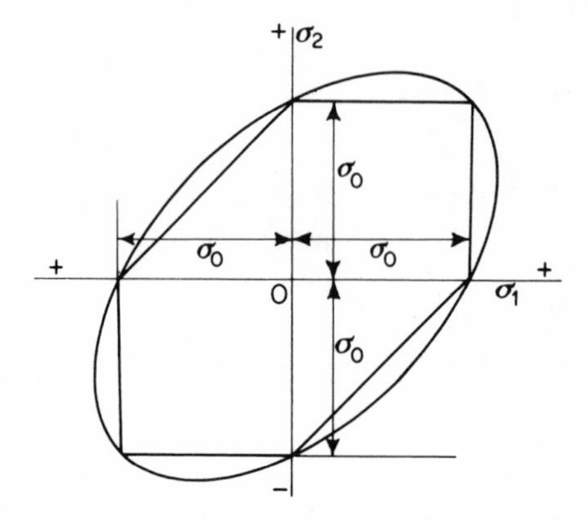

Fig. 6.5 Comparison of Tresca's (hexagon) and Huber-Mises (ellipse) yield criterions for plane stress.

After substituting this value of σ_2 in (6.5a) and simplifying, we obtain

$$\sigma_1 - \sigma_3 \leq \frac{2}{\sqrt{3}}\,\sigma_o \tag{6.7}$$

or

$$\sigma_1 - \sigma_3 \leq 1.157\sigma_o$$

Since the maximum effect of stress σ_2 occurs when $\sigma_2 = (\sigma_1 + \sigma_3)/2$, the allowable shear stress for these conditions is higher by about 15% than that suggested by the Tresca theory.

The trace of the Huber–Mises yield surface on any one of the coordinate planes is an ellipse circumscribed on Tresca's hexagon, Fig. 6.5. The equation of the ellipse shown is obtained from (6.5a) by putting $\sigma_3 = 0$, which leads to:

$$\sigma_1^2 - \sigma_1\sigma_2 + \sigma_2^2 = \sigma_o^2 \tag{6.8}$$

This is Huber's yield criterion for plane stress. The semiaxes of the ellipse are $\sigma_o\sqrt{2}$ and $\sigma_o\sqrt{\frac{2}{3}}$. The smaller value also represents the radius of the cylindrical yield surface.

When $\sigma_1 = \sigma_2$, the limiting stress is reached at $\sigma_1 = \sigma_o$ according to either theory. However, when $\sigma_1 = -\sigma_2$, the energy theory (6.8) yields

$$\sigma_1 = \frac{\sigma_o}{\sqrt{3}} = 0.577\sigma_o \tag{6.9}$$

whereas the shear theory (6.3) shows

$$\sigma_1 = 0.5\sigma_o$$

Thus, when σ_1 and σ_2 are equal in both sign and magnitude, or if the stress state is uniaxial, the two criteria are identical. In all other cases,

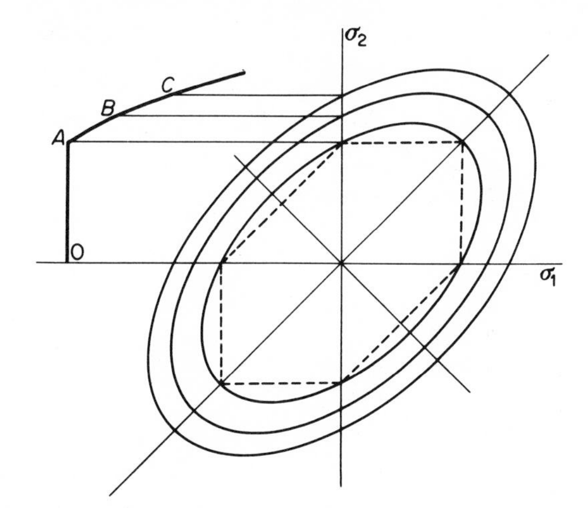

Fig. 6.6 Expansion of H-M ellipse caused by work hardening (*OABC*-stress-strain curve).

according to the Huber–Mises criterion, slightly higher loads are allowed without the danger of plastic deformation.

The existing experimental evidence slightly favors the distortional energy theory. However, the simpler shear theory is often employed because it is only in rare cases that the stress state is such that the two differ by the full 15 per cent.

Work-hardening causes the yield stress to rise, and the Huber–Mises cylindrical yield surface must expand accordingly. This is illustrated in Fig. 6.6 for a biaxial case where the smallest ellipse corresponds to the initial state while the two larger ones correspond to the two hardened conditions at points *B* and *C* on the stress–strain curve. The Tresca criterion would have been represented by a hexagon gradually expanding from its initial dashed form (see Fig. 9.9).

REAL AND IDEALIZED PLASTIC BEHAVIORS

It was stated in Chapter 4 that elastic properties of materials are structure-insensitive. Hence, the elastic modulus is almost constant for any one class of materials; e.g., steels, aluminum base alloys, etc. The plastic part of the curve, however, does depend on structure so that it is strongly affected by the physical nature of the particular metal. The strain-hardening rate may vary from very low to quite high, as illustrated by Fig. 6.7.

Problems in plastic deformation are handled in a number of different ways. If the amount of plastic flow is small, and comparable with the elastic strain, it is necessary to consider both simultaneously. Large plastic strains, as occur in metal forming or in mechanical testing, frequently permit the elastic part of the diagram to be ignored. A further simplification is

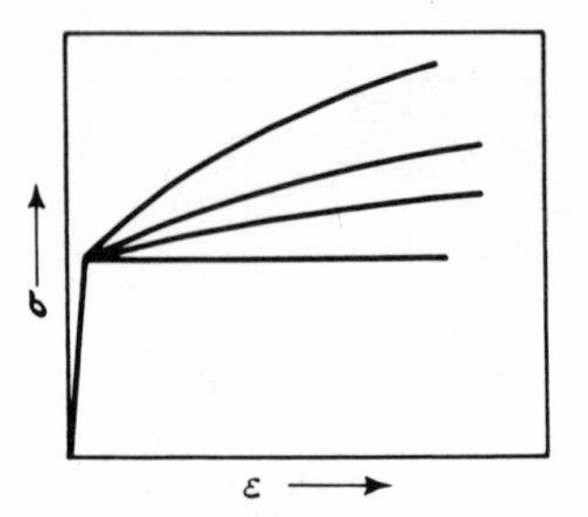

Fig. 6.7 Stress-strain curves of elastic-plastic substance showing different strain-hardening rates.

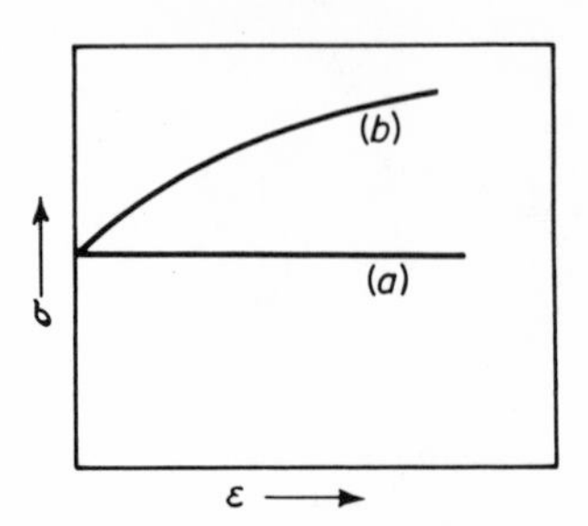

Fig. 6.8 Stress-strain curves of ideal (a) rigid-plastic and (b) rigid-work-hardening substances.

obtained by assuming the absence of strain-hardening, thereby arriving at the concept of a *rigid-plastic substance** which shows no deformation at all below the yield stress and a horizontal stress–strain relation beyond that point (Fig. 6.8). Although the idea of a substance with these properties appears academic, it has proved of considerable advantage in developing the formal framework of plasticity theory. Actually, in treating plasticity problems, other difficulties are encountered that require even more drastic assumptions to be made. The end results, while not quite satisfactory, are nevertheless valuable since they furnish a number of approximate analyses and solutions for otherwise impenetrable problems.

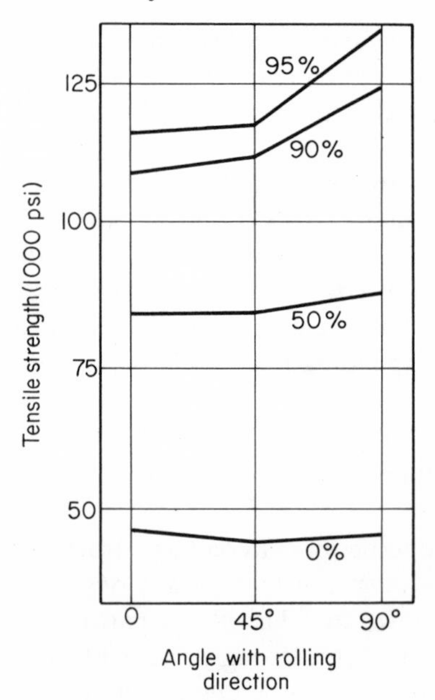

Fig. 6.9 Directional properties of cold rolled brass (70% Cu–30% Zn) strip. Reduced in thickness 0, 50, 90, and 95%. After M. Cook, Jl. Institute of Metals, **60,** 159, 1937.

Perhaps the most serious additional assumption is that the material is completely isotropic. It implies that the properties are entirely independent of the direction and of the straining path by which an existing condition was developed. In reality, metals that are cold-worked develop an anisotropy of mechanical properties (Fig. 6.9), and the principal manifestations of directionality will now be reviewed.

* Referred to as a " St. Venant body" by the rheologists. See Chapter 8.

PREFERRED ORIENTATIONS

The quantitative criteria governing the transition from the elastic to the plastic state apply strictly to isotropic materials only. Still, these laws are used with reasonable accuracy on metals which are composed of large numbers of individual crystals or grains, each of which is highly anisotropic. This is possible because the properties of the randomly and very small individual grains are "averaged out" within structural members. If an appreciable portion of the grains align themselves in certain preferred directions, however, averaging out does not occur, and the metal is said to have developed a *preferred orientation* or *texture*. This tendency for alignment occurs commonly in nature when objects assume relative positions while they are also moving as a group. For example, logs of wood dropped disorderly on a stream will gradually align themselves so that their axes will be parallel with the current.

In castings, columnar crystals grow in the direction of the maximum temperature gradient, so that their long dimension is perpendicular to the mold walls; these grains generally all have about the same crystallographic orientation because they grow faster in certain directions than in others. Similarly, in electroplating, textures are developed with the crystal growth normal to the surface being coated.

In crystalline materials, preferred orientations also develop from extensive plastic deformation. During deformation, while the grains are stretching, they also rotate so that certain crystallographic directions and planes gradually become aligned in a way that is characteristic of the type of deformation process used. For instance, in drawn wires, the preferred orientation develops relative to the wire axis, while in sheet rolling it is related to the rolling direction as well as to the rolling plane. A variety of textures obtained in beryllium after various combinations of rolling and extrusion is illustrated in Fig. 6.10. The strength of sheets with pronounced preferred orientation depends on the direction in which the test specimen is cut out of the sheet (Fig. 6.9), If such sheets are drawn into cylindrical cups, the weaker directions will show more extension than the stronger ones. This will produce two, four, six, or more undesirable *ears* on the finished product (Fig. 6.11a vs. b). Weaker and stronger directions in metals also show up in other forming processes. For example, when a cylinder was cut out of an aluminum plate and compressed, it did not maintain its circular cross section as shown in Fig. 6.11c. The preferred orientations of some metals formed in a number of ways are listed in Table 6.1.

Some metals, such as copper, show a *double fiber structure*. In this case a portion of the crystals tend to align themselves in one direction while the rest are oriented in another direction with respect to the specimen axis. Some examples of this are shown in Table 6.2.

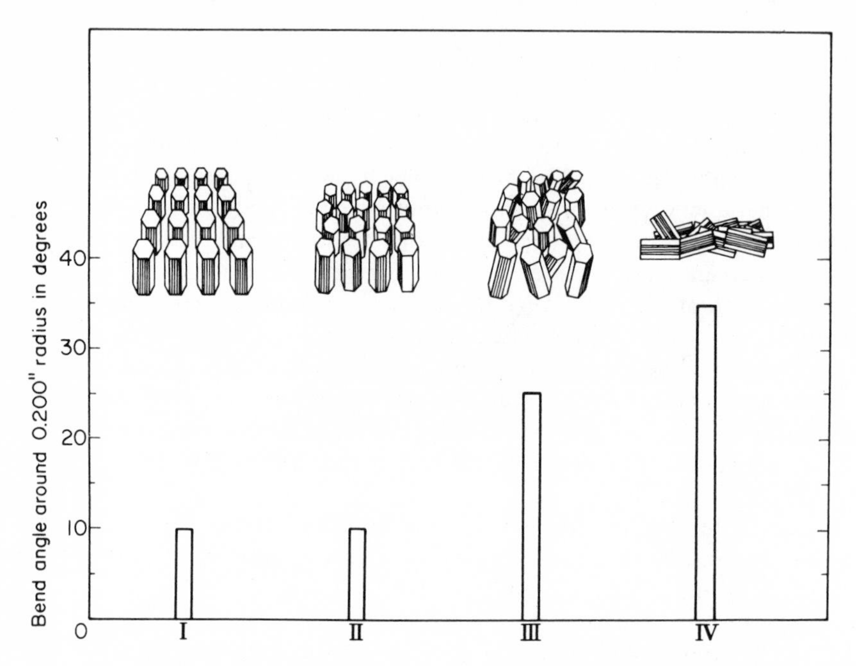

Fig. 6.10 Bend ductility of Be sheet correlated with texture. I = Extruded and transverse rolled. II = Hot pressed and random rolled to moderate reduction. III = Hot pressed and bi-directionally rolled to low reduction. IV = Extruded flat. Note the better ductility of relatively randomized structures III and IV. From J. Greenspan, G. A. Henrickson, and A. R. Kaufman. WADD, Tech Report No. 60–32 (1960).

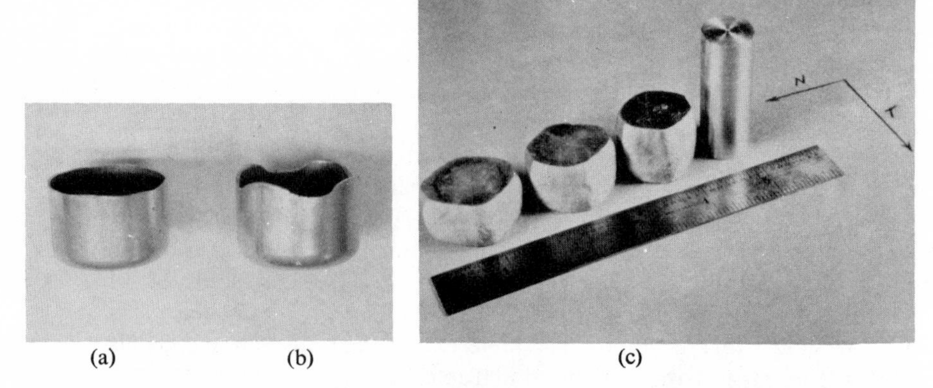

(a) (b) (c)

Fig. 6.11 Non-uniform flow of aluminum sheet and plate with preferred orientation. (a) Cup drawn from circular blank of randomly oriented sheet. (b) Same as (a) but with "ears" due to preferred orientation in the sheets. (c) Cylinder cut from aluminum alloy plate does not retain its circular cross section on compression. "N" indicates thickness direction of plate, and "T" indicates width direction. Cylinders from I. Rozalsky, *Trans. A.S.M.*, **47,** 77 (1955).

TABLE 6.1

TYPICAL PREFERRED ORIENTATION*

Metal	Forming Process				
	Drawing	*Compression*	*Torsion*	*Rolling*	
	Crystal direction that tends to become parallel with the specimen axis			Crystal plane and direction parallel to rolling	
				Plane	*Direction*
Al, Cu, Ni Au, Pb	[111] for Al [111] or [100] for others	[110]	[111]	(110)	[112]
Fe	[110]	[111]	[110] [112]	(001) (112) (111)	[110] [110] [112]
Mo, Ta, W	[110]	[111]		(001)	[110]
Mg, Zn, Cd, Zr				(0001)	[1120]

* After W. Boas, *An Introduction to the Physics of Metals and Alloys*. New York: John Wiley & Sons, Inc., p. 110 (1947).

TABLE 6.2

SOME DOUBLE FIBER STRUCTURES*

Metal	Percentage of Crystals	
	With [100] parallel to the wire axis	With [111] parallel to the wire axis
Aluminum	0	100
Copper	40	60
Gold	50	50
Silver	75	25

* After W. Boas, same as Table 6.1, p. 111.

The reason for different f.c.c. metals behaving differently in this respect is not known. Indeed, it is not possible as yet to predict from basic data the kind of preferred orientation that will develop after forming a specific metal in a particular manner. Approximate theories, however, were developed by Boas and Schmid (1931), G. I. Taylor (1938), and by Pickus and Mathewson (1938).

It is generally necessary to have a considerable amount of deformation to reveal preferred orientation. For instance, in rolling copper it becomes obvious only after an 80% reduction of thickness.

Due to differences in stress conditions, technical forming processes as a rule produce different textures on the outside and inside of the product. Because frictional shearing stresses or other constraints act at the surface, the principal stresses change their magnitude and direction with depth. In drawing, the center is deformed by a combination of axial tension and radial compression. The axial, and any two mutually perpendicular radial vectors, are thus the principal directions. At the periphery, however, even in the absence of friction, the principal directions are different (Fig. 6.12).

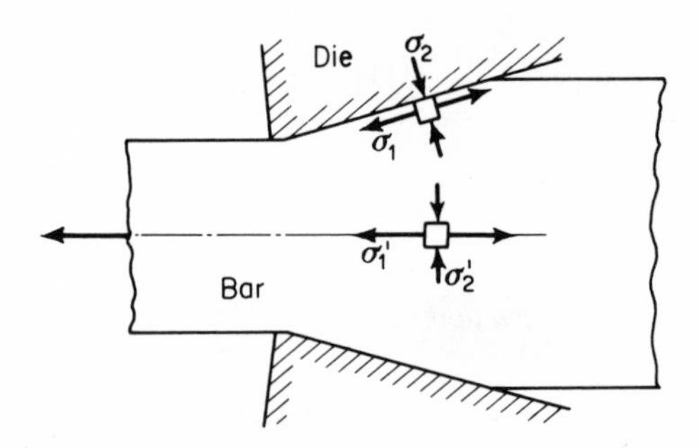

Fig. 6.12 Drawing of bar through a frictionless die. Principal stresses at surface (σ_1, σ_2) have a different orientation than those in the center. (σ_1', σ_2') The third principal direction is normal to the page.

THE BAUSCHINGER EFFECT IN PLASTIC POLYCRYSTALLINE MATERIALS

There is one other feature of plastic flow that results from the anisotropy of the individual grains and, specifically, from the variation of Young's modulus and the yield strength with orientation. This is the *Bauschinger effect* which is manifested by a lowering of the yield strength any time the direction of plastic deformation is reversed (Fig. 6.13).

When a force is applied to a crystalline body, all the grains change their lengths in the straining direction by the amount needed to maintain the

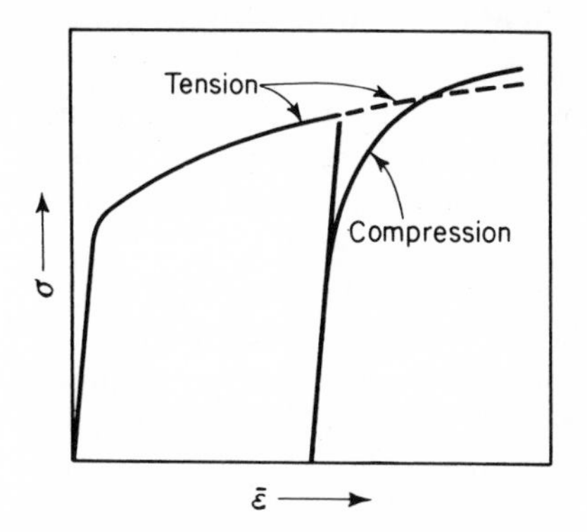

Fig. 6.13 Bauschinger effect showing lowered yield stress in compression following a tensile prestrain.

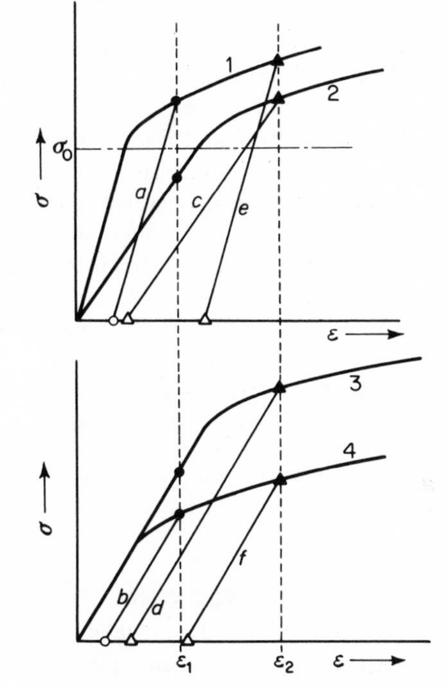

Fig. 6.14 The origin of low yield stress in cold worked metals on reversal of direction of straining. The unloading lines *a* to *f* are parallel to the appropriate elastic loading lines.

continuity of the material. Because neighboring grains have different orientations, however, the portion of their total strain that is elastic or plastic depends on the value of E and σ_o for the particular grains' orientation. Those with high elastic moduli and/or low yield strength will have the greatest fraction of their total flow made up of plastic deformation.

Consider, as an example, two small volumes in the metal, one containing adjacent grains with stress–strain curves 1 and 2, and the other having grains with the properties of 3 and 4 (Fig. 6.14). When an overall tensile strain ε_1 is applied, these grains deform to the points indicated by the four solid circles. To simplify the argument, the total strain is assumed to be the same in each grain. When the external force causing ε_1 is removed, each grain contracts. Since 2 and 3 suffered an elastic strain only, they will tend to assume their original length. Grains 1 and 4, however, were deformed plastically and will tend to contract along a and b, respectively, to the size shown by the open circles on the ε-axis. Since the other two crystals want to contract elastically still farther, they exert a residual compression on crystals 1 and 4. For this reason, when a compressive force is now applied to the specimen, this external force will be added onto the residual compression already on 1 and 4 so that a smaller applied force will be needed to produce plastic flow than would be the case if the member had not been predeformed. The same, of course, occurs in other similarly oriented grains. As a result, the overall yield stress in compression is reduced.

If all four crystals had been deformed plastically to strain ε_2 (solid triangles), a similar effect would have resulted. Here again grains 2 and 3 contracting along *c* and *d* tend to assume a shorter final length than 1 and 4 (open triangles) on release of the load.

It is this mechanism that is thought to reduce the yield strength observed in cold-worked metals when a load is applied to them in a direction opposite to the initial deformation. It will appear not only when compression follows extension and vice versa but also when torsion is followed by a reversed twist, or bending by bending in the opposite direction.

THE ASSUMPTION OF CONSTANT VOLUME AND USE OF NATURAL STRAIN

It is universally assumed in the analysis of plastic stress–strain relations that the volume of the material remains constant at all times; i.e.:

$$\varepsilon_V = 0. \tag{6.10}$$

Indeed, even for plastic strains involving five or tenfold extensions, ε_V does not exceed 0.25%. Advantage is taken of this volume constancy to simplify the mathematics of plasticity problems.

In defining the elastic volume strain ε_V as the sum of the three conventional strain components ε_1, ε_2, ε_3, their products could be ignored because of their relative smallness (3.10). In plastic deformation the individual strains can be large, often in excess of unity, so that their products are comparable with, or even exceed, their sum. This leads to cumbersome expressions in conventional strain units (see p. 56), but is very simply defined in terms of natural strain $\bar{\varepsilon}$ (3.4). For this purpose, consider a cube of side l_o being plastically transformed into a parallelepiped with sides l_1, l_2, l_3. If the volume does not change:

$$\frac{l_1}{l_o} \cdot \frac{l_2}{l_o} \cdot \frac{l_3}{l_o} = 1$$

Taking logarithms of both sides of the last equation, one obtains

$$ln\,(l_1/l_o) + ln\,(l_2/l_o) + ln\,(l_3/l_o) = 0,$$

or more conveniently

$$\bar{\varepsilon}_1 + \bar{\varepsilon}_2 + \bar{\varepsilon}_3 = 0. \tag{6.11}$$

Hence, if the volume remains constant during deformation, the sum of the three natural (normal) strains is zero.

For uniaxial tension, the strain in the direction of σ_1 is $\bar{\varepsilon}_1$, and the two lateral strains are $\bar{\varepsilon}_2 = \bar{\varepsilon}_3 = -\nu\bar{\varepsilon}_1$ (see p. 101). Because of the constancy of volume, $\bar{\varepsilon}_1 - 2\nu\bar{\varepsilon}_1 = 0$, making

$$\nu = \tfrac{1}{2} \tag{6.12}$$

so that Poisson's ratio for plastic flow is 0.5.

PLASTIC STRESS–STRAIN RELATIONS

In an attempt to establish a quantitative relationship between stress and strain in the plastic range, Levy, and later Mises, visualized an idealized isotropic substance. The straining conditions were such that the principal stress axes coincided with the principal axes of *strain increment* at all times.* By considering the elastic strains as negligible compared with the plastic deformation, the problem becomes one of a rigid-plastic, or rigid work–hardening substance (Fig. 6.8). They expressed the relation between stress and strain in the following, now generally accepted, form:

$$\frac{d\bar{\varepsilon}_x}{\sigma_x - \sigma_a} = \frac{d\bar{\varepsilon}_y}{\sigma_y - \sigma_a} = \frac{d\bar{\varepsilon}_z}{\sigma_z - \sigma_a} = \frac{d\gamma_{xy}}{2\tau_{xy}} = \frac{d\gamma_{yz}}{2\tau_{yz}} = \frac{d\gamma_{zx}}{2\tau_{zx}} = d\lambda \tag{6.13}$$

where $d\lambda$ is an instantaneous, but otherwise variable, proportionality factor, while σ_a is the mean stress (5.4). An alternative form of (6.13) is obtained by expressing σ_a through the three normal components:

$$\begin{aligned} d\bar{\varepsilon}_x &= \tfrac{2}{3}d\lambda[\sigma_x - \tfrac{1}{2}(\sigma_y + \sigma_z)] \\ d\bar{\varepsilon}_y &= \tfrac{2}{3}d\lambda[\sigma_y - \tfrac{1}{2}(\sigma_z + \sigma_x)] \\ d\bar{\varepsilon}_z &= \tfrac{2}{3}d\lambda[\sigma_z - \tfrac{1}{2}(\sigma_x + \sigma_y)] \end{aligned} \tag{6.14}$$

$$d\gamma_{xy} = 2\tau_{xy}\, d\lambda; \qquad d\gamma_{yz} = 2\tau_{yz}\, d\lambda; \qquad d\gamma_{zx} = 2\tau_{zx}\, d\lambda.$$

By the use of principal components, it is readily shown that

$$\frac{d\bar{\varepsilon}_1 - d\bar{\varepsilon}_2}{\sigma_1 - \sigma_2} = \frac{d\bar{\varepsilon}_2 - d\bar{\varepsilon}_3}{\sigma_2 - \sigma_3} = \frac{d\bar{\varepsilon}_3 - d\bar{\varepsilon}_1}{\sigma_3 - \sigma_1} = d\lambda \tag{6.15}$$

which, in effect, states that Mohr's circles for stress and plastic strain increment are identical except for the scale (Fig. 6.15).

The constant volume condition (6.10, 11) in an incremental form is

$$d\bar{\varepsilon}_x + d\bar{\varepsilon}_y + d\bar{\varepsilon}_z = 0 \tag{6.16}$$

and, to satisfy it, the zero point of the $d\bar{\varepsilon}$ axis in Mohr's diagram must coincide with σ_a (note that the difference between any two strain increments does not depend on the position of 0).

---

* Severe plastic torsion is a case where the directions of stress and strain differ (see Chapter 14).

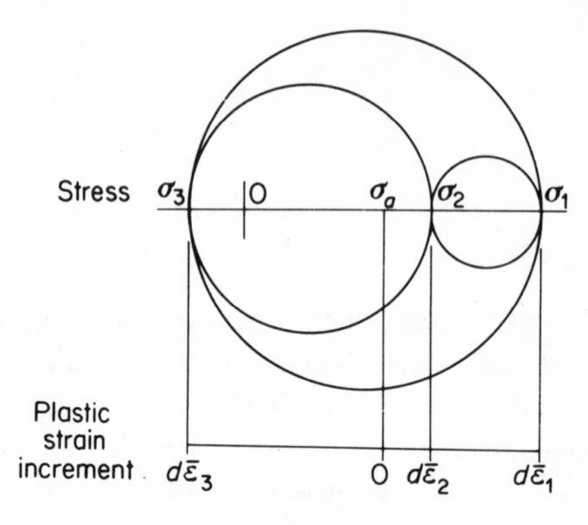

Fig. 6.15 The same Mohr circles are used to describe stresses σ and resulting strain increments $d\varepsilon$.

To find the factor $\frac{2}{3}d\lambda$, the *effective strain increment*, $d\bar{\varepsilon}_e$ defined by (6.17) is introduced

$$d\bar{\varepsilon}_e = \frac{\sqrt{2}}{3}\sqrt{[(d\bar{\varepsilon}_1 - d\bar{\varepsilon}_2)^2 + (d\bar{\varepsilon}_2 - d\bar{\varepsilon}_3)^2 + (d\bar{\varepsilon}_3 - d\bar{\varepsilon}_1)^2]} \qquad (6.17)$$

This term is structurally identical with (5.20b) for $\nu = \frac{1}{2}$ and using $d\bar{\varepsilon}$ instead of ε. Now the first three (6.14) equations are substituted successively in (6.17) using principal notations. Observing that $d\bar{\varepsilon}_1 + d\bar{\varepsilon}_2 + d\bar{\varepsilon}_3 = 0$, the following group of equations results:

$$\begin{aligned} d\bar{\varepsilon}_x &= \frac{d\bar{\varepsilon}_e}{\sigma_e}[\sigma_x - \tfrac{1}{2}(\sigma_y + \sigma_z)] \\ d\bar{\varepsilon}_y &= \frac{d\bar{\varepsilon}_e}{\sigma_e}[\sigma_y - \tfrac{1}{2}(\sigma_z + \sigma_x)] \\ d\bar{\varepsilon}_z &= \frac{d\bar{\varepsilon}_e}{\sigma_e}[\sigma_z - \tfrac{1}{2}(\sigma_x + \sigma_y)] \end{aligned} \qquad (6.18)$$

$$d\gamma_{xy} = 3\frac{d\bar{\varepsilon}_e}{\sigma_e}\tau_{xy}; \qquad d\gamma_{yz} = 3\frac{d\bar{\varepsilon}_e}{\sigma_e}\tau_{yz}; \qquad d\gamma_{zx} = 3\frac{d\bar{\varepsilon}_e}{\sigma_e}\tau_{zx}$$

The relationship between $d\bar{\varepsilon}_e$ and σ_e are obtained from a simple tension test.

PLASTIC STRESS–STRAIN RELATIONS IN "PROPORTIONAL" DEFORMATION

Since the plastic σ–$\bar{\varepsilon}$ relations are given above in an incremental form, calculation of total finite strains often calls for a complex integration. The situation is greatly simplified when the directions of successive strain increments coincide with those of the applied stresses, and the ratios of the strain increment components remain constant at all times. The latter case is defined by

$$\frac{d\bar{\varepsilon}_2}{d\bar{\varepsilon}_1} = k_1 \qquad \text{and} \qquad \frac{d\bar{\varepsilon}_3}{d\bar{\varepsilon}_1} = k_2 \tag{6.19}$$

where k_1 and k_2 are constants; this is referred to as *proportional straining*.

In view of (6.16),

$$k_1 + k_2 + 1 = 0 \quad \text{and} \quad k_2 = -(k_1 + 1).$$

Substitution in (6.17) produces

$$d\bar{\varepsilon}_e = \frac{2\, d\bar{\varepsilon}_1}{\sqrt{3}} \sqrt{(1 + k_1 + k_1^2)} \tag{6.20}$$

Equations (6.18) can then be used in a finite form

$$\frac{\bar{\varepsilon}_1}{\bar{\varepsilon}_e} = \frac{\sigma_1 - \frac{1}{2}(\sigma_2 + \sigma_3)}{\sigma_e} \quad \text{etc.} \tag{6.21}$$

and solved with the aid of (6.11).

Proportional straining actually takes place in a number of cases. For instance, in uniaxial extension of a long bar, the two transverse strains are related to the axial strain by a factor -0.5. In the expansion of a sphere, the radial (thinning) strain is -2 times the tangential extension, etc. In general, however, the use of (6.21), instead of (6.15) or (6.18), is not advocated unless conditions at least approximately fulfill (6.19).

REMOVAL OF THE EFFECTS OF COLD WORK, RECRYSTALLIZATION AND ANNEALING

Plastic deformation of a metal, carried out below about 0.35 to 0.4 of its melting point expressed in absolute units (°K.), increases its yield stress. This increase may be considerable. For example, soft copper has a yield strength as low as 5,000 psi or even less. When drawn into a thin wire, σ_o may raise to 60,000 psi; a stainless steel with 18% Cr and 8% Ni can have its yield strength raised from about 30,000 to 170,000 psi or even more by cold rolling or drawing. This deformation also causes the metals to become less ductile, however.

These high strength properties may be valuable for certain purposes like spring materials or wire cables but will be unsuitable for applications where ease of forming, such as bending or stamping, is important. Since in the course of manufacturing or fabrication of various metal products, cold working is an inherent part of the production process, it is very often necessary to remove the work–hardening and restore the original soft condition. The process employed for this purpose is called *annealing*.

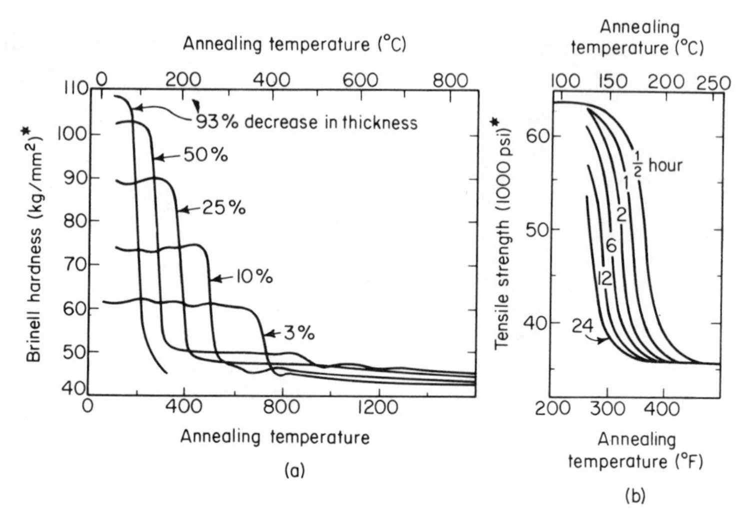

Fig. 6.16 Recrystallization characteristics of copper shown by the effect of amount of cold deformation, annealing temperature and time on the deformation resistance of copper (from C. H. Samans, *Engineering Metals and Their Alloys*, Macmillan, 1949, (a) after W. Koester (b) after W. A. Alkins and W. Cartwright). *Brinell hardness and tensile strength are measures of deformation resistance (see Chapters 10 and 12).

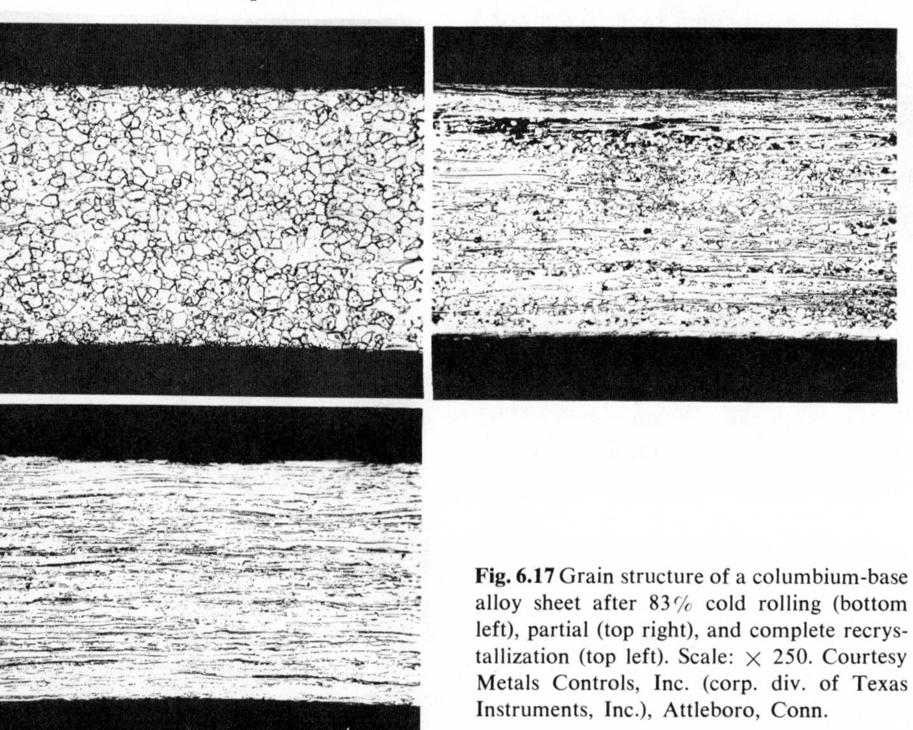

Fig. 6.17 Grain structure of a columbium-base alloy sheet after 83% cold rolling (bottom left), partial (top right), and complete recrystallization (top left). Scale: × 250. Courtesy Metals Controls, Inc. (corp. div. of Texas Instruments, Inc.), Attleboro, Conn.

Annealing consists of heating the cold worked metal to a relatively high temperature, holding it there for a certain length of time, followed by slow cooling. A lower temperature, or a smaller amount of initial cold work, requires a longer time to bring about a complete resoftening of the work-hardened structure. The effect of prior cold work, temperature, and time of heating are summarized in Fig. 6.16.

The principal physical phenomenon which causes this change is called *recrystallization* and it implies the formation, and quite often the subsequent growth, of new, unworked, and hence soft grains within the parent work-hardened structure.

As is seen in Fig. 6.17, the grains which became flattened and elongated by progressive cold working are replaced, during recrystallization, by new, equiaxed crystals.

HOT WORKING

If metals are deformed above their recrystallization temperature, strain-hardening, from a practical viewpoint, does not occur. Deformation of this type, especially when carried out at high rates, is known as *hot-working* because it is usually, but not necessarily, associated with elevated temperatures. Because of its low melting point, lead, for example, is hot-worked even at room temperature. A metal such as molybdenum, is cold-worked (i.e., it strain-hardens), even when deformed at a red heat, because of its high recrystallization temperature (> 1,100°C).

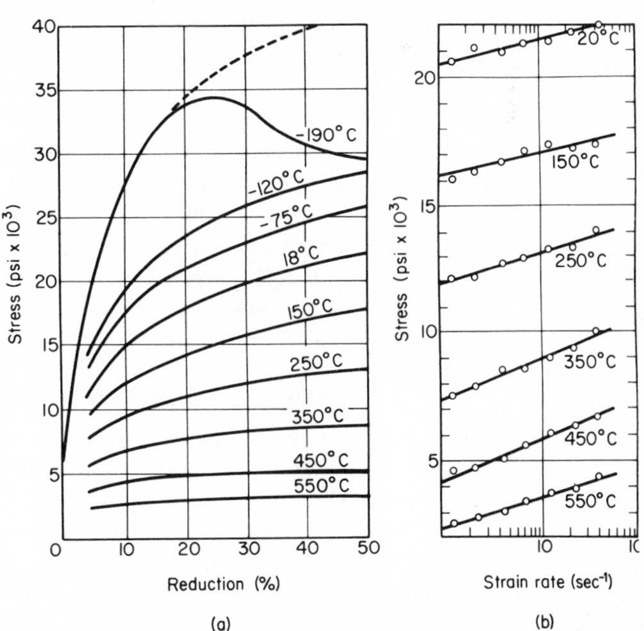

Fig. 6.18 (a) Effect of temperature on the stress-strain curve for aluminum. Strain rate = 4.38/sec. (b) Effect of strain rate on the stress required to compress aluminum to 40% reduction at various temperatures. (From J. F. Alder and V. A. Phillips *J. Inst. of Metals*, **83,** 80, 1954.)

The resistance of metals to plastic deformation generally falls with temperature (Fig. 6.18). For this reason large and massive sections are always worked hot by forging, rolling, or extrusion.

Metals display distinctly viscous characteristics at sufficiently high temperatures, and their resistance to flow increases at high forming rates (Fig. 6.18). This occurs not only because it is a characteristic of viscous substances (see Chapter 8) but also because the rate of recrystallization may not be fast enough to catch up with the rate of working.

PROBLEMS

1. Prove that the Huber-Mises yield criterion in plane strain is $\sigma_1 - \sigma_3 = (2/\sqrt{3})\,\sigma_o$.

2. Express the above criterion in terms of $\sigma_e = f(\sigma_o)$ where $\sigma_e = g(\sigma_1, \sigma_2, \sigma_3)$.

3. A circular blank (see sketch) of thickness t is deformed plastically by inwardly directed radial drawing. Calculate the radial and tangential stresses, assuming a yield stress σ_o and a stress free surface of the blank.

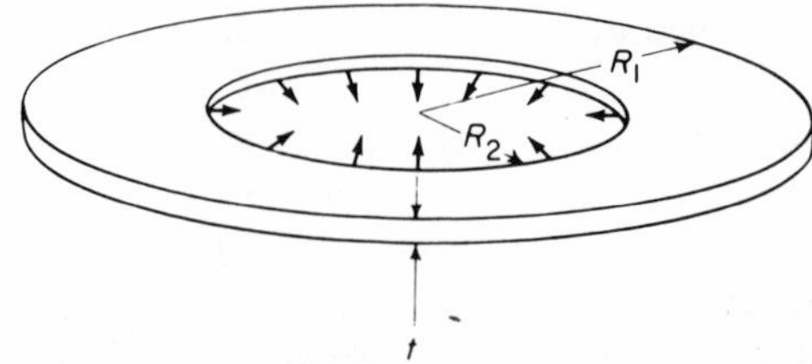

4. A closed-end tubular vessel with 3 in. O.D. and 0.1 in. thick wall is pressurized internally until it yields. Using Tresca's criterion, calculate the inside pressure at that moment, assuming $\sigma_o = 100{,}000$ psi.

5. A rigid - plastic cube is subjected to σ_1 on one pair of opposite faces and to $\sigma_2 = -\frac{1}{2}\sigma_1$ on another. Calculate the principal shear stresses and the ratios of the three strain increments at the moment of yielding if $\sigma_o = 50{,}000$.

6. What would be the effect of a superimposed 30,000 psi pressure on the strains?

7. Speculate how deformation will proceed when yield is reached in the presence and/or absence of strain-hardening.

8. A thick-walled steel tube of 6 in. O.D. and 2.5 in. I.D. is under 40,000 psi internal pressure. The periphery was layerwise machined off while the tube was under pressure, and yielding started when the O.D. decreased to 4.8 in. Determine the yield stress of the tube metal.

REFERENCES FOR FURTHER READING

(R6.1) Crandall, S. H. and N. C. Dahl, (eds.) *An Introduction to the Mechanics of Solids*. New York: McGraw-Hill Book Company, 1959.

(R6.2) Hoffmann, O. and G. Sachs, *Introduction to the Theory of Plasticity for Engineers*. New York: McGraw-Hill Book Company, 1953.

(R6.3) Johnson, W. and P. B. Mellor, *Plasticity for Mechanical Engineers*. London: D. Van Nostrand Co., Ltd., 1962.

7

PLASTICITY OF CRYSTALLINE MATERIALS

The quantitative treatment of plastic deformation in Chapter 6 was based on the assumption of an isotropic and "featureless" material. Nevertheless, the reader's attention had to be drawn to certain important modifications of this simplified concept. The Bauschinger effect and preferred orientation were shown to result from the microscopic anisotropy that comes from the crystalline structure of metals. To gain an insight in the mechanism of deformation, one must again turn to the individual grains and their internal structure and examine the nature of the atomic movements that result in plasticity. Only in this way can the onset of yielding, work-hardening, and other phenomena of plasticity be explained.

The anisotropy of single crystals is inherently very pronounced. With the aid of x-ray diffraction, the orientation of the crystallographic planes or direction of a crystal can be determined relative to some external easily identifiable axis or plane. For example, the standard projection discussed in Chapter 4 (Fig. 4.13) is one means of describing the angle between a specimen axis (normally the tensile direction) and any crystallographic element. Thus the motions of each of these sets of planes can be followed in the course of deformation.

Large single crystals can be produced by several techniques. Having a number of differently oriented crystals of the same material, one can systematically observe the effect of orientation on the mode of deformation and the stresses involved.

Many experiments that have been carried out show that plastic deformation occurs by *slip* on certain crystallographic planes. These generally are the planes which possess the highest atomic density; i.e., the largest number of atoms per unit area. Accordingly, slip occurs in h.c.p. crystals on ((0001)) planes, in f.c.c. on ((111)), and in b.c.c. on ((110)). Slip is also selective with respect to direction and is always along a line of highest atom density. The trivial explanation for this behavior can be found in the fact that atoms in close-packed planes are tightly bound into parallel layers, while the interaction between layers is relatively weak in view of the large distance between them. A more rigorous explanation for the fact that slip prefers the close-packed planes and direction is given in the discussion of Eq. (7.9). The slip planes and directions have been established for single crystals of the three basic structural groups. The combination of a slip plane and a direction is known as a *slip system.*

SLIP SYSTEMS

Slip Systems of Hexagonal Close-Packed Crystals

Typical of this group are magnesium, zinc, titanium, and beryllium. The close-packed slip directions here are the three base diagonals *DOA*, *BOE*, and *FOC* in Fig. 7.1a. The most common slip plane is the close-packed hexagonal basal plane (0001). Together they constitute only three slip systems, and this is responsible for the limited ductility of magnesium, zinc, or cadmium at ambient temperatures. At elevated temperatures, however, additional slip planes begin to operate—such as the pyramidal plane ABO' or the prismatic plane $AA'B'B$ with indices $(10\bar{1}1)$ and $(10\bar{1}0)$, respectively. This results in greatly improved plastic properties of Mg upon moderate heating (150–200°C).

In certain metals of this group, such as Zr and Ti, these extra slip planes are active even at room temperature, thereby permitting considerable plastic deformation by cold rolling or drawing of sheet and wire without annealing.

Slip Systems of Face-Centered Cubic Crystals

This group includes copper, aluminum, silver, nickel, and a number of others. The close-packed planes are the octahedral or ((111)) planes which have the same number of atoms per unit area as the basal planes in the h.c.p. group. The closest linear packing is along the diagonals [[110]], Fig. 7.1b. Although there are eight octahedral planes in a cubic crystal, only four, intersecting at one point, need be considered since each of the other four is parallel to one of the planes in the first group. Likewise, the total number of [[110]] directions is six but, of these, only three lie in each of the ((111)) slip planes. With the aid of (4.5), we find that the $(11\bar{1})$ plane, for example, contains the [101], [011], and $[1\bar{1}0]$ directions, thus producing three slip systems. It follows that these are $4 \times 3 = 12$ systems along which slip can propagate with equal ease.

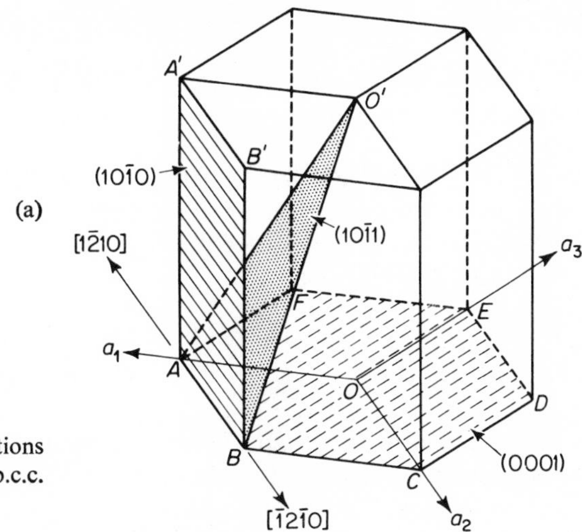

Fig. 7.1 Slip planes and directions in (a) h.c.p. (b) f.c.c. and (c) b.c.c. lattice.

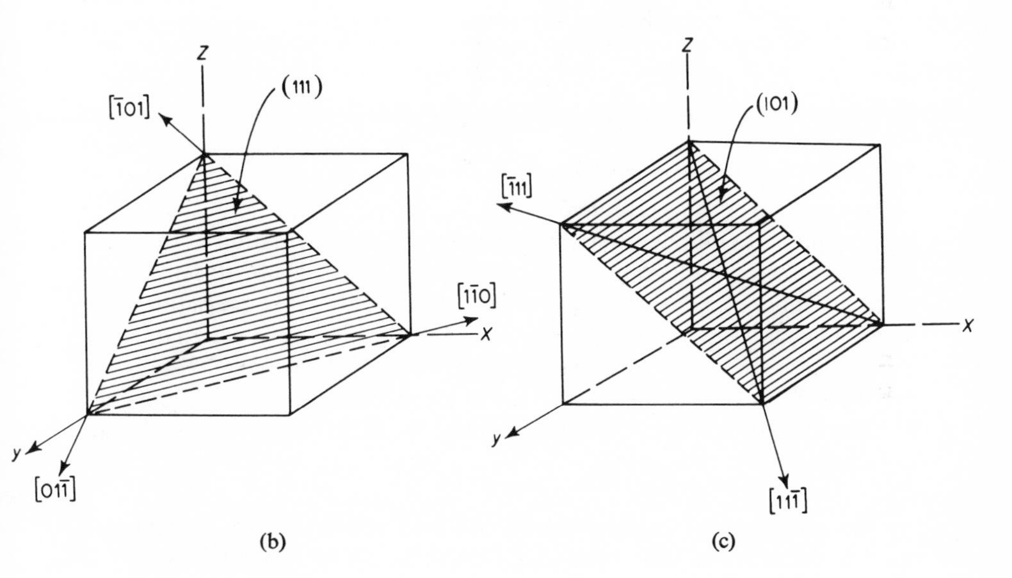

Slip Systems of Body-Centered Cubic Crystals

In b.c.c. metals like α-iron, molybdenum, or tungsten, the atomic density is highest on the ((110)) planes and along the [[111]] directions, Fig. 7.1c. However, the planar packing in this case is not as dense as in the h.c.p. and f.c.c. structures so that the direction becomes the controlling factor while several planes may be capable of slipping. For instance, in iron the highest density ((110)) planes, as well as ((112)) and ((123)), can become operative.

In general, knowing the type of slip plane and the number of slip directions, N_d, in the plane, we can calculate the number of slip systems, N_s, from

$$N_s = N_p \cdot \frac{N_d}{2} \tag{7.1}$$

where N_p is the multiplicity of the particular plane in the system.

For cubic crystals, the multiplicity, or number of equivalent planes (including those parallel to each other), is related to the Miller indices according to the following table:

Plane Type	*hOO*	*hhh*	*hhO*	*hkO*	*hhl*	*hkl*
Multiplicity	6	8	12	24	24	48

With the aid of (4.5) and (7.1), it is found that, in α-iron, the ((110)) planes contain two [[111]] directions, whereas the ((112)) and ((123)) planes each contains one of them. Hence,

$$N_s = \frac{(12 \times 2 + 24 \times 1 + 48 \times 1)}{2} = 48 \text{ slip systems.}$$

THE MECHANICS OF SLIP IN CRYSTALS

The Critical Resolved Shear Stress, τ_{ro}

Since slip in crystals is limited to certain planes and directions, the applied stress required to initiate plastic flow depends on the orientation of the stress relative to the crystallographic axis of the crystal. The h.c.p. crystals are an extreme example of this since they ordinarily slip only along the basal plane. If this plane is either normal or parallel to the direction of an external tensile or compressive stress (usually identical with the geometrical axis of the crystal), the shear stress on the plane is zero and plastic deformation is impossible. The metal then deforms elastically until it fractures. However, when the angle between the two is close to 45°, the yield strength, σ_o, will be at a minimum or nearly so.

The qualification, "nearly," had to be used above because we have still ignored the effect of the crystallographic direction of slip, *M* in Fig. 7.2, which generally does not coincide with the direction *K* of the maximum τ component on the particular plane. The shear stress component acting in the slip direction, τ_r, is conventionally described as a function of the angles between the axis of pull and the slip direction, λ, and this axis and the normal to the slip plane α (Fig. 7.2). The axial stress on the slip plane is $\sigma \cos \alpha$, and its shear component in the slip direction is

$$\tau_r = \sigma \cos \alpha \cos \lambda \tag{7.2}$$

The product of the two trigonometric terms in (7.2), which determines whether or not the orientation is favorable for slip, is known as the *Schmid factor*. The value of τ_r at the onset of plastic yielding is called the *critical resolved shear stress*, τ_{ro}, and it represents an important physical characteristic of a crystal. Although the yield stress, σ_o, measured in the axial direction varies within wide limits, τ_{ro} remains constant. In practice τ_{ro} is determined from measurements of σ_o which varies inversely with $\cos \alpha \cos \lambda$, with a minimum at $\cos \alpha \cos \lambda = 0.5$.

While changes of slip plane orientation α may cause very large variations in the apparent tensile yield strength of h.c.p. metals, this is not true with regard to the direction of slip within the plane, because a slip direction occurs every 60°. Consequently, the extreme effect of direction will be produced when ρ changes 30° ($\cos 30° = 0.866$), corresponding to a change of σ_o of not more than 14%.

The effect of crystal orientation on σ_o in f.c.c. and b.c.c. metals is much less pronounced than in the h.c.p. class because, with the large number of slip systems, it is impossible to find an orientation in which the resolved shear stress is zero on all potential slip planes at the same time.

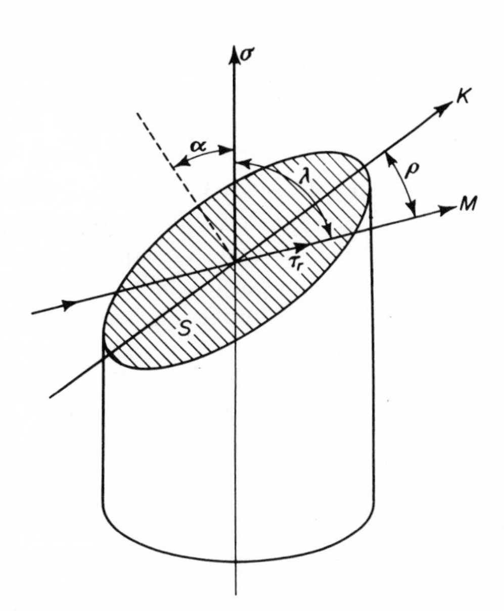

Fig. 7.2 Slip plane *S*, direction of maximum shear stress, *K*, and slip direction, *M*.

Slip Bands

Slip in single crystals does not begin simultaneously on all the slip planes of a system but only along some of them. It spreads suddenly by a relative displacement of the two neighboring layers by a distance of 1,000 Å. or more, stops suddenly, then starts again on a parallel plane about 100 Å away,

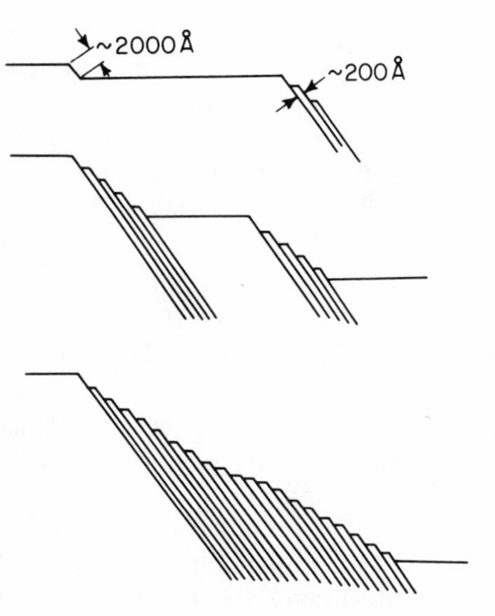

Fig. 7.3 Formation of laminar slip band by stepwise slip. From R. D. Heidenreich, in *Cold Working of Metals, Am. Soc. Metals*, 1949.

again ceases, and so forth. R. D. Heidenreich made electron microscope observations on single crystals of aluminum which had been polished prior to deformation and obtained the picture shown schematically in Fig. 7.3.

The cessation of slip on one plane is caused by work-hardening, whereupon slip is transferred to the next one, and so on. The density of these slip packets or *slip bands* increases as deformation proceeds. They can be observed through an optical microscope on the surface of crystals (including individual crystals in polycrystalline materials) because the sloping surfaces within the slip band "terraces" that separate two undistorted portions of the crystal do not reflect the vertical incident light beam into the microscope. They appear, therefore, as dark lines crossing the crystal.

The density of the slip bands increases with progressive deformation until they cannot be resolved with an ordinary microscope.

Reorientation Resulting from Slip. Multiple Slip

When a f.c.c. crystal is deformed by pure ("homogeneous") shear and suffers no extraneous constraints forcing it to assume a preferred shape, slip will occur on one set of planes only, and the orientation of the crystal relative to the direction of the acting force will not change with progressive deformation (shown schematically in Fig. 7.4a). However, when the crystal is pulled between coaxial grips or is subject to compression between parallel plates, the individual lamellae of the metal between the slip areas, together with the slip planes, will be forced to gradually rotate from their initial positions (Fig. 7.4b). In tension, the slip plane and direction tend to align themselves with the axis of the pull. In compression, they tend to become normal to the compression axis.

If the initial and final lengths of the specimen are l_o and l_1, and the angles between the slip plane and the direction of the applied force, before and after extension, are α_o and α_1, it can be shown that

$$l_o \sin \alpha_o = l_1 \sin \alpha_1 \tag{7.3}$$

As the active slip plane gradually rotates away from the position of high τ, other possible slip planes are rotated to a more advantageous position and begin to slip. This can result in a "pendulum" action, whereby two *conjugate slip systems* operate alternately, or even simultaneously when the shear stress components are equal on both planes. *Multiple slip* is the term applied to a condition where there are two or more intersecting slip systems (Fig. 7.5).

The h.c.p. lattice also rotates in the same fashion, but when it has only one family of slip planes, multiple slip is impossible. In b.c.c. iron, on the other hand, the multiplicity of slip planes is such that even small amounts of rotation will switch the slip from one plane to the next. Small structural irregularities will amplify this effect and, as a result, the slip bands acquire a wavy appearance (Fig. 7.6).

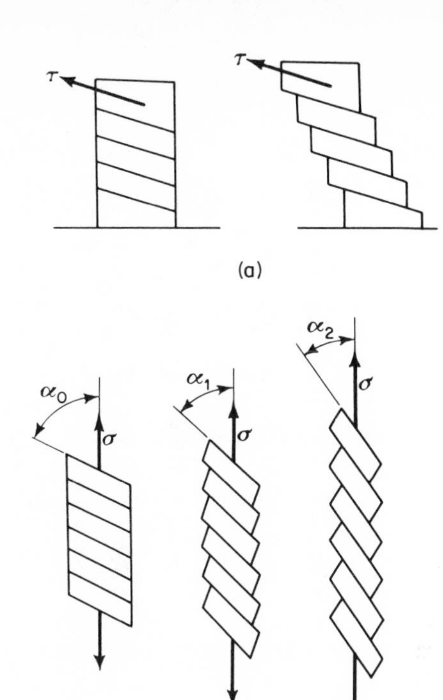

Fig. 7.4 Deformation of single crystal by: (a) homogenous shear (no rotation), (b) extension between grips (showing rotation).

Fig. 7.5 Cross slip bands in cold worked polycrystalline brass (70% Cr–30% Zn), 4500 ×. From C. S. Barrett, in *Cold Working of Metals*, *Am. Soc. Metals*, 1949.

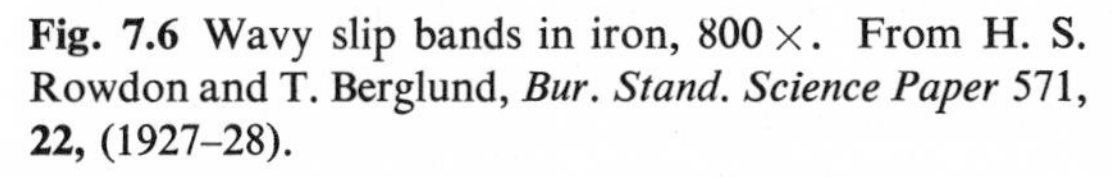

Fig. 7.6 Wavy slip bands in iron, 800 ×. From H. S. Rowdon and T. Berglund, *Bur. Stand. Science Paper* 571, **22,** (1927–28).

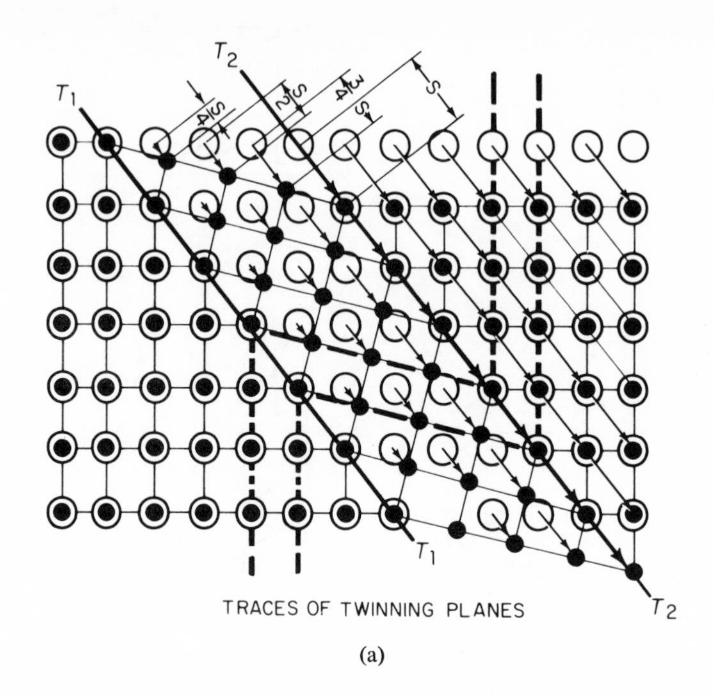

(a)

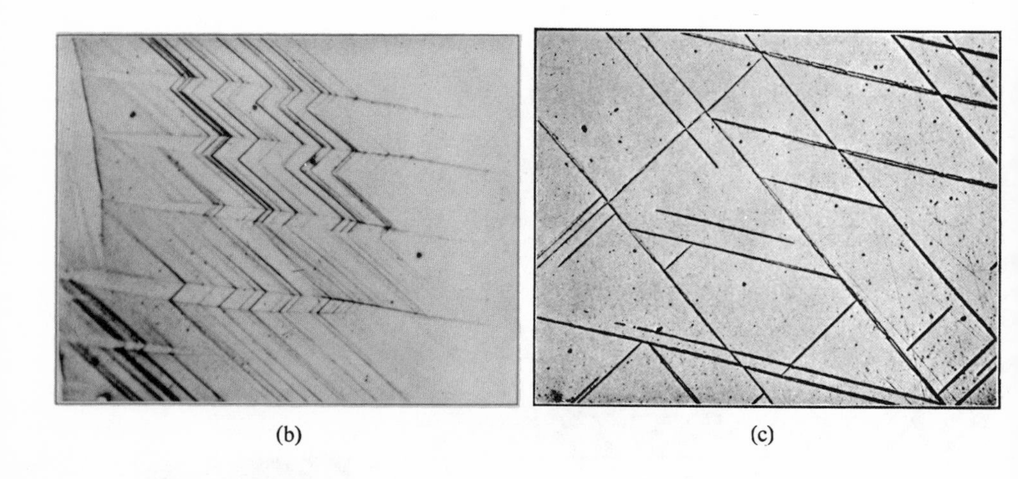

(b) (c)

Fig. 7.7 (a) Twin produced by a pack of atom planes between T_1 and T_2 shifting from ○ to ● positions. Note the symmetry of atom positions relative to a twinning plane. (T_1 or T_2). (b) Magnesium grain first twinned, then polished, etched, and caused to slip on basal planes. Second deformation was compression perpendicular to the first, 500 ×. Note the equivalence of the zig-zag steps with the heavy dashed lines in (a). Each zig-zag line is a slip band as shown in Fig. 7.3. (From C. S. Barrett, in *Cold Working of Metals*, *Am. Soc. Metals*, 1948. (c) Twins (Neumann bands) in a b. c. c. iron-silicon alloy. (From C. S. Barrett, in *Cold Working of Metals*, *Am. Soc. Metals*, 1948.

TWINNING

Deformation by Twinning

Twinning is a type of shear deformation that differs from slip. It occurs when each layer of atoms on one side of a certain plane, the *twin plane*, undergoes a fixed shift with respect to its neighbor. As a result the two parts of the twinned crystal represent mirror images of each other. The total displacement of each plane within the twinned portion of the crystal is proportional to its distance from the plane of symmetry (Fig. 7.7a).

There are no "in between" positions, and the process takes place instantly, often accompanied by a distinct clicking sound like the "cry" of tin when it is bent. Twins are generally local and have a lenticular appearance (Fig. 7.7b). The suddenness of twinning is reflected in the shape of the stress–strain curve which shows extensive jagging wherever twinning occurs. Twinning is very common in iron deformed by impact, or at low temperature in which case the twins are narrow and known as *Neumann bands* (Fig. 7.7c). In b.c.c. metals, the twinning plane is (112) and the direction $[11\bar{1}]$.

Since twinning results in a change of orientation, it can play an important role in the deformation of h.c.p. metals which have a rather limited capacity for plastic flow. The reformed orientation may put potential slip planes into a position associated with a high shear stress, thereby promoting further plastic flow. The twinning parameters in this structure are $(10\bar{1}2)$ and $[10\bar{1}\bar{1}]$.

For many years it was thought that deformation twins do not occur in the f.c.c. lattice. However, more recent experimental evidence indicates that they do at very low temperatures.

Strain Produced by Twinning

Schmid and Wassermann analyzed the shear deformation associated with twinning and the resulting dimensional changes. They showed that the amount of shear varies with the type of lattice and, in h.c.p. metals, it depends on the c/a ratio. The elongation* produced by complete twinning of a Zn crystal with $c/a = 1.856$ is 7.39%. This is a relatively small strain compared with what can be accomplished by slip.

DISLOCATION THEORY OF PLASTIC DEFORMATION

Shear Stress Requirements for Slip in a Perfect Crystal

Figure 7.8 represents a side view of the two neighboring close-packed planes in a perfect crystal, which are expected to slip relative to one another in the direction indicated by the arrows. In the positions shown, the atoms

---

* For details of calculation, see C. F. Elam, *The Distortion of Metal Crystals* (London: Oxford University Press, 1935), pp. 44–45.

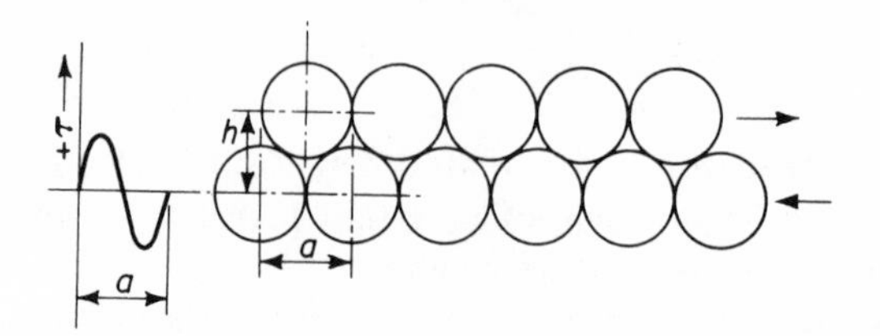

Fig. 7.8 Variation of shear stress, τ, required to slip one row of atoms over another by a distance of one lattice spacing, a.

are in equilibrium and, in order to reach the next equilibrium positions, the entire upper row will have to move by a distance, a. The shear stress, τ, varies periodically with period a as one row slides over the other. We can assume that τ varies according to a sine law (although other assumptions are possible and indeed desirable for improved accuracy of the shear stress estimate). The value for τ is given by $C \sin(2\pi x/a)$—where C is a constant, and x is the shear displacement or the distance over which the upper atom row moves with respect to the lower one. Since, for small displacements, $\tau = G\gamma$, $\gamma = x/h$, and $\sin(2\pi x/a) \approx 2\pi x/a$, $\tau \approx C(2\pi x/a) = G(x/h)$ from which $C = Ga/2\pi h$. The equation for the stress, therefore, becomes

$$\tau = \frac{Ga}{2\pi h} \sin\left(\frac{2\pi x}{a}\right) \tag{7.4}$$

and τ reaches a maximum at $\sin 2\pi x/a = 1$, i.e., at $x = a/4$. Hence, the lattice becomes unstable and yields when

$$\tau_{max} = \frac{Ga}{2\pi h} \tag{7.5}$$

Since a and h are usually similar numbers, the yield stress in shear of a single crystal should be about one-sixth of its shear modulus. The actual critical resolved shear stress, however, is only between 10^{-2} to 10^{-4} of the theoretical value.

By using more refined approximations, the calculated τ_{max} value is reduced by a factor of 5. This still leaves a discrepancy of two orders of magnitude between the experimental and calculated values.

This extraordinary disparity has forced a revision of the thinking on the mechanism of slip. It became obvious that, in a perfect crystalline lattice, colossal forces would be required to accomplish even a slight plastic deformation in a commercial metal like steel, aluminum, or even tin. To account for the relative ease with which slip takes place in crystals, it was necessary to assume that they contain certain structural defects called *dislocations*. Each dislocation is a disregistry that exists along a line over thousands of lattice spacings. It disturbs the perfect order of atoms in its surroundings, resulting in high internal stresses. Because this system of stresses about the dislocation determines its characteristics, this stress field will be discussed before considering the general properties of these line defects.

Mechanical Models of Edge and Screw Dislocations

Although the concept of submicroscopic dislocations was not introduced until 1934, models of dislocations were described early in this century by V. Volterra and others who investigated the internal stresses in "multiply-connected" elastic bodies. Of the various dislocations that have been described, the so-called edge and screw types are essential for interpreting plastic flow in crystals.

Both of these dislocations are produced in an elastic body by: (1) cutting it part way along a plane (Fig. 7.9), (2) subjecting the newly formed faces to a rigid shear displacement, b, in the surface of the cut, and (3) rejoining the faces in the new position to again produce a continuous solid. Depending on the direction of the displacement, an *edge dislocation* (Fig. 7.9a) or a *screw dislocation* (Fig. 7.9b) is formed. In both cases the cut terminates along z–z, the *dislocation line.* However, in the first case, the displacement b is normal to the dislocation line, whereas, in the second, it is parallel to it.

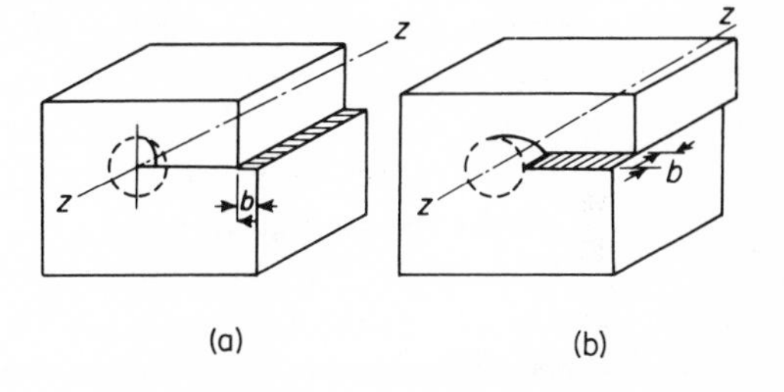

These models can only describe the displacement b that occurs at some distance from the dislocation line z–z. In the immediate vicinity of z–z, displacements of magnitude b are not possible. To see why this is so for the edge dislocation, consider a point on the cut surface at a distance b from z–z. If the displacement, starting at the outside and continuing to z–z, were constant, it would require the material between the aforementioned point and the dislocation line to be compressed to zero length. This means that $\varepsilon = \infty$ and, consequently, requires an infinitely large stress which, of course, is not possible to produce. Similarly, for the screw dislocation, the shear strain $\gamma = b/r$, (r = distance from a point on the plane of the cut to z–z) will increase infinitely when $r \to 0$.

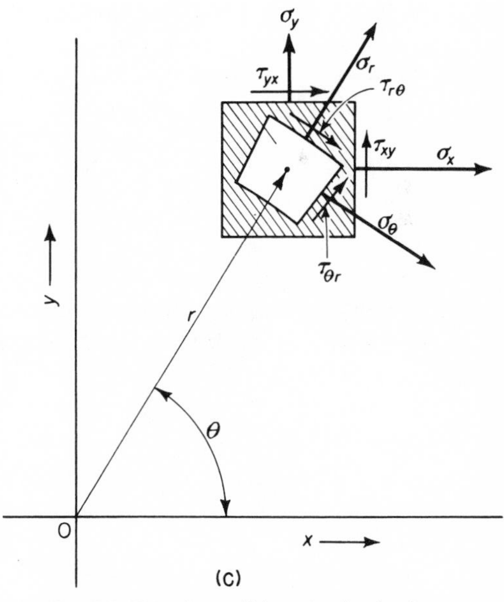

Fig. 7.9 Forming a dislocation in elastic solid by a planar cut and displacement b. (a) Edge dislocation. (b) Screw dislocation. (c) Stress components in polar (r, θ) and Cartesian (x, y) coordinates.

This circumstance complicates the mathematical treatment of dislocations. Luckily, it does not prevent the understanding of their physical functions in crystals.

Dislocations control the plastic properties of crystals by virtue of the system of internal stresses that exist in their surroundings. Since the strain histories that produced the dislocations are known, one can anticipate, in at least a general fashion, what these stresses should be. In forming the edge dislocation, the portion of the block above the cut in Fig. 7.9 was compressed. When the external compressing force is removed, this upper part will expand, pulling with it the initially stress free lower half. An equilibrium will finally be established with the upper half still in compression and the lower under residual tension. There will also be a shear stress in the plane of the "healed" cut. In view of the continuity of the body, the stresses will be symmetrical with respect to a vertical plane passing through z–z. That is, the same stress condition results on compressing the material above, or extending it below, the cut by the amount b.

Since deformations along one direction invariably give rise to lateral strains, an edge dislocation is also associated with strains and stresses in the direction normal to the plane of the original cut. These stresses can be evaluated by methods of the theory of elasticity, and the appropriate expressions are simpler in polar than in Cartesian coordinates.

Denoting the radial distance of a point from z–z by r, and the angle between the plane of the "cut" and the radius by θ, the stress components (Fig. 7.9c) for an edge dislocation are

$$\sigma_r = \sigma_\theta = -\frac{Gb}{2\pi(1-\nu)}\frac{\sin\theta}{r}$$

$$\tau_{r\theta} = \frac{Gb}{2\pi(1-\nu)}\frac{\cos\theta}{r} \tag{7.6a}$$

If desired, the σ_x, σ_y, and τ_{xy} components can be calculated from $x^2 + y^2 = r^2$ and $\tan\theta = y/x$. Hence, the stresses in Cartesian coordinates are

$$\sigma_x = -\frac{Gby(3x^2 + y^2)}{2\pi(1-\nu)(x^2+y^2)^2}$$

$$\sigma_y = \frac{Gby(x^2 - y^2)}{2\pi(1-\nu)(x^2+y^2)^2} \tag{7.6b}$$

$$\tau_{xy} = \frac{Gbx(x^2 - y^2)}{2\pi(1-\nu)(x^2+y^2)^2}$$

The screw dislocation carries only shear stresses, $\tau_{z\theta}$ ($\tau_{\theta z}$), on planes normal to or passing through z–z. This shear stress is

$$\tau_{z\theta} = \frac{Gb}{2\pi r} \tag{7.6c}$$

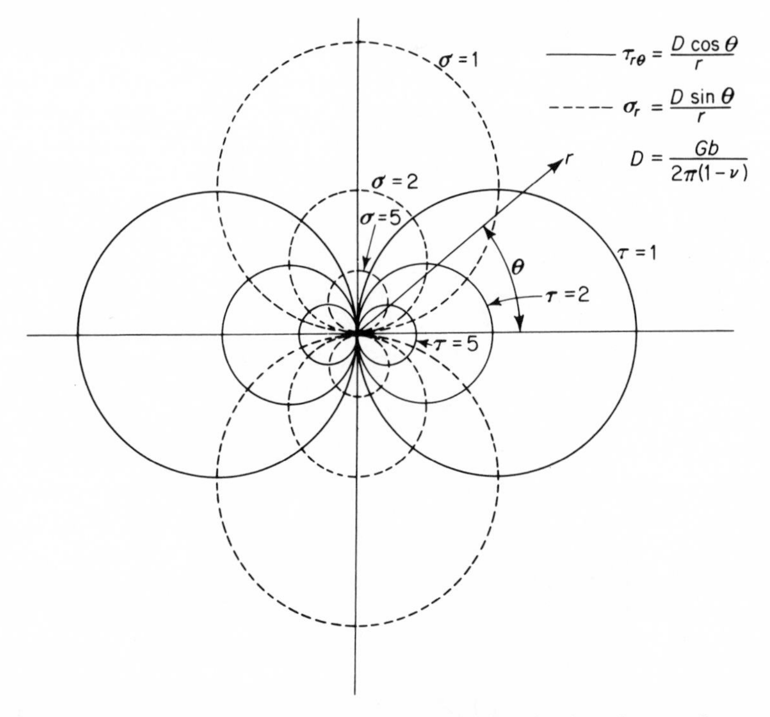

Fig. 7.10 Constant stress lines around an edge dislocation in polar coordinates.

Because of radial symmetry, $\tau_{z\theta}$ does not depend on θ, while $\tau_{r\theta} = \tau_{rz} = 0$.

For any selected radial direction from z–z, the normal and shear stresses produced by the dislocations decrease inversely with the distance r. The stress fields around dislocations are seen to be rather extensive and not negligible, even at some distance from their centers (Fig. 7.10). For this reason dislocations would be expected to interact with each other, much like electrical or magnetic disturbances do.

Edge Dislocations in Crystals

A crystal is built up of a large number of individual atoms and in this respect differs from the ideal mechanical continuum. Nevertheless, by making a partial cut between two neighboring atom planes and then shifting the upper or lower half of the crystal one atomic spacing relative to the other half, an edge dislocation is created just as for the elastic medium. A perfectly rigid displacement of all the material above the cut relative to that below it would create the same difficulty as in the case of the continuous model: the sheet of atoms adjacent to, and extending vertically from, the bottom of the cut z–z would have to be squashed to a zero thickness. To avoid this

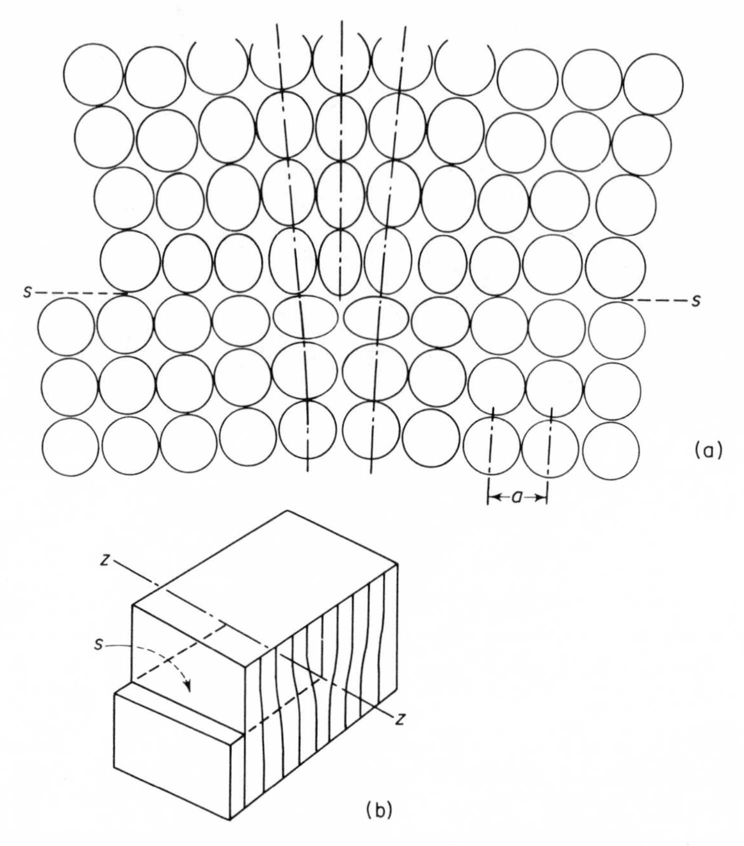

Fig. 7.11 Edge dislocation in cubic crystal. (a) Arrangement of atoms. (b) identification of elements (s—slip plane).

difficulty, the displacement must terminate not at a single line of atoms but over an area of a few atomic diameters wide in the vicinity of z–z. This "flattens" the atoms ahead of the displaced volume in the direction of the displacement and stretches them out in the direction normal to it. When the force that caused the displacement is removed, a certain readjustment of the individual atoms occurs, producing a final configuration as shown in Fig. 7.11a. Line z–z is again a *dislocation line*, and the plane of the cut is the slip plane (Fig. 7.11b).

As a result of the operation just described, the following situation developed:

(a) There is one extra half-sheet of atoms above the slip plane, causing the upper half of the crystal to be in compression and the lower in tension;

(b) a residual shear stress exists along the slip plane, its value decreasing with the distance from the dislocation line (Fig. 7.10);

(c) it is evident that the stress field and distribution of atoms, caused by forcibly inserting the extra half-plane into the crystal above the slip plane, could also have been produced by removing a half-plane from below it. If the extra half-plane were added in the lower half, the same situation would have arisen except that the signs would be reversed.

According to an accepted convention, the edge dislocation is considered positive when the extra plane is above the slip plane and is denoted by the symbol ⊥; if the extra plane extends from the slip plane downwards, the dislocation is negative, symbol ⊤ (Fig. 7.12).

Some dislocations develop when crystals grow from the melt while others are created by plastic deformation. Practically all mono- and polycrystalline materials contain numerous dislocations. Their formation is suppressed, however, in certain minute single crystals of only about 1 micron diameter. These crystals, known as *whiskers*, exhibit extremely high strength properties as shown in Table 8.1.

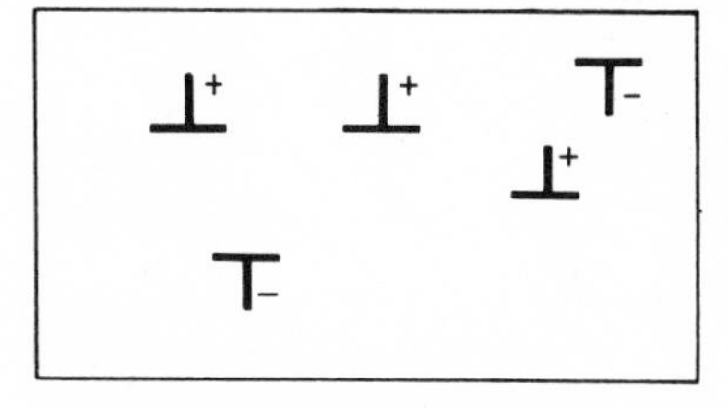

Fig. 7.12 Convention regarding sign of edge dislocations.

TABLE 8.1*

STRENGTH OF WHISKERS

Metal	*Diameter (microns)*	σ_{max} kg/mm^2	τ_{max} kg/mm^2	$\frac{\tau_{max}}{G}$	$\frac{\tau_{theoretical}}{G}$
Fe	1.60	1,340	364	0.060	0.033–0.19
Cu	1.25	300	82	0.022	0.033–0.13
Ag	3.80	176	72	0.031	0.033–0.13

* After S. S. Brenner, *Journal of Applied Phys.*, **27** 1484, (1956).

Screw Dislocations in Crystals

In order to produce a screw dislocation in a crystal, it is again slit to line *z–z*, but instead of displacing the two halves in the direction normal to *z–z*, the displacement *b* is parallel to it (Fig. 7.9b). This results in the configuration shown in Fig. 7.13. The structure resembles a spiral ramp or staircase rising helically around the vertical dislocation line *z–z*. In view of the discontinuous nature of the crystal structure and the voids it inherently contains, the mathematical dilemma of the infinite shear strains at *z–z* does not arise here.

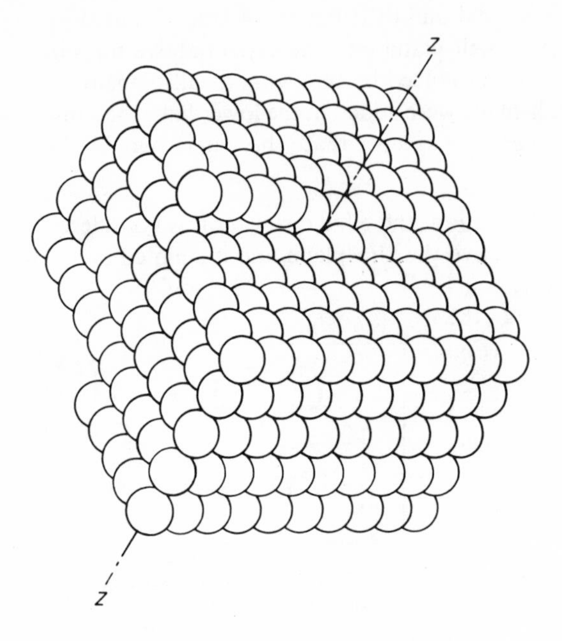

Fig. 7.13 Arrangement of atoms in a screw dislocation.

The Burgers Vector and Energy of Dislocations

A dislocation of either type can be described by a vector, **b**, showing the direction and magnitude of displacement undergone by the slipped portion of the crystal. This vector, introduced by Burgers, is obtained as follows: In a crystal containing a dislocation, an atom or point A is selected at some distance from the disturbed area (Fig. 7.14). One then moves from this point y_1 atomic spacings in the $+y$ direction, then x_1 spacings along the x-axis, then $-y_1$ and $-x_1$ units in the directions opposite to the former, thus forming a loop, known as a *Burgers circuit*. In a perfect material (Diagram a), the circuit closes, ending at the starting point A.

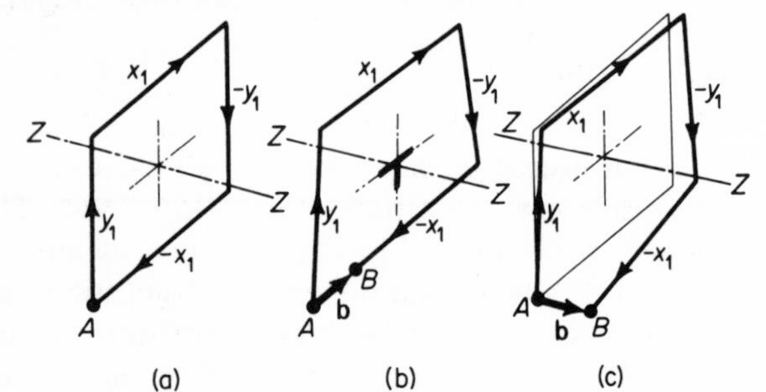

Fig. 7.14 Burgers circuit and Burgers vector, **b**:
(a) In perfect lattice.
(b) Around edge dislocation.
(c) Around screw dislocation.

The presence of an edge dislocation at z–z (diagram b) will result in a failure of the loop to close, the missing link AB being exactly **b**. A screw dislocation will cause a similar closure deficiency, except that now (diagram c) $AB = \mathbf{b}$ is parallel to z–z instead of being normal to it as in (b). In either case the vector $AB = \mathbf{b}$ is known as a *Burgers vector*, and it is spoken of as the *dislocation strength.* The vector connects two successive positions of atom equilibrium and is independent of the path taken around a particular dislocation line. If, as is very often the case, b represents one lattice spacing then vector **b** characterizes a *dislocation of unit strength.*

The relative directions of the dislocation line and Burgers vector give an easy means for distinguishing edge and screw dislocations. If the vector and line are perpendicular, the dislocation is an edge type; if these are parallel, it is a screw type. Note that the edge dislocation is defined by a line and a perpendicular direction which fixes it in a plane. The screw, on the other hand, is completely defined by just two parallel directions; i.e., the dislocation line and its Burgers vector. We shall see that edge dislocations can move only when both the dislocation line and its Burgers vector lie in one slip plane, while a screw dislocation has much greater freedom. The latter may, for example, glide on one slip plane, then shift to some intersecting slip plane, and continue to glide with equal ease. This should not be surprising since Eq. (7.7) showed the stresses about a screw dislocation to have rotational symmetry.

In dealing with dislocations and Burgers vectors, the discussion was purposely restricted to simple cubic structures, with the imperfections extending along the sides of the unit cell, in order to keep the concept as simple as possible. Of course, dislocations occur in directions other than this one. Hence, to have a notation that represents the direction as well as the strength of Burgers vectors, the Miller directional indices are used in conjunction with the length of the vector. Remembering that, because Miller's indices are proportional to their direction cosines, their components can be used directly to add or subtract various Burgers vectors. For example, vectors along a cube edge, a face diagonal, a body diagonal, and a diagonal stretching across two coplanar cube faces, which are identified in Fig. 7.15, are written as $\mathbf{b}_1 = a[100]$, $\mathbf{b}_2 = a[110]$, $\mathbf{b}_3 = a[111]$, and $\mathbf{b}_4 = a[021]$, respectively. The symbol "a", denotes that the component vectors extend in the x, y, and z directions by the number of unit lattice spacings indicated by the consecutive numerals (indices) in the brackets (see Fig. 4.11a). It is also possible to define B-vectors in fractions of "a," for instance, $a/2$ [024] means that the terminal point is reached by making two $a/2$ motions along "y," followed by four along the "z" axis. This is equivalent to a [012]. For example, to add $\mathbf{b}_2$ to $\mathbf{b}_4$, each of the corresponding components are added separately; i.e., $\mathbf{b}_2 + \mathbf{b}_4 = a[110] + a[021] = a[131]$. Of course, in adding or subtracting the components, identical unit vectors must be used;

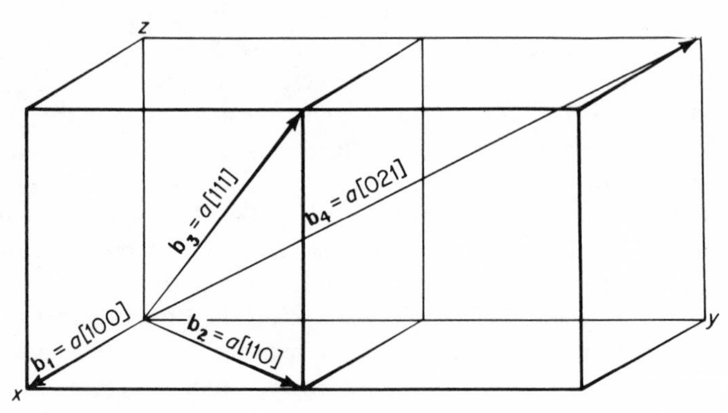

Fig. 7.15 Examples of Burgers vectors for four different directions in a cube lattice.

e.g., to add $a[110]$ to $a/2[131]$, the first vector is expressed as $a/2[220]$. The sum of the two then is $a/2[351]$.

Energy of Dislocations

By distorting and thereby straining the perfect lattice, dislocations increase the internal energy of crystals. The strain energy introduced by the dislocation can be calculated. The work used to form it consisted of moving a part of the crystal against a linearly increasing shear resistance acting in the slip plane. This elastic energy of formation is all stored in the crystal and is equal to

$$\frac{1}{2}\int_{r_o}^{r_1}(\tau_{r\theta}b)\,dr$$

where $\tau_{r\theta}$ comes from Eq. (7.6) and $b = |\mathbf{b}|$. On integrating for $\theta = 0$, the energy of an edge dislocation is found to be

$$W = \frac{G}{4\pi(1-\nu)}\,b^2\left(ln\,\frac{r_1}{r_o} - 1\right) \tag{7.7a}$$

and that of a screw,

$$W = \frac{G}{4\pi}\,b^2\left(ln\,\frac{r_1}{r_o} - 1\right) \tag{7.7b}$$

The meanings of r_o and r_1 must now be clarified. Inasmuch as the strains (and stresses) in a Volterra dislocation approach infinity near the dislocation line (see p. 164), it follows that $W \rightarrow \infty$ when $r_o \rightarrow 0$. This is avoided in the continuum model by providing a hole along the dislocation line (Fig. 7.9) so that r_o = the hole radius. In crystals, because of the available open spaces and the compressibility of atoms, infinite strains do not arise even at the dislocation center. Values—such as 10^{-7} centimeter (= 10 Å) for r_o and 1 cm. for r_1—have been used in these calculations by Cottrell.

Thus, the energy of a dislocation of strength **b** is found to be proportional to b^2, whereas the energy of one twice this strength is proportional to $4b^2$. Since the stability increases as the strain energy decreases, a dislocation of more than unit strength will tend to dissociate into two or more dislocations of lesser strength. For instance, a dislocation of 2**b** strength will probably split up into two, each of **b** strength because

$$b^2 + b^2 = 2b^2 < (2b)^2. \tag{7.8}$$

Resolution of Burgers Vector into Edge and Screw Components

A dislocation is produced whenever slip occurs over only part of a crystal. The slipped areas (shown crosshatched in Fig. 7.16) are all separated from the undeformed sections of the crystal by a dislocation line (shown in detail in Figs. 7.11 and 7.13). If this line is straight and normal to its Burgers vector, the dislocation is an edge type as shown diagrammatically in (a) and (b). Again, if the dislocation line were straight and had a Burgers vector parallel with it, it would be a pure screw. In many cases, however, the dislocation line is curved so that, over most of its length, it is neither perpendicular nor parallel to its Burgers vector (Fig. 7.16c). Indeed, in the most general case, the slipped region with Burgers vector **b** may be wholly surrounded by undeformed material [diagram (d)]. A condition of this type is very common and is actually basic to the mechanism by which slip occurs; hence, the nature of the boundaries between slipped and unslipped areas will now be analyzed.

Fig. 7.16 Straight and curved dislocation lines as boundaries between slipped and unslipped portions of crystals.

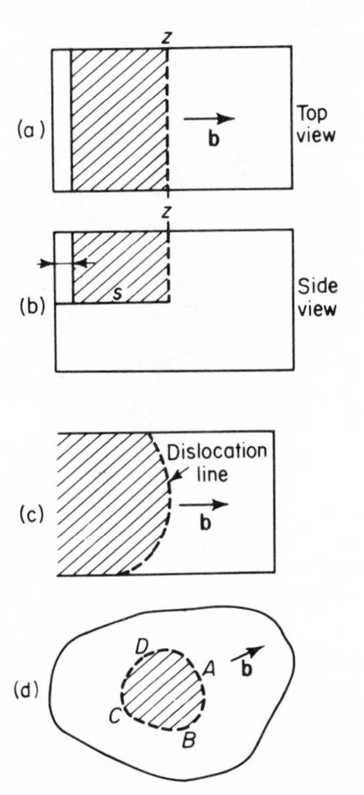

Let us first consider a somewhat simplified case (Fig. 7.17) where a prism, $ABCD$ $A'B'C'D'$, has slipped on plane s–s by one atomic spacing in the direction of **b**, thus forming a positive edge dislocation, e, along $B'C'$ (because an extra plane of atoms is now crowded in *above* $B'C'$). At the same time the half-sheet above $A'D'$ moved to the right, leaving a vacant "slot" in its wake and thus producing a negative dislocation. Details of these displacements are seen in Fig. 7.17b where the bounding planes of the slipped region are 15 and 5. To form the two dislocations, the portions of these planes above slip plane

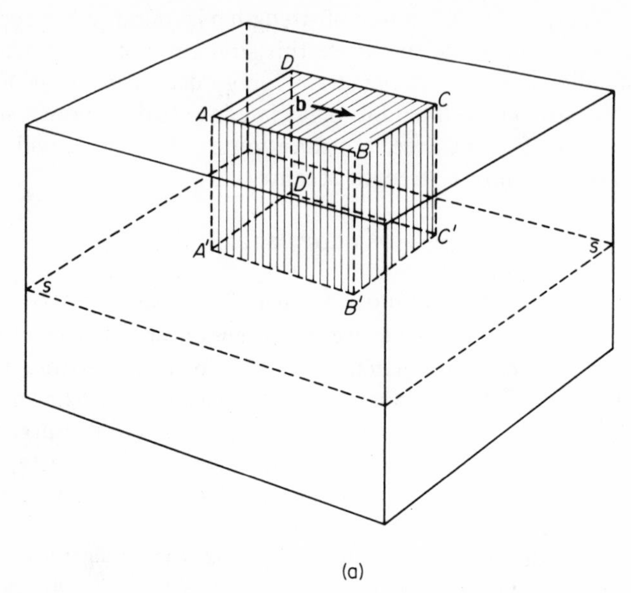

Fig. 7.17 Dislocation structure surrounding block slipped by Burgers vector **b**. (a) Three-dimensional view. (b) Details of side view showing edge dislocation. (c) Details of top view on slip plane showing screw dislocation.

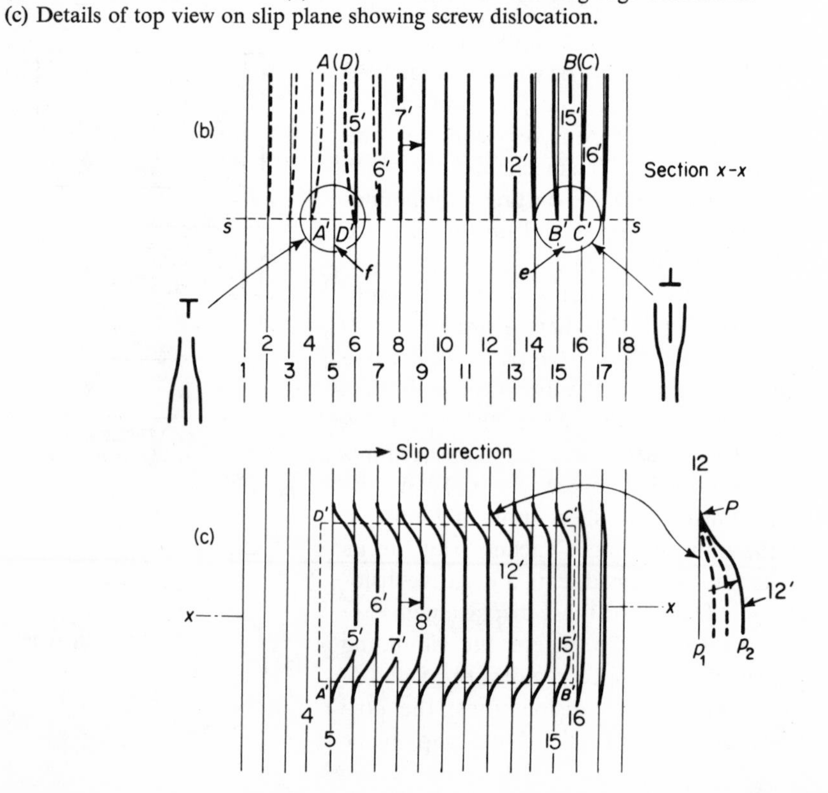

s-s assume new positions 15′ and 5′, the dislocations being e and f. When the shear force is removed, the displaced atom planes will assume new equilibrium positions (shown dashed at f). The configuration of the crystal along $A'D'$ and $B'C'$ is quite simple since it consists of pure edge dislocations.

A different condition has arisen along $A'B'$ and $C'D'$, Fig. 7.17c. The fringes of the shifted planes; i.e., their sections 5′–15′ at points P perform a hinge or scissor-like motion around the atomic verticals along planes $ABB'A'$ and $CDD'C'$. For example, when the slipping section 12′ of plane 12 moves in the **b** direction, its fringe on $C'D'$ is pivoted at P (see inset) and moves from the original P–p_1 to the final P–p_2 configuration. With each plane following the same course, we get a screw dislocation with $C'D'$ as its axis. The same applies, of course, to the situation along $A'B'$. One of these screws is right-handed and the other left. In conclusion, it is seen that the boundary of a rectangular area which slipped in the direction of one of its sides consists of two edge and two screw dislocations. The former are normal and the latter parallel to the Burgers vector **b**.

In the above discussion the boundary of an enclosed slipped area was described. This can expand in either its edge or screw directions. If edge $B'C'$ moves forward, the slipped area increases in the direction of Burgers vector **b** while maintaining its width. Each increment of forward motion increases the length of the two boundary screws by adding consecutive threads. Increasing the width of the slipped area involves forward movements of the atoms outside the area bounded by $A'B'$ and $C'D'$. As successive atoms on any one numbered plane undergo the "hinging" action shown at P, the fringe moves to expand the slipped area. Thus, forward movement of atoms at the screw dislocation line causes the line to move perpendicular to the direction of atom motion. Just as the moving edge leaves a screw dislocation in its wake, the sidewise gliding screw axis extends the edge dislocation at its terminals.

In the more general case shown in Fig. 7.16d, the boundary will consist of pure edge dislocations at A and C and of screw dislocations at B and D. The intermediate section will represent "mixed" dislocations of a partly edge and partly screw character. The relative contributions of both can be determined by resolving the Burgers vector at each point in directions parallel and normal to the boundary. Obviously, the edge component will be **b** at A and C, zero at B and D, and will vary between these extremes in the intermediate positions. Read constructed a schematic drawing of the layers of atoms immediately above and below the slip plane s–s for a curved dislocation (shown in Fig. 7.18).

Stress Required to Move a Dislocation

The displacement required to move a dislocation is indeed small. Looking at the "extra" plane 4′ (Fig. 7.19), it is seen that a very small movement of its lowest atoms to the right will suffice to line it up with the highest atoms of

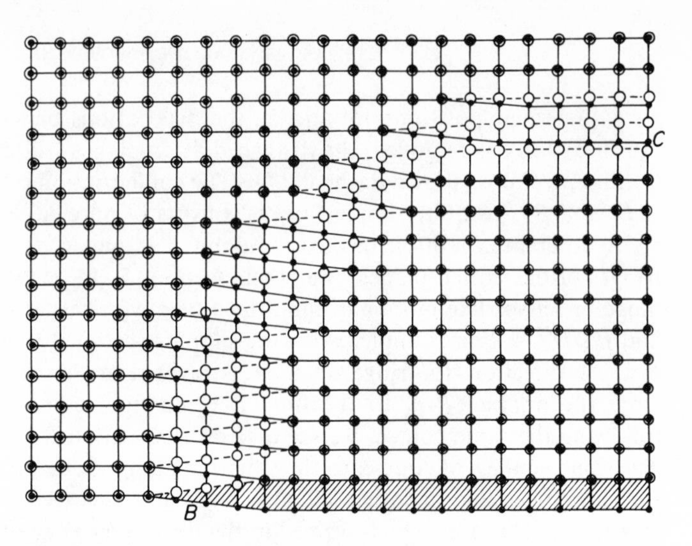

Fig. 7.18 Arrangement of atoms at a curved dislocation. Letters *B* and *C* refer to Fig. 7.16d. Closed circles represent the atomic plane above the slip plane; open circles represent the atoms just below. From W. T. Read, Jr., (R7.5).

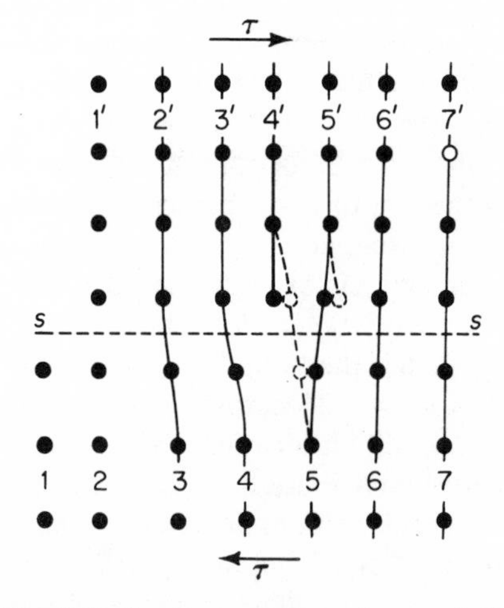

Fig. 7.19 Small movement of atom half-plane (4′) shifts the dislocation to 5′.

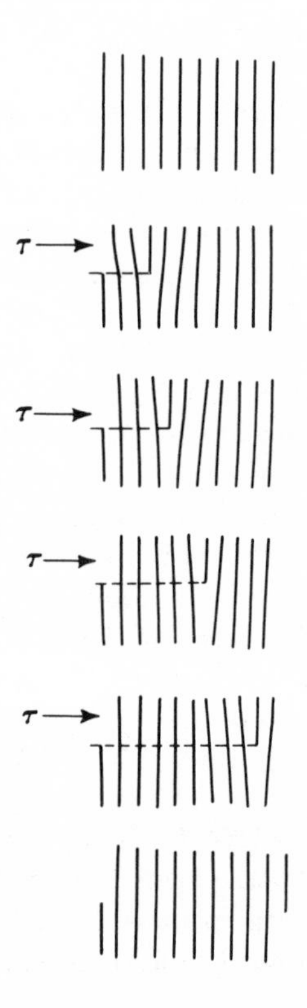

Fig. 7.20 Motion of a single dislocation across lattice causing one unit of slip.

half-plane 5. This will cause half-planes 4′ and 5 to form a single atom sheet, while severing half-plane 5′ from its mate below slip plane *s–s*. Hence, the (edge) dislocation line moved from its first position between 4 and 5 to a new one between 5 and 6. By repeating this procedure again and again, the dislocation will traverse the entire crystal leaving it on an outside surface. Several stages of such motion are shown diagrammatically in Fig. 7.20. When the dislocation leaves the crystal, a step is produced on the end face. As a result, the upper half of the crystal moved or slipped over the lower half by one atomic spacing, thereby producing a unit amount of permanent or plastic deformation.

The forces required to move the dislocation represent only a small fraction of those needed to move similar atoms from one equilibrium position to the next in a perfect lattice. Calculations by R. Peierls (1940) and F. R. N. Nabarro (1947) gave a shear stress to move the dislocation of the order of

$$\tau = \frac{2G}{1-\nu} \exp\left\{\frac{-2\pi h}{a(1-\nu)}\right\} \tag{7.9}$$

where h = distance between planes,
and a = distance between atoms in the slip direction.

The value of τ is strongly affected by Poisson's ratio and $h{:}a$. For slip on ((001)) planes in a cubic lattice, when $h/a = 1$, and $\nu = 0.3$, $\tau = 3.6 \times 10^{-4}\,G$. For slip on ((110)) in close-packed crystals, $h/a = \sqrt{2}$, and the appropriate τ is then about a hundred times less. On the contrary, when $h < a$ (closely spaced but loosely packed planes), τ is very high and slip on them is highly improbable. Thus expression (7.9) gives a formal justification for slip occurring on close-packed planes.

Extensive Slip Caused by Single Dislocation. Dislocation Loops

The mechanism just discussed explains how a dislocation facilitates a *unit* slip over a distance of a single atomic spacing. After performing this minute function, the dislocation leaves the crystal. In reality, very extensive plastic deformations are observed, and it is unthinkable that a sufficient number of dislocations exist in a crystal to account for this on the basis of one dislocation per one slip unit. It was necessary, therefore, to conceive of a mechanism whereby large amounts of slip of the order of thousands of atomic spacings can be generated with the aid of a single dislocation. This mechanism is provided by the *Frank-Read source or generator* shown schematically in Fig. 7.21 and by the photograph in Fig. 7.22.

It requires an edge dislocation to extend only part way through the crystal. For instance, the "inserted" atom sheet in diagram 7.21a is $BB'zz'$ which extends halfway through the crystal. The dislocation lines, Bz and $B'z'$, must be unable to slip, either because they are "sessile" (see p. 181) or are

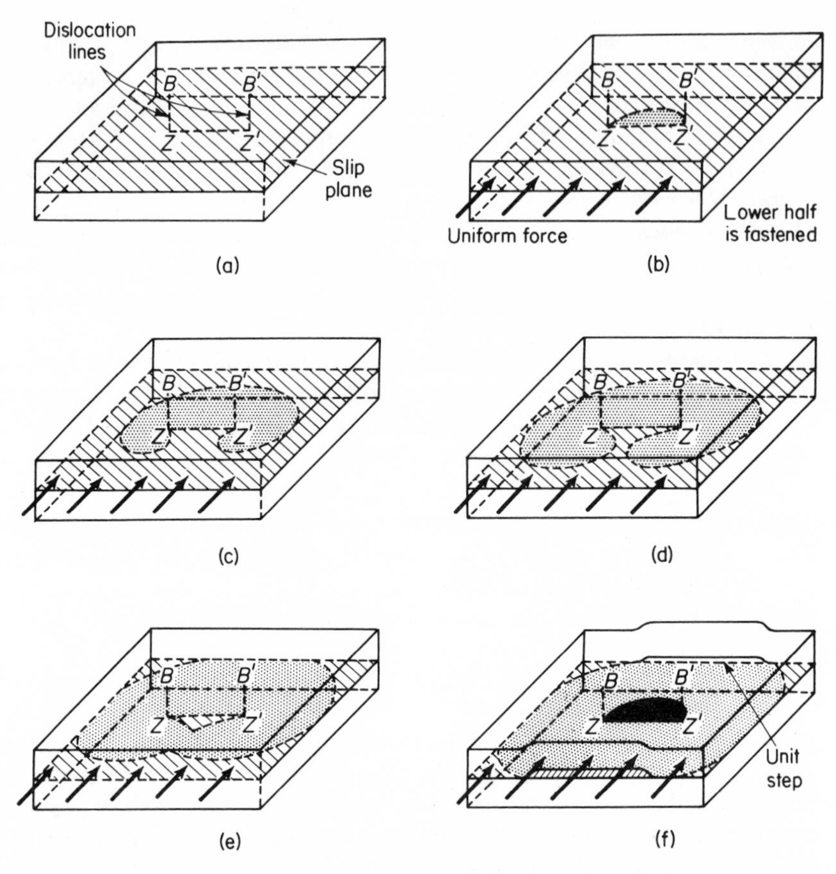

Fig. 7.21 Frank-Read mechanism for extensive slip from a single source. From A. G. Guy, *Elements of Physical Metallurgy*, 2nd ed., Addison-Wesley Publishing Co., 1960.

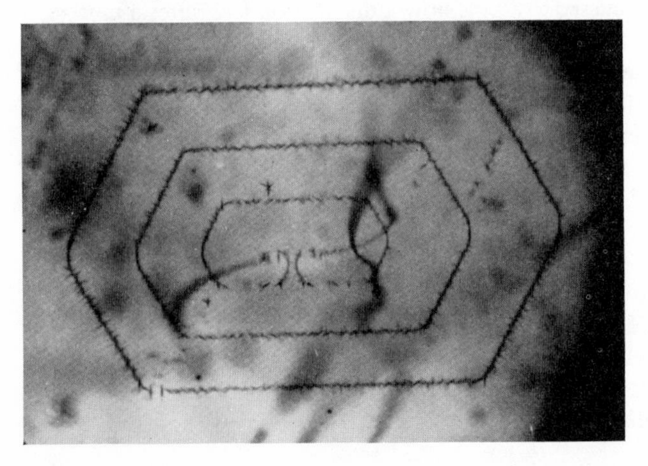

Fig. 7.22 Dislocation loops spreading out from a Frank-Read source in a silicon crystal. From W. C. Dash, in *Dislocations and Mechanical Properties of Crystals*, J. Wiley and Sons, Inc., 1957.

pinned down by the presence of foreign atoms. When a force is now applied to the upper half of the crystal, as shown in (b), the dislocation $zz'BB'$ bulges out and begins to move forward, gradually traversing the entire slip plane by a sweeping motion around the stationary anchors Bz and $B'z'$. As a result we find that the material above the slip line has moved one atomic spacing forward, by forming a protrusion on its front, and leaving a step in its wake. These are shown partly formed in diagram (f), but they eventually extend across the entire width.

This process caused, in effect, the upper half of the crystal, together with the anchor lines and the dislocation which did not vanish, to move forward by one atomic spacing with respect to the unslipped lower part. If the force is still applied, this process will repeat itself [indicated by black bulge in (f)] until the Bz and $B'z'$ anchors reach the far edge of the lower crystal portion. The dislocation then disappears and further slip ceases.

The parts of the sweeping dislocation that are closest and farthest from the reader are of the edge type, whereas its sides represent screw components. The edge and screw are both moving in the direction of the Burgers vector, but the former moves perpendicularly to and the latter parallel with the dislocation line.

In this way a limited number of anchored dislocations produce considerable slip. It is not even essential for the slip generating dislocations to be pinned on two sides; one anchored end will suffice for this purpose, with the free end of the dislocation sweeping the slip surface like the hand of a clock.

When a generator operates according to the scheme in Fig. 7.21, and the area to be swept is large compared with the length of the pinned section BB', the "tail ends" of the expanding bulge come so close to one another that the two screw elements in it which are of opposite signs (i.e., left- and right-handed) are mutually attracted and annihilate themselves, leaving a completely closed dislocation ring (Fig. 7.21d and e). This is a natural course of events since the disappearance of the two screws decreases the total strain energy. The closed ring keeps expanding until it disappears on leaving the crystal. It is apparent that several such rings can exist at the same time and experimental evidence to that effect was produced by Dash (Fig. 7.22).

Multiplication of Dislocations by Transfer of Slip to Parallel Planes (Cross Slip)

When a screw segment of an expanding dislocation loop on a plane encounters an obstacle in its path, it may readily deflect onto an oblique plane. This condition is illustrated in Fig. 7.23a, where a screw segment of a *F–R* source was deflected from plane A to B and then again to plane C. Upon traversing along B, the screw left two edge dislocations "e" (as explained on page 173). However, as long as the original loop on A, the edges e on B, and

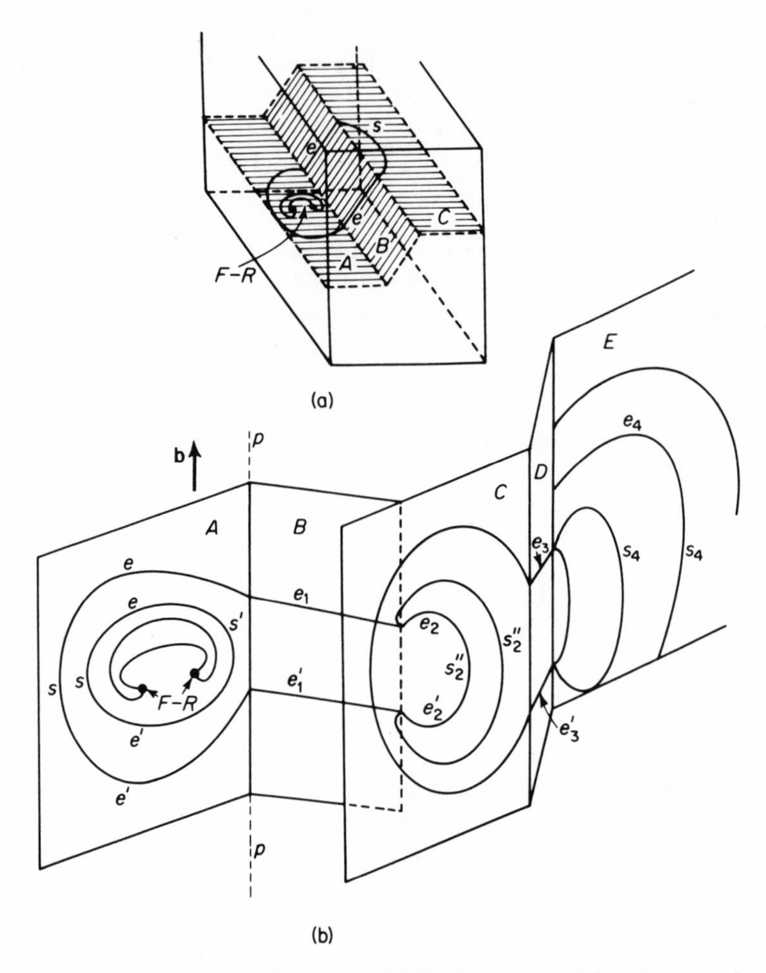

Fig. 7.23 Bypassing obstacles and multiplication of Frank-Read sources by double cross-slip as suggested by Orowan and Koehler: (a) General layout. (b) Details. *F-R*—source, *e*—edge, *s*—screw dislocation.

the screw segment *s* on *C*, all travel at the same speed, the *Z*-shaped loop behaves quite normally.

On the other hand, if the edge sections e_1–e_1' in Fig. 7.23b were stationary, or moving slower than the rest of the loop, their intersections with plane *C* would become a generator for new loops on *C*. The same can be repeated again by a part of screws *s* being diverted to *E* via *D*. These sources on planes *C* and *E* are similar to Frank–Read generators, except that it is a screw rather than an edge dislocation that is pinned.

This mechanism is referred to as multiplication by cross-slip (or *double cross-slip*) and probably enables dislocations to bypass particles of a precipitate in the matrix.*

Partial Dislocations, Stacking Faults, Glissile and Sessile Dislocations

So far it has been implied that slip always involves a movement of atoms from their initial sites to equivalent neighboring sites so that the crystal structure looks the same before and after slip occurred. Because two dislocations with short Burgers vectors may introduce less strain energy into the lattice than one dislocation whose length is the vector sum of the other two (see Eq. 7.8), slip in some cases does not leave an undisturbed lattice. This can occur in the f.c.c. system where atoms slipping on ((111)) may find new equilibrium positions different from those which correspond to the original lattice arrangement.

In Chapter 4, the ((111)) plane was represented by contacting spheres, and the complete f.c.c. system was built up by placing identical layers of such spheres, one on top of the other, so that the relative positions of atom centers on successive layers were always given by identical vectors (Fig. 7.24a). This packing was indicated by the symbol △△△ . . . , and was contrasted with the h.c.p. structure in which odd layers are stacked one way and even layers another (Fig. 7.24b), leading to a packing of △▽△▽△ . . . . Slip in the f.c.c. system occurs on the ((111)) planes in the [[110]] direction, as shown by a single atom on the second plane moving from position I to II in Fig. 7.24c and d. Specifically, the depicted motion is in the $[\bar{1}10]$ direction and could be produced by an edge dislocation formed by inserting an extra plane of $(\bar{1}10)$ atoms above or below the plane of the page. A corner of the inserted plane is shaded in 7.24d. A Burgers circuit around the dislocation line, z_1–z_1, identical with the z axis, yields a Burgers vector equal in length to the distance between I and II. This length is only one half of the unit cell diagonal so that its Burgers vector, $\mathbf{b}_1 = \frac{1}{2}a[\bar{1}10]$. If all the second layer atoms travel from position I to II (diagram c), the lattice is identical before and after slip. However, this same total movement would require less energy if made in the two steps from position I to IV to II. In the vector triangle below Fig. 7.24c, these paths are compared. In the I–IV–II case, the Burgers vectors are $\mathbf{b}_2 = \frac{1}{6}a[\bar{1}2\bar{1}]$ and $\mathbf{b}_3 = \frac{1}{6}a[\bar{2}11]$, which reduces the strain energy of the lattice from being proportional to $a^2/2$ to proportional to $a^2/3$. Further, $\mathbf{b}_2$ and $\mathbf{b}_3$ are of the same sign, and, as shown in a later section, this caused them to repel each other, resulting in an extended area of the lattice between z_1–z_1 and z_2–z_2 (Fig. 7.24e), in which the slipped atoms are stacked in the Y' rather than in the Y positions. On examining the stacking of the layers

* D. Thomas and J. Nutting, *Jl. Inst. Metals*, **86**, 7 (1957). (Appendix by P. B. Hirsch.)

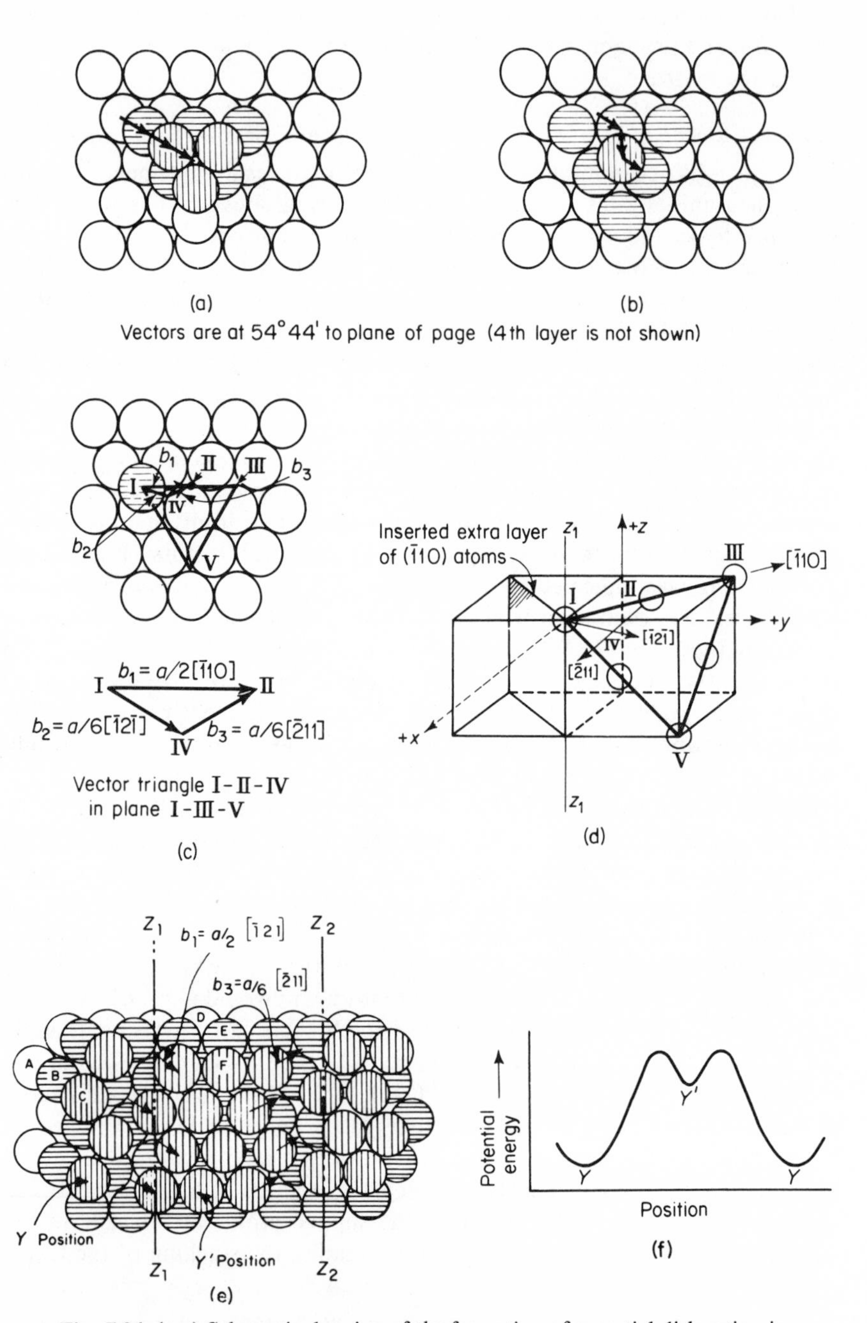

Fig. 7.24 (a–e) Schematic drawing of the formation of a partial dislocation in a f.c.c. lattice. (f) Relative energy of atoms in the Y and Y' positions shown in (e).

of atoms in 7.24e, one finds a f.c.c. lattice outside the area bound by z_1–z_1 and z_2–z_2. In going from the bottom layer of atoms,—e.g., atom A to atom B in the next layer and then to atom C (in a Y position in the third layer)—the translation vectors are all identical producing the △△△△ . . . structure shown in 7.24a. Within the slipped area, however, in going from atom D to E to F, the vector direction changes on alternating layers, indicating the local existence of some △▽△▽ . . . or h.c.p. structure.

Although the "normal" Y sites correspond to the minimum potential energy of the atoms, sites Y' are also positions of relative equilibrium (Fig. 7.24f). This small region, in which the f.c.c. structure changed to h.c.p., is termed a *stacking fault*.

Dislocations, such as z_1–z_1 and z_2–z_2 that do not produce a full lattice translation and, as a consequence, result in a stacking fault are termed *partial dislocations*. Specifically, these are *Shockley partial dislocations* formed by making a cut parallel to a ((111)) plane, moving the material on one side of the cut by $\mathbf{b}_2$, or $\mathbf{b}_3$ and healing the cut to again form an uninterrupted crystal. Any dislocation, full or partial (including the Shockley type), whose Burgers vector is in the slip plane, and hence is able to move, is called *glissile*. Other types of dislocations, whose Burgers vectors are not in the slip plane, can also be introduced into the lattice. These dislocations cannot move and hence are termed *sessile*. For example, if a part of one of the close-packed planes in Fig. 7.25 is removed, and the sides of the "slot" are brought together again, partial dislocations are formed at the slot perimeter. The Burgers vector, in this case, is perpendicular to the close-packed planes so that slip, according to Eq. (7.9), is extremely improbable. For this reason the dislocation remains sessile.

Fig. 7.25 A Sessile dislocation is formed when part of a layer of close packed atoms is removed from the lattice as shown in (a). When the lattice collapses (b) a dislocation is formed around the slot perimeter with Burgers vector **b** perpendicular to the close packed plane. After F. R. N. Nabarro.

STRAIN- (OR WORK-) HARDENING

The description of plastic flow in crystals based on the presence of dislocations and dislocation generators, as presented up to this point, does not account for strain-hardening; i.e., the gradual increase in resistance to deformation with increasing plastic strain. This resistance to flow should be constant as long as dislocations or their sources are present in the crystal. When they are all used up, however, flow or yield stress should abruptly increase to its calculated value (7.5) which is about $G/6 \approx 0.07\, E$. Actually, the stress–strain curve climbs steadily in the plastic range, and the various ways by which this can be accounted for in terms of dislocation movements will now be described.

A. *Interaction between Dislocations on Parallel Slip Planes*

A dislocation produces in its surroundings a stress field which, in edge dislocations, is determined by Eqs. (7.6b). The fields of neighboring dislocations interact, and some of these interactions are easily visualized. For example, a positive and negative edge type on a common slip plane will tend to attract and annihilate each other. Each of these was initially surrounded by stress fields of the same magnitude, as given by (7.6), but of opposite signs. Their annihilation, in reducing these stresses to zero, reduces the strain energy of the lattice.

There is a great variety of other types of interactions. A particularly important case is one of two dislocations on parallel planes (Fig. 7.26). Calculations which will not be reproduced show that the force per unit length of the dislocation line, P_x, exerted by dislocation A on B, is $\tau_{xy}b$. The shear stress caused by A at point B is τ_{xy}, and $\mathbf{b}$ is the amount of relative shift involved in forming the dislocation. Force P_x acts in the slip direction x and it determines the interactions between the two dislocations. If A and B are of the same sign, they repel each other along x when $x > y$, but attract when $x < y$. For $x = y$ and $x = 0$, P_x is zero, but only for $x = 0$ is the equilibrium stable. Because of this, dislocations of the same sign tend to align themselves as shown in Fig. 7.27. Positioning of this type, although stable, does require some additional energy to be introduced into the lattice so that aligning of the dislocations occurs in crystals after plastic deformation (to introduce dislocations) has been followed by heating (to supply the energy for the dislocations to move until alignment is reached). The two parts of the crystal (in Fig. 7.27) are divided by the aligned dislocations so that their orientations differ from a fraction to a few degrees. Boundaries of this type are called *tilt boundaries* and they divide the individual grains into *subgrains*. Because the aligned dislocations are in a position of minimum energy, it is difficult to move them. Hence, they are effective barriers to the free motion of gliding dislocations, thereby causing an increase in the crystal's yield strength. It is also possible to form a sub-boundary that consists of

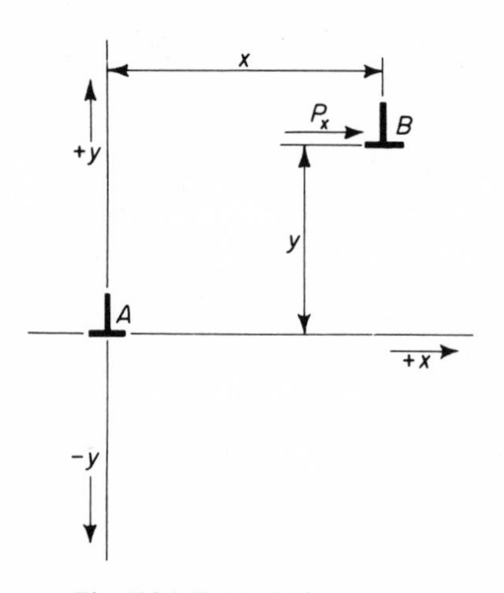

Fig. 7.26 Force (P_x) exerted by dislocation A on B.

an array of screw dislocations (see Fig. 7.28). These are known as *twist boundaries*.

The process of forming subgrains is called *polygonization.* The latter occurs rapidly during *recovery* which proceeds at temperatures below that needed for recrystallization.

Fig. 7.27 A tilt boundary is formed in a single grain when edge dislocations align themselves one on top of the other. From W. T. Read, Jr., (R7.5).

Fig. 7.28 Twist boundaries formed by a network of screw dislocations. Schematic drawing. From W. T. Read, Jr., (R7.5).

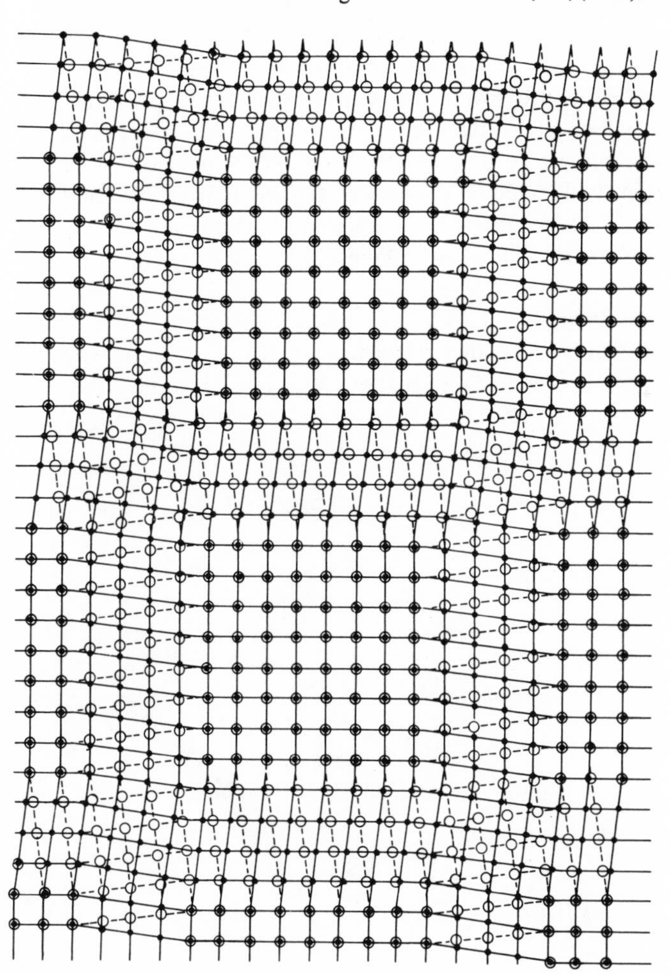

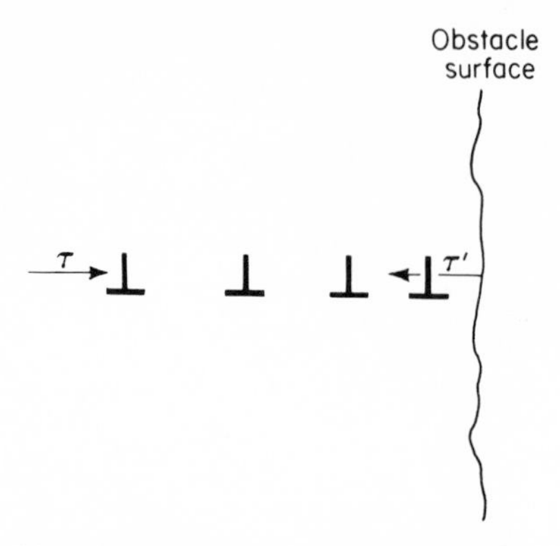

Fig. 7.29 Pile-up of edge dislocation against an obstacle (grain boundary, subboundary, etc.).

If an edge dislocation moves towards one which is stationary (e.g., being an element of a sub-boundary) on the same or nearby slip plane, and both dislocations are of the same sign, the repulsive force that builds up when the two approach may balance the force which causes the first dislocation to move. The dislocation will then come to standstill and so will others following it, thereby creating a *dislocation pileup* (Fig. 7.29). The increasing spacings result because the stress fields around each dislocation are additive, so they amplify the repelling forces.

If these dislocations originate from one source, the "back pressure" may grow to a point at which the source will cease to operate. The applied stress will now have to increase to a level which will set in motion other sources less advantageously located with respect to the maximum shear stress direction, and so on. These additional dislocation sources may operate on planes inclined to the already "blocked" planes.

B. *Combining of Dislocations*

Because of rotation of the slip planes away from high τ directions, slip is transferred from the initial plane to other planes which come into positions

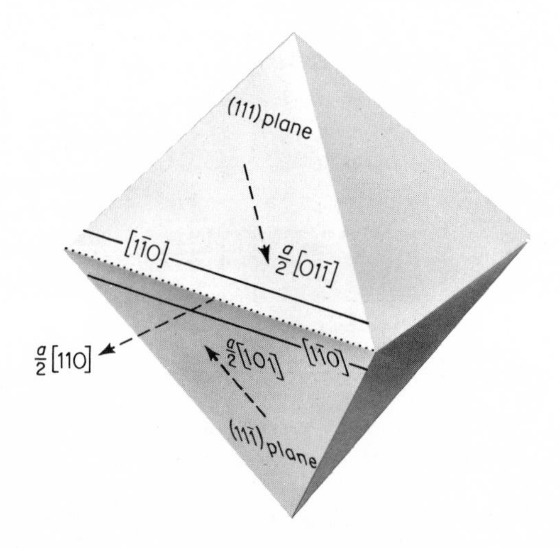

Fig. 7.30 Lomer reaction involving the intersection of $a/2$ $(01\bar{1})$ and $a/2$ (101) to form $a/2$ (110).

of high Schmid factors (see p. 157), producing conjugate slip. In the f.c.c. system, such slip results in rapid strain-hardening. This is believed to occur by what is known as a *Lomer reaction.* The concept first assumes slip on two intersecting ((111)) planes, such as the (111) and (11$\bar{1}$) in Fig. 7.30. The dislocation lines in this example are along the (1$\bar{1}$0) directions, and the Burgers vector of the top dislocation is $\frac{1}{2}a[01\bar{1}]$, while the one on the bottom face is $\frac{1}{2}a[101]$. These two dislocations run together and meet at the dotted [1$\bar{1}$0] edge of the octahedron in Fig. 7.30. They now combine to form a new dislocation, according to the reaction,

$$\tfrac{1}{2}a[01\bar{1}] + \tfrac{1}{2}a[101] \rightarrow \tfrac{1}{2}a[110] \qquad (7.10a)$$

We now have an edge dislocation since its Burgers vector and dislocation line are perpendicular. They lie in the (001) plane, however, which is not a slip plane, and, hence, it cannot move easily. According to Cottrell, this dislocation becomes even more stable by decomposing, according to the reaction,

$$\tfrac{1}{2}a[110] \rightarrow \tfrac{1}{6}a[11\bar{2}] + \tfrac{1}{6}a[112] + \tfrac{1}{6}a[110] \qquad (7.10b)$$

in which case it is known as a *Cottrell-Lomer extended dislocation* or "lock."

C. *Intersection of Dislocations*

When one dislocation cuts across another one lying on an intersecting plane, either one or both initially straight dislocation lines acquire a Z-like break or *jog*. To understand how jogs arise, consider a cubic structure containing two initially perpendicular edge dislocations (Fig. 7.31).

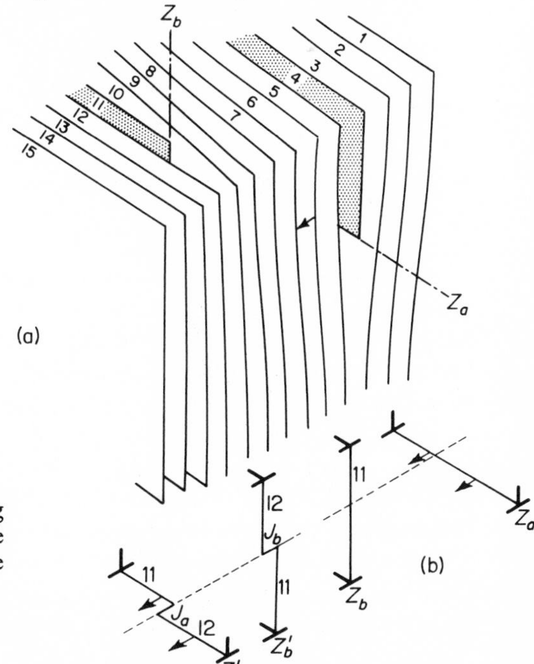

Fig. 7.31 Jogs J produced by intersecting dislocations Z_a and Z_b. The relative positions prior to and after crossing are shown in (b) and (c), respectively.

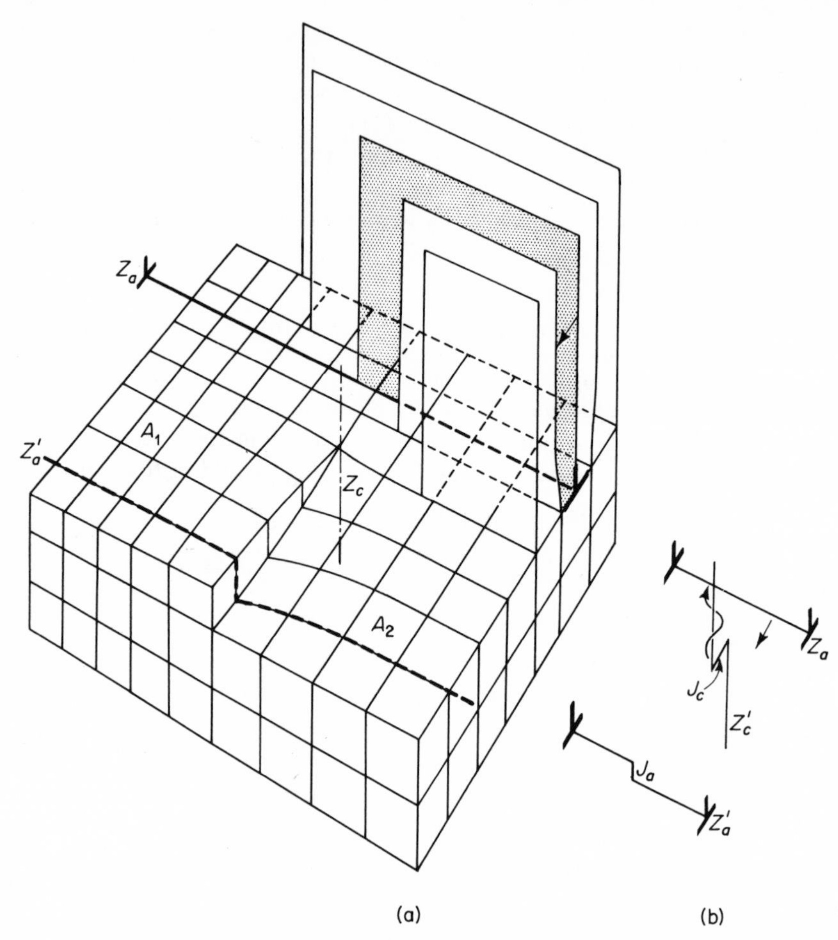

Fig. 7.32 Jogs J_a and J_c produced by edge dislocation Z_a cutting through screw Z_c.

Let dislocation Z_a (diagram a) glide in the direction of its Burgers vector (arrow) until it reaches plane 10. Planes 1 to 9 are now complete, and 10 becomes the half-plane above line Z_a. When the dislocation now moves one more lattice spacing forward, its left-hand part (looking from plane 15) will move to 11 while the right-hand end will move to plane 12. Hence, by crossing Z_b, dislocation Z_a has developed a jog, J_a, and now has the form Z_a' (see transition $Z_a \rightarrow Z_a'$ in diagrams b and c). Upon further movement, the "distorted" dislocation Z_a' retains the jog and can be described as 12-J_a-13, 13-J_a-14, etc. The length of the jog is evidently equal to the Burgers vector.

What about dislocation Z_b which was tacitly assumed to stand still? Let it again be assumed that Z_a is on plane 10. In the next motion of Z_a, the extra half-plane with which it is associated pushed the top half of plane 11 to position 12, while the bottom half of 11 is left stationary. Hence Z_b, which was initially all vertical, has formed a horizontal jog, J_b (diagram c), at the same time that J_a was formed.

It is also readily found that, had Z_a been in the same horizontal plane but at 90° to the direction shown in Fig. 7.31, its intersection with Z_b would have caused a jog on Z_b but not on itself. Note that the jog on one of the two crossing dislocations is always in the direction of the Burgers vector of the other one. In the case of Fig. 7.31, the jogs and their Burgers vectors have the same direction, so they are screw dislocations one lattice spacing long. There is no reason why jogs like these should interfere with the motion of edge dislocations of which they are a part.

The situation is different, however, if one or both of the intersecting dislocations is a screw. An edge dislocation, Z_a, is shown cutting through a screw, Z_c, in Fig. 7.32. When one part of the edge glides along A in the direction of the rising "ramp," A_1, and the other along the descending surface, A_2, line Z_a acquires a vertical jog and becomes Z'_a. By reasons identical with those that caused Z_b to acquire a jog in the preceding example, Z_c also obtains one. Again, since Z'_a and jog J_a are normal to their Burgers vector, they readily glide together. On the other hand, if Z_c were now to move, its jog (an edge dislocation one lattice spacing long) could slip easily only in the direction of its Burgers vector which is parallel to the screw axis. If $Z'c$ were to sweep laterally, Jc would be dragged in a direction normal to its vector, creating a line of vacancies or displacing a line of atoms from the normal to interstitial positions (see Section D). On the intersection of two screw dislocations, edge jogs are formed on both so that this same obstruction to movement occurs.

D. *Nonconservative Motion and Climb of Dislocations*

A motion of a dislocation in a direction other than that determined by the usual slip parameters (i.e., plane and direction of easy slip) is termed *nonconservative*. Consider, for example, Fig. 7.33, where dislocation Z_a at A is forced to move in a direction inclined under angle α to the slip plane.

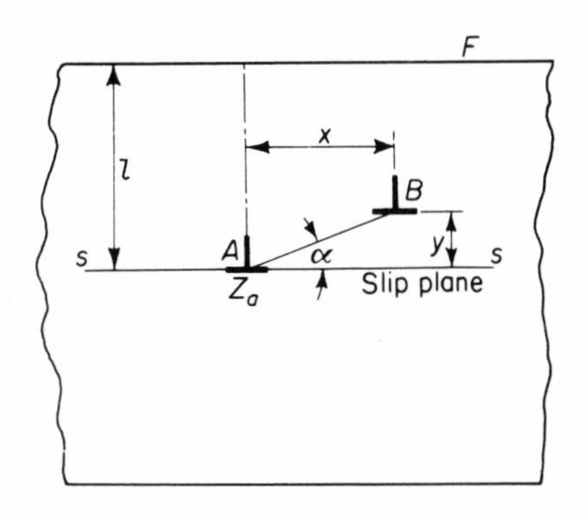

Fig. 7.33 Movement of a dislocation in a plane other than the slip plane, s, requires removal of atoms or acquisition of vacancies from the moving half-plane.

Upon reaching position B, the half-plane above Z_a is shorter by a length, $y = x \tan \alpha$, so that the now redundant atoms must either have been removed from the crystal or shifted into interstitial positions. If the dislocation moves so that the partial plane gets larger rather than shorter, an insufficient number of atoms is available to fill all the lattice sites, so vacancies are formed.

The movement of either vacancies or atoms from normal into interstitial positions, or vice versa, involves mass transfer. As such, it is a diffusion process. The diffusion rate D increases exponentially with temperature, according to

$$D = Ae^{-\Delta H/RT}. \tag{7.11}$$

A and ΔH are material constants. R is the gas constant (1.987 cal per gram-mol per degree Kelvin), T is the absolute temperature. ΔH, which has the dimension of cal per gram-mol, is the *activation energy* (for diffusion). It is a measure of resistance to atom movement and may be regarded as an energy barrier which the atoms must overcome before they can move into equilibrium positions. It is for this reason that jogs, such as J_c (Fig. 7.32), exert a drag on dislocation movement.

The mechanism illustrated in Fig. 7.33 enables an edge dislocation to move out of its original slip plane into a second slip plane. This process is called *climb* and becomes increasingly important as the temperature is raised. In view of the high diffusion rates at high temperatures, climb can occur under moderate or quite low stresses.

E. *Other Obstacles to Motion of Dislocations*

The dislocation interactions that were described in the previous sections explain how strain-hardening can occur in any crystal. No requirements were placed on its purity. Actually, all constructional metals contain impurities or alloys in various amounts. These foreign atoms interfere with free movement of the dislocations and contribute greatly to the strength of actual materials. However, the effect of the foreign atoms on dislocation movements depends on the manner in which they are distributed in the parent crystal, and this distribution is easily controlled. Hardening, both by the simple addition of foreign atoms and by heat treating, is discussed in textbooks dealing with physical metallurgy.

PHENOMENA ASSOCIATED WITH YIELDING

Dislocation Theory of the Discontinuous Yield Point

Some metals like copper, gold, or magnesium, show a gradual and almost imperceptible transition from the elastic to the plastic range. Others, the most common of which are iron-carbon alloys (steels), display a discontinuous jagged yield point (Fig. 7.34, curve 1). When a soft steel specimen is pulled so that its total elongation is within the length of the "platform," Y, it will be found that only a portion of the material deforms permanently while the

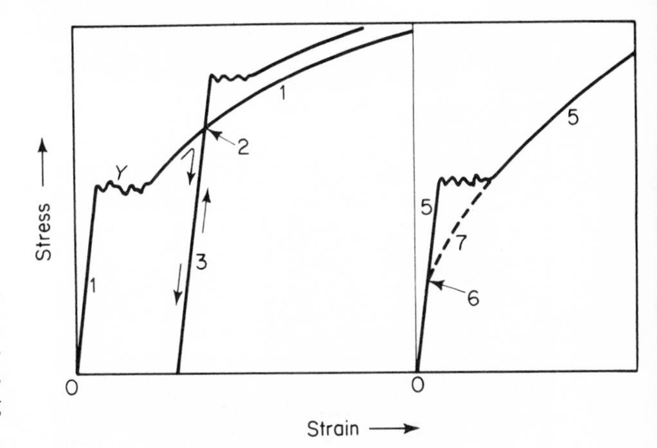

Fig. 7.34 Discontinuous yielding of iron containing carbon or nitrogen. Specimen 1 unloaded at 2, aged and retested to produce curve 3. Discontinuous yield in 5 eliminated by alloying with Ti or Cb, curve 0-6-7-5.

rest remains elastic. The elongation measured is actually a sum of elastic plus plastic deformation in one portion of the specimen and a purely elastic one in the remainder. If the tension test is performed on a soft steel sheet, the localized plastic deformation leads to the formation of peculiar surface patterns known as *Lüders lines* or *stretcher strains* (Fig. 7.35). When the total deformation exceeds Y, and reaches the continuously rising right portion of curve 1 in Fig. 7.34, the stretcher strains vanish.

The discontinuous yield of iron is caused by the atoms of carbon and nitrogen it contains. If these are completely eliminated or are tied up chemically as stable carbides and nitrides, the jagged yield point disappears. In either case, the shape of the stress–strain curve changes from 0–5–5 to 0–7–5, with the new lowered yield point now at 6 (Fig. 7.34).

Although attempts were made prior to the advent of the dislocation theory to explain the nature of the discontinuous yield, none was really satisfactory. A concept widely accepted at present is due to Bilby and Cottrell; briefly, it is as follows:

Because carbon and nitrogen atoms are much smaller than iron atoms, they will be present in steel as interstitials. Thermal fluctuation will cause them to drift from "tight" into "expanded" interstitial positions. Obviously, the largest voids are right beneath the bottom of the extra atom planes in edge dislocations. The drifting C and N atoms will gather in the spacious troughs at the dislocation. By interacting with the nearest atoms of the half-plane, the interstitials hold the plane in position and prevent it from moving under a low shear stress. When the stress reaches a sufficiently high level, the dislocation is suddenly torn away from the carbon "atmosphere," which was anchoring it, and propagates in the usual way under a much lower stress. When the deformation continues, the force increases due to strain-hardening until it reaches a value sufficient to cause a breakaway of dislocations in another part of the specimen, and so forth. This accounts for the jagged appearance of the yield discontinuity.

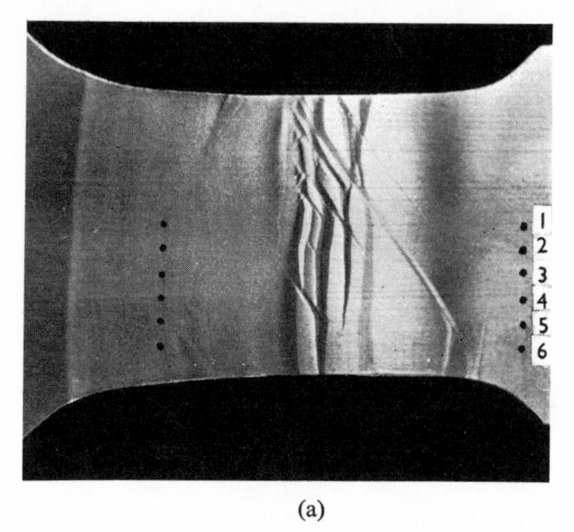

(a)

Fig. 7.35 Lüder's markings on soft steel: (a) Tensile specimen (R. Chadwick and W. H. L. Hooper). (b) Stamped sheet part (B. B. Hundy). From *J. Inst. Metals*, **81**, 744–5, (1953)

(b)

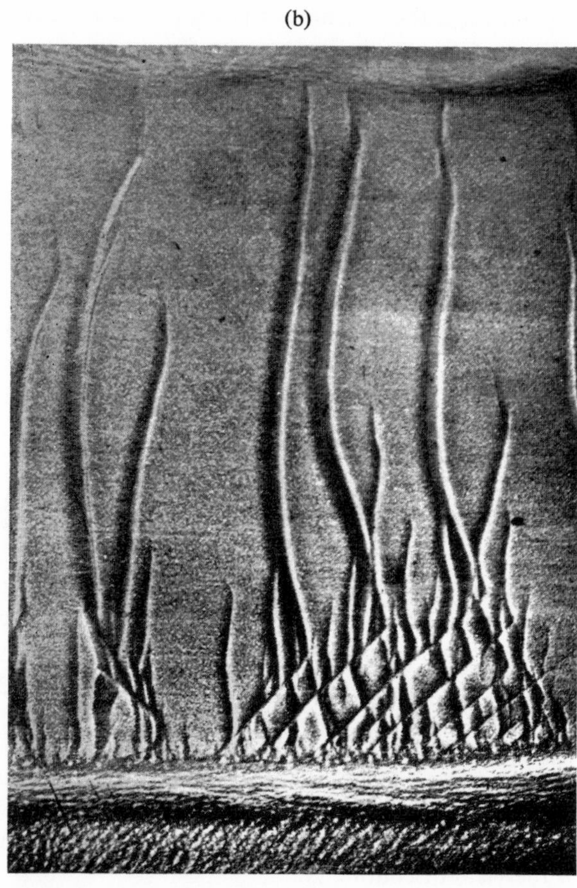

When all the dislocations anchored by the solute C and N atoms are freed, plastic flow becomes uniform. However, if straining is interrupted, say, at 2 (Fig. 7.34), and the metal is left for some time at room or higher temperature, the yield discontinuity gradually returns (curve 3). The rate at which this occurs was found to agree with the rate of diffusion of carbon or nitrogen in iron. This return of the yield point is one aspect of *strain-aging*. Discontinuous yielding occurs also in molybdenum, certain aluminum, copper, and other alloys, but was investigated in greatest detail on low-carbon steel. Its occurrence creates difficulties in stamping products, such as automobile body panels, sauce pans, and similar objects, where the presence of stretcher strains is highly objectionable.

Reversed Deformation and the Bauschinger Effect in Single Crystals

Plastic deformation (e.g., by tension) will leave a number of dislocation pileups like the one in Fig. 7.29. These pileups create internal stress which results in local regions of high strain energy. The stress, and hence the associated energy, is lowered if the pileup is allowed to expand. This will not occur spontaneously, unless the temperature is high enough to substantially increase the mobility of the atoms composing the crystal. However, if the direction of the applied force is reversed, in this case by compressing the crystal following initial extension, the tendency of these pileups to expand will assist the external forces in starting plastic deformation. As a result, the yield stress in compression will be reduced below the level of the final stress reached during preceding extension on lines similar to the Bauschinger effect in Fig. 6.13.

The Yield Strength of Polycrystals from a Dislocation Viewpoint

Grain boundaries must consist of disarrayed atoms and, therefore, represent serious obstacles to slip propagation. In fact, calculations show that the stress required to force a dislocation across a boundary is likely to be so high that the probability of slip propagating from one grain to the next in this way is remote. On the other hand, the stress concentration at the tip of a slip band may be high enough to activate dislocation sources in the neighbor grain across the boundary.

There were several attempts to compute the yield stress of polycrystals from the properties of single crystals, with or without making use of dislocation concepts. Some of the early attempts in this direction simply endeavored to find a mean yield strength for a statistical distribution of single crystals with random orientations. Such calculations have the defect that, once a random distribution of orientation is assumed, the yield stress should be independent of grain size. This is not true.

N. J. Petch, who worked on the quantitative aspects of brittle fracture in polycrystalline iron (b.c.c.), found a simple relation between the grain size

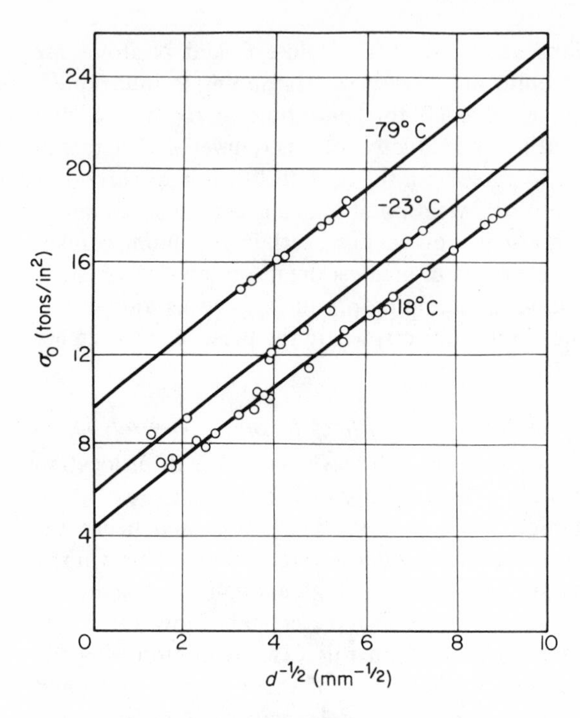

Fig. 7.36 Linear relationship between yield strength (σ_0) and reciprocal square root of grain size [$d^{-(\frac{1}{2})}$] for iron at three different testing temperatures. From N. P. Petch, *The Ductile-Cleavage Transition in Alpha-Iron*, in *Fracture*, see (R4.3).

and the lower yield stress (see Chapter 10) in ingot iron and low carbon steel. By making certain rather liberal assumptions regarding the arrangement of dislocations within the grain, and the forces acting on them, he was able to derive the following equation to describe the line in Fig. 7.36:

$$\sigma_o = \sigma_o' + kd^{-\frac{1}{2}} \tag{7.12}$$

where σ_o' = the intercept on the σ axis
k = a temperature-dependent constant
d = mean grain diameter

DIRECT EXPERIMENTAL EVIDENCE OF DISLOCATIONS

It is a noteworthy fact that, while the theory of dislocations started about 1933, attempts to prove their existence directly became successful many years later. A variety of techniques was developed for this purpose, including etching (which attacks preferentially the disordered dislocation sites) and electron microscopy.

The magnification of a modern electron microscope is such that lattice spacings of the order of 10 angstroms can be directly resolved. By transmission microscopy, which uses foils about 100 angstroms thick, it is possible to show happenings in depth and it is not confined to surfaces. Figure 7.37

Fig. 7.37 Dislocations in an iron-phosphorus alloy. Magn. 60,000×. From E. Hornbogen, *Trans. Am. Soc. Metals*, **56,** 16 (1963).

Fig. 7.38 Dislocation loops in high-purity aluminum. Magn. 35,000×. From R. M. J. Cotterill and R. L. Segal, *Phil. Mag.*, **8,** 1105 (1963).

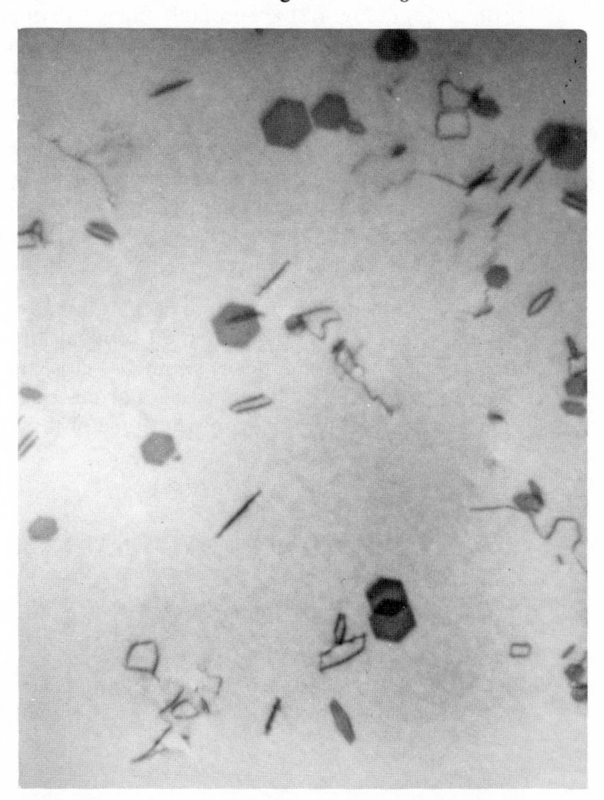

represents dislocation loops observed by this technique in an Fe-1.8% P alloy after 1% deformation by rolling. They clearly emanate from a grain boundary. Figure 7.38 shows numerous dislocation loops in high-purity aluminum. The hexagons are lying on ((111)) planes with edges along the [[110]] directions. They are edge dislocations which delineate stacking faults.

PROBLEMS

1. Prove that ((110)), ((112)), and ((123)), all contain [[111]] directions.
2. Crystals grow easiest by addition of atoms to existing incomplete layers. When a layer is full, the next layer must be started by a nucleation process which is energetically much less favorable. How can the presence of screw dislocations facilitate the growth of a crystal?
3. Develop a rule for defining the type of a screw dislocation in a dislocation loop by placing your right or left hand along the edge line with the thumb pointing in the direction of the Burgers vector.
4. A closed dislocation loop of length L is shaped as a (i) circle, (ii) square. Which of the two is more stable from the energy viewpoint?
5. Jogs J_a and J_b in Fig. 7.31 are unstable and will disappear as soon as the "trailing" part of the dislocation of which they are a part is moved forward. The same is true of jog J_c but not of J_a (Fig. 7.32). Explain why, and draw a configuration of two edge dislocations where a stable jog is formed on at least one of them.
6. A moving edge dislocation encounters in its path a chain of inclusions spaced several tens of lattice constants apart and extending substantially parallel to the dislocation line. Explain the stages (using sketches) of the dislocation approaching the chain, passing through it, and, finally, leaving it behind. Assume that double-cross slip is not involved.
7. Splitting of perfect dislocations into partials occurs only when the energy balance is favorable. However, the formation of a stacking fault uses some energy to form an interface. The latter is about 200 ergs/cm^2 in Al, 80 in Au, and only 13 in 18% Cr–8% Ni stainless steel. For cross-slip to occur, a recombination of the partials is required. With which one of these metals will this occur most readily?

REFERENCES FOR FURTHER READING

(R7.1) Chalmers, B., *Physical Metallurgy*. New York: John Wiley and Sons, Inc., 1959.

(R7.2) Cottrell, A. H., *Dislocations and Plastic Flow in Crystals*. Oxford: The Clarendon Press, 1953.

Cottrell, A. H., *The Mechanical Properties of Matter*. New York: John Wiley & Sons, Inc., 1964.

(R7.3) Dieter, G. E., Jr., *Mechanical Metallurgy*. New York: McGraw-Hill Book Company, 1961.

(R7.4) McLean, D., *Mechanical Properties of Metals*. New York: John Wiley and Sons, Inc., 1962.

(R7.5) Read, W. T., Jr., *Dislocations in Crystals*. New York: McGraw-Hill Book Company, 1953.

8

TIME-DEPENDENT DEFORMATION—RHEOLOGICAL MODELS AND INTERNAL FRICTION

In the preceding chapters strain was treated as a function of stress alone. No mention was made of the time over which loads acted, the implication being that mechanical performance is time independent. In reality, however, a steel tube operated at, say, 900°F. in a rocket booster could safely carry a much higher pressure than the same tube used under identical conditions of temperature and environment in a steam boiler. The reason for this lies in the difference in life expectancies of the two pieces of equipment. In the first case, life is measured in minutes, and in the second, in years. A lower pressure is mandatory in the boiler because the tube deforms slowly with time at the elevated temperature. Although this may result in a negligible amount of permanent deformation during the few minutes of life of the rocket, the same rate of distortion over the life of the boiler would be excessive. Time dependent permanent deformation of this type is shown by all construction materials and is called *creep*. Although the term often has a generic meaning including all time-dependent deformation, its use in this text is restricted to flow that occurs at constant load (or stress). In the harder metals, such as steel, titanium, or cobalt alloys, creep need be considered only at elevated temperatures, although there are some exceptions; e.g., heavily loaded high strength steel springs may take a permanent set after long times even at room temperature. On the other hand, soft metals, like lead, tin, or very pure aluminum, creep appreciably at ambient temperatures. Creep is even more important in amorphous materials like polymers and glass; these are shaped by processes involving mechanisms that are more time dependent than slip.

VISCOUS FLOW AND NEWTONIAN LIQUIDS

Viscous flow is another term applied to time dependent deformation. Like creep, it is frequently used as a generic term, but in this text it will identify flow that obeys specific laws. The equations of viscosity are similar to those of both elastic and plastic flow. Elastic strain is proportional to the applied stress (see Eq. (5.2a)), while for viscous flow, the deformation *rate* is proportional to the applied stress.

Viscous materials are assumed to be "perfect" fluids, or *Newtonian liquids*. These materials, by definition, exhibit a deformation rate that is exactly proportional to the applied stress, are incompressible, and possess no elastic characteristic. Actually, all materials are somewhat compressible and display some elastic response to an external stress. In "thin" liquids, such as water, an elastic behavior is easily detected only in compression, but in other viscous materials, such as glass, tensile elastic characteristics are also obvious, especially at room or lower temperatures. Nevertheless, many materials, including glass at high temperatures, approach the behavior of a Newtonian liquid.

As already pointed out, viscous behaviors are defined in terms of strain rates rather than in terms of strains, and the same relationships that exist between strains and displacements (Chapter 3) are found between strain rates and velocities. Hence, by analogy with Eqs. (3.2) and (3.5), the six strain rate components are:

$$\dot{\varepsilon}_x = \frac{\partial \dot{u}}{\partial x}, \qquad \dot{\varepsilon}_y = \frac{\partial \dot{v}}{\partial y}, \qquad \dot{\varepsilon}_z = \frac{\partial \dot{w}}{\partial z},$$
$$\dot{\gamma}_{xy} = \frac{\partial \dot{v}}{\partial x} + \frac{\partial \dot{u}}{\partial y}, \qquad \dot{\gamma}_{yz} = \frac{\partial \dot{w}}{\partial y} + \frac{\partial \dot{v}}{\partial z} \qquad \text{and} \qquad \dot{\gamma}_{zx} = \frac{\partial \dot{u}}{\partial z} + \frac{\partial \dot{w}}{\partial x} \tag{8.1}$$

The dot in each case represents differentiation with respect to time. Hence, since u, v, and w are the displacement components in the x, y, and z directions, respectively, $\dot{u} = du/dt$, $\dot{v} = dv/dt$, and $\dot{w} = dw/dt$ are the velocities in these same directions. The dimension of strain rate is 1/time (e.g. inches per inch per second = 1/sec).

The simplest type of viscous flow is one in which the particles of the material move in straight parallel lines. Flow of this type, which is shown for a two dimensional model in Fig. 8.1, is termed *parallel flow*.

Using the symbols for strain rate given in Eq. (8.1), and the definition of

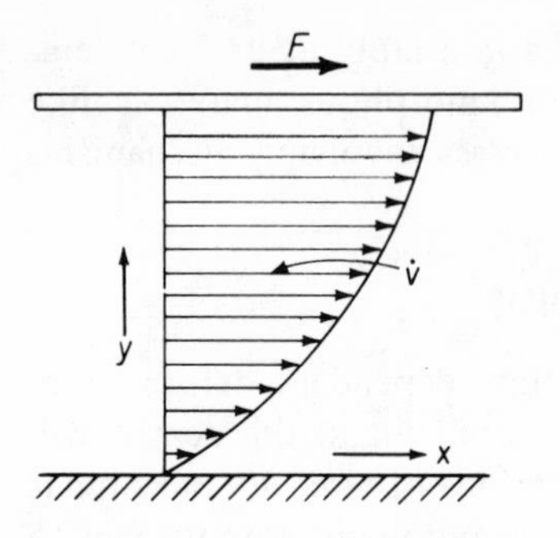

Fig. 8.1 Velocity distribution in laminar flow.

a Newtonian liquid as one in which the rate of strain is proportional to the applied stress, the relationships between the various strain rates and the corresponding stress components are

$$\dot{\varepsilon}_x = \frac{1}{\phi}\sigma_x, \quad \dot{\varepsilon}_y = \frac{1}{\phi}\sigma_y, \quad \dot{\varepsilon}_z = \frac{1}{\phi}\sigma_z \quad \text{and}$$
$$\dot{\gamma}_{xy} = \frac{1}{\eta}\tau_{xy}, \quad \dot{\gamma}_{yz} = \frac{1}{\eta}\tau_{yz}, \quad \dot{\gamma}_{zx} = \frac{1}{\eta}\tau_{zx} \tag{8.2}$$

The proportionality constants, ϕ and η, are defined as the *tensile viscosity* and *shear viscosity* respectively. The dimensions of viscosity are seen to be

TABLE 8.1*

VISCOSITY OF SOME COMMON SUBSTANCES

Substance	*Shear Viscosity η in Poise*
Air (20°C.)	1.86×10^{-4}
Water (20°C.)	0.010
Mercury (20°C.)	0.0156
Castor oil (20°C.)	7.2
Pitch (15°C.)	2×10^{11}
Pitch (50°C.)	1.5×10^{7}
Glass (20°C.)	10^{22}
Glass (575°C.)	1.1×10^{13}
Lava (flowing)	4×10^{4}

* From (R8.4)

the product of stress and time. One of the most common units of its measurement is the *poise* or *centipoise* (1 poise = 100 centipoise = 1 dyne-sec/cm²). Viscosity values given in the literature are usually for shear. Table 8.1 gives η for some common substances.

The condition of incompressibility of a Newtonian liquid, expressed in terms of normal strain rates, is

$$\dot{\varepsilon}_x + \dot{\varepsilon}_y + \dot{\varepsilon}_z = 0 \tag{8.3}$$

This equation, also known as the "continuity equation of hydrodynamics," is analogous to Eqs. (6.11) and (6.18) which defined the constancy of volume during plastic flow.

The application of normal stress (e.g., σ_x) to a viscous body produces an extension in the direction of its application and a contraction in the two lateral directions. To satisfy (8.3), the latter must occur at one half the rate at which the body is extending so that

$$\dot{\varepsilon}_y = \dot{\varepsilon}_z = -\frac{\dot{\varepsilon}_x}{2} = -\frac{\sigma_x}{2\phi} \tag{8.4a}$$

Similar equations result when the load is applied in the y or z directions:

$$\begin{aligned} \dot{\varepsilon}_z = \dot{\varepsilon}_x &= -\frac{\dot{\varepsilon}_y}{2} = -\frac{\sigma_y}{2\phi} \\ \dot{\varepsilon}_x = \dot{\varepsilon}_y &= -\frac{\dot{\varepsilon}_z}{2} = -\frac{\sigma_z}{2\phi} \end{aligned} \tag{8.4b}$$

Now, by combining the appropriate terms in (8.4), one can express the normal strain rate in any of the three directions in terms of the three stresses: $\sigma_x, \sigma_y, \sigma_z$. This procedure results in equations analogous to the generalized form of Hooke's law for an isotropic elastic body (5.2a); i.e.,

$$\begin{aligned} \dot{\varepsilon}_x &= \frac{1}{\phi}\,[\sigma_x - \tfrac{1}{2}(\sigma_y + \sigma_z)] \\ \dot{\varepsilon}_y &= \frac{1}{\phi}\,[\sigma_y - \tfrac{1}{2}(\sigma_x + \sigma_z)] \\ \dot{\varepsilon}_z &= \frac{1}{\phi}\,[\sigma_z - \tfrac{1}{2}(\sigma_y + \sigma_x)] \end{aligned} \tag{8.5}$$

Unlike the elastic case in which two material constants were required to describe deformation, only one constant is needed for Newtonian liquids because of their incompressibility. Hence, Eq. (8.5) is structurally identical with the plasticity equation (6.14).

Again by analogy with plasticity, the incompressibility condition, $\nu = \frac{1}{2}$, transforms the elastic equation (5.5), where $E = 2(1 + \nu)G$ into

$$\phi = 2(1 + \tfrac{1}{2})\eta$$

or

$$\phi = 3\eta \tag{8.6}$$

That is, the viscosity in tension is three times that in shear.

Note the similarity between the strain rate vs. stress relationships in Newtonian flow, with the strain–stress equations for elastic and plastic deformation, by replacing the elasticity terms in Table 8.2 with the corresponding plasticity or viscosity symbols.

TABLE 8.2

CORRESPONDING ELASTICITY, PLASTICITY, AND VISCOSITY TERMS

Elasticity	u	v	w	ε_x	ε_y	ε_z	γ_{xy}	γ_{yz}	γ_{zx}	ν	E	G
Plasticity	u	v	w	$\bar{\varepsilon}_x$	$\bar{\varepsilon}_y$	$\bar{\varepsilon}_z$	—	—	—	$\frac{1}{2}$	—	—
Viscosity	$\dot{u}$	$\dot{v}$	$\dot{w}$	$\dot{\varepsilon}_x$	$\dot{\varepsilon}_y$	$\dot{\varepsilon}_z$	$\dot{\gamma}_{xy}$	$\dot{\gamma}_{yz}$	$\dot{\gamma}_{zx}$	$\frac{1}{2}$	ϕ	η

MECHANISM OF NEWTONIAN FLOW

Liquids and amorphous "solids" that display stress–strain behaviors similar to Newtonian liquids possess short range, but not long range, order (Chapter 4). It is apparent that such a structure would be formed by clusters of closely packed structural units; viz., atoms, molecules, or segments of molecules that are separated by gaps or "holes" from neighboring clusters with different orientations. Actually, as stated in connection with glass, the exact structure of amorphous materials is difficult to ascertain, and, consequently, the structural unit movements that lead to viscous flow in these substances are not as yet understood in a completely satisfactory manner. The most generally suggested mechanism, however, is based on a relative movement of the structural units and neighboring holes, as proposed by Eyring and his co-workers for low molecular weight liquids.

The moving structural units shall be referred to as molecules in this section, and, indeed, whole molecules are thought to move about in liquids or low molecular weight polymers. In high polymers, segments of long molecules, consisting of about 20 to 30 atoms along the chain backbone, are the movable units, and it takes many of these movements before the relative position of entire molecules is changed. Only the movement of complete molecules can be considered viscous flow. The individual jumps cause viscoelastic deformation which is discussed in the following section. In other cases, it is the individual atoms near holes that move into the voids, leaving a gap for another unit, etc. This movement is not without restrictions, however, since the initial site of the molecule is its equilibrium position, and a certain activation energy is required to make it jump from this position into the neighboring hole. As the amplitude of thermal vibration of the molecule increases with temperature, the activation energy required to move the molecule into a new site decreases. In addition, the number of holes is also temperature dependent and is related to the effect of temperature on volume. This molecular movement is simply diffusion, and the probability of a molecule moving in one direction is as great as its probability of moving in any other direction so that it does not result in a body shape change. If, however, external stress is applied, the movement is biased in the direction of the stress, so the energy required to move in this direction is decreased. As a consequence, there is a greater number of jumps in one direction than in any other, thus resulting in a shape change of the body. The two temperature dependent factors, the extra energy needed for jumping over the barrier into neighboring holes and the density of holes, are lumped together as an *activation energy for viscous flow*. This energy term is related to the viscosity by

$$\frac{1}{\eta} = Ae^{-\Delta H_a/RT} \tag{8.7}$$

or

$$\eta = A^{-1}e^{\Delta H_a/RT}$$

where A = constant
ΔH_a = activation energy for viscous flow
R = gas constant
T = temperature

Although a plot of $\log \eta$ vs. $1/T$ is not actually a straight line for real materials, excepting over a small range of temperatures, Eq. (8.7) is generally used to describe the temperature-viscosity relationship for most substances.

THE SOFTENING POINT AND THE APPARENT SECOND ORDER TRANSITION POINT

The Softening Point

When an amorphous material is slowly cooled from a truly liquid condition, it remains soft and pliable over some span of temperature and, at least near the top temperatures of this range, it acts like a liquid whose viscosity is approximately given as a function of temperature by Eq. (8.7). At the low end of the temperature range, the material's viscosity is so high that its response to stress is completely elastic and, if the elastic range is exceeded, it breaks; i.e., it is brittle or glassy. The transition from a glassy state to a soft pliable one occurs over a rather narrow range of temperatures, generally designated as the *brittle point* or *softening point*. Actually, this temperature is not a material constant and depends both on the length of time over which the substance is held at temperature prior to testing (so long as it is not held at temperature for a long enough time to reach equilibrium) and on the rate of testing.* The relationship between stress and

* A specific procedure for defining the "softening point" in glass as the temperature at which it will deform under its own weight is given by A.S.T.M. (A.S.T.M. Desig. C338–57).

strain for materials that can deform elastically and viscously at the same time is the sum of the elastic and viscous responses to stress. Hence, if a shear stress is applied, the resulting shear strain rate is

$$\frac{d\gamma}{dt} = \frac{1}{G}\frac{d\tau}{dt} + \frac{1}{\eta}\tau \tag{8.8}$$

where $(1/G)(d\tau/dt)$ is the elastic strain rate written in terms of stress (5.3), and $(1/\eta)\tau$ is the viscous component (8.2).

The measure of relative magnitudes of viscous to elastic response is given by the ratio η/G. It has the dimension of time and is sometimes referred to as the *transition time*, Ŧ. We shall see in a following section that this ratio is also referred to as the *relaxation time* in certain cases or *retardation time* in others. Although G is a function of temperature, it varies so slowly, compared with η, that it can be considered constant so that

$$\text{Ŧ} \approx A' e^{\Delta H_a/RT} \tag{8.9}$$

The stress-strain relationships displayed by a material, and the classification of whether it is primarily elastic, viscous, or a combination of the two, depend on the length of time over which the body is stressed. This interval is known as the *time scale* of the observation, ⓣ, and it becomes most meaningful when compared with Ŧ.

Alfrey defines the softening point as the temperature at which Ŧ = ⓣ and uses this to distinguish glassy from liquid behaviors. Hence, on cooling, a material changes from liquid to glassy when $\eta = G$ ⓣ. It is common to consider viscous and elastic behaviors on a time basis. If an experiment involves slow strains, Ŧ ≪ ⓣ, and there is sufficient time for a biased type of diffusion to take place, the material acts as though it were completely viscous. If, on the other hand, Ŧ ≫ ⓣ, the elastic response dominates the behavior; if Ŧ ≈ ⓣ, the body displays both viscous and elastic effects. Since E for glass at room temperature $\approx 10 \times 10^6$ psi (Table 5.1), and Poissons ratio is 0.25, $G \approx 4 \times 10^6$ psi. From Table 8.1, η for glass $= 10^{22}$ poise $\approx 15 \times 10^{16}$ psi–sec. From this, Ŧ $= \eta/G = (15 \times 10^{16})/(4 \times 10^6) \approx 4 \times 10^{10}$ sec. Hence, it is obvious for any reasonable testing time, ⓣ, at room temperature, Ŧ ≫ ⓣ so that glass can be considered completely elastic.

The Second Order Transition Point

At about the temperature of the softening point, materials undergo changes that in some respects resemble primary phase transformations even though a new structure is not actually formed. At the temperature of a primary phase change (e.g., from a crystalline to an amorphous structure), the primary thermodynamic variables—volume, internal energy, etc.—abruptly change as shown in Fig. 8.2 for one type of rubber at about +10°C. Near the softening point there is a sudden change in slope of these same variables. This slope change that occurs at about −70°C in Fig. 8.2 is

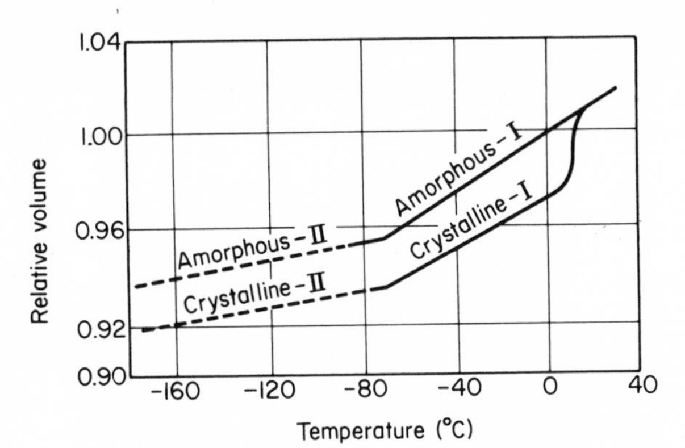

Fig. 8.2 Volume vs. temperature of purified rubber in the vicinity of a primary phase transition (11°C) and a second order transition (−72°C). After Bekkedahl

termed an *apparent second order transition*, and the temperature at which it happens is the *transition point*. Of special interest in the study of mechanical properties is the fact that the slope of the viscosity-temperature curve also changes at the transition point. Unlike the softening point, the transition point is a material constant if, before the measurements are made, the substance is allowed to come to equilibrium at the testing temperature. In high viscosity materials, however, this equilibrium is difficult to obtain.

VISCOELASTIC PROPERTIES

By considering two components to the shear strain rate, Eq. (8.8) satisfactorily describes low molecular weight liquids and inorganic glass. Amorphous linear high polymers, however, undergo three modes of straining. First, they experience ordinary elastic deformation consisting of a temporary stretch or compression of the valence bonds and moderate changes in the valence angles. Like all elastic atom movements, these are transitory and vanish immediately on release of load. In addition, they also deform by the wriggling of chain segments from one position into an adjacent hole, resulting in *retarded* or *configurational elasticity*, which is manifested by a delay in elastic response on the application or release of the applied load. This is generally referred to as a "rubbery behavior" since it takes a certain amount of time for a suddenly stretched rubber band to assume its final or equilibrium shape. This is neither a truly elastic or viscous response, but a combination of the two that requires both a modulus and a time dependent term for its description. Indeed, only high polymers show this mode of deformation to an appreciable extent. The ability to undergo large amounts of retarded elasticity, and their large molecular weight, are the two distinguishing features of this class of substances.

The third mode of deformation entails a number of individual jumps along the length of a molecule so that the entire molecule effectively wanders through the body. This component of deformation is an irreversible viscous

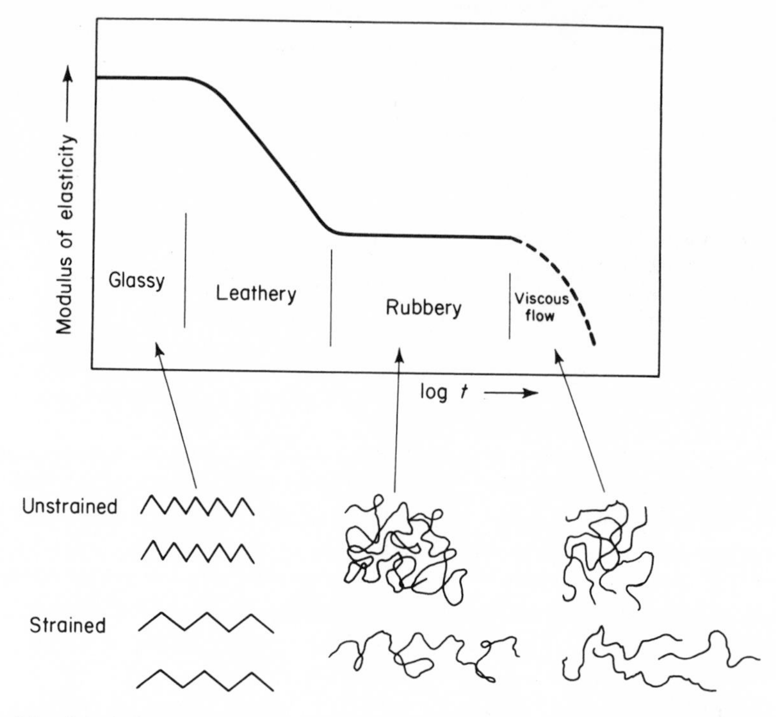

Fig. 8.3 Variation of modulus of a typical linear amorphous polymer and schematic representation of molecular movement as the strain rate is changed. Temperature held constant.

flow approximated by Eq. (8.2). Although both cross-linked and linear polymers can show retarded elasticity, the former do not flow viscously. As stated in Chapter 4, on heating these to a temperature high enough to allow for viscous flow, primary valence bonds are broken, causing these materials to decompose.

The complete spectrum of mechanical behaviors of high polymers, as well as the molecular movements associated with each type of deformation, are shown schematically in Fig. 8.3 as a function of time. The modulus of elasticity, or stiffness, is high—of the order of 10^{10}–10^{11} dynes/cm^2 when straining is very rapid (i.e., Ⓣ is very short). Under this testing condition, the materials show Hookean elasticity and are brittle, much like inorganic glass at room temperature. At slower strain rates or longer times, their stiffness decreases and they become leathery. Still slower straining leads to a rubbery behavior, and, if the material is a linear polymer, it will eventually undergo viscous deformation.

Figure 8.3 represents such behavior at constant temperature. Increasing the test temperature moves the curve to the left. If the time were held constant,

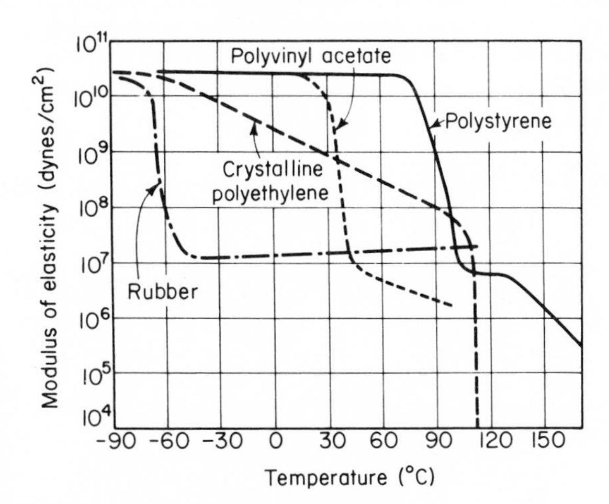

Fig. 8.4 Effect of testing temperature on the modulus of four typical high polymers. Strain rate held constant. From A. V. Tobolsky, *Scientific American*, **197**, No. 3 (1957).

and the temperature varied, the modulus would change as shown in Fig. 8.4 for some typical polymers. At a sufficiently high temperature, rubber decomposes while the other three materials flow viscously.

TIME-DEPENDENT DEFORMATION OF CRYSTALLINE MATERIALS

From a macroscopic viewpoint, creep of crystalline materials is similar to the time-dependent deformations of high polymers—as shown by a comparison of the creep curves for a steel at high temperature and methyl methacrylate at room temperature in Fig. 8.5. The mechanisms of deformation in these two materials must be different, however, since their atomic structures are dissimilar. Permanent deformation of crystalline materials was shown in Chapter 7 to result from dislocation movement, and this must have some time dependency.

In spite of a vast number of experimental studies of creep of crystalline solids, and especially of metals, understanding of its exact mechanism is still far from complete. There are a number of reasons for this. First, the constructional metals which are generally the subject of these studies have complex microstructures that may not be stable under the conditions of use. Hence, the mechanisms that account for deformation are intertwined with physico-chemical changes that take place in the material while it is being strained. Second, from an applied viewpoint the major interest is in very long time properties at elevated temperatures, and these may be impractical to duplicate in laboratory studies.

Nevertheless, a substantial amount of work has been done on creep mechanisms, generally for times much shorter than those of greatest commercial interest. These investigations indicate that the major difference between short and long time deformation is that the latter allows time for

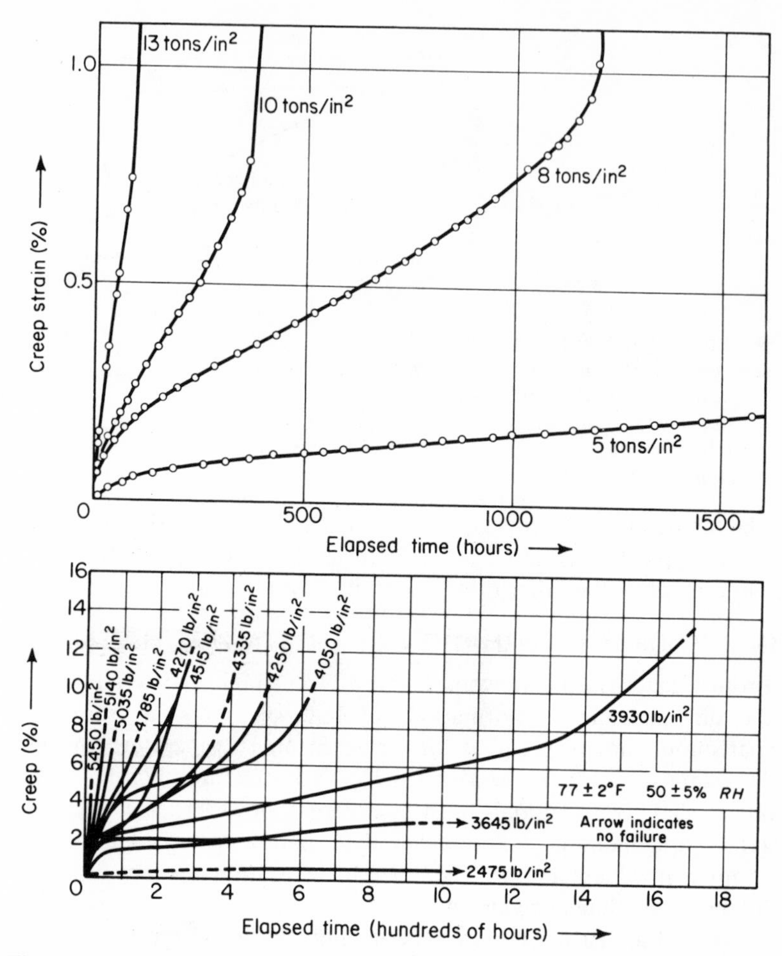

Fig. 8.5 Comparison of creep-time curves for steel at 600°F and methyl methacrylate at room temperature. After A. H. Sully, *Progress in Metal Physics*, Pergamon Press, **6,** 1956.

dislocation climb. It will be recalled from Chapter 7 that dislocations move easily on the slip plane until they encounter some obstacle or barrier in their path. These can be surmounted if the stalled dislocation can move to a new plane by climbing (Fig. 7.33). Climb occurs either if vacancies or "holes" in the lattice diffuse to the compression side of the dislocation or if extra atoms diffuse to the tension side. Either of these atom movements, as shown schematically in Fig. 8.6, moves the dislocation to an adjacent atom plane, and the hindrance to slip that was found on the first plane is avoided so that slip can again proceed. Like all diffusion processes, climb

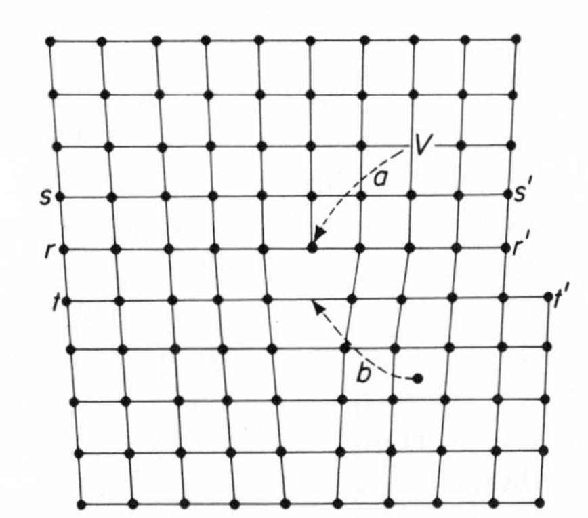

Fig. 8.6 Dislocation climb. If a vacancy, V, diffuses to the compression side of the dislocation as shown by arrow (a), the dislocation moves from plane r-r' to s-s'. If an interstitial atom moves to the tension side, climb occurs in the opposite direction, i.e., from plane r-r' to t-t'.

is a function of time and temperature (see Eq. (7.11)) so that one would expect the activation energies for creep and diffusion to be the same. Evidence substantiating this is shown in Fig. 8.7.

The ease with which metals deform at elevated temperatures over long periods, as compared with their short time rigidity at the same temperature, is a serious problem in the design of many structures. This is especially true of power generating equipment—such as boilers, turbines, etc.—which operates at elevated temperatures. Because of its enormous commercial importance, the phenomenological features of creep are discussed in more detail in Chapter 16.

Fig. 8.7 Correlation between activation energies for creep and self-diffusion. From J. E. Dorn, *J. Mech. Phys. Solids* **31,** 85 (1955).

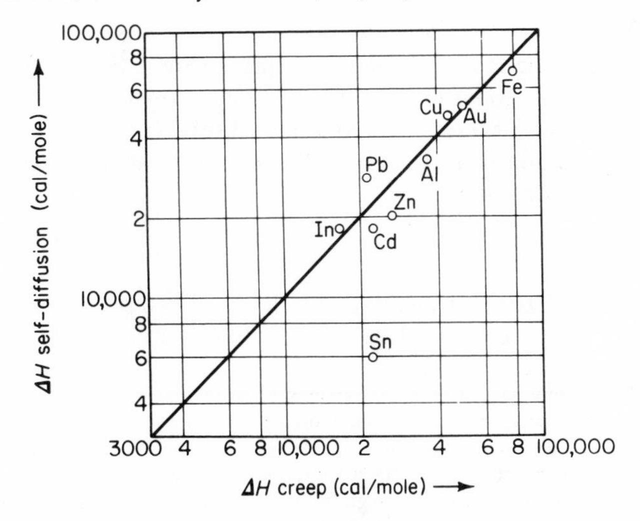

RHEOLOGICAL MODELS

The many different modes by which materials, and especially high polymers, may deform makes a description of their mechanical properties very difficult. This is particularly true under conditions in which different time dependent mechanisms operate simultaneously.

Substances that exhibit a mixture of different types of deformation cannot be defined simply as elastic, viscous, etc. These materials, of course, can be described by equations representing their stress–strain characteristics, and, if these equations are developed with the help of simple mechanical models (as is done by rheologists), the physical significance of the equations can be easily visualized.

By representing each of the basic types of deformation by a model which conforms to a relatively simple equation, complex behaviors can be described by combinations of the model elements. The requirement for these *rheological models* is that they exhibit the same relationship between force, elongation, and time that stress, strain, and time do in the real body. Either normal or shear stresses and strains can be represented by equations based on these models, although their physical picture involves normal forces and extensions only.

Rheological models do not necessarily describe the complete life of a real substance. For example, each of the strain–time curves in Fig. 8.5 is made up of three segments: an initial length over which the strain rate decreases with time; a second portion during which the strain rate is constant with time; and, finally, a length over which creep accelerates, leading eventually to fracture. The model analog of creep only duplicates the first portion of these curves.

Basic Model Elements

The first basic model element is a spring (Fig. 8.8a) and it represents a perfect elastic or *Hookean substance.* All the energy supplied to this element is stored as strain energy. Its extension (displacement), δ, is instantaneous on the application of force F (Fig. 8.8b), and the two are related by M, a proportionality constant:*

$$F = M\delta \tag{8.10}$$

The second basic element is a dashpot, Fig. 8.8c, which dissipates all the energy supplied to it. This is an analog of a perfectly viscous fluid or

* In this section, we shall use the symbols M for elastic modulus, β for viscosity, and δ for model extension, rather than the more commonly used E or G for modulus, η for viscosity, and ε for extension. The latter four symbols have been used in the earlier parts of the text to represent parameters with specific dimensions such as psi, etc. As used here, the units are quite different; e.g., M has the dimensions of force per unit of length.

Newtonian substance. In this case it is the rate of strain that is proportional to the stress, and the proportionality constant is the viscosity β of the dashpot

$$F = \beta\dot{\delta} \tag{8.11a}$$

If $\delta = 0$ at time $t = 0$, then at time t

$$\delta = \frac{Ft}{\beta} \tag{8.11b}$$

The model extension as a function of time on abrupt application and release of load is shown in Fig. 8.8d.

The third of the basic models, the *St. Venant substance*, is represented by a block that resists being pulled because of the friction, F_o, between it

Fig. 8.8 Basic rheological elements and their force-time-displacement curves (a) and (b) Hookean, (c) and (d) Newtonian, (e) and (f) St. Venant. The St. Venant body represents a yield strength that is time independent.

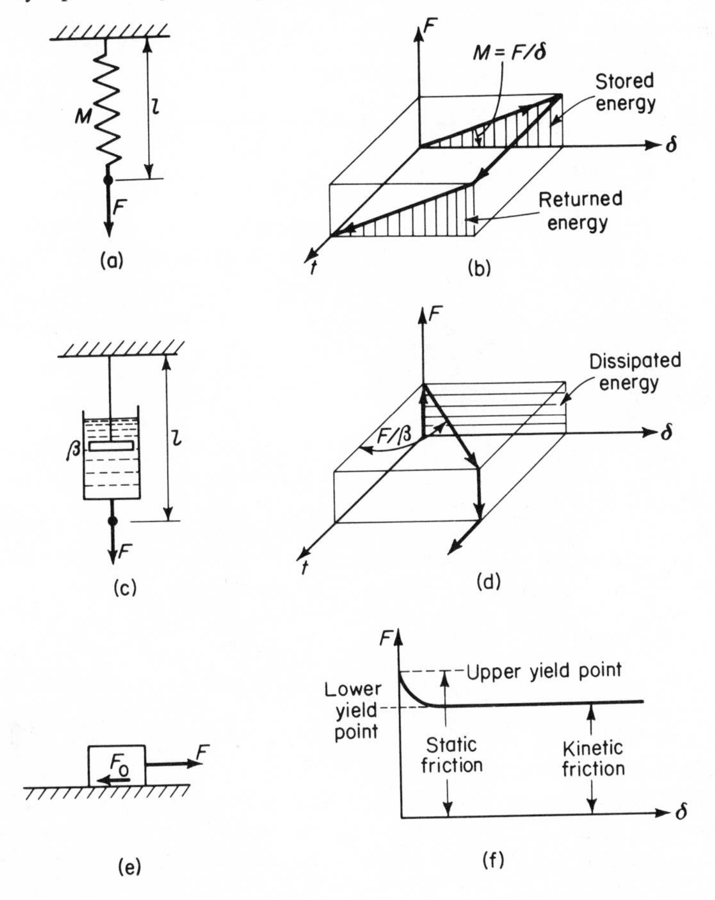

and the horizontal tabletop on which it rests (Fig. 8.8e). If a force $F < F_o$ is applied, there is no displacement. A sufficiently large force $F > F_o$ moves the weight, and, when this force is removed, the motion stops. This model is unrealistic, however, since a constant force, once it overcame friction, would cause the body to move faster and faster. For this reason, a St. Venant body is always used in conjunction with other elements to avoid the acceleration and, in these combinations, it represents a yield strength which is time independent. A force-extension curve, ignoring acceleration for this model, is shown in Fig. 8.8f. Note that this diagram has the same shape as that proposed for the rigid-plastic substance in Fig. 6.8a if the static and kinetic frictions are equal.

Compound Models

By various combinations of the basic elements, the flow behavior of almost any real material can be simulated. The differences between various materials are represented by changing the constants of the springs, dashpots, and blocks. *Viscoelastic* deformation is approximated by combinations of springs and dashpots connected in various ways. The simplest combination is a *Maxwell element*, consisting of a spring and dashpot in series (Fig. 8.9a). The applied force, F, is the same on both elements, and the model extends by an amount that is equal to the sum of the extensions of each element. Hence,

$$\delta = \delta_1 + \delta_2 \qquad \delta_1 = \frac{F}{M} \qquad \text{and} \qquad \dot{\delta}_2 = \frac{F}{\beta}$$

where δ = extension of complete model
δ_1 = extension of spring
δ_2 = extension of dashpot,

and

$$\dot{\delta} = \frac{\dot{F}}{M} + \frac{F}{\beta} \tag{8.12a}$$

If, at $t = 0$ and $\delta = 0$, a constant force is applied, the spring immediately extends by an amount F/M, and the dashpot extends with time according to Eq. (8.11b) so that the Maxwell element, after time "t", has elongated by

$$\delta = \frac{F}{M} + \frac{Ft}{\beta} \tag{8.12b}$$

as shown in Fig. 8.9b. On release of load, the spring contracts while the extension of the dashpot remains.

The ratio of the shear viscosity, η, to the shear or rigidity modulus, G, has been defined in an earlier section as the transition time $\mathcal{T}$. This same ratio, when applied to a body that behaves like a Maxwell element, is termed

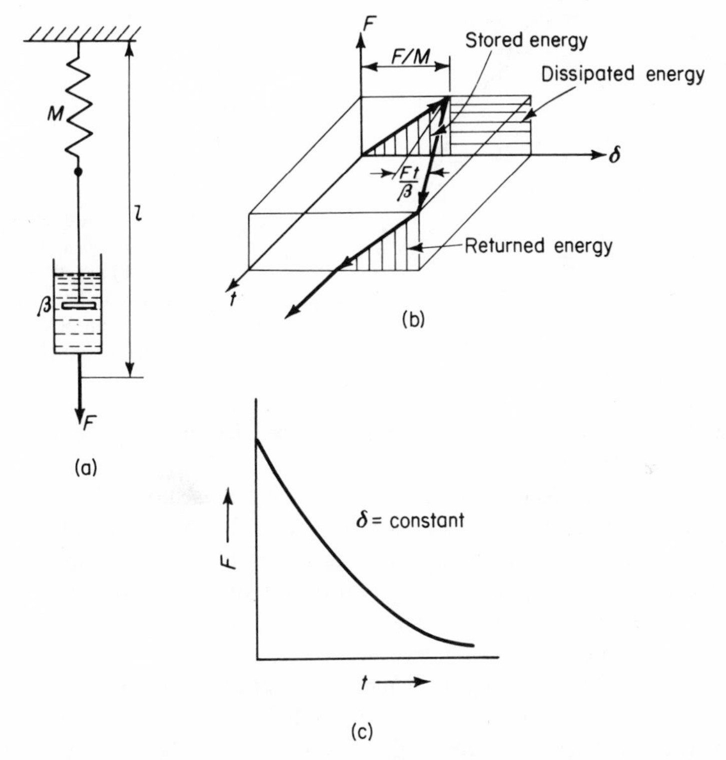

Fig. 8.9 (a) Maxwell element. (b) Force-time-displacement relationship for the element on loading and unloading Relaxation curve when δ is applied at $t = 0$.

the *relaxation time*. The symbol, $\mathcal{T}_m$, will be used for the ratio of β/M and thus (8.12b) can be written as

$$\delta = \frac{F}{M} + \frac{Ft}{\mathcal{T}_m M} \tag{8.12c}$$

where $\mathcal{T}_m$ is its relaxation time.

If the system is suddenly extended by δ_o, the force required to maintain this extension as a function of time is found by Eq. (8.12a). In this case $\dot{\delta}$ is zero, and $F = M\delta_o$ at $t = 0$, since the spring can respond instantly to the applied force while the dashpot acts like a rigid unit on an abrupt extension. Therefore, on integrating Eq. (8.12a) and using the limits:

$$\text{at } t = 0, \qquad F = M\delta_o \text{ and}$$

$$\text{at } t = t, \qquad F = F, \text{ one obtains}$$

$$F = M\delta_o e^{-Mt/\beta} \qquad \text{or} \qquad F = M\delta_o e^{-t/\mathcal{T}_m} \tag{8.12d}$$

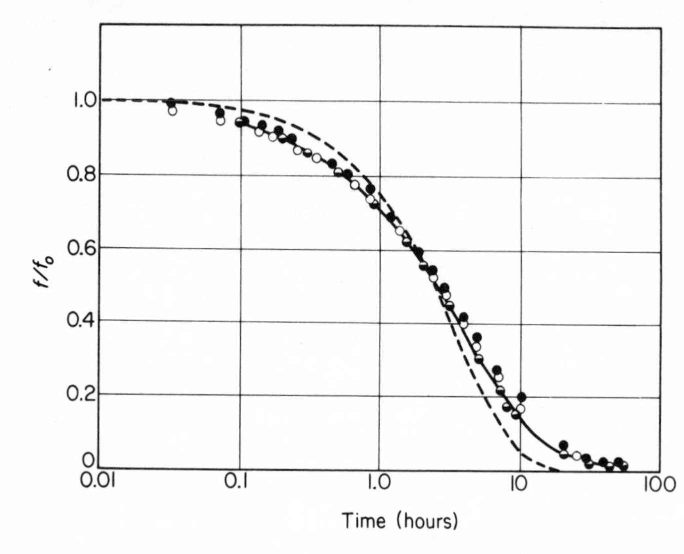

Fig. 8.10 Relaxation of stress in polysulfide rubber H-II at various elongations, and plot of exp $(-t/T) = 3$ hrs. (This figure differs from 8-9c because time is plotted on a logarithmic scale.) From (R4.6).

In the time $t = T_m$, the force F relaxes from its initial value of $M\delta_o$ to $M\delta_o/e$ (Fig. 8.9c).

Some polymers do indeed relax in this manner, as shown for a particular rubber in Fig. 8.10, but a single Maxwell element can only represent a very limited number of real materials since most of the relaxation in the model occurs within a time span equal to two orders of magnitude. Further, the analogy is only applicable to relaxation, not extension. Consequently, some other spring and dashpot combinations will be examined. The next simplest model considers the two basic elements in parallel (Fig. 8.11a). This behavior was described independently by Kelvin in 1875 and Voigt in 1890. Hence, the combination is referred to as a *Kelvin* or *Voigt body*. It is this arrangement that is required for configurational or rubber elasticity so that one expects to find at least one parallel spring and dashpot combination in any description of high polymers. The extension δ of a Kelvin body is equal

Fig. 8.11 (a) Kelvin element. (b) Force-time-displacement relationship for the element on loading and unloading.

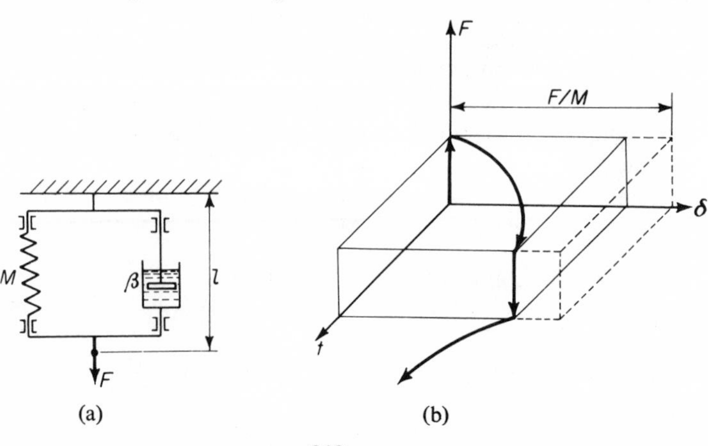

to the extension of either the spring or dashpot

$$\delta = \delta_1 = \delta_2 \tag{8.13a}$$

The force on the body equals the sum of the forces on the two elements.

$$F = F_1 + F_2 = M\delta + \beta\dot{\delta} \tag{8.13b}$$

or using the symbol $T_k = \beta/M$ which, in the Kelvin body, is known as the *retardation time*,

$$(1/\beta)F = \dot{\delta} + \delta(1/T_k) \tag{8.13c}$$

When the Kelvin model is stressed, part of the deformation energy is stored in the spring while the remainder is dissipated in the dashpot. If at time $t = 0$, $\delta = 0$, and a force F is applied, the extension, as a function of time, is found by integrating Eq. (8.13b) or Eq. (8.13c), using as limits the fact that $\delta = 0$, at $t = 0$, and $\delta = \delta$, at $t = t$; thus

$$\delta = \frac{F}{M}(1 - e^{-t/T_k}) \tag{8.13d}$$

Hence, δ increases exponentially with time, approaching F/M, the amount that the spring would have extended in the absence of the dashpot (Fig. 8.11b). On release of load, the model contracts exponentially approaching zero.

In order to more closely approximate the behavior of viscoelastic material, a Kelvin and Maxwell body are placed in series (Fig. 8.12a). This makes it possible to approximate all the deformation modes found in high polymers; the spring and dashpot of the Maxwell body correspond to Hookean and Newtonian behaviors, while the configurational elasticity is given by the Kelvin element. The extension-time curve for the system, if a constant force F is applied at $t_1 = 0$ and removed at t_2, is shown in Fig. 8.12b and given by Eq. (8.14a)

$$\delta = \underbrace{\frac{F}{M_1}}_{\text{Hookean extension}} + \underbrace{\frac{F}{M_2}(1 - e^{-t/T_k})}_{\text{Kelvin extension}} + \underbrace{Ft/\beta}_{\text{Newtonian extension}} \tag{8.14a}$$

where F = applied force
M_1 = modulus of Hookean element
M_2 = modulus of Kelvin spring
T_k = retardation time of Kelvin element
t = time
β = viscosity of Newtonian element

The best correlation between models and the extension of viscoelastic materials consists of a Maxwell element in series with a number of Kelvin

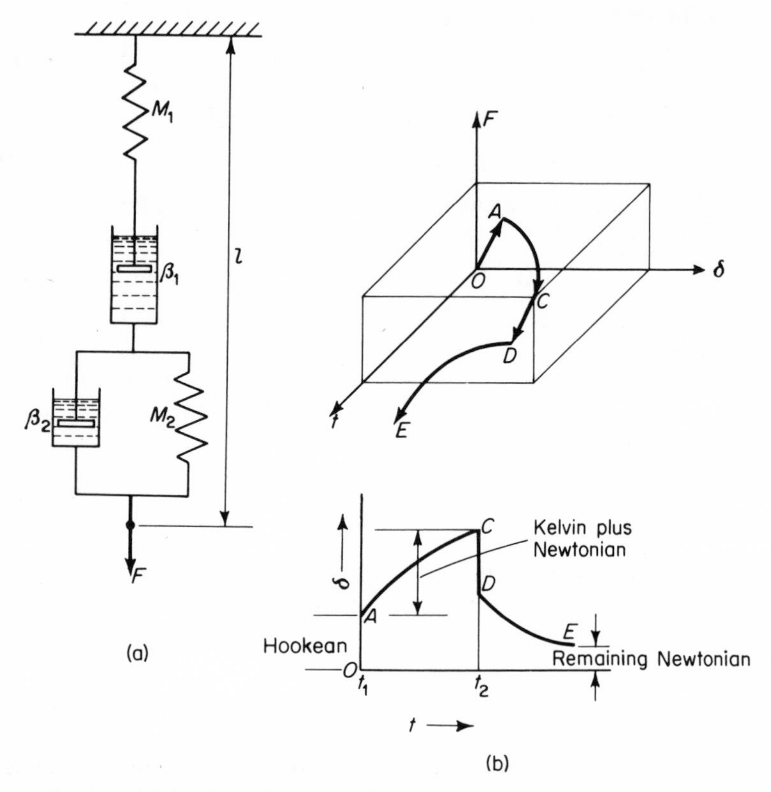

Fig. 8.12 (a) A Maxwell element in series with a Kelvin element simulates to the deformation of high polymers. (b) Force-extension time curve of (a).

elements (shown in Fig. 8.13). With this large number of individual elements, it is possible to pick a spectrum of moduli and viscosities to duplicate almost any real behavior. The improved accuracy in describing a real material by a multitermed expression (i.e., Kelvin elements in series), as compared with a single termed expression (one element), is shown by a practical example in Chapter 16. Naturally, as the model becomes more complex, the mathematical analysis becomes more involved. The most realistic analogy of relaxation of polymers is provided by a number of Maxwell elements in parallel.

Many materials, and certainly metals, frequently show a combination of elastic–plastic deformation. The elastic–plastic body, known as *Prandtl substance*, is represented by St. Venant and Hookean bodies in series (Fig. 8.14). Again ignoring the acceleration of the St. Venant body when F_o is exceeded, this model is analogous to metals in which $\sigma_o = F_o$.

To describe the rheological properties of paint, Bingham used a parallel St. Venant-Newtonian combination in series with a Hookean element (Fig.

8.15). This model, known as a *Bingham substance*, suffers very little deformation with slight loads (less than F_0) even at long times. A load in excess of F_0 on the model causes an extension that is a function of the viscosity of the dashpot. It is interesting to note that, if paint did not act like a Bingham substance, it could not be readily brushed onto a vertical wall. Paint should flow easily under the force supplied through the brush and also flow after application so as to obliterate the brush marks. Both of these requirements are met by a liquid with a low viscosity. However, if it were a Newtonian substance, and had properties that one normally thinks of as "liquid," it would run off the wall under its own weight before it could dry. This would be especially true if it had a low enough viscosity to make it spread easily. In actual practice, however, when the thickness of the paint layer is sufficiently small, its own weight does not exceed its yield strength so that it effectively does not flow at all.

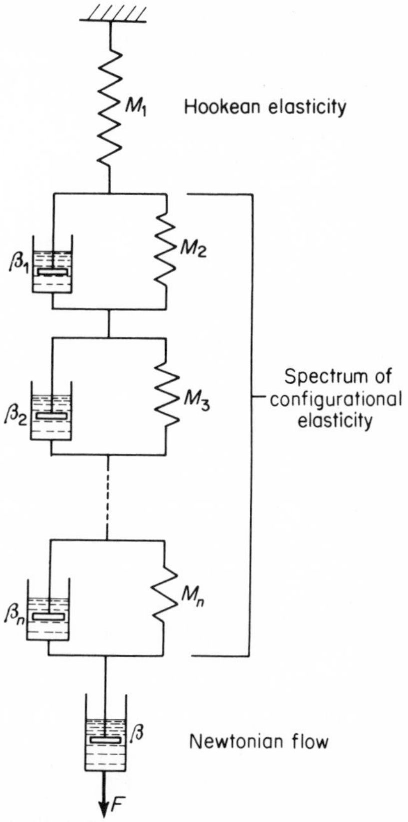

Fig. 8.13 Best correspondence between models and real high polymers requires a spectrum of Kelvin elements.

Rheologists have greatly simplified their model nomenclature by using a "shorthand" known as *rheological formulas*. According to this system, each of the model elements is referred to by its initial; e.g., H for Hookean, N for Newtonian, and St V for St. Venant. The coupling between elements is indicated by a vertical line (|) for parallel connections, and a horizontal one (—) for series. Hence, the Maxwell body is written as M = N—H, the Kelvin as K = N | H, and the Bingham as H—(St V | N). In series coupling, the force on the various elements is the same, and, in parallel coupling, the extensions are the same. The various element combinations

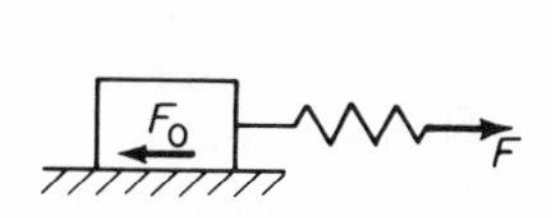

Fig. 8.14 Prandtl body corresponds to elastic plus plastic flow.

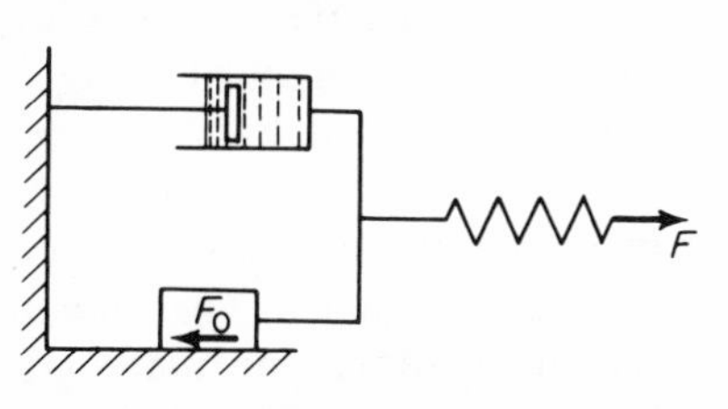

Fig. 8.15 A Bingham body analogous to paint.

that have been discussed to date are shown in Table 8.3. Some deformation patterns are not only described by the name of their model analogs but by other descriptive names as well. Where these have been introduced, they are added to the Table.

TABLE 8.3

COMPOUND RHEOLOGICAL MODELS

Model Name	*Symbol*	*Descriptive Name*
Maxwell	M = N—H	Elastico-viscosity
Kelvin	K = N | H	Retarded elasticity
Bingham	B = H—(St V | N)	Plastico-dynamic
Letherich	L = K—N	Elastic sols
Jeffrey	J = N | M	Relaxing gels
Schwedoff	Schw = H—(St V | M)	Plastic gels
Poynting-Thomson	PTh = H | M	Anelasticity (standard linear solid)
Burgers	Bu = M—K	
Trouton-Rankine	TR = N—PTh	
Schofield-Scott-Blair	Sch Sc B = Schw—K	

The most complex combination of elements so far assembled to describe a real material is the grouping designated as a Schofield-Scott-Blair body, used as an analogy for flour dough.

In spite of the handiness of these models, they have one serious shortcoming in that they can only demonstrate behaviors in which each element shows a linear response to loading. That is, all Hookean elements are such that if force F_1 causes extension δ_1, force $2F_1$ causes an extension of $2\delta_1$. If the dashpots extend by an amount δ_2 from force F_2 in time t, the extension within this time interval is doubled if F_2 is doubled. An example of a nonlinear model is given in Chapter 16.

DAMPING CAPACITY OR INTERNAL FRICTION AND ANELASTICITY

We have seen that, on the application of a force or displacement to these models, the applied energy may be stored in the spring as elastic energy or dissipated by the dashpot as heat. Hence, if force is applied and abruptly removed from a mass connected to a spring and dashpot combination, the spring, since it alternately stores and then releases potential energy, will oscillate. The moving mass alternately stores and dissipates kinetic energy so that the total energy surges back and forth between potential and kinetic forms. The dashpot on the other hand absorbs the kinetic energy of the moving mass so that it attenuates the motion. The result is that the oscillating model will slowly come to rest with time. All solids display this ability to dissipate vibrational energy to varying degrees, and this property is referred to as *damping capacity* or *internal friction*. This energy dissipation occurs even though the body is so completely isolated that losses to its surroundings are negligible. Since damping occurs even in metal bodies at low amplitudes of stress, Hooke's law cannot be rigorously applied to real materials. Ideal Hookean bodies dissipate no energy since their behavior shows no component equivalent to the dashpot. Hence, for real materials, stress and strain are always somewhat out of phase. This phase difference, for most structural materials, is so small that it may safely be ignored, as was done in Chapter 5, but it is just this slight phase shift that allows damping to occur.

Because of damping, the sound emitted from a struck metal object gradually dies out, and the great difference between various materials in this respect is seen by comparing a struck piece of lead with a tuning fork. The block in an automobile engine is made of high damping capacity material to minimize noises transmitted to the body. A similar consideration is made in choosing materials for machine tool parts, such as lathe beds to reduce "chatter" (the vibrational pattern sometimes superimposed on smooth machined parts). On the other hand, bells are made of metals with low damping capacity.

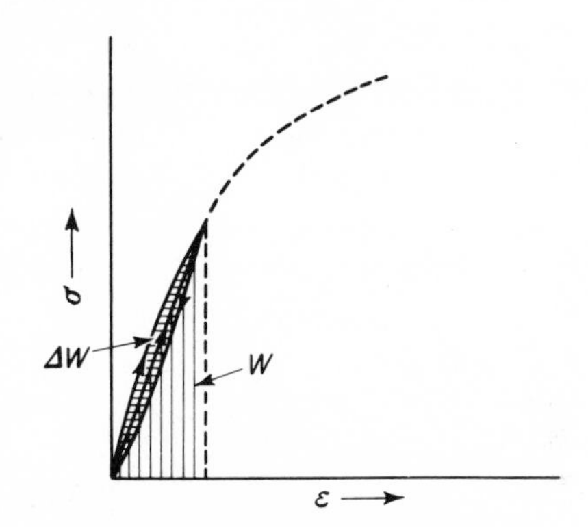

Fig. 8.16 Specific damping is defined at ratio of energy dissipated, ΔW, to total energy per cycle, W. (Width of ΔW band highly exaggerated for metal.)

The tendency to dampen or suppress vibrations through dissipation of vibratory energy is of importance in high speed machinery, as well as in structures like bridges or airplanes exposed to vibrations. Very large and dangerous cyclic stress and strain amplitudes can arise in them when the frequency of the exciting source approaches their natural frequency (resonance or near resonance conditions).

The ability of a material to dissipate vibrational energy can be expressed in a variety of ways. If, when the body is vibrating at its natural frequency, W is the total vibrational energy per cycle, and ΔW is the energy dissipated per cycle, the damping capacity, or, more accurately, the specific damping capacity, D, is given as the ratio of ΔW to W as shown in Fig. 8.16; i.e.,

$$D = \frac{\Delta W}{W} \tag{8.15}$$

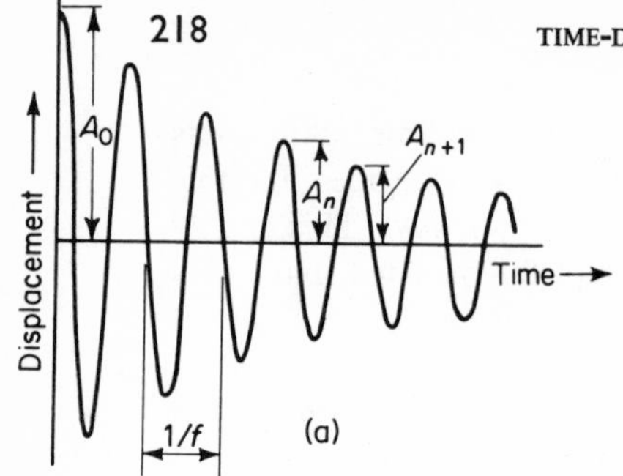

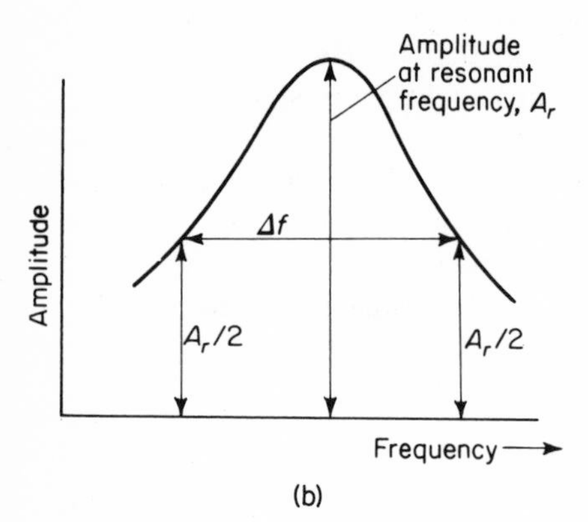

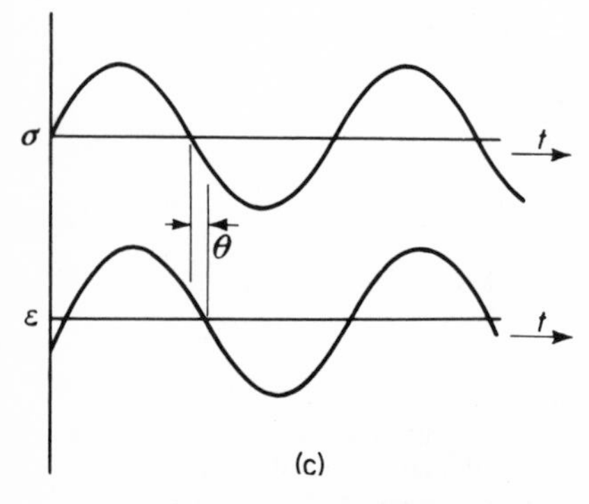

Fig. 8.17 Three methods for defining damping capacity. (a) According to Eq. (8.16). (b) According to Eq. (8.17). (c) According to Eq. (8.18b).

In engineering practice, damping is generally expressed as per cent so that D is multiplied by 100.

If no outside source of energy is supplied to the system that is vibrating at its natural frequency, f, and the amplitude of vibration decays from A_o to A_n in t seconds (Fig. 8.17a), the damping capacity is

$$D = \frac{2}{ft} \ln (A_o/A_n) = 2 \ln (A_n/A_{n+1})^* \quad (8.16)$$

Still another means for measuring damping consists of noting how rapidly the amplitude of vibration decreases as the frequency is moved off resonance. If a fixed amount of energy per cycle is supplied to a body, a maximum amplitude of vibration occurs at the resonance frequency, and the amplitude decreases as one moves from this frequency. The breadth of the resonance curve, Δf (Fig. 8.17b), is taken as the frequency range, on either side of resonance, over which the amplitude decays to half its resonance value. Hence,

$$D = \left(\frac{2\pi}{\sqrt{3}}\right)\left(\frac{\Delta f}{f}\right)^* \quad (8.17)$$

So far no distinction was made between the damping properties of materials at high and at low stresses. The basic difference between the two is that measurements of internal friction (vibration decay rate) are made at very low stresses which are not expected to cause any structural changes, such as strain-hardening, in the tested substance. Thus, it is customary to speak of internal friction when the experiment is conducted to probe into the basic properties of matter. In engineering applications on the other hand, the vibratory stress level is often high enough to affect the properties of the material, and one's interest is then directed toward the ability of the

* Equations (8–16, 17) only apply when the damping is small and independent of amplitude over the studied range. In (8.17) the damping must also be independent of frequency.

material to either sustain or dampen the vibrations rapidly. The term then used is *damping capacity*.

C. Zener has popularized the term *internal friction*, (R8.7) defined as the tangent of the phase angle θ between the impressed load and resulting displacement (Fig. 8.17c). Since θ is always small for metals, the tangent is equal to the angle so that

$$I.F. = \tan\theta = \theta \tag{8.18a}$$

It can be shown that the damping capacity and the angle are related by

$$D = 2\pi \tan\theta \tag{8.18b}$$

when θ is small. Hence, by the use of Equations (8.15, 16, 17),

$$I.F. = \frac{\Delta W}{2\pi W} = \frac{\ln(A_n/A_{n+1})}{\pi} = \frac{\Delta f}{f\sqrt{3}} \tag{8.19a}$$

In fatigue studies, the symbol Q^{-1}, borrowed from the theory of electrical resonance, is frequently used:

$$Q^{-1} = \frac{\Delta f}{f\sqrt{3}} \tag{8.19b}$$

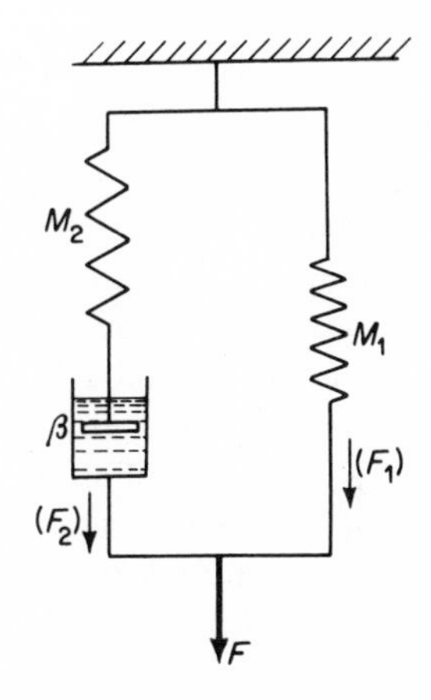

Fig. 8.18 Poynting-Thomson Model used to represent Anelasticity.

The Poynting-Thomson model (Fig. 8.18) is the basis for analyzing the characteristics of internal friction. If a vibrating force, F, is applied to the model shown in Fig. 8.18, the force is distributed so that F_1 is supported by the Hookean element while F_2 is taken up by the Maxwell element.

Now, we may write the displacement rate, $\dot{\delta}$, for the left-hand (Maxwellian) part of the model as the sum of the rate at which the spring length changes, plus the rate at which the piston travels in the dashpot.

$$\frac{\dot{F}_2}{M_2} + \frac{F_2}{\beta} = \dot{\delta} \tag{a}$$

For the Hookean element

$$F_1 = M_1\delta \tag{b}$$

and

$$\dot{F}_1 = M_1\dot{\delta} \tag{c}$$

We also know that

$$F = F_1 + F_2 \tag{d}$$

and

$$\dot{F} = \dot{F}_1 + \dot{F}_2 \tag{e}$$

Substituting (b), (c), (d), and (e) in (a) gives

$$\frac{\dot{F} - M_1\dot{\delta}}{M_2} + \frac{F - M_1\delta}{\beta} = \dot{\delta} \tag{f}$$

or

$$M_2F + \beta\dot{F} = M_1M_2\delta + \beta(M_2 + M_1)\dot{\delta} \tag{g}$$

By dividing (g) by M_2, and regrouping our constants, we arrive at

$$F + \tau_F\dot{F} = M_R(\delta + \tau_\delta\dot{\delta}) \tag{8.20}$$

Hence, the PTh body can be described by three constants, M_R, τ_F, and τ_δ. The modulus M_R is known as the *relaxed elastic modulus* since it is the ratio of force to displacement after all relaxation has taken place (i.e., under static loading conditions). The other two constants are relaxation times which include τ_δ—the relaxation of stress at constant strain, and τ_F—relaxation of strain under constant stress.

If the rate at which the applied force and displacement change is so great that no relaxation can occur,

$$dF = \frac{M_R\tau_F}{\tau_\delta}\,d\delta = M_u\,d\delta \tag{8.21}$$

Hence, the *unrelaxed elastic modulus*, M_u, is measured under conditions of high vibrational frequency.

Damping capacity or internal friction may now be defined in terms of these constants. First, however, note that, when the applied force or displacement varies very slowly or very rapidly, the relationship between F and δ is given by M_R or M_u, respectively. The material in either of these cases acts like a Hookean element and hence dissipates no energy; i.e., its damping capacity is zero. Physically, the basis for this is seen in examining the model in Fig. 8.18. At very low frequencies, the dashpot offers no resistance to the applied load so that the model effectively consists of a single Hookean element identified as M_1. At very high frequencies, on the other hand, the dashpot acts like a rigid body so that the element is described by the characteristics of the two springs, M_1 and M_2.

When the period of vibration becomes close to the relaxation time of the model, however, the dashpot can absorb energy, making the displacements lag the applied force, and this gives rise to damping. In this case, strain is no longer a single valued function of stress. Instead, the stress–strain curve becomes an ellipse. Zener has shown the damping capacity to be given by

$$D = \frac{\Delta W}{W} = 2\pi\,\frac{M_u - M_R}{M}\,\frac{\omega\tau}{1 + (\omega\tau)^2} \tag{8.22}$$

where M = geometric mean of M_u and M_R, i.e., $M = (M_u M_R)^{\frac{1}{2}}$
T = geometric mean of T_F and T_δ
and ω = angular frequency.

Equation (8.22) reaches a maximum value—i.e., the stress–strain ellipse has its largest area—when $\omega T = 1$. The width of this peak is approximately equal to a ratio of frequencies of about 100:1, as shown in Fig. 8.19.

Experimentally, it is difficult to vary ω over this range of 100:1 to study a single peak, and it is even more difficult to vary it over thousands of megacycles if a complete frequency spectrum is to be studied. This dilemma is avoided by taking advantage of the fact that Eq. (8.22) is symmetrical in ω and T. Since T is often a function of e^t, where t is temperature, the experimental procedure is simplified by varying t instead of ω.

There are a great many mechanisms that provide peak values of internal friction as a function of ω. For example, Zener has developed the curve in Fig. 8.20 in which six peaks, each associated with a different mechanism, are found over the range of 10^{-14} to 10^4 cps. All of the mechanisms that make materials behave like a PTh body have been grouped together by Zener and called *anelastic* phenomena. Further, because Eq. (8.20) could be written in the form

$$C_1 F + C_2 \dot{F} = C_3 \delta + C_4 \dot{\delta}$$

in which C_1—C_4 are constants, this model is seen to represent the most general linear relation between F, δ, and their time derivatives. Hence, the model is sometimes referred to as the *standard linear solid*, and anelastic effects are those associated with such a solid. Anelasticity also encompasses effects other than internal friction. For example, if a stress is suddenly applied to a real elastic body, it does not immediately show its complete

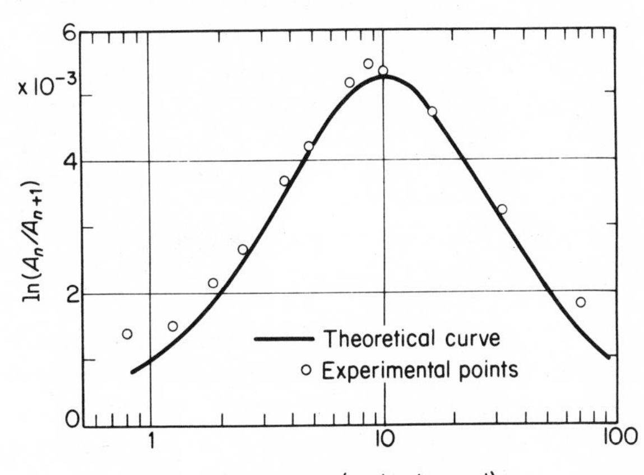

Fig. 8.19 A typical internal friction peak. From H. Kolsky, *Stress Waves in Solids*, Clarendon Press, Oxford, 1953.

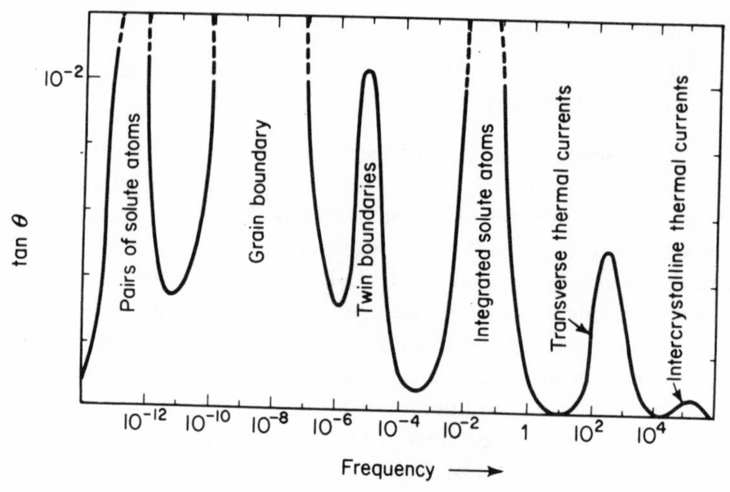

Fig. 8.20 A typical curve of damping capacity over a spectrum of frequencies. From C. Zener, *Elasticity and Anelasticity of Metals*, U. of Chicago Press, 1948.

response as an ideal Hookean element. Instead, a small fraction of the total displacement occurs only after some time. An identical behavior is found on abrupt unloading. This is known as the *elastic aftereffect* and is one more manifestation of anelasticity. On the other hand, not all internal friction is anelastic. Only those types which are independent of the amplitude of displacement during vibration can be considered in this category. This includes stress relaxation across grain boundaries, stress induced atomic ordering, and thermoelastic effects. It is the latter phenomenon (viz., a heating of a metal on compression and cooling on tension) that was discussed at the end of Chapter 5. Notice, incidentally, that thermoelastic effects are consistent with the model in Fig. 8.18 and Eq. (8.22) so far as internal friction is concerned. If a metal is vibrated slowly, the piece does not dissipate energy and it remains in thermal equilibrium with its surroundings. If it vibrates rapidly, the temperature changes are adiabatic, and it dissipates no energy. Between these two limits, however, it gives up and absorbs heat from its surroundings, accounting for high values of internal friction.

MECHANISMS FOR THE FORMATION OF INTERNAL FRICTION PEAKS

Various mechanisms, in addition to the thermoelastic effect, cause internal friction peaks in metals, as discussed in detail by Entwistle (R8.3). One source of peaks among those he lists is the "stress-induced ordering of solute atoms." Two types of peaks are classified in this group. One, the *Snoek peaks* or *Snoek effect*, results from the ordering of interstitial atoms.

The occurrence of these can be demonstrated by considering the possible positions that interstitial atoms, carbon, for example, can take in body-centered cubic iron. These are shown by the x, y, and z points in Fig. 8.21. The letter identifying each point indicates the direction of greatest strain produced in the lattice if the position is occupied. The energy in the lattice is minimized when the interstitial atoms occupy the largest voids. Hence, the application of a tensile stress in the x-direction causes the carbon atoms to prefer the expanded x sites, and, when an alternating stress is applied, the carbon atoms would move toward and away from these positions resulting in a strong peak. Note the similarity of this physical movement with the model of Fig. 8.18. Snoek's peaks have been found in iron-carbon, iron-nitrogen, iron-hydrogen, and possibly in iron-boron solutions as well as in carbon, nitrogen, and oxygen in other b.c.c. transition elements. The phenomenon has been used in studying diffusion, solubility, and precipitation.

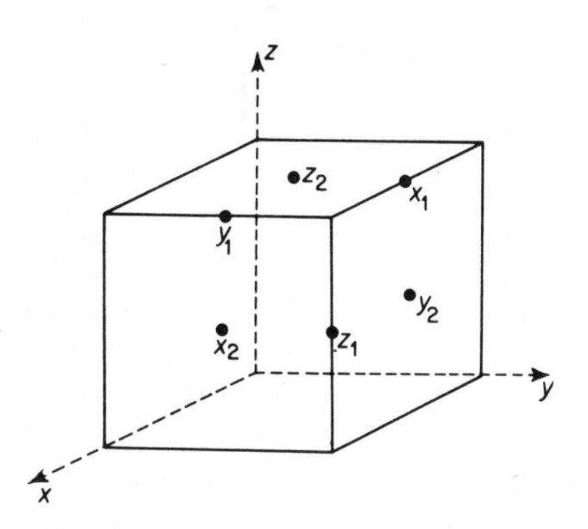

Fig. 8.21 Positions that a carbon atom can take in b.c.c. iron.

Zener peaks arise in internal friction vs. temperature curves of substitutional solid solution alloys. Such a peak occurs in α-brass single crystals at 420°C. and is thought to result from a change in degree of ordering brought about by the applied stress.

A second general class of internal friction peaks were those associated with crystal boundaries. Kê compared the internal friction vs. temperature curves for a single crystal and polycrystalline aluminum over the range from 0 to 500°C. and found the pronounced difference shown in Fig. 8.22. The

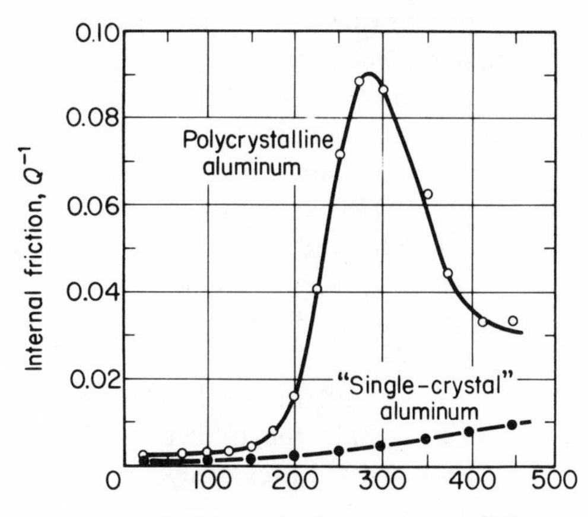

Fig. 8.22 Internal friction versus temperature curves for a single crystal in polycrystalline aluminum. From T. S. Kê, *Phys. Rev.*, **71**, 533 (1947).

temperature-frequency dependence of this peak suggests that it results from grain-boundary self-diffusion; i.e., interchange of position of atoms of the same kind. The relative sliding of adjacent crystals, that leads to the relaxation of the shear stress across the grain boundary, suggests a resemblance between the behaviors of a boundary and a film of viscous fluid which corresponds to the dashpot in Fig. 8.18.

Internal friction peaks can also be caused by dislocations. Because a study of these contributes to an understanding of the mechanisms of plastic flow, this class of peaks has attracted considerable attention. *Dislocation damping* originates from two types of phenomena: those that are primarily frequency dependent, and those that are primarily amplitude dependent. The amplitude-independent group gives rise to *Bordoni peaks* found at low temperatures in the face-centered cubic and possibly body-centered cubic, and in at least one hexagonal metal. They are not found in annealed materials but occur after even very small amounts of plastic deformation. There is no complete agreement on their cause, but they have been attributed to the resistance to dislocation movement or Peierls force described by Eq. (7.9). On the application of a stress, only part of the dislocation line moves, causing "kinks" consisting of short lengths of dislocations at the ends of the moved line (see Fig. 7.17). These kinks are of opposite sign so that they attract each other by a force that is inversely proportional to the distance between them. The applied stress however opposes this attraction and, when the kinks are far enough apart for their attractive force to be small, the applied stress can cause them to run in opposite directions, accounting for the anelastic strain.

Frequency dependent damping could also occur if dislocation loops undergo forced vibration under the action of an alternating stress. The damping will go through a maximum at the natural vibration frequency of the dislocation line.

An extension of this concept of a vibrating dislocation line can also be used to explain amplitude-dependent dislocation damping. The additional requirement is that the dislocation lines must be pinned at some positions along their lengths. The pinned dislocation represents a series of short segments which, because of their reduced lengths, are stiff. Hence the stress for any amount of movement is large. After the dislocation breaks free from its pinned points, it acts like a long, more flexible line that can be moved back the same distance with far less force. On returning to its initial position, it becomes repinned so that the different stress–strain relations, in moving away from and toward its initial location, causes mechanical hysteresis.

Other mechanisms for the formations of peaks have also been described. These include the influence of magnetic domains and the thermoelastic effect discussed previously.

PROBLEMS

1. A 2 in. long by 2 in. diam. piston fits into a 2.01 in. diam. cylinder. An oil with a viscosity $\eta = 0.5$ poise separates the piston and cylinder walls. How much force is required to keep the piston in motion at a velocity of 100 ft./min.? (Assume the thickness of the oil film is uniform over the periphery of the piston, and that the flow is linear.)

2. Assume that the viscosity of glass is linear with temperature, and the relationship is defined by the two points in Table 8.1. The elastic constants are $E = 25 \times 10^6$, and $\nu = 0.25$, for glass at room temperature. Assume that these constants are temperature independent. Plot the transition time, T, vs. temperature. According to Alfrey's definition, what is the softening point in one week?

3. Compare the δ vs. time curves for 1, 2, 3, and 4 Kelvin elements in series. The values M and T are as follows:

Element No.	M	T
1	30×10^6	1000
2	20×10^6	100
3	10×10^6	10
4	5×10^6	1

   What can you state about the curve shapes?

4. Explain why the H, N, and K bodies are simply special cases of the PTh body.

5. Draw a Bingham body model and write an equation for δ in terms of M, β, and t.

6. Calculate the damping capacity of the polyethylene whose tensile stress–strain curves are shown in Fig. 10.15.

REFERENCES FOR FURTHER READING

(R8.1) Alfrey, Turner, Jr., *Mechanical Behavior of Polymers*. New York: Interscience Publishers, Inc., 1948.

(R8.2) Ehring, F. R., editor, *Rheology, Theory and Applications*, Vol. II. New York: Academic Press, Inc., 1958.

(R8.3) Entwistle, K. M., *Metallurgical Reviews*, Vol. 7, No. 6 (1962), 175.

(R8.4) Jaeger, J. C., *Elasticity, Fracture and Flow*. 2nd ed. London: Methuen and Co., Ltd., 1962.

(R8.5) Mark, H. and Tobolsky A. V., *Physical Chemistry of High Polymer Systems* (2nd ed.). New York: Interscience Publishers, Inc., 1950.

(R8.6) Reiner, M., "Rheology," in *Handbook of Physics*, Condon and Odishaw (eds.). New York: McGraw-Hill Book Company., 1958.

(R8.7) Zener, C., *Elasticity and Anelasticity of Metals*. Chicago: U. of Chicago Press, 1948.

9

FRACTURING

Deformation of any type—i.e., elastic, plastic, viscous, etc.—can be terminated by a separation of the material into two or more parts. For example, on increase of load on the aluminum alloy wire and the glass strand discussed in the Introduction, they continuously stretch—the former elastically and then plastically, the latter only elastically—until they fracture; these fracture points are indicated as F_1 and F_2 (Fig. 9.1). Their ordinate values represent the *fracture* or *rupture stress*, which is defined as the breaking load divided by the cross-sectional area at fracture. If the ordinate denotes the breaking load divided by the cross-sectional area before straining, it is referred to as the *fracture* or *rupture strength*. The latter terms are applicable only when the specimen cross section changes an insignificant amount before fracture as is the case for glass or cast iron. The abscissa of these same points is called the *fracture strain* or, if only the plastic strain is considered, it is referred to as the *ductility* of the material. All of these terms imply a substantially uniaxial stress state during deformation and fracturing unless otherwise stated.

Fig. 9.1 Stress-strain curve of glass and aluminum alloy showing fracture, F_1 and F_2 after elastic and after elastic plus plastic deformation.

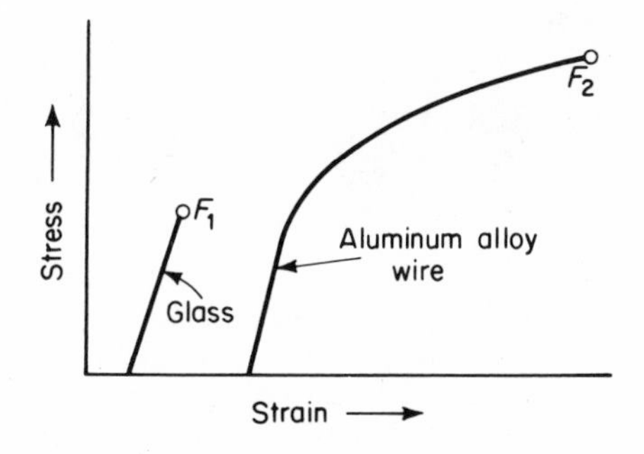

If the material undergoes an appreciable amount of permanent

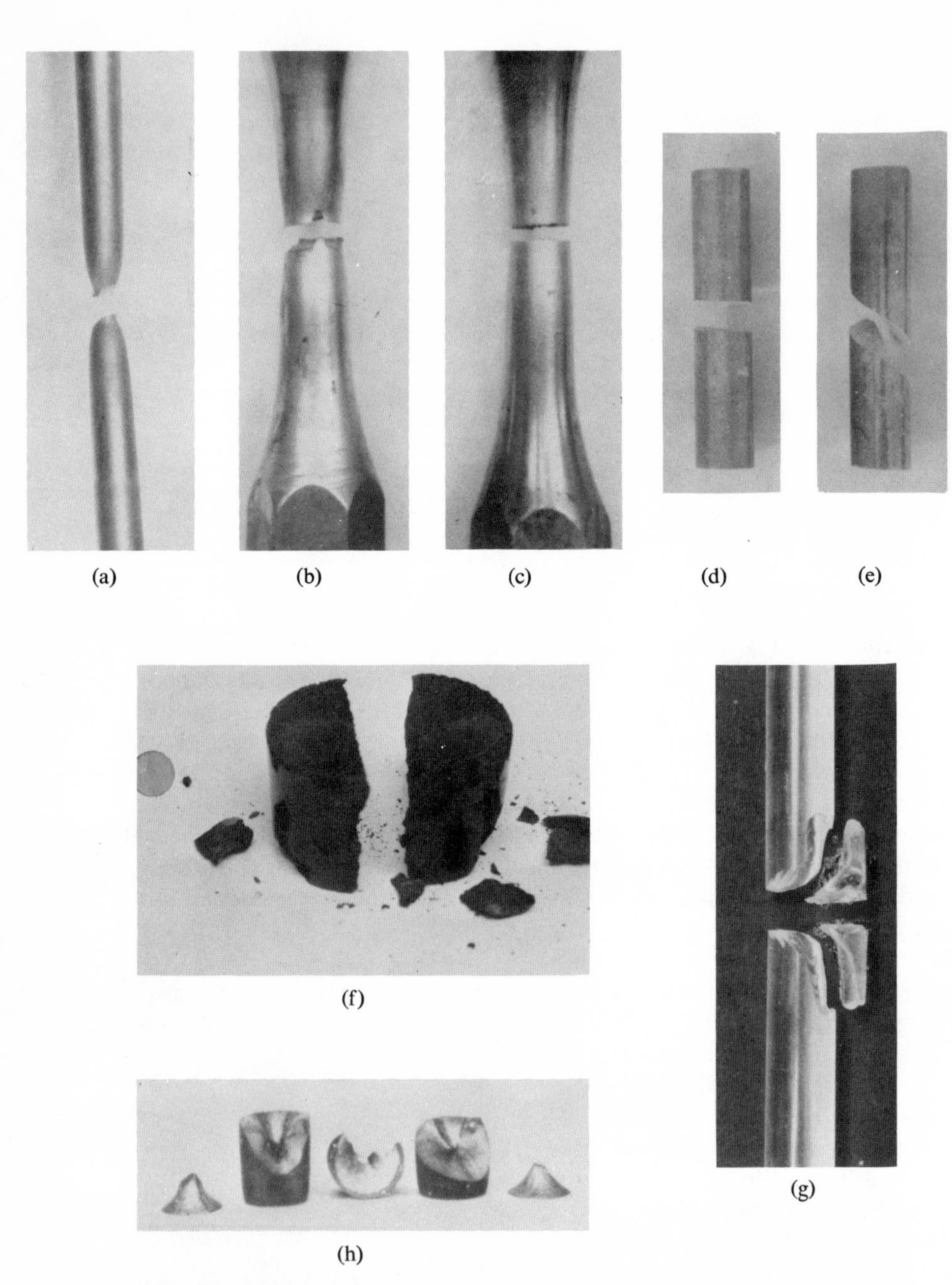

Fig. 9.2 Photographs of a variety of types of fracture. (a) Brass, tensile. (b) Ductile steel, tensile. (c) Brittle steel, tensile. (d) Chalk, bending. (e) Chalk, torsion. (f) "Bakelite," compression. (g) "Plexiglas," bending. (h) Leaded brass, compression.

deformation prior to fracture, it is said to be *ductile*; if this prior deformation is small, the material is *brittle*. The appearance of fractures varies considerably. Some typical examples are shown in Fig. 9.2. "Ductile" and "brittle" are relative terms. A mild steel rod that ruptured after straining a few per cent would be classed as brittle because normally it would be expected to stretch 25–30%; yet, a normally very brittle ceramic such as a magnesia crystal would be considered ductile had it withstood a few per cent permanent deformation before breaking.

One of the most perplexing phenomena of fracturing is that some materials that generally are ductile become brittle under certain service conditions. As an example of the seriousness of this problem, the bridge girder in Fig. 9.3 was split in half by a crack that macroscopically appeared

Fig. 9.3 Fracture in span of Kings Bridge in Melbourne, Australia. (a) External view of span. (b) Internal view of span. (c) Close-up of fractured girder. Courtesy of Melbourne and Metropolitan Board of Works, Victoria, Australia.

to propagate through a material that was completely brittle. Yet, if a test bar made from this same girder were pulled apart at the temperature at which the beam cracked, it would have displayed a substantial ductility.

In discussing fracture we shall have to go through the same procedure as used in dealing with plastic flow and consider the problem on the macroscopic, microscopic and atomic levels. In the first place, a number of empirical laws will be introduced in order to predict fracturing behaviors under complicated stress systems from data obtained on simple tension or compression specimens. This will be followed by a description of the proposed mechanisms of atomic movements and energy conditions that lead to fracture in order to explain the macroscopic observations. Both of these aspects of fracturing are less completely understood than is the case for plastic flow.

FRACTURE MODES

Fracturing is a complex phenomenon and it is difficult to design laboratory tests for evaluating the significant variables. One of the complicating factors is the variety of ways or *modes* in which solids break. In crystalline materials, fractures most commonly are intragranular (transcrystalline); i.e., go through the grain. In some cases, however, fracture follows along the grain boundaries. This latter is a special type that occurs either on slow straining at elevated temperatures or under the combined action of stress and a corrosive environment. Some materials, however, break intergranularly because they contain a soft or else a brittle phase in the grain boundaries. In this chapter we shall mainly be concerned with fractures that occur from the action of continuously increasing loads (the effect of alternating loads in causing rupture is discussed in Chapter 18).

Ruptures may occur in polycrystalline materials by an internal slipping apart of individual grains. The fracture surface hence consists of cross-sections of grains exposed by the shearing process. In this case, the separation surface is identical with the slip plane. Another common and frequently dangerous type of fracture occurring in b.c.c. and h.c.p. metals, as well as in many inorganic nonmetallics, consists of cleaving by splitting along a crystallographic plane, the *cleavage plane*. This occurs similarly to the way layers of mica are peeled; it is because diamonds are readily cleaved that they can be split into the familiar gem shapes. There is a critical resolved normal stress for cleavage of single crystals much like the critical resolved shear stress for slip. The cleavage strength, however, can only be considered a material constant so long as fracture occurs without an appreciable amount of prior plastic flow. Unlike τ_{ro}, the cleavage strength of crystals is more or less independent of temperature. The curve in Fig. 9.4 is calculated, assuming a critical normal stress, while the points were experimentally obtained at two different testing temperatures. The cleavage and shear (slip) planes in a crystal may or may not be identical. The cleavage planes and the critical normal stress for some metals and rock salt at various temperatures are shown in Table 9.1.

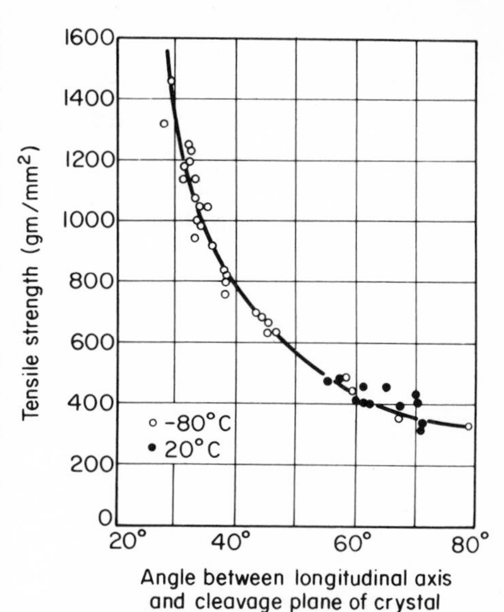

Fig. 9.4 Dependence of cleavage fracture strength on orientation for bismuth crystals. Notice that the cleavage strength is independent of temperature. From W. Georgieff and E. Schmid, *Z. Physik*, **36**, 759, (1926).

Intuitively one would expect a larger amount of deformation or energy expenditure with a shear type than with a cleavage fracture. This is a correct inference when one is concerned only with the energy dissipated during the time that the fracture is extending. However a large or small amount of deformation may precede either of these fracture types. For example, physical changes may occur in metal during large plastic strains with sufficient embrittling influence to produce a cleavage fracture after the extensive deformation.

Whether a particular fracture surface is the result of slipping apart or cleavage of individual grains is recognizable from the fracture appearance, provided the break is "clean"; i.e., the fracture surfaces do not rub against each other after the separation occurred. In mild steel, the former results in a dull gray, fibrous surface (Fig. 9.5a). These fractures are called *fibrous*. Cleavage fractures (Fig. 9.5b), on the other hand, are brighter and show

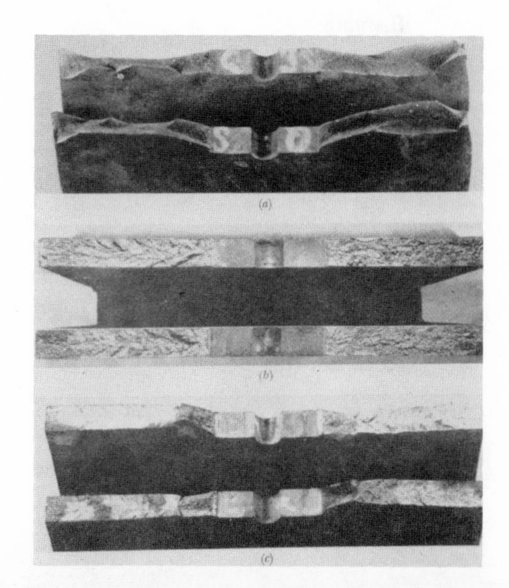

Fig. 9.5 Photographs of fractured steel specimens showing the fracture appearance for the following modes of fracture: (a) Shear, (b) Cleavage. (c) Mixed shear and cleavage. From (R9.4).

highly reflective spots, each of which is a cleavage plane of a single grain. As the specimen is rotated, the light and dark patterns on the fracture surface shift because the cleavage planes are randomly oriented, as are the crystals in a typical specimen. Of course, it is not necessary that a specific fracture be completely of one kind. The specimen in Fig. 9.5c shows a mixed cleavage and fibrous fracture.

FRACTURE STRESS AS FUNCTION OF STRAIN

Since there is one resolved stress for slip by shear and another one for cleavage, it seems logical that the possible reactions to an applied load compete with each other. After some initial elastic strain, the response to an additional stress may be plastic flow, viscous flow, cleavage, or any other response, depending on which requires the lowest stress. This led P. Ludwik to suggest in 1927 that, in addition to the easily observed stress–strain curve for flow (shown in Fig. 9.1), there exists a fracture stress vs. strain curve that defines the fracture stress at every value of strain as shown in Fig. 9.6.

TABLE 9.1

CRITICAL NORMAL STRESS FOR BRITTLE CLEAVAGE OF SINGLE CRYSTALS*

Metal	*Cleavage Plane*	*Temperature °C.*	*Critical Normal Stress, kg. per sq. mm.*
Zinc (0.03 per cent cadmium)	(0001)	−80	0.19
	(0001)	−185	0.19
	$(10\bar{1}0)$	−185	1.80
Zinc + 0.13 per cent cadmium	(0001)	−185	0.30
Zinc + 0.53 per cent cadmium	(0001)	−185	1.20
Bismuth	(111)	20	0.32
	(111)	−80	0.32
	$(11\bar{1})$	20	0.69
Antimony	$(11\bar{1})$	20	0.66
Tellurium	$(10\bar{1}0)$	20	0.43
Magnesium	$(10\bar{1}2)$		
	$(10\bar{1}1)$		
	$(10\bar{1}0)$		
α-Fe	(100)	−100	26
α-Fe	(100)	−185	27.5
Rock salt (dry)	(100)		0.44

* From (R4.2).

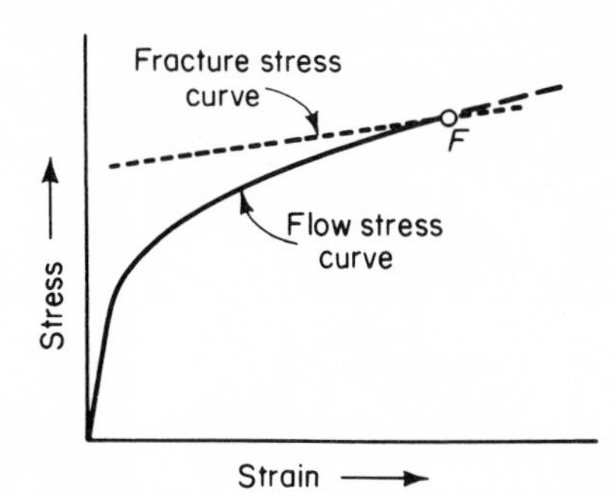

Fig. 9.6 Proposal of Ludwik suggesting that fracture occurs at a point, *F*, where flow stress and fracture stress vs. strain curves intersect.

Fig. 9.7 Relative positions of flow stress, cleavage fracture stress, and shear fracture stress vs. strain curves to produce: (a) Ductile shear failure. (b) Ductile cleavage failure. (c) Brittle shear failure. (d) Brittle cleavage failure.

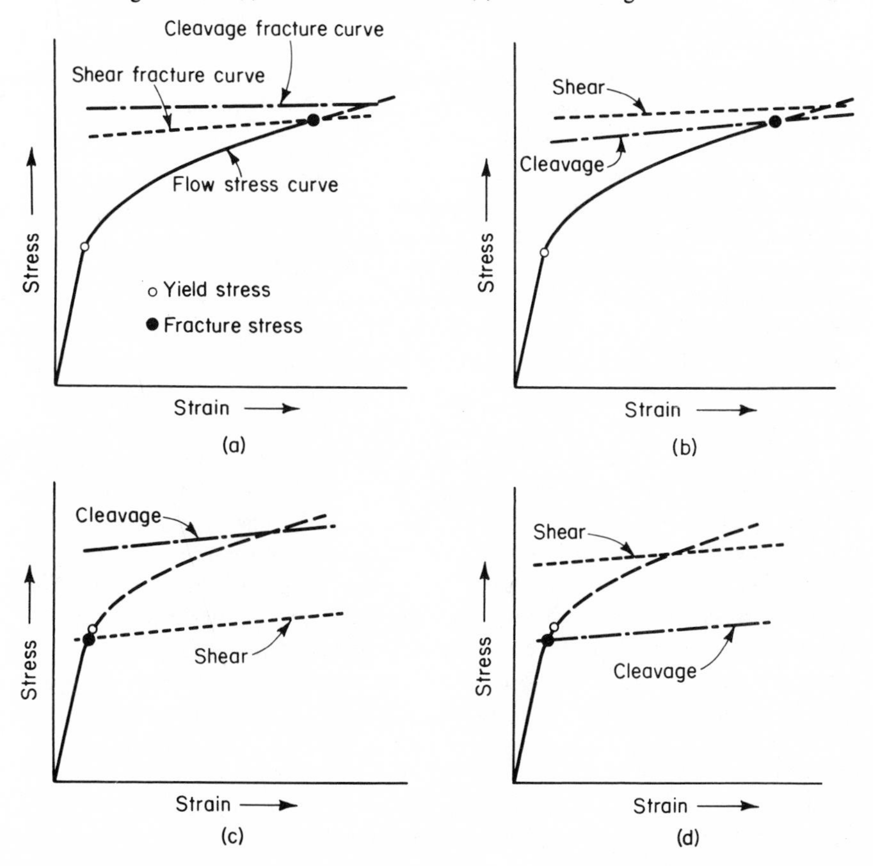

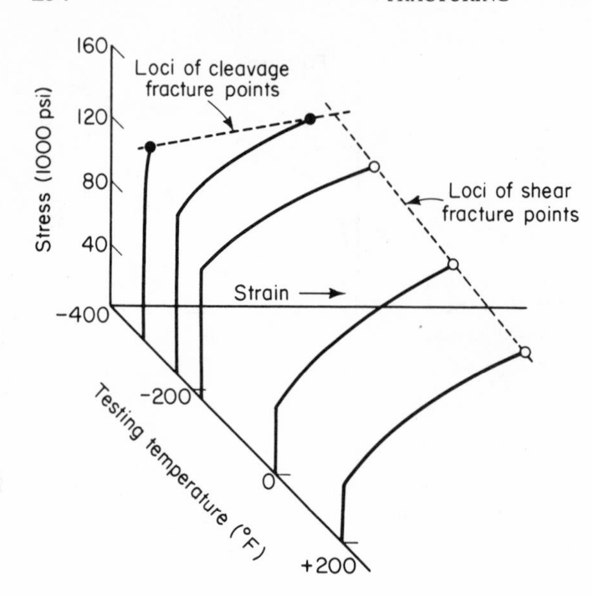

Fig. 9.8 Stress-strain curves at a variety of temperatures showing the fracture changes from shear to cleavage. (Stress-strain curves from R9.4.)

So long as this curve lies above the flow curve, the body deforms on loading. As the flow stress increases with strain, the flow and fracture curves eventually intersect and the material breaks. Unfortunately, unless some testing variable such as temperature or stress state is changed, only this single point can be obtained on the fracture stress–strain curve.

Following Ludwik's original suggestion, N. N. Davidenkow and F. F. Wittman pointed out in 1937 the necessity for having two fracture stress–strain functions, one for each of the two fracture modes, since separation could be either by shear or cleavage. By a combination of three stress–strain curves, one for flow and two for rupture, any type of fracture—ductile or brittle with either cleavage or shear separation—is possible (Fig. 9.7). According to this concept, a completely brittle fracture occurs if the fracture stress, σ_f, is less than the yield strength, σ_o. This fracture stress–strain concept is helpful in describing the influence of some testing variables. For instance, when mild steel is tested at continuously lower temperatures, the fracture mode changes from shear to cleavage as shown in Fig. 9.8. The fracture stress vs. strain and temperature curves are shown dashed in this figure. Note that the ductility barely changes over the temperature range where fracture is by shear. With the onset of cleavage, however, the ductility is abruptly reduced. It is because cleavage fracture can produce such a brittle or unsafe behavior in steel that a study of this transition is so important.

A number of objections have been raised to the use of a fracture stress–strain function. First, as explained above, the curve cannot be determined experimentally since only one fracture point can be observed as a function of strain so long as all other variables are held constant. This inability to locate the curve by direct measurements is a serious limitation, but even finding it experimentally would contribute little to explaining the underlying mechanism. Actually, the concept itself may be incorrect since it implies that metals may fracture with zero plastic strain, and this appears never to occur.

LAWS OF FRACTURING

Just as a number of empirically developed "laws" have been proposed to relate the yield stress to the stress state, so attempts have been made to predict the fracture stress under multiaxial stress states from the measured fracture stress in uniaxial tension. These, however, have been found to be less reliable than their counterparts such as the energy of distortion criterion for plastic flow. Unlike the two commonly used plasticity criteria, each of the fracturing laws predicts widely different behaviors as stress states are varied.

Maximum Shear Stress Theory

The fracturing law that appears to best represent the effect of stress state for plane stress of ductile metals is an extension of the Tresca criterion for plastic yielding. When one of the three principal stresses vanishes, this criterion is represented by a hexagon (Fig. 6.3). As the body hardens during straining, the hexagon expands because the yield stress, σ_o, continuously increases (Fig. 9.9). When the expansion reaches the point at which the uniaxial yield stress, σ_o, equals the uniaxial fracture stress, σ_f (the point

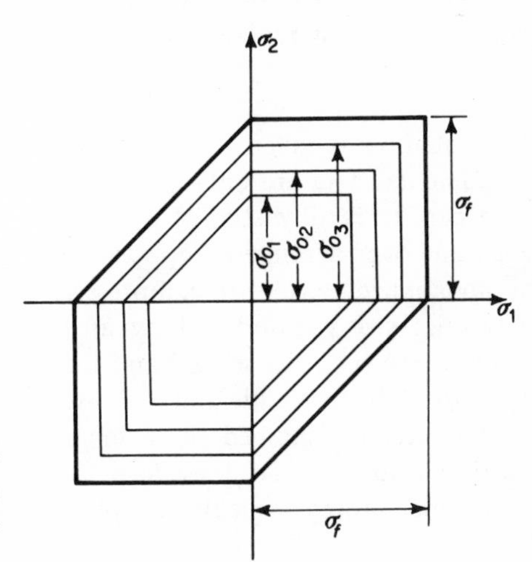

Fig. 9.9 Initial yield strength of member is σ_{o_1}; as it hardens the yield strength increases to σ_{o_2}, σ_{o_3}, etc. When $\sigma_o = \sigma_f$ fracture occurs.

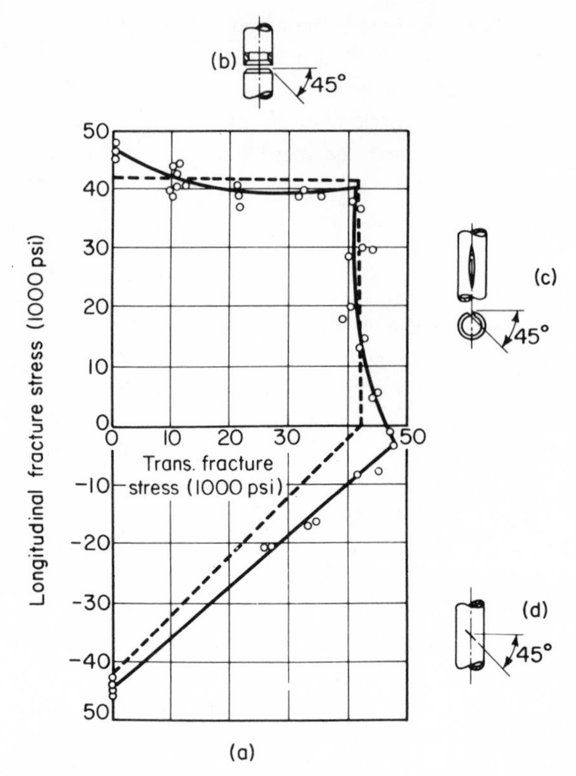

Fig. 9.10 (a) Fracture stress of magnesium tubes (J-1 alloy) as ratio of σ_1/σ_2 varies. Dashed curve is Eq. (9-1). Cusps in experimental data are thought to result from anisotropy. (b), (c), and (d) are drawings of fractured specimens tested on horizontal, vertical, and sloping portions of curve, respectively. From J. E. Dorn, in *Fracturing of Metals*, *ASM* (1948).

represented by the intersection of the two curves in Fig. 9.6), the resultant hexagon (outlined heavily in Fig. 9.9) describes the proposed law. Hence, the maximum shear stress criterion for fracturing would be written identically with (6.2) by replacing σ_o with σ_f:

$$-\sigma_f \leq \sigma_1 - \sigma_3 \leq \sigma_f \tag{9.1}$$

The equation is given in this fashion rather than that of (6.1b) because experimental data are only available for plane stress. This particular stress state can be readily studied on tubular specimens in the first and fourth quadrant over a range of σ_1/σ_2 ratios. The longitudinal stress is supplied by an applied tensile or compressive load along the tube axis while the tangential stress is induced by an internal pressure. Tests under radial compression (i.e., the second and third quadrant) are avoided because they cause the tube to buckle and collapse prematurely.

Test results obtained on a magnesium alloy are shown in Fig. 9.10. Further evidence that these are shear fractures is seen in the fracture appearance, three typical examples of which are sketched in this same figure.

An equation probably exists that defines the fracture stress as a function of all three principal stresses, ($\sigma_f = f[\sigma_1, \sigma_2, \sigma_3]$). It represents a surface analogous to the hexagonal prism in Fig. 6.2, but its exact shape is not known at present. It is reasonable to assume, however, that it differs from the one in Fig. 6.2 since fracture can occur when $\sigma_1 = \sigma_2 = \sigma_3$, although this could not be a shear fracture.

Maximum Normal Stress Theory

Many materials, especially those subject to cleavage, break when the applied normal stress reaches some critical value, σ_f; i.e.,

$$\begin{aligned} \sigma_1 &\geq \sigma_f \\ \sigma_2 &\geq \sigma_f \\ \sigma_3 &\geq \sigma_f \end{aligned} \tag{9.2}$$

The three dimensional diagram representing this condition is shown in Fig. 9.11. The figure consists of three crosshatched planes, none of which intersects the axes on the compressive side, suggesting that materials cannot rupture when subject to compressive stresses alone. Such a stress state is difficult to realize other than for purely hydrostatic compression. The application of any other compressive system, such as uniaxial compression, results in secondary tensile or shear stresses which may lead to fracture. This is discussed in greater detail in Chapter 11.

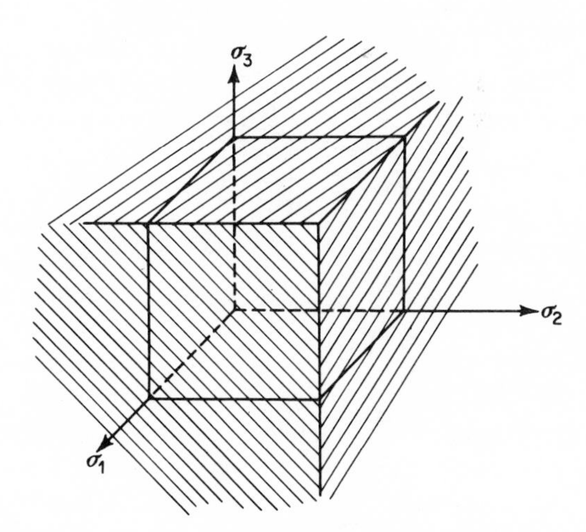

Fig. 9.11 Three-dimensional diagram illustrating the maximum normal stress theory of fracture.

Mohr's Strength Theory

At about the turn of the century, O. Mohr proposed a "failure" law for predicting either the yield strength or fracture stress of a material as a function of stress state. Like most of the workers in this field at the time, he was trying to develop design criteria so that he considered either permanent deformation or fracture as a "failure." His concept was initially presented in a form that was really a generalization of the maximum shear stress yield criterion to include those substances whose strength in tension and compression differed. Much later (1934) the hypothesis was extended by A. Leon to predict cleavage fractures as well.

Mohr proposed that a material fails when either the shear stress across a plane is greater than some critical magnitude or when σ_1 exceeds a certain limiting value. The critical τ, however, depends on the normal stress acting across the same plane. Hence,

$$\tau = f(\sigma) \tag{9.3}$$

The use of Mohr's strength law, just like the use of his stress circle (see page 37), requires that one plot the stresses in σ, τ coordinates, with normal stresses along the abscissa and shear stresses along the ordinate. A typical plot of this type, known as *Mohr's envelope*, is shown in Fig. 9.12. It consists of two branches which are symmetrical about the σ axis and diverge toward the compression side. The shaded area between the curves contains safe combinations of normal and shear stresses while that outside represents failures. The higher permissible shear stresses to the left are merely an expression of the well established fact that many materials, and especially brittle ones, have a greater resistance to compressive than to tensile forces.

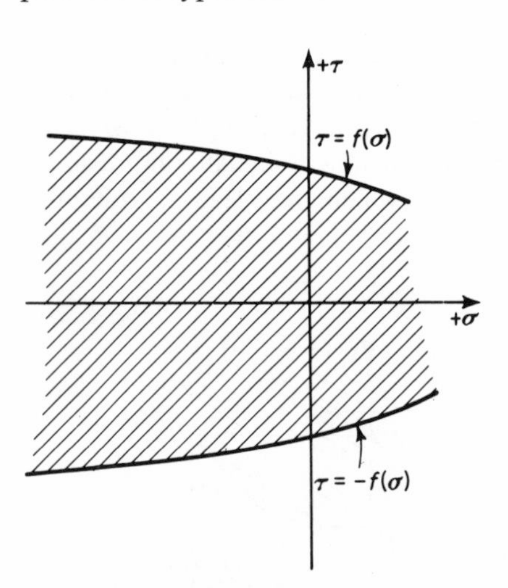

Fig. 9.12 Mohr's envelope. Safe stresses lie within shaded area.

As expected, there is a relationship between Mohr's envelope and the stress circle obtained at the most highly stressed point in the body. To show the connection between these two, the case of three-, as well as two-dimensional stress states will be considered, even though Mohr's circle was only discussed for biaxial stress in Chapter 2. Fortunately, only a few points on the figure, representing the three dimensional stress state, are used in connection with the envelope and their significance is easily seen when the two dimensional case is understood. Even where none of the three principal stresses is zero, Mohr's representation can still be shown on a plane. This is done by drawing three circles with diameters $(\sigma_1 - \sigma_2)$, $(\sigma_2 - \sigma_3)$, and

$(\sigma_3 - \sigma_1)$. The largest of these, $|\sigma_3 - \sigma_1|$, just encompasses the other two (Fig. 9.13) and is called the *large principal stress circle.** The maximum shear stress, τ_m, is equal to $(\sigma_3 - \sigma_1)/2$ (see Eq. (2.18)) which is the radius of the large circle. Mohr, like Tresca, assumed that the intermediate stress was unimportant so that in drawing the enveloping curve, AB or $A'B'$ in Fig. 9.13, only the large stress circle need be used to locate tangents, P and P'. If a series of circles is drawn for a point in the body and each circle is large enough to cause yielding under the particular stress state it represents, their envelope gives the condition for plastic flow. If these circles are of a size representative of fracture rather than flow, AB and $A'B'$ define the fracturing conditions. Mohr's envelope has found most use in dealing with brittle materials such as rocks. These substances will sustain greater shear stresses if the mean stress $\sigma_a < 0$ than if $\sigma_a > 0$. Mohr's theory has been used in geology, for example, to explain faulting in the earth's crust. If the material were such that it had the same strength (yield or fracture) in tension and compression, which is the case for yielding of ductile metals, Mohr's envelope would be represented by a pair of horizontal τ lines. The "failure" hypothesis then reduces to Tresca's criterion for yielding and a maximum shear law for fracturing.

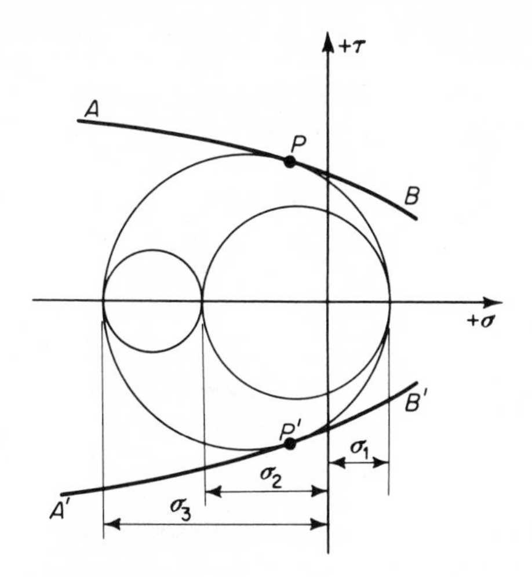

Fig. 9.13 Mohr's circle for a three-dimensional stress state and the "failure" envelope.

The envelope, $\tau = f(\sigma)$, can be defined if the fracture stress circle for three different stress states—e.g., tension, torsion, and compression—is known. Mohr's circle in uniaxial tension (see Fig. 2.12) has its center at $\sigma_1/2$, for torsion the center is at the origin ($\sigma_1 = -\sigma_3$, Fig. 2.13), and for compression it is at $\sigma_3/2$. These three circles and a typical envelope for a material are illustrated in Fig. 9.14. According to this diagram, fracture in uniaxial compression does not occur on the plane of maximum shear stress, τ_m, at 45° from the direction of σ_1, but at some angle $\phi < 45°$. This is found to be the case in materials like marble. Rocks and other brittle materials cannot be readily tested in tension or torsion, however, so that a number of different triaxial compressive stress states are usually used to define the envelope.

* For a more complete discussion of the stress circle for triaxial stresses, see (**R**9.3).

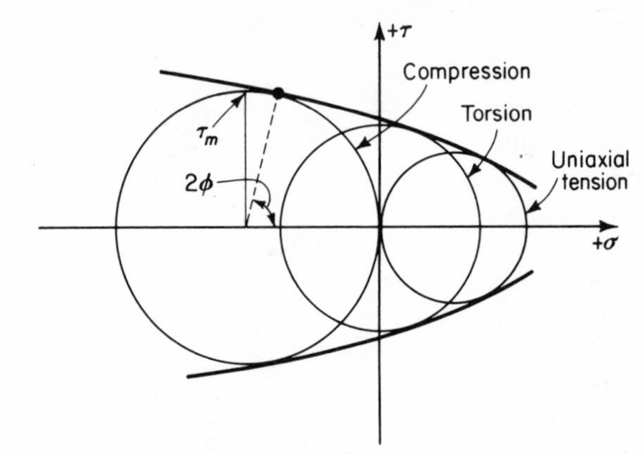

Fig. 9.14 Mohr's envelope for a material can be defined by means of the stress circles for uniaxial tension, torsion, and compression.

A useful Mohr envelope is given by a pair of straight lines (Fig. 9.15) in which

$$\tau = \pm(\tau_1 - \mu\sigma) \tag{9.4}$$

τ_1 is the fracture shear stress in the absence of normal loads. The direction between the maximum principal stress and the normal to the fracture plane for this case is the same for any stress state:

$$\phi = \frac{\tan^{-1}(1/\mu)}{2} \tag{9.5}$$

The term, $\mu\sigma$, is similar to a friction force across the fracture plane so that the proportionality constant, μ, is sometimes considered a coefficient of internal solid (Coulomb) friction. Eq. (9.4) has been found to represent the fracturing characteristics of rocks quite well.

It is more difficult to attach physical significance to the ends of Mohr's envelope than it is to its middle portion. The envelope appears to be open to the left since it could only cross the negative σ-axis if fracturing resulted when $\sigma_1 = \sigma_2 = \sigma_3 < 0$, and neither flow nor fracture appears to occur under this condition. (Actually, A. A. Griffith showed that some materials do fracture with only negative stresses acting, but these stresses cannot be equal.) Mohr's envelope is perpendicular to the positive σ axis at its crossing point, and the distance between the apex of the envelope and the origin of the coordinate system appears to be significant so far as

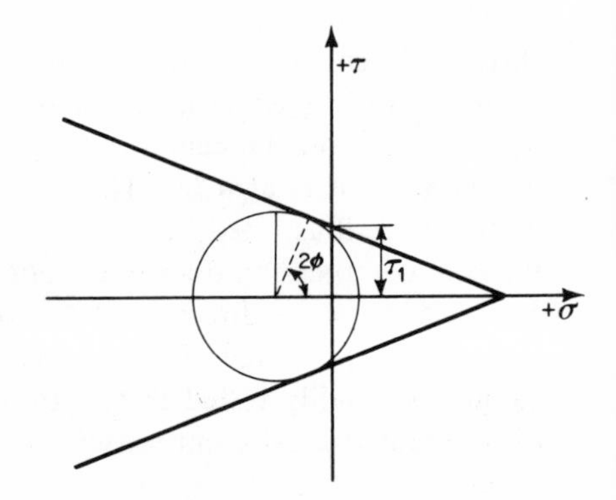

Fig. 9.15 A simple form of Mohr's envelope, given by two straight lines, has been found useful in describing rocks.

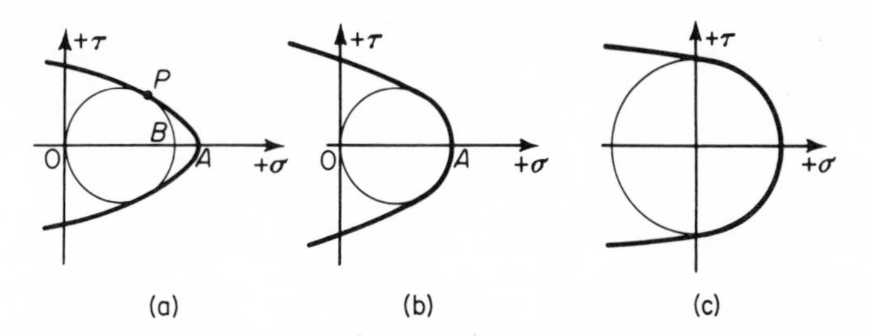

Fig. 9.16 Three different ways for Mohr's envelope to cross the positive axis.

predicting whether fracture will be shear or cleavage for a particular stress state. If the distance OA in Fig. 9.16a is greater than the diameter OB of Mohr's circle for uniaxial tension, a tensile specimen of the material will fracture in shear since the stress circle touches the envelope at point P. If the material is such that its Mohr's circle in tension has a diameter OA equal to the distance from the origin to the apex (Fig. 9.16b), a body of the material stressed in tension will form a cleavage separation surface normal to σ_1. The material described by 9.16a will break only by cleavage if a triaxial tensile stress is applied. Such a material might be steel at room temperature. If the envelope has a very large curvature at its apex (Fig. 9.16c), both tension and torsion would produce cleavage fracture as is the case for cast iron and marble.

Bridgman's Fracture Theory

P. W. Bridgman has made a significant contribution to the understanding of fracture by developing data on the fracturing characteristics of materials under very high pressures. His results (shown for a typical steel in Fig. 9.17) indicate that increasing superimposed hydrostatic pressure increases the fracture stress. This might imply a fracture criterion of the type:

$$\frac{\sigma_f + \sigma_2 + \sigma_3}{3} = \text{c}$$

or

$$\sigma_f = c' - (\sigma_2 + \sigma_3) \tag{9.6}$$

However his test results were not in complete agreement with such a "law."

FRACTURE STRESS AND FRACTURE DUCTILITY

In discussing laws of fracturing, only one coordinate of the fracture point, the fracture stress, was considered. No mention was made of ductility, the other coordinate, although it is as much a function of stress state as is the fracture stress. There is good reason for this: established methods for designing structures cannot make direct use of ductility values, only of stress.

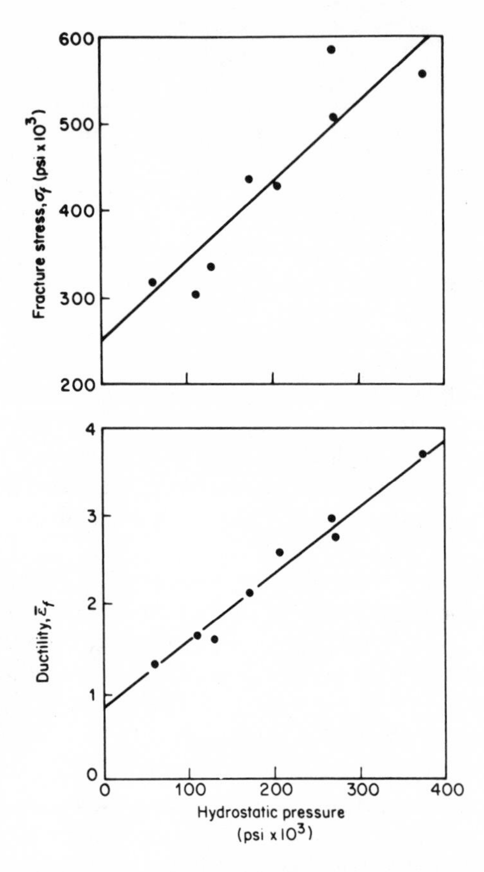

Fig. 9.17 Fracture stress and ductility of a medium carbon steel as a function of applied pressure. Data after P. W. Bridgman,

Yet ductility is very important since it is this factor that generally determines the "safety" of a structure. In many constructional parts, localized stresses developed during service may build up to dangerous levels; e.g., around sharp corners, at the root of threads, near keyways, etc. If the material from which the part is built has sufficient ductility under the conditions of use, it can permanently deform, redistributing these stress peaks. At present, materials are tested by such means as notched impact tests (Chapter 15), and the results are used for comparison with other materials that have performed satisfactorily in similar applications. Rather than use ductility in these evaluations, we often use *toughness*, where the latter is defined as the amount of energy that a member can absorb prior to fracturing. The toughness of the aluminum alloy wire described in Fig. 9.1 could be given as the area under the stress–strain curve. In units of inch-pounds per cubic inch, this would have the value of

$$W_t = \int_o^{\bar{\varepsilon}_f} \sigma \, d\bar{\varepsilon} \tag{9.7}$$

where $\bar{\varepsilon}_f$ = natural fracture ductility (strain)
σ = flow stress in psi.
In order to integrate (9.7), σ would have to be written as $f(\bar{\varepsilon})$. Means for doing this are described in Chapter 10, Eq. (10.6).

The fracturing laws described above ignored the complication that not only the fracture stress but also the ductility varies with stress state. No difficulty would be expected from this as long as the ductility is low. However, fracture laws lose their significance when applied to materials such as high strength steels ($\sigma_o > 250{,}000$ psi) which are brittle under one stress state and ductile under another. Because the ductility varies with stress state, the amount of strain-hardening is also changing. Hence, the fracture stress is determined on the virgin material for one stress combination while in others the material may have undergone considerable additional work-hardening. An extreme example of this is found in Bridgman's test already mentioned (Fig. 9.17).

MECHANISM OF FRACTURE

Although studies of fracturing on a macroscopic level have been continuing for many years, it is considerably more recent that reasonable mechanisms which may lead to rupture have been proposed. In order to describe these, it will be assumed that the solid has no imperfections and it breaks simply by a normal separation of planes of atoms. Using this model, the fracture strength can be calculated by following the arguments proposed by Orowan in 1949 and 1955. This discussion proceeds along lines very similar to those used in Chapter 7 to find the yield strength of an ideal crystal. The starting point now is a differentiation of the energy vs. distance-between-atoms curve shown in Fig. 4.2c, resulting in a stress-distance curve (Fig. 9.18). The interval a is the spacing between atoms in their equilibrium positions. To move the atoms from these locations requires an increasing stress up to the distance Δa where the stress reaches its maximum value, σ_m. (Note that for normal separation of layers of atoms,

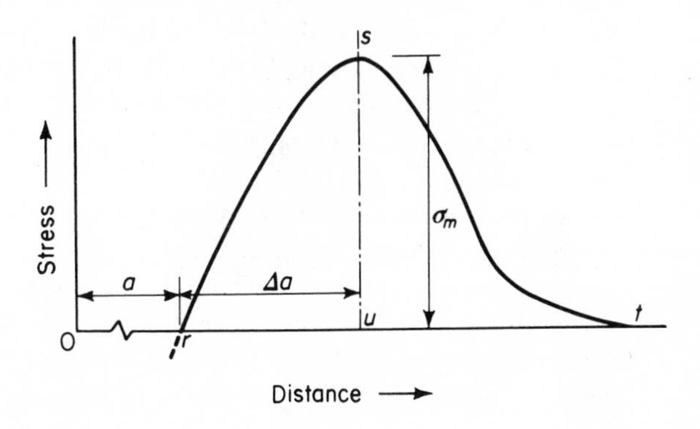

Fig. 9.18 Stress needed to move an atom away from its neighbor.

the force-distance curve is not periodic as was the case for slip.) Movement beyond Δa then becomes increasingly easier. The area under curve *rst* is the energy needed to produce fracture over a unit of cross section. Fracturing produces two new surfaces, each having an energy W_α per unit area. If one assume areas *rsu* and *stu* to be equal, and *rs* a straight line up to σ_m, the sum of the elastic areas under the two halves of the curve is (Δa) (σ_m). The distance Δa can be written as a strain, $\varepsilon = \Delta a/a$, so that the energy becomes equal to $a\varepsilon\sigma_m$ or $\sigma_m^2\ a/E$, where E is Young's Modulus. When rupture occurs, this energy term is balanced by the surface energy of the two sides of the crack, $2W_\alpha$. Hence

$$\frac{a\sigma_m^2}{2E} = W_\alpha$$

or (9.8)

$$\sigma_m = \sqrt{\frac{2EW_\alpha}{a}}$$

On applying (9.8) to a typical metal, σ_m is found to have a value of about 4×10^6 psi which is 20 to 1000 times higher than the generally observed strength. The strength of the "whiskers" described on p. 167, however, does approach this theoretical value.

Griffith Cracks

This wide discrepancy between calculated and observed fracture strengths implies that there is something in ordinary material that concentrates the average, relatively low, applied stress so that over some small regions the theoretical value of fracture stress is reached. A. A. Griffith was a pioneer, not only in pointing out the importance of subtle stress concentrations but also as the first to interpret the fracture phenomenon by applying the technique used in the preceding paragraph; i.e., equating the strain energy that is released by cracking to that needed for the formation of the new surfaces. His work was described in 1920–21 (R9.2) and is considered a classic in fracturing studies. He assumed that the brittle amorphous materials with which he worked (glass, fused silica) contained many small invisible cracks, and that fracture resulted when the applied stress caused these to grow to macroscopic size.

To find the amount of strain energy released by the crack as it propagates, the magnitude of the stress at the tip of the crack must be known. When a load is applied to a body in which there are discontinuities, such as Griffith's cracks, the stress has a much higher than average value near the tip of a crack. Inglis, using the techniques of the theory of elasticity, had calculated the stress distribution near the end of the major axis of an elliptical hole

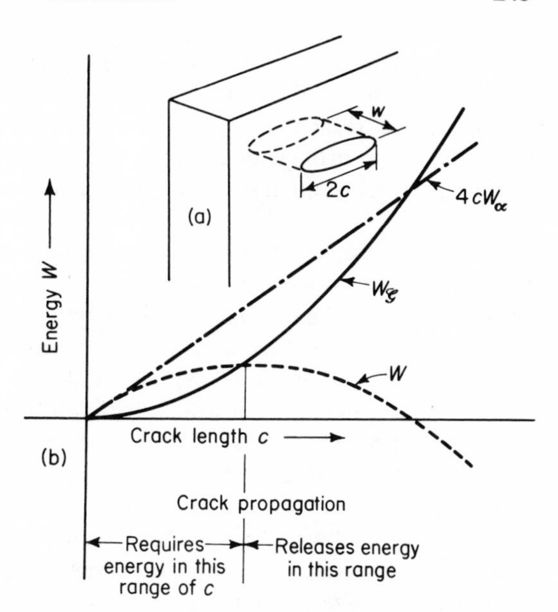

Fig. 9.19 (a) Griffith Crack. (b) The energy W required to propagate a crack; first increases and then decreases as the crack becomes longer.

even before Griffith's work. Griffith assumed the cracks in glass to be slots with a flat elliptical profile (Fig. 9.19a), and used Inglis' equation to determine the strain energy released as initially present cracks grew. This strain energy released, $W_\mathscr{G}$, was

$$W_\mathscr{G} = \frac{\pi(1 - \nu^2)\sigma^2 c^2}{E} \tag{9.9}$$

where $2c$ = length of a crack that extends completely through a unit thick plate ($w = 1$)
E = Young's Modulus
ν = Poisson's Ratio

The surface energy of a crack of unit width and $2c$ units long is $4cW_\alpha$. Hence the energy for its propagation, the difference of this quantity and Eq. (9.9) (Fig. 9.19b) is

$$W = -\frac{\pi(1 - \nu^2)\sigma^2 c^2}{E} + 4cW_\alpha \tag{9.10}$$

The crack will propagate spontaneously when $dW/dc = 0$, giving

$$\sigma = \sqrt{\frac{2EW_\alpha}{\pi(1 - \nu^2)c}} \tag{9.11}$$

i.e., the stress to cause the crack to grow is proportional to $c^{-\frac{1}{2}}$. Griffith checked his theory by artificially producing cracks in sheets of silicate glass and he found that $\sigma^2 c$ was a constant ($= 2EW_\alpha/\pi(1 - \nu^2)$) which is in agreement with Eq. (9.11). His experimental value for $\sigma^2 c$ and the strength of the glass before artificially cracking it enabled him to estimate the size of the largest cracks initially in the glass. These were found to be about 10^{-4} inches long.

In Griffith's calculations and experimental work, the cracks went completely through thin sheets. More recent calculations (by Sneddon)* assuming an internal, penny shaped crack did not greatly differ from Griffith's results.

Although it may become necessary to modify Griffith's hypothesis of pre-existing cracks as more experimental data are collected, his original suggestion is generally consistent with all the mechanical characteristics presently attributed to glass. For example, in addition to the fact that glass is weak and brittle, its strength also varies greatly from specimen to specimen and this, of course, is consistent with the concept of a distribution of flaws. Further, glass shows a pronounced section size effect; i.e., its strength decreases as the test specimen cross section is increased (Griffith showed an appreciable size effect in glass fibers with diameters varying from 0.13 to 4.2 mils). This size dependence is thought to result because pieces with small cross sections are less likely to contain large flaws than are large specimens. There is some question about the interpretation of these tests, however, since thin fibers and thick ones undergo a somewhat different strain and temperature cycle during their production. Nevertheless, a section size effect also shows itself as the specimen length is increased. A striking experiment of this type was described by O. Reinhober who fractured a silica fiber, fractured the fragments, again tested the fragments of the fragments, etc., and found that each generation became progressively stronger—presumably by eliminating the largest flaw in each test. Direct evidence of surface flaws in glass has also been obtained by "decorating" the glass with sodium vapor. In this technique, the sodium deposits on the cracks make them visible so that one can see flaws that ordinarily are invisible. Not only are the flaws found, but their size does indeed correlate with the strength of the glass.

Calculations of theoretical fracture strengths, involving energy balance equations between the strain energy released by cracking on the one hand and the energy of the newly formed surface on the other, have been widely applied even to such diverse materials as rocks and liquids. Application of this concept to metals, or even to relatively glass-like materials such as polymethyl methacrylate ("Plexiglas" or "Lucite") requires some modification, however, because these do not break in a completely brittle fashion

* I. N. Sneddon, Proc. Phys. Soc. London, **187**, 229 (1946).

to produce a plane fracture surface. Even in those cases where rupturing can be described as brittle in a macroscopic sense, there is always some plastic flow in the volumes bordering the fracture surface. Metals, for example, show a "river" fracture surface (Fig. 9.20a). This roughness is thought to result from tearing between parallel cleavage planes (Fig. 9.20b) and the localized flow requires an appreciable amount of energy. Irwin (1948) and Orowan independently proposed that the energy irreversibly consumed as plastic flow per unit area, W_p, must be added to the surface energy term, W_α. However, W_p is orders of magnitude larger than W_α so that the latter can be neglected and Eq. (9.8) then becomes

$$\sigma \approx \sqrt{\frac{(EW_p)}{c}} \tag{9.12}$$

This is sometimes referred to as the Griffith-Irwin-Orowan theory. Felbeck and Orowan tested this equation by artificially introducing cracks into mild steel plates and found the fracture stress was proportional to $c^{-\frac{1}{2}}$, and W_p had a value that was close to that obtained by other methods of evaluating it.

Fig. 9.20 (a) Cleavage step "river pattern." 3% Si-Fe single-crystal cleavage surface. Cleaved at 78°K. Direction of crack propagation is from top to bottom of the photograph. (b) Schematic representation of cleavage step formation between two parallel cleavage cracks on differing crystal planes. From J. R. Low, Jr., in *Fracture*. See (R4.3).

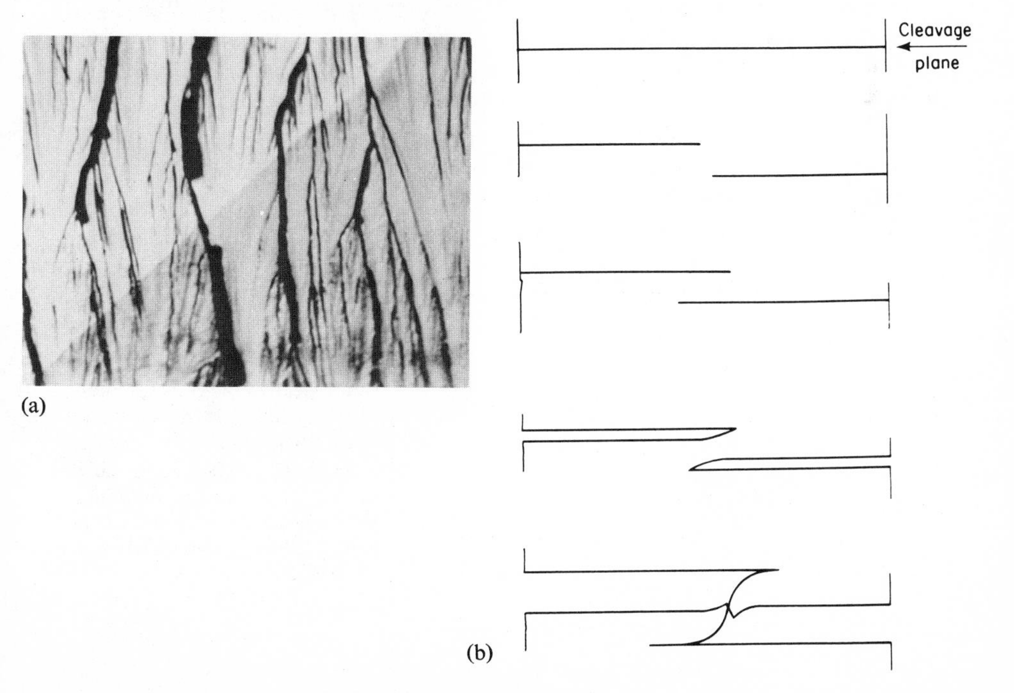

FRACTURE THEORIES BASED ON DISLOCATION MOVEMENTS

Griffith's equation, as applied to a truly brittle material [Eq. (9.12)], and its modification to account for energy used in nonelastic deformation [Eq. (9.12)], does much to aid our understanding of the low fracture stress of the common structural materials. Many of the characteristics that have been learned about the fracturing of metal, however, are not explained by Eq. (9.12) other than by making numerous assumptions regarding W_p. For example, cleavage occurs in b.c.c. and some h.c.p., but not in f.c.c. metals. Furthermore, the ductility of these first two classes of metals is very sensitive to the temperature of deformation. They become brittle at low temperatures while f.c.c. metals generally do not. This tendency toward brittleness is enhanced by the presence of interstitial elements (see Chapter 4) such as carbon and nitrogen.

A mechanism is needed to more fully explain the fracturing characteristics of metals as well as those of ductile cubic ionic crystals. It will be necessary to consider cleavage and fibrous fractures separately since these are thought to be caused by different mechanisms. The first of these can result in catastrophic failures as shown by the bridge failure in Fig. 9.3. For this reason cleavage cracks have been the most intensively studied. Cleavage is not a single continuous event but rather a two step process consisting of first, initiation, and then propagation of a crack.

The permanent deformation that always appears to precede fracturing, even in the most brittle metals, is thought to be the process by which cracking is started. There is substantial evidence indicating the need for some plastic flow before rupture can occur in metals. It follows, therefore, that crack-free metals should not fracture at a stress below their yield strength. This is supported by Low's experiments which showed that a steel broke brittlely in tension, or flowed in compression, at identical stresses irrespective of grain size (Fig. 9.21). In addition, the Griffith cracks that would be required in constructional metals to reduce the theoretical strength to the observed strength would have to be too large to have been undetected. (Their calculated lengths, according to E. Parker, are of the order of 0.1 inch for ductile

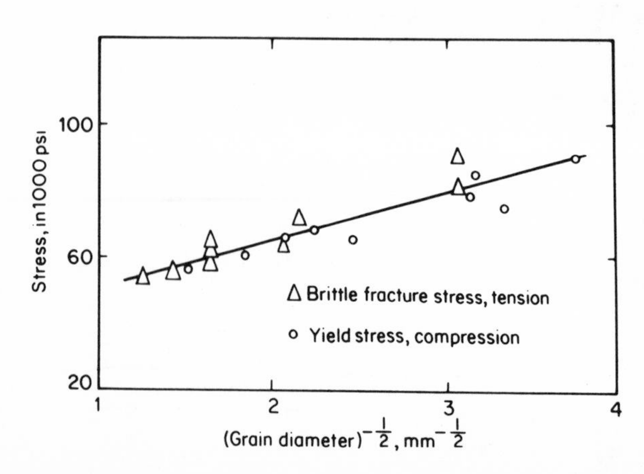

Fig. 9.21 Effect of grain size on yield strength in compression, and brittle fracture stress in tension, on mild steel at −196°C. After J. R. Low, Jr., "Deformation and Flow of Solids," *I.U.T.A.M. Madrid Colloquim*, Springer-Berlin, 1956, 60.

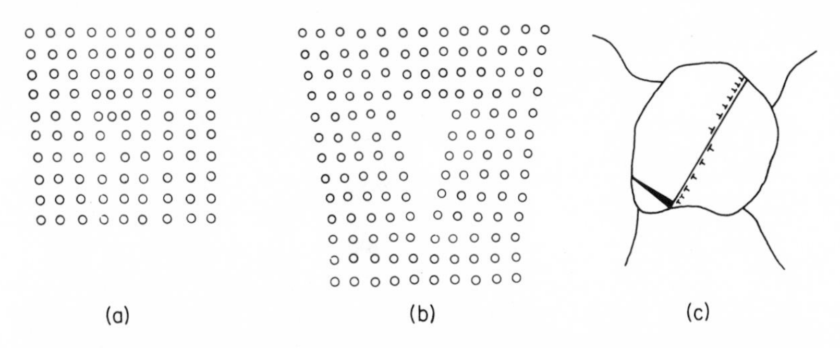

Fig. 9.22 (a) Edge dislocation with Burgers vector **b**. (b) Coalescence of three dislocations Burgers vector 3 **b**. (c) Microstructure of a crack due to a dislocation pile-up at a grain boundary. (d) Schematic drawing of slip band intersection showing a dislocation piled up and crack. (e) Crack at the intersection of slip bands in a MgO single crystal. The arrows alongside the bands indicate the shear within the bands. Magnification: 500. From W. G. Johnston, *Phil. Mag.*, **5**, Ser. 8, 407 (1960).

metals as compared with 10^{-4} for glass.) Although it may be possible for such cracks to exist in some cases where one has a multiphased microstructure, entrapped foreign materials, etc., in most instances cracks of this size would not go undetected. The following section is concerned with the initiation of cracks in reasonably perfect samples. In fabricated structures, on the other hand, cracks of sizes in excess of that required, according to Parker, are likely to occur in the process of manufacturing.

C. Zener (1948) pioneered the concept that fracturing was a natural outgrowth of plastic flow and, indeed, stated that flow was necessary for the formation of cracks. He suggested that rupturing resulted from dislocations piling up at an obstacle. In Chapter 7 this was also stated to be one cause for strain-hardening. The difference in the case now considered is that after piling up, the dislocations coalesce. According to Zener, the coalesced pileup acts like a freely slipping crack that moves on the slip plane under the action of a shear stress. His proposal can be most simply visualized by considering a coalescence of edge dislocations. A single edge dislocation (Fig. 9.22a) looks much like a small crack on an atomic scale and, when a number of these run together so that there is no space between them (Fig. 9.22b), it becomes easy to imagine the beginnings of a microcrack.

Ten years later A. N. Stroh pointed out that the stress concentration around a pileup could be large enough to initiate and propagate a crack on a cleavage plane. If the obstacle to dislocation movement were a grain boundary, the microstructure of the piled-up group and crack would look like that in Fig. 9.22c. Cracks that do appear to form by this method have been found in magnesium oxide crystals. Slip bands may serve as obstacles to dislocations moving on intersecting slip bands. A schematic drawing of

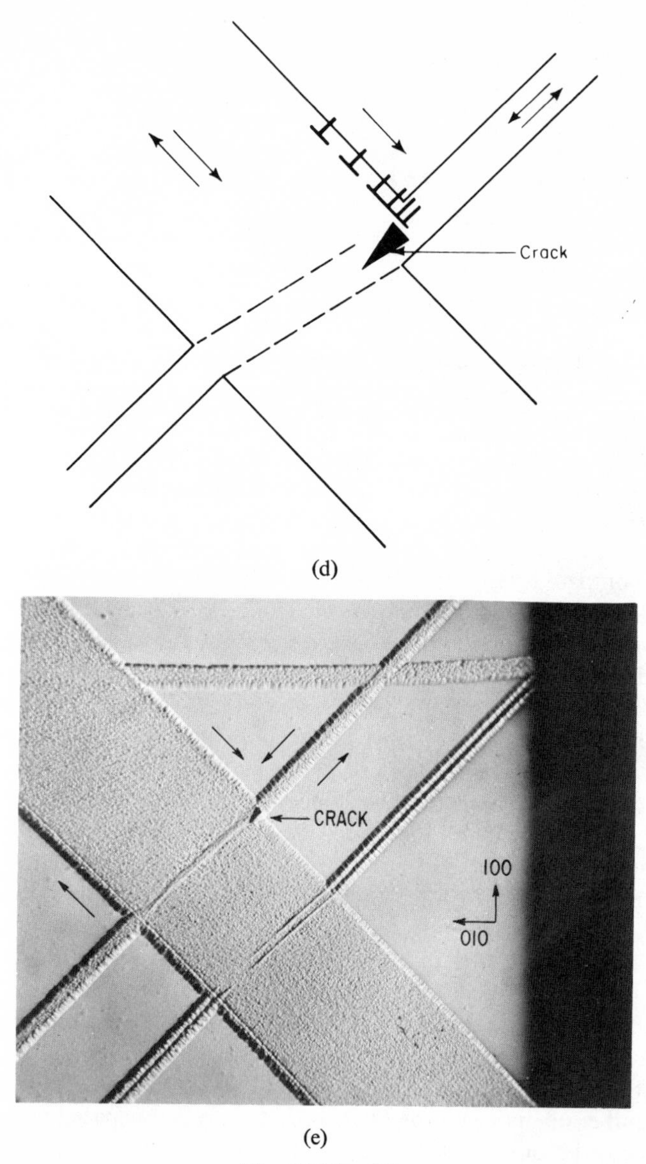

(e)

Fig. 9.22 (con't.)

this proposed mechanism is shown in Fig. 9.22d, and a crack formed this way is shown in (e). Not only grain-boundaries and slip bands, but other obstacles such as twin interfaces, may prohibit glide as well. At low temperatures steel deforms by twinning easier than by slip. This throws many obstructions in the path of gliding dislocations and hence contributes to the low temperature brittleness of steel.

A. H. Cottrell has proposed still another means for dislocations on intersecting planes to glide together and convert to "cavity dislocations." Evidence of this type of crack has also been found in magnesium oxide crystals, and the process is expected to occur in b.c.c. metals although it has not been verified by experiment as yet. Cavities in this case are formed on the (001) cleavage plane by the intersection of one dislocation with Burgers vector $a/2[\bar{1}\bar{1}1]$ slipping on the (101) plane, and another with Burgers vector $a/2[111]$ slipping on the $(10\bar{1})$ plane (Fig. 9.23). The cavity formation can be written as

$$\frac{a}{2}[\bar{1}\bar{1}1] + \frac{a}{2}[111] \rightarrow a[001] \tag{9.13}$$

It is interesting in this instance that, if a tensile stress, σ, had been applied to a single grain in a polycrystalline metal as shown by the arrows in Fig. 9.23, gliding would occur in the maximum shear direction and cracking would occur at 45° to slip; i.e., on a plane subjected to maximum normal stress. This is significant since after the crack attains some critical size it propagates as a Griffith crack under the influence of normal, not shear, stress. The coalescence of intersecting glide dislocations to form cavities is possible in the f.c.c. as well as in b.c.c. metals. According to Cottrell, however, coalescence releases elastic energy only in the b.c.c. system so that the cavity is stable just for this structure.

One serious flaw in the concept of fracture resulting from dislocation pile-ups, as outlined above, is that cracks, even the length of a single grain,

Fig. 9.23 Cavity dislocation formed on the (001) plane by glide dislocations on the (101) and $(10\bar{1})$ planes according to Eq. (9-16). (a) Three dimensional view. (b) Side view.

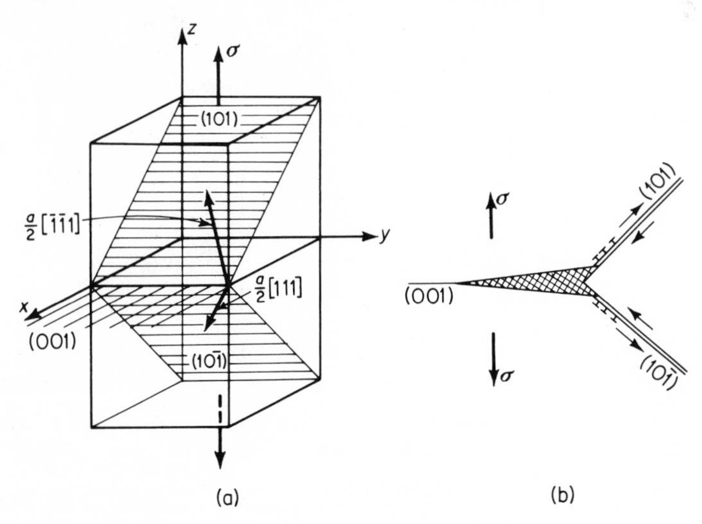

which are frequently found, would not be expected to result solely from the action of an applied stress. Any stress concentration about a pileup should cause additional plastic flow; i.e., activate new dislocations as discussed in Chapter 7. Hence, we must assume that, in addition to an applied stress, cavity dislocations can only develop if the elastic energy concentrated about the dislocations is not dissipated by plastic flow of nearby material. Consequently, crack initiation requires that slip be blocked. The conditions required for blocking are discussed in the following section. Other micromechanisms for cleavage fracture have also been proposed (see, for example, R9.1), but neither these suggestions nor those outlined above are consistent with all that is known about fracture.

TRANSITION TEMPERATURE

One phenomenon of metal fracturing that is of great technical importance, and which can be rationalized by dislocation concepts, is the *transition temperature*. This is actually a temperature range in which the ductility or toughness changes abruptly as shown for a steel in Fig. 9.24. The transition temperature is a function of stress state and of strain rate for strain-rate sensitive materials. The more positive the value of the mean stress, σ_a, the greater the tendency for embrittlement. The crack that split the girder (Fig. 9.3) was able to propagate across this large structure because the transition temperature of the ordinarily ductile steel was above ambient.

A. H. Cottrell and N. J. Petch (R9.1) have independently presented similar tentative fracturing laws to explain this behavior on the basis of dislocation pileups. We shall briefly review one of these (by Cottrell) since the resulting equation is consistent with much that is known about cleavage fracture. This development requires first that one write Griffith's equation so as to equate the energy of a dislocation pileup with the surface energy of a crack. The energy of the pileup is given by the product of the applied stress and the displacement between faces of the crack. Hence, by referring to a unit so that stress and force are equivalent,

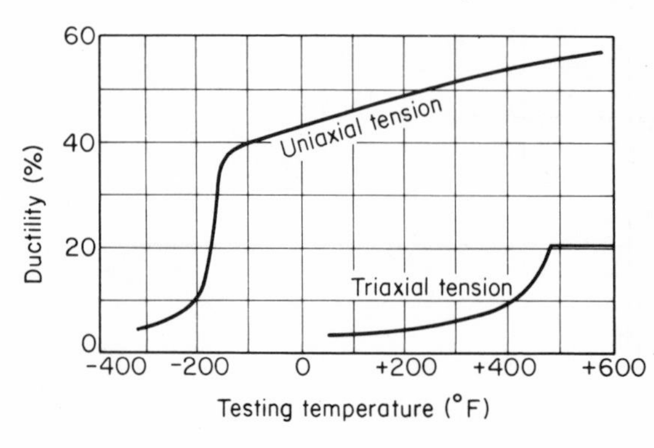

Fig. 9.24 Transition temperature of a steel tested in uniaxial tension and in triaxial tension. From E. J. Ripling, in "Symposium on Effect of Temperature on Brittle Behavior of Metals," *ASTM Spec. Tech. Pub.*, No. 158.

$$\sigma na = 2W_\alpha \tag{9.14}$$

where σ = applied stress
n = number of dislocations
a = interatomic spacing

The energy is the product of σ and the total displacement, na.
The magnitude of the total Burgers vector can also be expressed as

$$na = \frac{\tau - \tau_i}{G} d \tag{9.15}$$

where $\tau - \tau_i$ = "effective" shear stress (τ_i is a constant referred to as the "friction stress").
d = length of band containing the dislocations and can be taken as one half the grain diameter in a polycrystalline material.

Further, by analogy with (7.12), the yield stress in shear is

$$\tau_0 = \tau_i + k_y d^{-\frac{1}{2}} \tag{9.16}$$

If the fracture and yield stress are approximately equal, (9.16) can be re-written with the aid of (9.17)

$$na \approx \frac{k_y d^{\frac{1}{2}}}{G} \tag{9.17}$$

Crack propagation, according to Cottrell and Petch, is more difficult than crack initiation; consequently, the competitive processes on the application of a load are yielding and crack propagation. According to Griffith, crack propagation is controlled by the normal stress σ rather than shear stress τ. Now (9.17) may be substituted into (9.14) as long as the equality is corrected for the fact that (9.17) was based on shear stress while (9.14) involves normal stresses. The correction is made by introducing factor β which equals twice the ratio of shear to normal stress for the particular stress state considered; e.g., $\beta \approx 2$ for torsion, 1 for tension, etc.

$$\sigma k_y d^{\frac{1}{2}} = \beta G W_\alpha \tag{9.18}$$

If the left hand side of (9.18) is smaller than the right, ductile yielding results from the application of a load. If the reverse is true, the behavior is brittle. Notice that (9.18) makes fracturing a function of grain size (d), stress state (β), and temperature through temperature sensitive k_y. (Petch, on the other hand, states that temperature influences fracture by affecting W_α in (9.18) and τ_i in (9.16) rather than k_y.)

The development of (9.18) does not account for the occurrence of transition temperatures in b.c.c. and h.c.p. metals and its absence in f.c.c. The crystal system of the metal becomes important when one adds the requirement that slip must be blocked in order that dislocations pile-up to the extent required for cleavage fracture. Blocking in b.c.c. metals is associated with Cottrell interstitial atmospheres (see p. 189). When one dislocation breaks away from its atmosphere, an entire avalanche of dislocations can glide behind it on the slip plane. When these encounter an obstacle, their high velocity does not allow the initiation of new dislocation sources which would relax the stress by producing plastic flow, and a crack is formed instead. In h.c.p. metals, cracking rather than initiation of new sources occurs because of the limited number of slip planes in the material.

DUCTILE FRACTURE

Because of its commercial implications and catastrophic nature, cleavage has been more extensively studied from a dislocation viewpoint than fibrous fracture. However, ductile rupture appears to be the easier of the two to understand. The face-centered cubic metals, with the exception of those containing brittle networks on a microscopic scale, etc., are not only ductile at room temperature but, unlike commercial b.c.c. and many h.c.p. metals, they do not become embrittled by low temperatures. It has been suggested that fracturing of these is initiated at inclusions; such as oxides trapped during solidification, precipitated brittle phases, etc. Fractures initiated by these foreign particles cannot grow in the direction normal to the applied stress by the release of elastic strain energy as do Griffith cracks. Instead, they elongate in the direction of straining and, as the deformation continues and the space between these voids becomes smaller and smaller (Fig. 9.25), fracture occurs. H. C. Rogers (1959) stated that metal grains adjacent to the voids, or surface grains at the bottom of a notch, simply glide apart like single crystals. This continuously exposes new grains to the crack or void. These in turn have a free surface and glide apart, etc. According to this mechanism, very pure metals should not fracture until their cross section is reduced to zero. This has been found to be the case.

In commercial metals the flaws that control fibrous fracturing are not randomly oriented. During processing, such as rolling or drawing, inclusions and other flaws are lined up in the direction of the maximum extension. For this reason, their fracture properties are highly anisotropic. The fracture stress and ductility of commercial sheet, plate, or bars in the transverse direction may be less than half that in the rolling or drawing direction. H. W. Swift in 1939, and other investigators more recently, twisted rods by various amounts after which they were stretched to fracture. The original torsional prestrain oriented the flaws along a helical path so that the subsequent tensile fracture surfaces were also helical, and the fracture stress and ductility were lowered, (Fig. 9.26). If the prestrained rods were untwisted before pulling, the tensile fracture appearance was much like that of rods

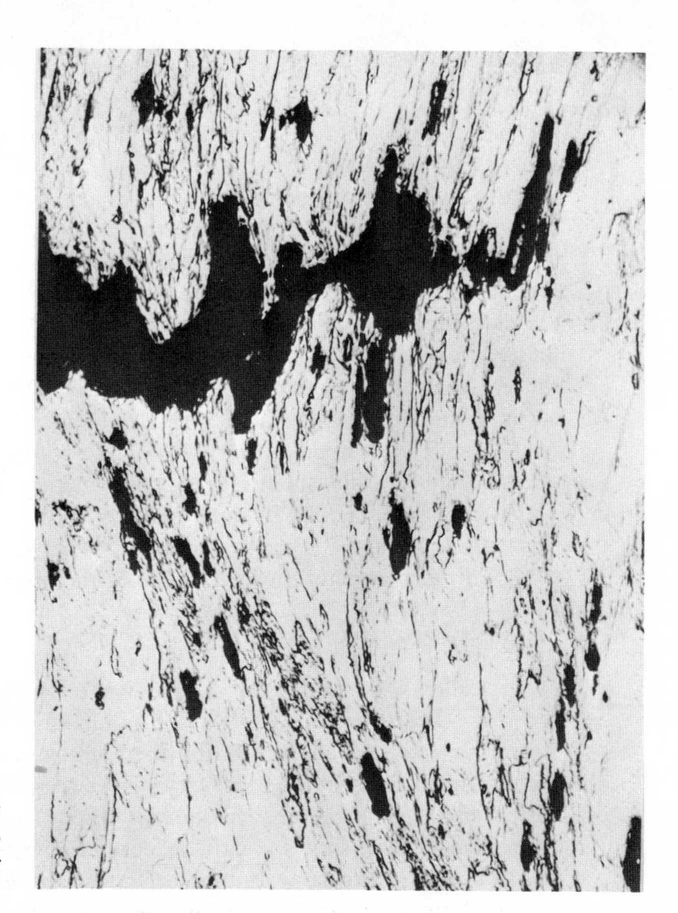

Fig. 9.25 Crack growth by coalescence of cavities. Courtesy of J. I. Bluhm, U.S. Army Materials Research Agency.

Fig. 9.26 Tensile fracture along a helical path in a pre-twisted steel tube.

(or tubes) that were not prestrained at all, and the fracture stress and ductility were recovered.

A second type of ductile fracture, a *shear fracture*, sometimes occurs on surfaces acted upon by large shear stresses; e.g., in machining metals. This type of failure has much in common with the fibrous type in that it seems to originate at initially present flaws, but differs from it in two respects: first, the deformation involved is generally restricted to a volume very close to the separation surface, and, second, the fracture propagates very rapidly. Indeed, some shear fractures seem to be a result of plastic instability and occur so rapidly that the deformation may be adiabatic; e.g., in punching out holes. The heat generated by the large plastic deformation over a very small volume locally softens the metal, making the process autocatalytic. This heat is large enough in some cases to produce a phase change in metal. Martensite, a phase formed by quenching steel from temperatures in excess of 1300°F, is frequently found near shear fractured surfaces of this material.

DELAYED FRACTURE (STATIC FATIGUE)

In dealing with the fracture mechanism of glass, it was assumed that this and other brittle materials always contain flaws that serve as stress concentrators and this accounted for their low fracture strengths. Actually, freshly drawn glass fibers have very high strengths and presumably few or no flaws. Glass, however, ages quickly and, after a few minutes exposure to the atmosphere, shows greatly reduced strength. The water vapor in the air is thought to react with the surface of the glass to form cracks that grow with time. When a load is maintained on a glass sample that is exposed to the water vapor in air, surface flaws grow relatively slowly until they reach the size of Griffiths cracks and then fracture occurs catastrophically. This time-dependent fracturing is termed *delayed fracture* or *static fatigue*. If the glass is baked out and tested in an inert atmosphere or vacuum, aging does not occur, and its strength becomes independent of time under load.

If the mechanism of delayed fracture is slow growth of a crack, the corrosion of stressed glass must be highly directional. A small surface crack that is attacked by a corrosive agent will have its sharp radius at the crack bottom rounded out if the surface were attacked at the same rate in all directions. There are reasons to believe that corrosion follows the path normal to the maximum tensile stress because this is the most expanded part of the lattice from which sodium ions can migrate (p. 94). Hence, the crack stays sharp as it continues to grow.

Another reason that has been proposed to explain the time-dependent strength of glass is that the crack has its surface energy (W_α) lowered by adsorbing the products of corrosion. The latter, of course, could work in conjunction with an anisotropic growth of the crack.

Not only glass, but also some metals, have been found to show delayed fracture. Steel will absorb large amounts of hydrogen; e.g., during electroplating.

In this hydrogen charged state, its fracturing characteristics become time-dependent. Petch and Stables attribute this to hydrogen being adsorbed on the fracture surface of the steel and a consequent lowering of the surface energy.

FRACTURE SURFACE MARKINGS (FRACTOGRAPHY)

Only in special cases can fracturing be watched while it is happening. Much that has been learned about the process is the result of devising mechanisms probable consistent with the observed markings and texture of the fracture surface. Indeed, in attempting to learn the cause of service failures, fracture markings are frequently the most helpful aid.

The chronology of ductile fractures as discussed on p. 254 would lead one to expect these to form a very rough surface due to stretching of the metal between the microvoids or flaws. The fracture of the steel tensile specimen shown in Fig. 9.27 (see also Fig. 10.9c) is the most common type of tensile failure found in ductile polycrystalline metals and, because of its characteristic appearance, is known as a *cup-and-cone fracture*. The flat bottom part of the fracture is the section over which the voids grow to form the first internal crack. As this crack expands, and the crack tip approaches the surface, the stress state at the crack tip changes from triaxial to biaxial. With this change in stress state, the metal becomes weaker in shear than in tension, accounting for the smooth inclined edges of the fracture.

Brittle materials also exhibit fracture surfaces that are partially rough and partially smooth but for a completely different reason. Brittle fractures

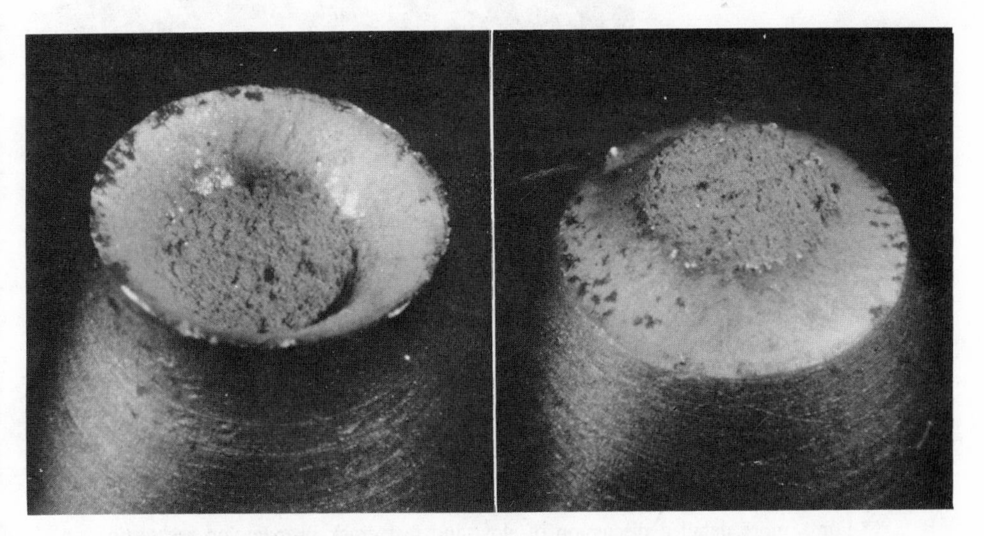

Fig. 9.27 Cup and cone fracture in a ductile steel tensile bar. (Light and dark spots on shear area are points where surfaces touched when halves were put together for measuring elongation.)

(a)

(b)

Fig. 9.28 (a) Surface of fractured glass rod. Fracture started at flaw on surface (top) and formed mirror during slow cracking, 27X. (b) Mirror in epoxy originating at scratch on surface, 54X. From (a) N. M. Cameron, University of Illinois, *T. and A. M. Report*, No. 242, 1963 and (b) C. E. Feltner, ibid, No. 224, 1962.

generally occur in two stages. While the crack is small, it expands at an increasing rate to its limiting velocity.* Glass fractures show a sharp boundary when the velocity attains this value. During slow growth the fracture surface is extremely smooth, forming a *mirror* which originates at the point of crack initiation (Fig. 9.28). The mirror, at least in glass, can be used as an internal stress gauge. By measuring its size, and using the

* For a more detailed discussion of slow and fast crack propagation see section on 'Fracture Mechanics", starting on p. 404.

techniques described in Chapter 15, one is able to calculate the stress at which fast fracture started.

Fast fracture in any material is limited to a velocity that is equal to about $\frac{1}{2}$ the speed at which a shear stress wave travels through the solid. This limits the rate at which stored strain energy can be dissipated in forming the new surfaces. In order to release as much strain energy as possible, additional fracture surfaces are formed. For this reason the fast crack in glass may branch into a number of surfaces by "forking," which explains the roughness or *hackle* of the fracture beyond the mirror in Fig. 9.28a.

The epoxy fracture, Fig. 9.28b, appears similar to that of glass, but its cause is somewhat different. Epoxy may show a slow fracture region that is followed by a mirror. The hackle also appears because the crack velocity attains such a high value in this strain-rate sensitive material.

Fast fracture in some cases also produces smooth surfaces for still another reason. A typical crack surface in a brittle plastic frequently shows parabolic-markings (Fig. 9.29). These are thought to result from the formation of a new fracture surface ahead of the main crack. The secondary crack starts from a point so that it forms a circular front as it grows. The full circle cannot be developed, however, because the linear main crack front is right behind the secondary one and it is the intersection of the two that causes the observed parabolic shape. The marking is visible because it is on a different level from the primary crack. A secondary crack leading the main fracture front is shown in Fig. 9.30 for balsa wood.

A great many studies have been made of the fine details of fracture surfaces of crystalline materials. This discipline is generally referred to as *fractography*. The fracture appearance of polycrystalline materials varies considerably, depending on the amount of permanent flow adjacent to the surface as well as on the fine structure of the material (see Fig. 9.20). Large differences are, of course, expected near the grain-boundaries as compared with the crystalline material within the individual crystals. Even such fine details as screw dislocations are evidenced in fracture appearance. Because

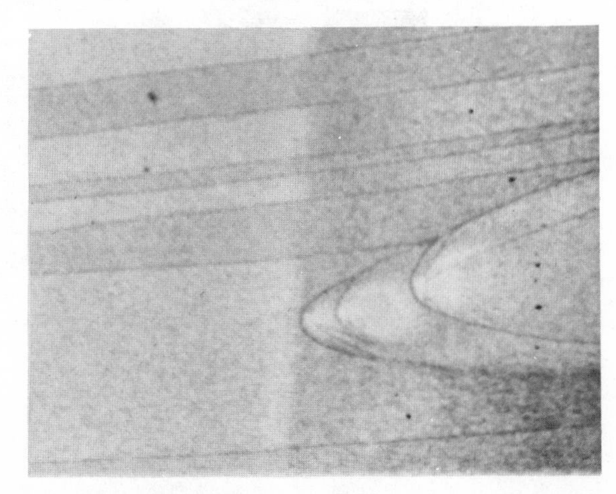

Fig. 9.29 Parabolic markings in the fast fracture surface of methyl-methacrylate, 120X. Courtesy I. Wolock, U.S. Naval Research Laboratory.

Fig. 9.30 Main crack front growing from a slit in a piece of balsa wood. Secondary crack is shown ahead of main one, (approximately 8X). Courtesy of E. Wu, Dept. Ther. and Applied Mech., University of Illinois.

Fig. 9.31 Fracture appearance and direction of crack propagation in polycrystalline iron. From J. R. Low, Jr., in (R9.1).

of the structural complexity, the crack does not even necessarily propagate in a single direction, as shown in Fig. 9.31.

PROBLEMS

1. Assume that a metal flows according to the Tresca criterion, while its fracturing behavior is described by the maximum normal stress theory. Compare the deformations to which it can be subjected in tension, torsion, and compression, assuming that all processes produce homogeneous deformation and the stress–strain curve between σ_o and σ_f is given by $\varepsilon = \sigma \times 10^{-6}$.

2. Describe an experiment you might use for finding the fracture stress vs. strain curve.

3. High pressure cylinders fracture in shear. In light of Eq. (9.4), what steps would you take to minimize such failures?

4. According to Bridgman's Fracture Law, what sort of working process (i.e., stress state) would you use to convert a short stubby cylinder of a brittle metal to a long slender one?

5. The fracture stress of steel is lower when the metal is tested under liquid lithium. Assuming that Fe is not soluble in Li, how would you explain this?

6. If W_α for glass $= 300$ ergs/cm^2, and, statistically, the largest crack in a glass rod equals 10 per cent of its diameter, plot σ vs. fiber diameter for fibers less than 0.1 inch diam.

7. If the flaws in polycrystalline metals are of the order of one grain long (1×10^{-3} inch), calculate the fracture stress of steels ($E = 30 \times 10^6$ psi $W_\alpha =$ 1,200 ergs/cm^2) if all the supplied energy is used for forming the fracture surface.

8. If the steel in (7) has a $\sigma_f = 200{,}000$ psi, what is the value of W_p? Compare this with $W_\alpha = 1{,}200$ ergs/cm^2.

9. According to Eqs. (9.16–19), what could you do to a steel to improve its resistance to fracture in a particular service requirement?

REFERENCES FOR FURTHER READING

(R9.1) Averbach, B. L., D. K. Felbeck, G. T. Hahn, and D. A. Thomas, (eds.) *Fracture*. New York: Technology Press and John Wiley and Sons, Inc., 1959.

(R9.2) Griffith, A. A., "The Phenomena of Rupture and Flow in Solids," *Phil. Trans. Roy. Soc.*, London, A 221 (1920), p. 163.

(R9.3) Nadai, A., *Theory of Flow and Fracture of Solids*, Vol. 1. New York: McGraw-Hill, Inc., 1950.

(R9.4) Parker, E. R., *Brittle Behavior of Engineering Structures*. New York: John Wiley and Sons, 1957.

(R9.5) Rostoker, W., J. M. McCaughey, and H. Markus, *Embrittlement by Liquid Metals*. New York: Reinhold Publishing Corp., 1960.

(R9.6) Tipper, C. F., *The Brittle Fracture Story*. Cambridge: Cambridge University Press, 1962.

(R9.7) Walton, W. H., (ed.) *Mechanical Properties of Nonmetallic Brittle Materials*. London: Interscience Publishers, Inc., 1958.

(R9.8) *Fracture of Metals*. Metals Park, Ohio: Am. Soc. Metals, 1948.

part IV

MECHANICAL BEHAVIORS: PROPERTIES AND TESTS

10

TENSILE PROPERTIES

In developing many of the concepts presented in Part II of this book, it was often necessary to use tensile stress–strain curves. Tensile tests are of great importance in determining basic material characteristics. These tests serve not only as a basis for acceptance, but their results also are used to evaluate the performance of materials both in regard to their permissible loading in service and their ease of formability. These properties are measured by applying a continuously increasing load to a smooth specimen of the material and noting the resulting extension. Both Young's modulus, E, and the yield strength, σ_o, are obtained in this way. The broad usage of this test is shown by the fact that specifications have been written for tensile specimens for practically all structural materials ranging from steel to mortar (Fig. 10.1 on pages 266–267). The relationship between uniaxial tensile stress and strain is one of the most basic curves in engineering. Its most important features and their meaning are discussed in this chapter.

DEFINITIONS OF YIELD STRENGTH

The type of deformation produced by a tensile load may continuously change as the load is increased. At low loads the response is essentially elastic. With larger loads, metals and other ductile crystalline solids flow plastically, and the stress that was just enough to cause plastic flow was referred to as the yield stress, σ_o. This offhand reference to σ_o may have led the reader to believe that it is an easily identified point on the stress–strain curve, marking a discontinuity between elastic and plastic flow. This however, is only occasionally found to be the case. In general, the transition

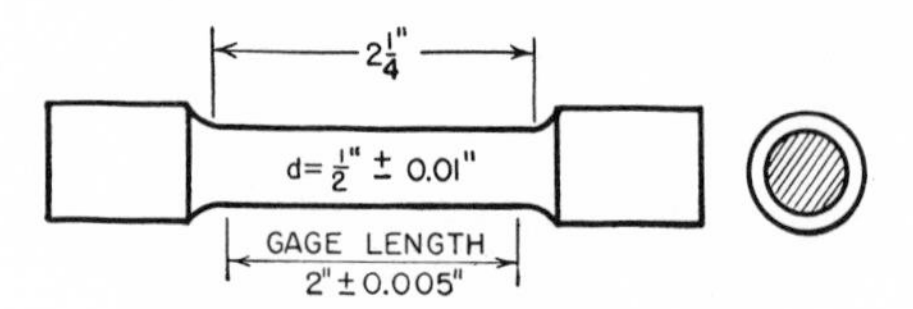

ROUND SPECIMEN WITH 2 IN. GAGE LENGTH
FOR DUCTILE METALS

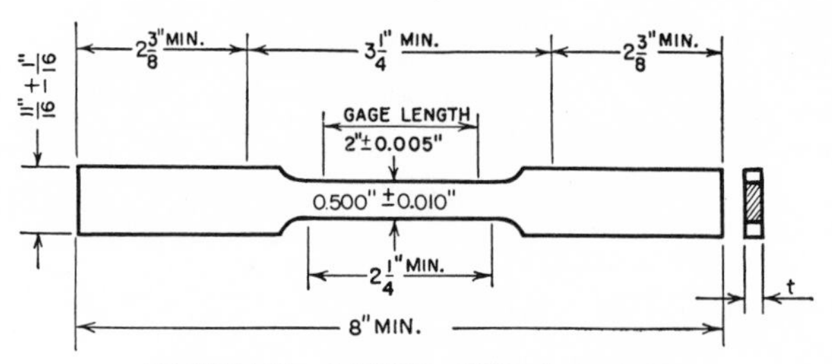

RECTANGULAR SPECIMEN WITH 2 IN. GAGE
LENGTH FOR PLATES AND SHEETS WITH
t = 0.01 TO 0.50 IN. FOR DUCTILE METALS

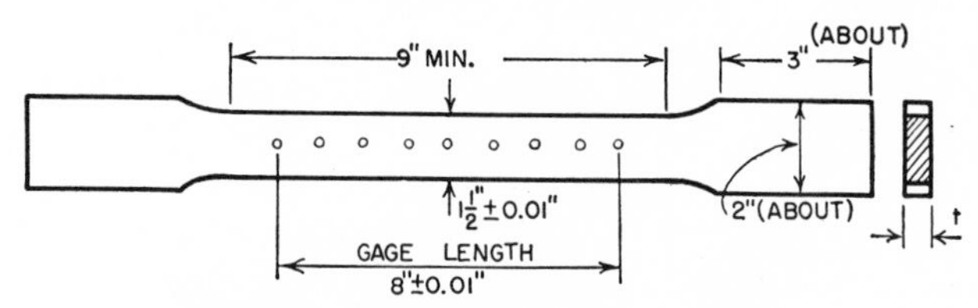

RECTANGULAR SPECIMEN WITH 8 IN. GAGE
LENGTH FOR PLATES AND SHEETS WITH
$t > \frac{3}{16}$ IN. FOR DUCTILE METALS

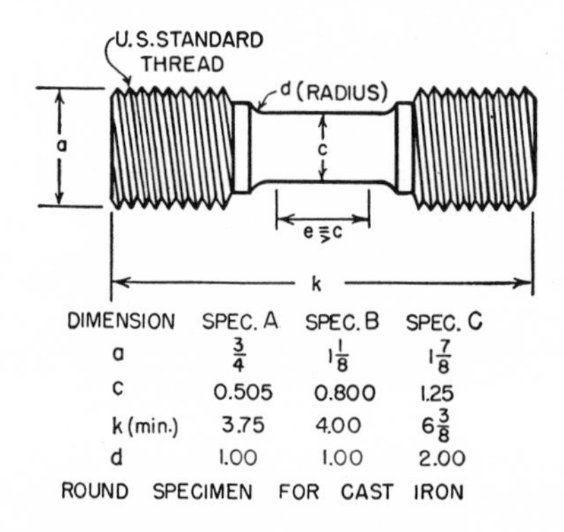

DIMENSION	SPEC. A	SPEC. B	SPEC. C
a	$\frac{3}{4}$	$1\frac{1}{8}$	$1\frac{7}{8}$
c	0.505	0.800	1.25
k (min.)	3.75	4.00	$6\frac{3}{8}$
d	1.00	1.00	2.00

ROUND SPECIMEN FOR CAST IRON

(a) SPECIMENS FOR METALS

Fig. 10.1 Common types of standard tension specimens. Left-hand group are for metals; right-hand group for non-metals. From J. Marin, *Engineering Materials, Their Mechanical Properties and Applications*, Prentice-Hall, Inc., 1952.

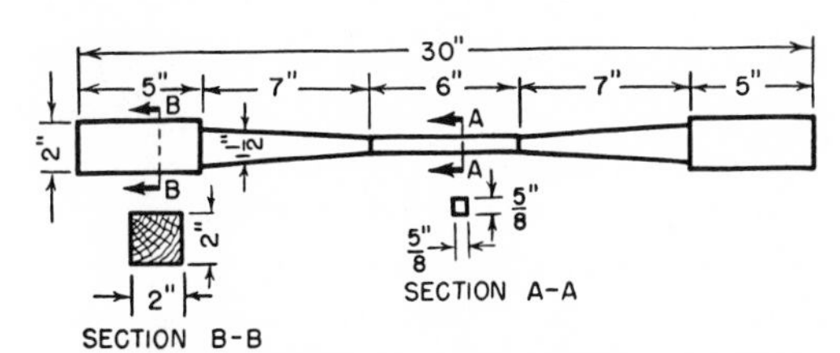

WOOD SPECIMEN FOR TEST PARALLEL TO GRAIN

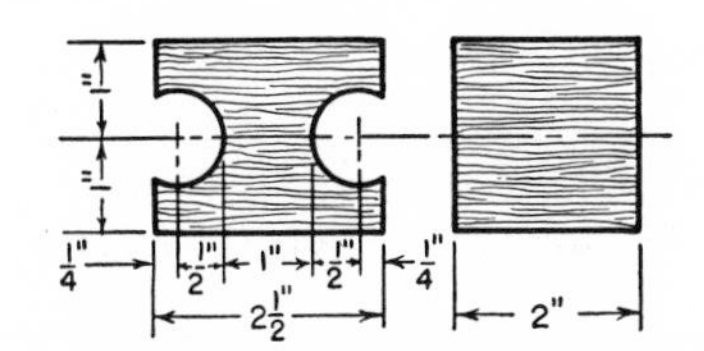

WOOD SPECIMEN FOR TEST PERPENDICULAR TO GRAIN

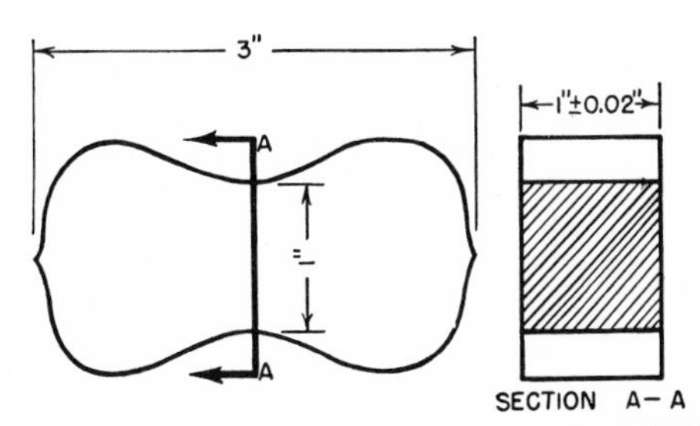

SPECIMEN FOR CEMENT MORTAR AND GYPSUM PRODUCTS

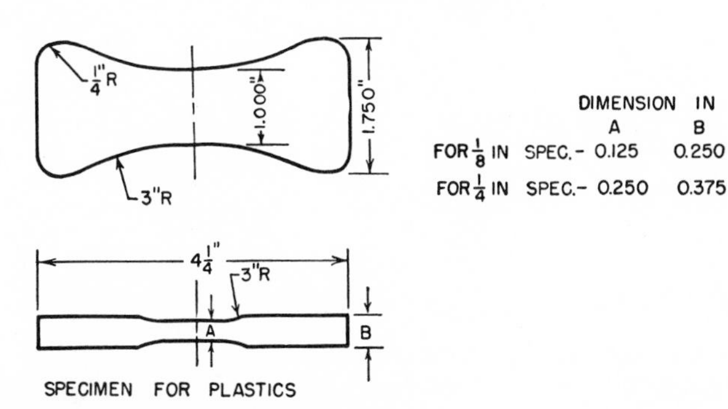

SPECIMEN FOR PLASTICS

(b) SPECIMENS FOR NON-METALS

from the elastic to the plastic range is gradual and almost imperceptible, making it difficult to locate the end of the elastic range. For this reason, the yield strength is usually found by procedures that make it possible to exactly and repeatably define points on the stress–strain curve that are close to, but do not actually represent, the elastic-plastic boundary.

The most obvious means for finding σ_o is to make use of the characteristics of elastic deformation as discussed in Chapter 5. Two features distinguished this type of deformation: (1) it was transient; i.e., the strain vanished on removal of stress, and (2) the strains were proportional to stress. If advantage is taken of the transient nature of elasticity, we can define the *elastic limit*, σ_{el}, as the largest stress that will not cause any permanent set. To determine this quantity, the specimen dimensions are first accurately measured and then the piece is successively loaded and unloaded. The largest load that did not cause a measurable permanent deformation is taken as the elastic limit. The second characteristic of elasticity is used to find the *proportional limit*, σ_{pl}. This is the largest stress at which the stresses and strains are still related by Hooke's Law. Although in design engineering it is usually assumed that stress is proportional to strain over the complete elastic range, the discussion

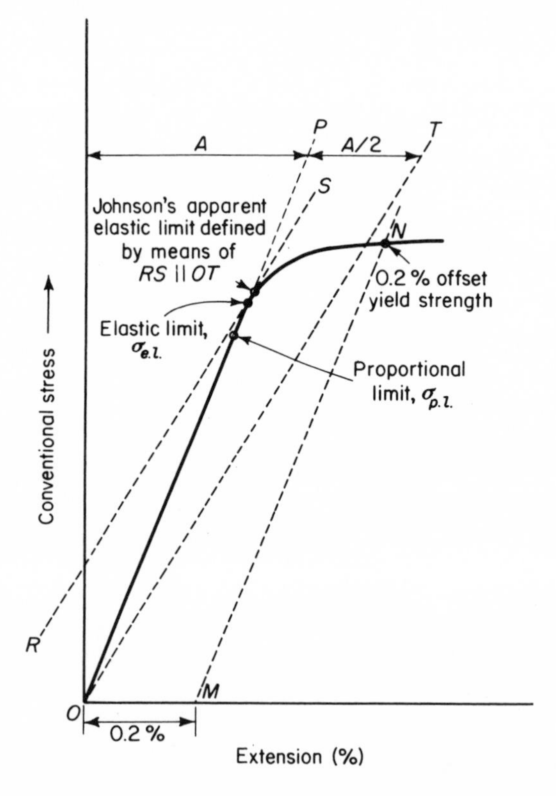

Fig. 10.2 A variety of methods for measuring yield strength.

of damping in Chapter 8 reveals that this is not quite the case. Because of this slight but systematic deviation from Hooke's law, the elastic limit is somewhat higher than the proportional limit for most metals. They have about the same value, however, for low and medium carbon steels.

Neither the σ_{el} or σ_{pl} method for defining σ_o is used in routine testing since their apparent values depend to a large extent on the accuracy of the measuring apparatus and the stress or strain intervals at which readings are taken. As the measuring equipment becomes increasingly sensitive both these values decrease. The elastic limit and proportional limit are schematically compared with some other measures of yield strength in Fig. 10.2.

One procedure that is used to approximate the proportional limit defines a quantity known as *Johnson's Apparent Elastic Limit*. This is found by locating the stress at which the slope of the stress–strain diagram has two thirds the value of Young's modulus. In practice one draws a line with a slope equal to $2E/3$ from the origin of a stress-strain curve (line OT in Fig. 10.2). The position at which a line parallel to this one (RS) is tangent to the stress–strain curve gives the value of Johnson's apparent elastic limit. Even this technique does not assure good reproducibility since it is difficult to find the exact point of tangency; consequently, it is not as commonly used for locating the elastic-plastic boundary as the other methods discussed below. Nevertheless, Johnson's limit is the common method for specifying yield strength in certain industries such as wire.

More often, the end of the elastic range is approximated by compromise methods that are based on the assumption that some small amount of permanent deformation is tolerable. The most widespread of these methods in this country is the so-called offset procedure used to find the stress at which the permanent deformation is equal to 0.2 per cent. It takes advantage of the fact that the elastic stress–strain curve, in unloading, is parallel with the initial elastic line. Hence if a strain of 0.2 per cent is measured on the abscissa (Fig. 10.2), and a line with slope E drawn through this point, it will intersect the stress–strain curve at a plastic strain of 0.2 per cent, yielding the value for the *0.2 per cent offset yield strength*. Other values of offset may also be used; e.g., 0.1 per cent, if less plastic deformation can be tolerated. Because of this, the amount of offset should be specified with any offset yield strength figure (e.g., $\sigma_{o(0.2\%)}$). Rather than refer to these values as yield strength, they are sometimes called *proof stress*, especially in Great Britain where frequent use is made of either 0.1% or 0.5% offsets.

Some metals, such as soft copper, show virtually no region in which Hooke's law is applicable (Fig. 10.3). Hence, it is impossible to draw an elastic line so that the yield strength is arbitrarily defined at some total strain. A value of 0.5 per cent total strain is frequently used.

It was pointed out in Chapter 7 that in still other metals, the most important example being mild steel, the elastic range is abruptly terminated by

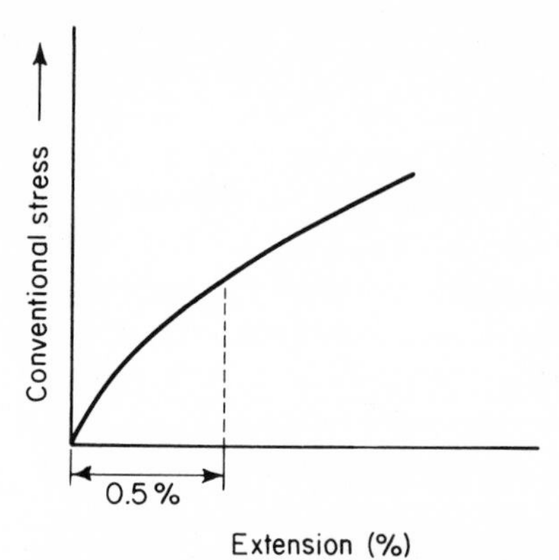

Fig. 10.3 The yield strength of metals, such as soft copper, that show no straight portion in their stress-strain curve, is arbitrarily defined at some total strain.

a drop in the load from the highest value of elastic stress to a lower value required for initiating plastic flow (Fig. 10.4).* In these metals, plastic extension starts in some small volume of metal at the reduced load (point *B* in Fig. 10.4). The load then remains more or less constant until the entire

* Although this is a practical elastic-plastic boundary, some plastic strain can be detected before the drop occurs.

Fig. 10.4 Typical stress-strain curve for a mild steel.

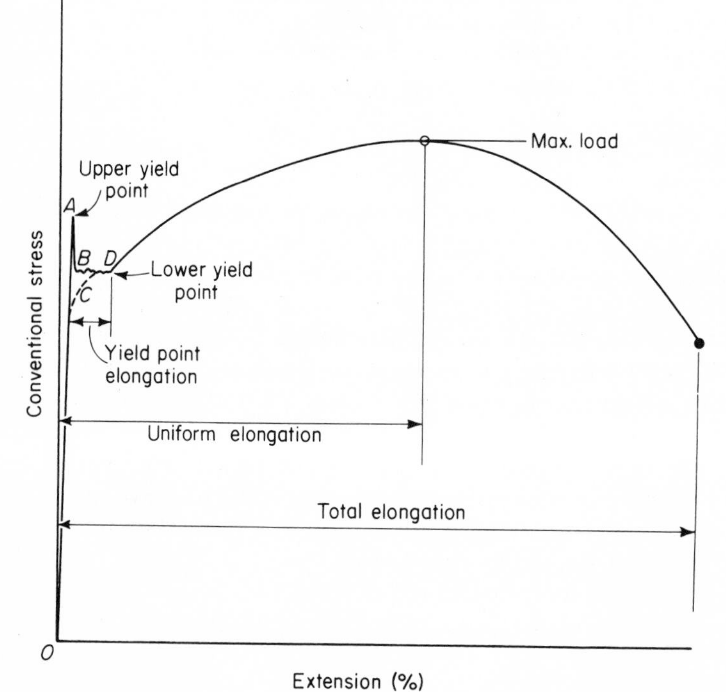

test section has suffered this same amount of extension, point D, and only then does the load increase with further deformation so that it is a relatively simple matter to determine the beginning of the plastic range. The average stress at this horizontal portion of the stress–strain curve is known as the *lower yield point* or, somewhat more ambiguously, the *drop-of-beam yield strength*. The first term distinguishes this stress from the *upper yield point*, which is the highest "elastic" stress reached during the test (point A in Fig. 10.4). In some cases the upper yield point is higher than any other part of the stress–strain curve, including the plastic part over which strain-hardening has occurred. The upper yield has little importance as a material property because it depends to a large extent on the testing procedure. If the specimen is mislaigned during testing, the upper yield may be suppressed as shown by the dashed curve in Fig. 10.4. The stress–strain curve in this case is given by OCD rather than $OABD$.

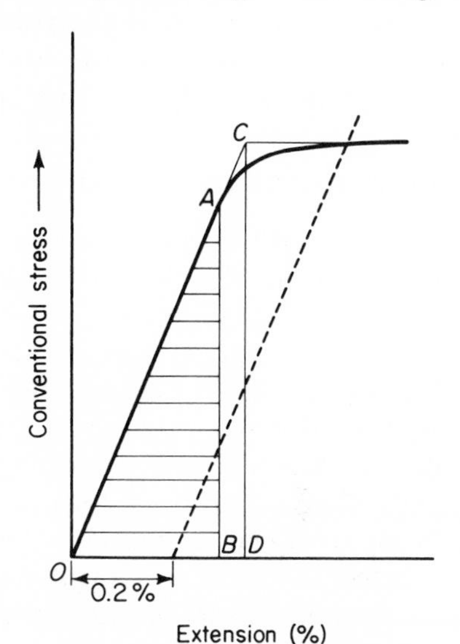

Fig. 10.5 Modulus of resilience is given by area under elastic portion of stress-strain curve. Using σ_{p1}, its value is OAB. Using 0.2% offset yield strength, the resilience equals area OCD.

The discontinuous flow that is found at the yield point is known as a *yield jog*, and the length of the substantially horizontal plastic range, BD, is the *yield point elongation* or *yield jog elongation*. As stated in Chapter 7 (see p. 189), this results in Lüder's lines or stretcher strains during sheet forming operations.

RESILIENCE

Resilience is a measure of the amount of elastic energy a material can store, and the *modulus of resilience*, W_p, is the energy stored per unit volume of metal. Hence, it is equal to the energy under the stress–strain curve up to the proportional limit; i.e.,

$$W_p = \tfrac{1}{2}(\sigma_{pl})(\varepsilon_{pl}) = \frac{\sigma_{pl}^2}{2E} \qquad (10.1)$$

where σ_{pl} = proportional limit
ε_{pl} = strain at proportional limit.

Although the modulus of resilience is an energy term and thus should have the dimensions of a product of force and distance, the units of (10.1) are seen to be pounds per square inch in English units. Actually this is a simplification of inch-pounds per cubic inch of material. Again, because σ_{pl} is difficult to determine, W_p is frequently calculated by using any of the previously described values for yield strength in the place of σ_{pl}. This only introduces a small error as shown in Fig. 10.5. In the latter case, W_p is synonymous with W_0 from Eqs, (5.8) and (5.16 ante).

TENSILE STRENGTH AND DUCTILITY

Beyond the elastic range metals flow plastically, and, because of strain-hardening, increasing loads must be applied if deformation to continue within this interval. While this hardening occurs, however, the cross-sectional area of the specimen decreases. If the metal is sufficiently ductile, the rate at which the cross section diminishes eventually exceeds the rate of strain-hardening. From this point on, the load required for additional flow decreases so that the load-extension curve goes through a maximum (Fig. 10.4). After this maximum load is reached, the deformation becomes highly localized, producing a *neck* as shown by the brass rod in Fig. 9.2a. Whether the material is one that necks or not, the largest load sustained by the specimen—other than the upper yield point—divided by the original cross-sectional area is known as its *tensile strength*, *ultimate strength*, or *ultimate tensile strength*, S_u*; i.e.,

$$S_u = \frac{\text{maximum load}}{\text{cross-sectional area before straining}} = \frac{F_m}{A_o} \tag{10.2}$$

The tensile strength is not really a stress value since the load is not divided by the area on which it acts. However, it acquired a considerable stature because of the ease with which it is calculated and by a century old habit. Data in terms of tension loads, divided by the initial undeformed cross section of the specimen, are referred to as *nominal* (or *conventional*) *stress*, *S*.

The amount of plastic extension undergone by a specimen at any stage of a tensile test is usually specified as *per cent elongation*.

$$e = \frac{l - l_o}{l_o} \times 100 = 100\left(\frac{l}{l_o} - 1\right) \tag{10.3}$$

where l_o = initial distance between two fixed points on the central, parallel portion of the specimen; it is referred to as *gauge length* and is standardized for various specimen types (see three top left items in Fig. 10.1).

l = gauge length at any stage of deformation.

---

* U.S. Government reports generally use the symbol F_{tu} for this quantity.

The term in parentheses was defined in Eq. (3.3) as conventional strain, ε, so that $e = 100\varepsilon$. The amount of permanent strain reached just at the onset of necking is known as *maximum uniform elongation or necking strain.*

When a neck forms, the extension ceases to be uniform and is confined to the neck where the specimen eventually breaks. When the distance between the gauge markings after fracture, l_f, is inserted into (10.3), the *percentage elongation at fracture*, e_f, is obtained. The term "elongation" in routine tensile testing actually means e_f.

The fracture elongation e_f gives no idea of the contraction of area suffered by the specimen at its thinnest section at the bottom of the neck. In metal forming operations, such as severe deep drawing of cups or boxes from sheet, the ultimate "thinning out" that can occur prior to fracture may represent the limiting factor. Thus, the maximum *reduction of area* or *contraction in area*, q, is an important measure of ductility. Its value is given by

$$q = \frac{A_o - A}{A_o} \times 100 = 100\left(1 - \frac{A}{A_o}\right) \tag{10.4}$$

where A_0 = original cross-sectional area
A = cross-sectional area at any instant of deformation.

The maximum theoretical value of q is 100 per cent. Although this is never reached, it is sometimes approached in tests on very soft aluminum or copper at elevated temperatures. Again, in tensile testing, the term, "reduction of area," denotes the reduction of area at fracture, q_f.

As long as elongation is uniform, and the gauge portion of the specimen is strictly cylindrical,

$$Al = A_o l_o;$$

$$\frac{l}{l_o} = \frac{A_o}{A} = \frac{100}{100 - q}$$

Substitution in (10.3) yields

$$e = \left(\frac{q}{100 - q}\right)100 \tag{10.5}$$

On the other hand, if the specimen necks, the elongation based on a short gauge length, located at the bottom of the neck, would be up to several times that obtained by using long gauge lengths.

TRUE STRESS–STRAIN CURVE

During plastic flow, the cross-sectional area of ductile specimens is progressively reduced. Hence, as straining proceeds, the nominal stress, S, based on the specimen's undeformed cross section, becomes more and more meaningless. A more realistic stress value is obtained by dividing the load by the true (instantaneous) area. A stress calculated on the basis of actual area is termed *true stress*, σ. If σ rather than S is used, tensile stress–strain curves do not exhibit a maximum (curve B in Fig. 10.6).

Although conventional strain values based on reduction in area or uniform elongation are not meaningless in the sense that conventional stress is, they have the disadvantage of "bunching up" the strain values at large deformations. For example, assume that we were to plot the yield strength of a ductile metal as a function of cold work by stretching. If the cross-sectional area were initially A_o, and it were stretched to an area, $A_1 = A_o/2$, followed by $A_2 = A_1/2$, $A_3 = A_2/2$, $A_4 = A_3/2$, etc., the values of q would

Fig. 10.6 True stress vs. natural strain and conventional stress-reduction in area curve for a $2\frac{3}{4}$% Si steel.

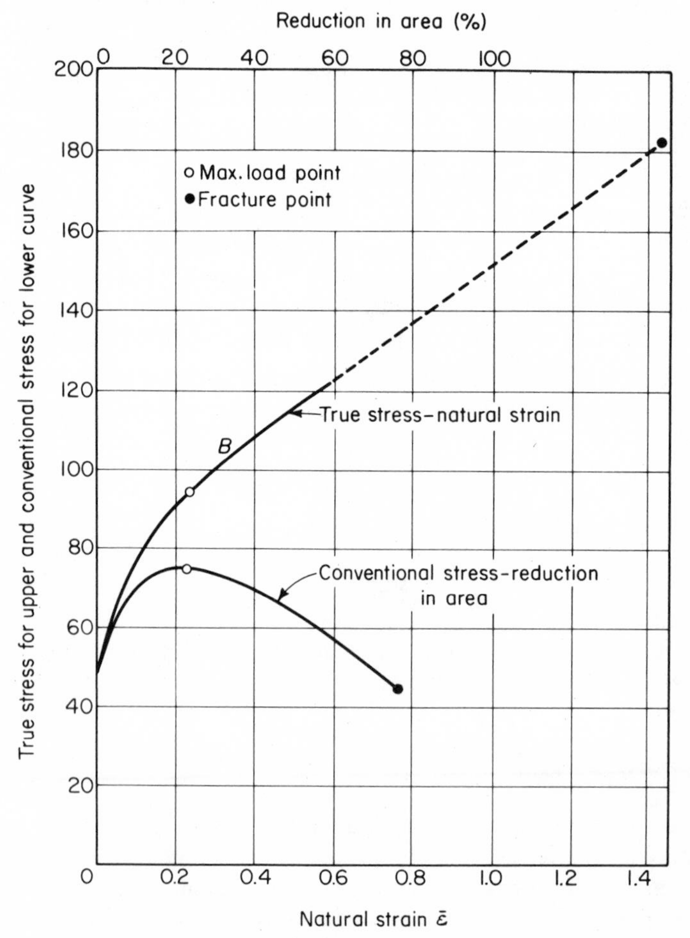

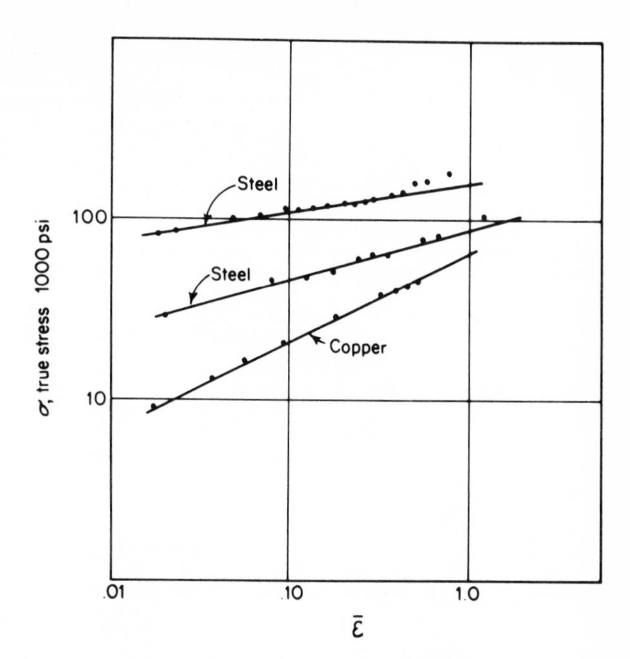

Fig. 10.7 True stress vs. natural strain on log-log plot for two steels and copper. K is intercept on $\bar{\varepsilon} = 1$ and n = slope. From L. R. Jackson, *Trans. AIME*, **171**, 622 (1947).

be 50, 75, 87.5, 93.75%, etc. Thus the differences in q are decreasing gradually from 50 to 6.25% even though each of these contraction increments is identical. If, on the other hand, contraction (or extension) is defined in terms of natural strain as given in Eq. (3.4a) (i.e., $\bar{\varepsilon} = ln\,(l/l_o) = ln\,(A_o/A)$) the increments are uniform, $\bar{\varepsilon} = 0.69$, 1.39, 2.08, 2.77 etc., the difference being $\ln 2 \approx 0.69$ in each case. Hence, a true stress-natural strain curve ((*B*) in Fig. 10.6) is the most useful type for studies in the plastic range. Of course, the bunching up that results from using conventional strain would also occur if true stress were plotted as a function of e, and it would have the added disadvantage of not taking into account all the deformation at the bottom of the neck.

With true stress and natural strain units, the plastic region of the tensile curves for most ductile metals can be approximated by a power function:

$$\sigma = K\bar{\varepsilon}^n \tag{10.6}$$

where K and n are material constants. The first of these, K, is the *strength coefficient* and is equal to the stress at a natural strain of $\bar{\varepsilon} = 1$. The second is the *strain-hardening exponent*. This relationship best fits the rapidly rising part of the curve and is applicable to a wide range of metals as shown by the log-log plot in Fig. 10.7.

It is often convenient to characterize the stress–strain properties of a particular material without reference to a diagram, and the constant (n) which specifies the rate of work-hardening has been widely used for this purpose. (Equation (10.6) permits numerical evaluation of (9.7).)

By taking advantage of the fact that necking occurs when the applied load goes through a maximum, the strain-hardening exponent can be shown to be equal to the necking strain. This is done by first taking the logarithm of (10.6) and differentiating both sides,

$$\frac{d\sigma}{\sigma} = n\,\frac{d\bar{\varepsilon}}{\bar{\varepsilon}} \tag{a}$$

The applied load, F, is equal to the cross-sectional area, A, times the stress, σ. The maximum load, or necking strain, occurs when $dF = d(A\sigma) = 0$. Hence $A\,d\sigma + \sigma\,dA = 0$, $A\,d\sigma = -\sigma\,dA$, or

$$\frac{d\sigma}{\sigma} = \frac{-dA}{A} \tag{b}$$

Using

$$\bar{\varepsilon} = ln\,\frac{A_o}{A}$$

we find

$$\frac{dA}{A} = -d\bar{\varepsilon} \tag{c}$$

Substituting (c) in (b) and then (b) in (a) yields

$$\bar{\varepsilon}_n = n \tag{10.7}$$

The necking strain in metals is of interest in many forming operations because $\bar{\varepsilon}_n$ rather than the total ductility establishes the deformation limits. For example, when sheets are formed by stretching, as is sometimes done for aircraft wing panels, the maximum stretch that will still result in a uniform section is the necking strain. For this reason both analytical and graphical methods have been developed for finding this value.

Again, using the fact that necking is a state of instability,

$$\frac{dF}{d\bar{\varepsilon}} = 0 \tag{d}$$

and the applied load,

$$F = \sigma_1 A \tag{e}$$

we find

$$\frac{dF}{d\bar{\varepsilon}} = \frac{d\sigma_1}{d\bar{\varepsilon}}\,A + \frac{dA}{d\bar{\varepsilon}}\,\sigma_1 = 0 \tag{f}$$

The instantaneous cross-sectional area, A, is related to the initial area by

$$A = A_o(1 + \varepsilon_2)(1 + \varepsilon_3) \tag{g}$$

and since

$$\bar{\varepsilon} = \ln(1 + \varepsilon)$$

and

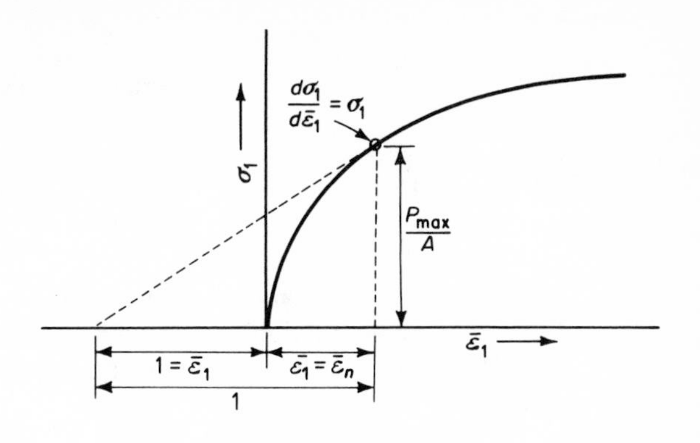

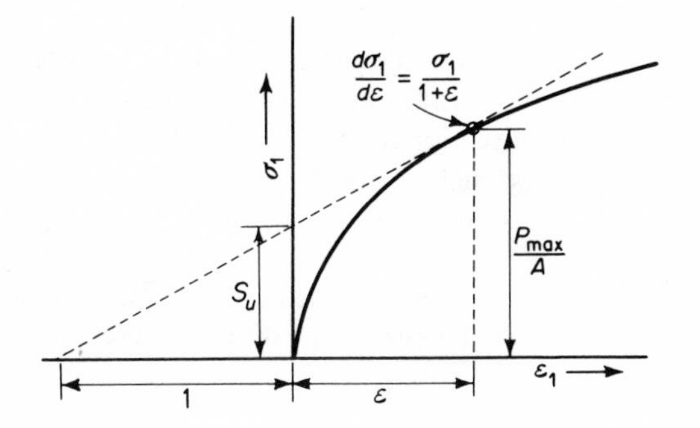

Fig. 10.8 Graphical methods for determining necking point. (a) True stress-natural strain. (b) True stress-conventional strain.

$$\varepsilon_1 = \varepsilon_2 + \varepsilon_3 \tag{6.11}$$

(g) can be rewritten

$$A = A_o e^{\bar{\varepsilon}_2 + \bar{\varepsilon}_3} = A_o e^{-\varepsilon_1} \tag{h}$$

Differentiating (h) with respect to $\bar{\varepsilon}$,

$$\frac{dA}{d\bar{\varepsilon}} = -A_o e^{-\varepsilon_1} = -A \tag{j}$$

and substituting (j) in (f) gives

$$\frac{d\sigma_1}{d\bar{\varepsilon}_1} = \sigma_1 \tag{10.8}$$

Thus, instability arises when the slope of the true stress vs. natural strain curve is equal to the stress. This value is readily found graphically (Fig. 10.8a).

If the stress–strain data were in the form of true stress vs. conventional strain (elongation as in (10.4) rather than contraction as in (10.6)), Eq. (b) with the aid of (c) is

$$\frac{d\sigma}{\sigma} = d\bar{\varepsilon} \tag{k}$$

Differentiating (3.4a) yields

$$d\bar{\varepsilon} = \frac{d\varepsilon}{1 + \varepsilon} \tag{m}$$

so that

$$\frac{d\sigma}{d\varepsilon} = \frac{\sigma}{1 + \varepsilon} \tag{10.9}$$

Again the necking strain is found graphically (Fig. 10.8b). This is known as Considère's construction and is simpler to carry out than the procedure of Fig. 10.8a. In this case the tangent is easily found since it intersects the abscissa at -1.

STRESS DISTRIBUTION IN NECK

After a ductile metal begins to neck in tension, the stress state in the neck is no longer uniaxial. The reason for this is seen by examining a series of cross-sectional elements within the neck (Fig. 10.9a). Each of these elements, such as disc No. 2, wants to extend by an amount that is dictated by the stress acting on it. In order to extend, however, its cross-sectional area

Fig. 10.9 (a) Necked tensile specimen. (b) Distribution of stresses within neck. (c) Photograph of crack in necked specimen.

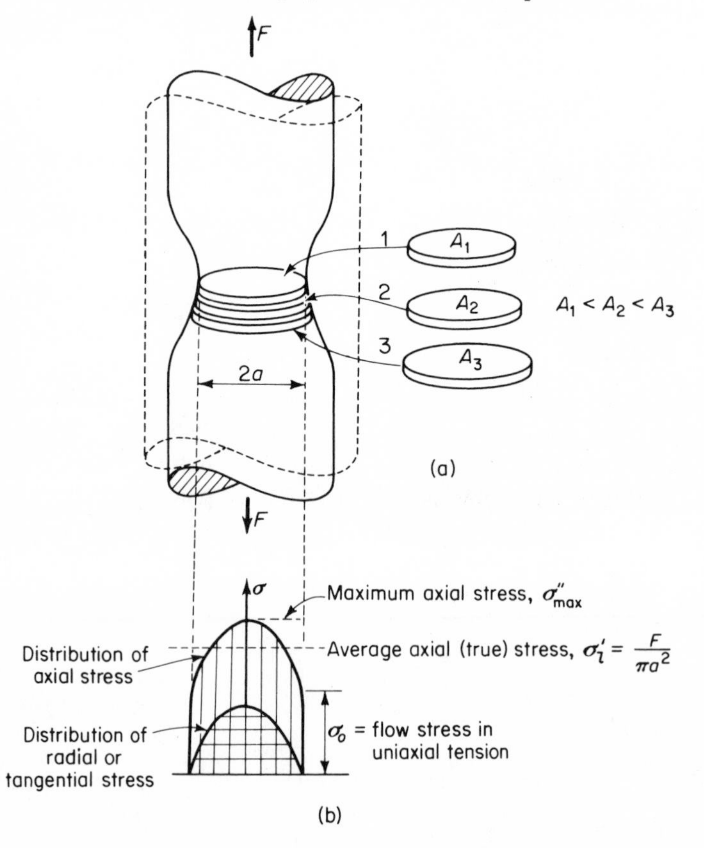

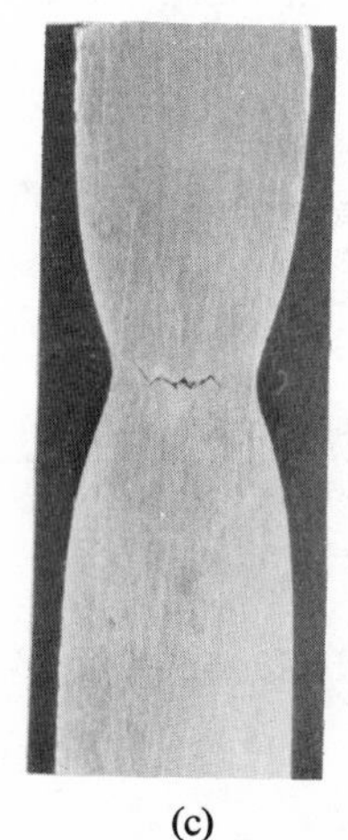

(c)

must become smaller. Since each disc has a different area supporting the same applied load, the stress on each element, and the consequent tendency of each to contract, differ. The discs must remain compatible with their neighbors so that disc No. 2 is restrained from contracting by disc No. 3; No. 2, on the other hand, imposes a restraint on No. 1, etc. The restraint manifests itself as radial and tangential tensile stresses that each disc exerts on its smaller neighbor.

N. N. Davidenkov and N. I. Spiridonova have determined the magnitude of these stresses, using an experimental approach. They analyzed the strains by noting the change in shape of initially equiaxed grains. P. W. Bridgman calculated the stresses analytically. Both found the stress to be distributed within the neck as shown schematically in Fig. 10.9b. On the surface of the specimen the radial stresses vanish, making the axial stress, σ_l, equal to the uniaxial flow stress, σ_o. As the radial and tangential stresses, σ_r and σ_θ, increase on traveling in from the surface, σ_l must also increase in order to maintain a condition of plasticity. At the center of the test bar the distinction between radial and tangential directions disappears, and these two transverse stresses must be identical at this point. Parker, Davis, and Flanigan made more direct measurements of the three principal stresses in a necked tensile bar and found that not only σ_l but the largest principal shear stress as well went through a maximum at the specimen centerline.

The material at the minimum cross section of a necked tensile specimen is subjected to a complex stress state so that curve B in Fig. 10.6, between the maximum load point and the fracture point, represents average axial stress values. The magnitude of the transverse stresses increases as the neck becomes deeper. Hence, for flow to continue, σ_l must also increase as the neck deepens to satisfy the condition of plasticity Eq. (6.1b). According to Bridgman's analysis, the flow stress, σ_o, can be calculated from the average axial stress, σ_l', the specimen diameter at the neck, $2a$, and the neck contour radius, R.

$$\sigma_0 = \sigma_l' \left[\frac{1}{\left(1 + 2\frac{R}{a}\right) ln \left(1 + \frac{1}{2}\frac{a}{R}\right)} \right] \qquad (10.10a)$$

Davidenkov and Spiridonova's expressions for σ_o and the maximum axial stress, σ_l'', are somewhat simpler.

$$\sigma_o = \frac{\sigma_l''}{1 + (a/4R)}$$

$$\sigma_l'' = \sigma_l' \frac{R + \frac{a}{2}}{R + \frac{a}{4}} \qquad (10.10b)$$

Because both the maximum normal and shear stresses occur at the axis of the bar, fracturing is initiated in necked specimens at this point. A large number of experiments in which necked test bars were stretched almost to complete fracture and then longitudinally sectioned proved this to be the case. A typical example is shown in Fig. 10.9c. Macroscopically, the fracture surface appears to be normal to the applied load. This led to the assumption that fracturing was controlled by a normal fracture stress law (see Chapter 9). Microscopic examination of these surfaces, however, showed them to consist of shear failures of individual grains. This is consistent with the comments in Chapter 9 wherein it was stated that ductile fracturing of metals generally occurs by shear. Of course, not all tensile bars neck. Whether or not necking occurs, the ratio of the breaking load to the minimum cross-sectional area is known as the *fracture stress*, σ_f.

EFFECT OF TESTING TEMPERATURE AND STRAIN RATE

The tensile properties of metals, and indeed of most materials, are temperature sensitive, and the ductility loss that abruptly occurs in some metals as the testing temperature is lowered (see Figs. 9.8 and 9.24) is particularly alarming. The temperature at which the abrupt change from ductile to brittle occurs is termed the *transition temperature*. It is usually studied by means of notch tests which measure toughness and, therefore, is discussed in more detail in Chapter 15. As stated in Chapter 9, this phenomenon occurs in b.c.c. and some h.c.p. metals but very rarely, and then only with special microstructures, in f.c.c. metals. Transition temperatures are elevated by

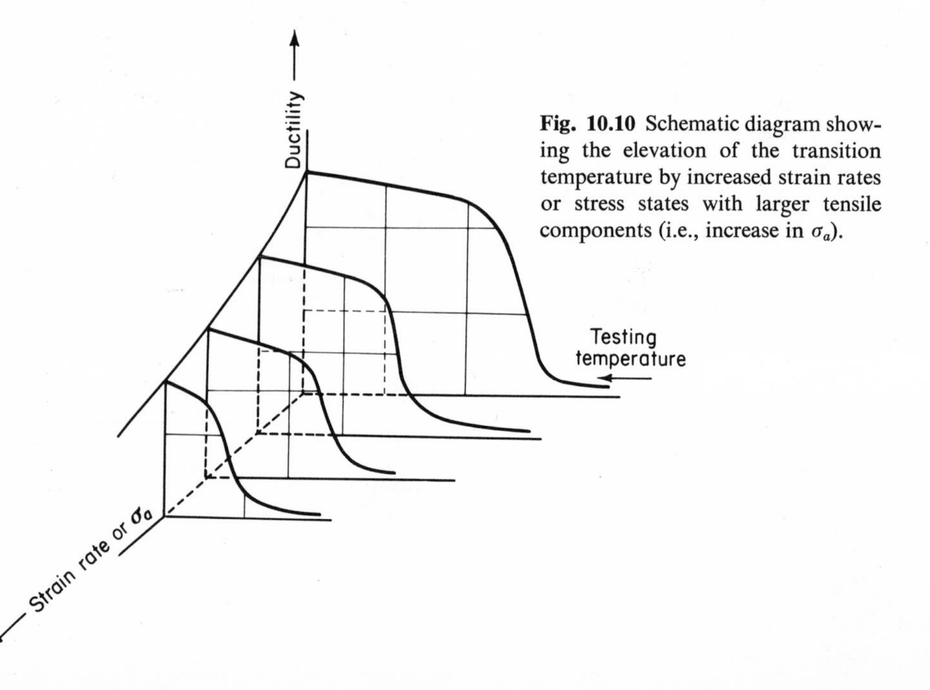

Fig. 10.10 Schematic diagram showing the elevation of the transition temperature by increased strain rates or stress states with larger tensile components (i.e., increase in σ_a).

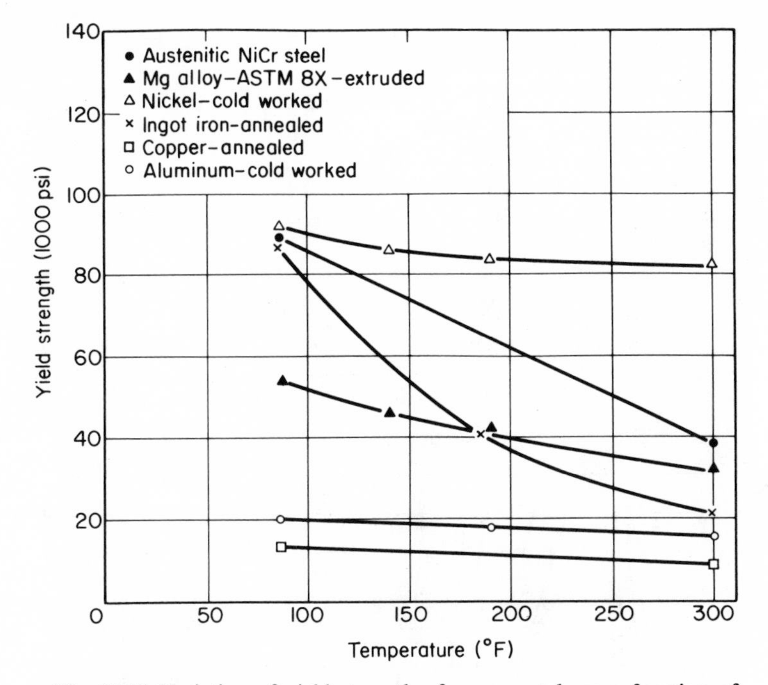

Fig. 10.11 Variation of yield strength of some metals as a function of testing temperature. From L. Seigle and R. M. Brick, *Trans. Am. Soc. Metals*, **40**, 813 (1948).

either increasing strain rates or by increasing the mean tensile stress component (Fig. 10.10). This point was made in discussing Eq. (9.19).

Not only ductility but also the strength properties are affected by low temperature. Because both the ultimate tensile strength and fracture stress may be influenced by ductility changes, only the yield strength can be considered as an independent strength property in describing temperature effects. Again, the yield strength of b.c.c. metals is more temperature-sensitive than the f.c.c. group (Fig. 10.11).

Strain rate has an inherent effect on tensile properties in addition to its influence on the transition temperature. The yield strength of all metals is raised with strain rate although the magnitude of $d\sigma_0/d\varepsilon$ varies greatly for different alloys (Fig. 10.12a). Tensile ductility on the other hand may increase, decrease, or remain reasonably unchanged with speed of testing. In addition to being material-dependent, the ductility changes depend on the specific manner in which it is measured; i.e. whether it is e_f or q_f. This is especially pronounced for the steel whose properties are shown in Figs. 10.12b and c. The apparent contradiction, that strain rate has a different influence on e_f than on q_f, results because the shape of the stress–strain curve is also altered by testing speed.

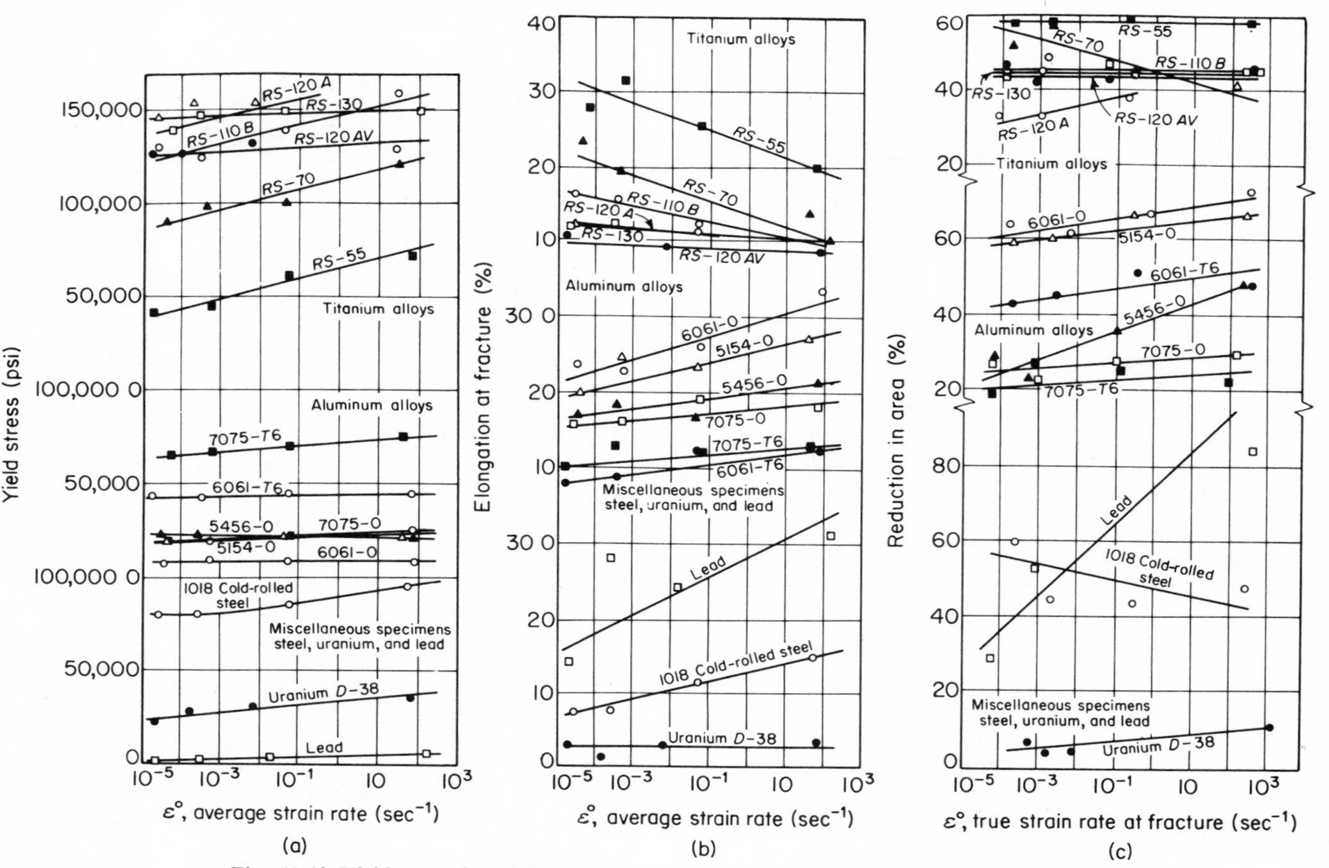

Fig. 10.12 Yield strength and elongation as a function of average strain rate and reduction in area vs. true strain rate. From R. F. Steidel and C. E. Makerov, *ASTM Bulletin*, **247**, 57 (1960).

EQUIVALENCE OF STRAIN RATE AND TEMPERATURE

Because increasing strain rate or lowering temperature both increase the resistance of materials to deformation, it is not surprising that attempts have been made to prove an equivalence of these two. A relationship of this type would be helpful, for example, in simulating service conditions where strain rates are too high to be easily duplicated in the laboratory, such as in studying the deformation of armor plate. Hollomon and Zener* proposed one such relation for metals,

$$P = \dot{\varepsilon}_e^{\Delta H/RT} \tag{10.11}$$

where P = a parameter
$\dot{\varepsilon}$ = strain rate
ΔH = activation energy
R = gas constant
T = temp. °K

This and other expressions aimed at the same purpose unfortunately have only a very limited range of validity.

TENSILE PROPERTIES OF POLYMERS

The manner in which smooth polymer rods react to a tensile load appears to vary enormously, depending on the specific polymer, its molecular weight, the testing conditions, etc. Nevertheless, all these more or less fit into a general pattern. Indeed, it is surprising to find even a similarity in the tensile behavior of some of the materials that are included in this class. Unlike metals whose atoms are always bonded together by identical forces in all directions and whose structure is completely crystalline, high polymers differ greatly on an atomic level (Chapter 4). Of the many polymers that exist, two types are of special interest from a mechanical viewpoint—plastics (both thermoplastic and thermosetting) and rubbers.

THERMOPLASTICS

Elastic Deformations. The thermoplastic materials may undergo very large extensions. As in metals, a variety of types of deformation may occur. When materials such as nylon or polyethylene are stretched, a stress–strain curve, much like that observed on mild steel, is obtained (Fig. 10.13). The curve shows a typical upper and lower yield point, and at large strains appears to undergo strain-hardening. The upper yield point in this case is a material constant and is generally referred to simply as the *yield point*. In polymer technology deformation that takes place at the lower yield point is called

* J. H. Holloman and C. Zener, *Trans. AIME*, **158**, 283 (1944).

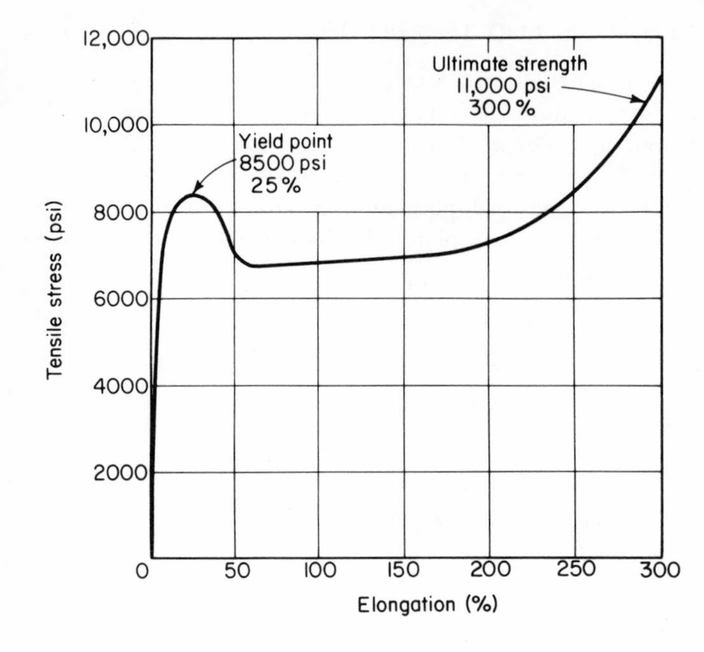

Fig. 10.13 Stress-strain curve for one type of Nylon ("Zytel" 101) at 73°F and 2.5% moisture. From "Designing with Du Pont Plastics," E.I. du Pont de Nemours and Co., Inc., Wilmington, Del. (1956).

cold drawing and will be discussed in more detail below. A closer examination of the curve shows that much of this similarity between crystalline materials and thermoplastics is superficial. The initial portion of the curve over which the stress rises rapidly with strain is not linear but exponential, and at a constant strain rate the "modulus" in this region, designated as E_p, is

$$E_p = E_o e^{-\alpha\varepsilon} \tag{10.12}$$

where E_p = slope of tangent at any point
E_o = slope at origin
α = material constant
ε = strain

From the form of Eq. (10.12) the constant, α, is seen to be a measure of curvature of the stress–strain curve. If α is small, the curvature is slight, and the reaction of the plastic to small applied loads would be similar to "hard" metals. Although α is a material constant, it does depend on testing temperature, strain rate, molecular weight, and degree of crystallinity. For example, when polymers are quenched from the amorphous temperature, they do not tend to crystallize so that the effect of crystallinity can be seen by comparing the stress–strain curves of quenched and annealed polyethylene (Fig. 10.14).

The extension of polymers by stresses below the upper yield point is elastic in that it is generally recoverable on the release of load. However, a large portion of the deformation is anelastic as shown by the loading and unloading curves in Fig. 10.15.

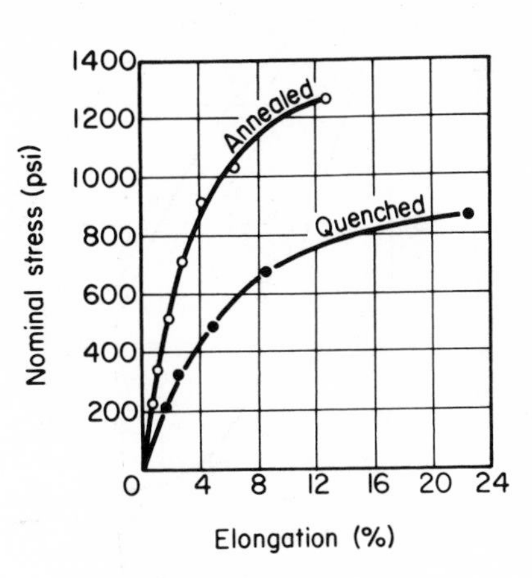

Fig. 10.14 The effect of different states of crystallinity on the stress-strain curve of low density polyethylene at 25°C. The annealed specimen was 0.035 in. thick; the quenched specimen, 0.037 in. thick. Both were 0.5 in. wide. Average strain rate was 3% per minute. From I. L. Hopkins and W. O. Baker, in *Rheology*, **3**, F. R. Eirich, ed., The Academic Press, N.Y. (1960).

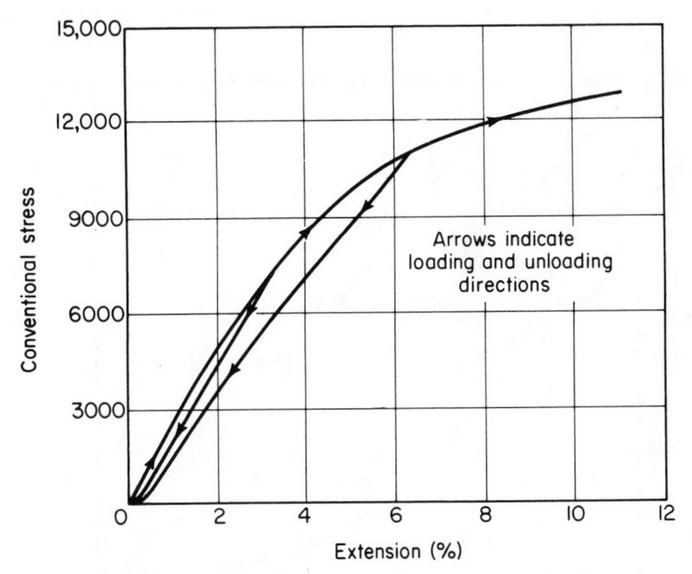

Fig. 10.15 Stress-strain hysteresis loops formed in linear polyethylene at 23°C (50% rel. humidity) showing an elastic aftereffect during loading and unloading.

In presenting the most used tensile properties of metals, viz., Young's modulus and the yield strength, there was no need to be especially concerned with the testing environment. Such is not the case with polymers. Many of these, such as nylon, are hygroscopic and both E and σ_o are affected by moisture (as shown in Fig. 10.16a and b). Some other plastics, such as the

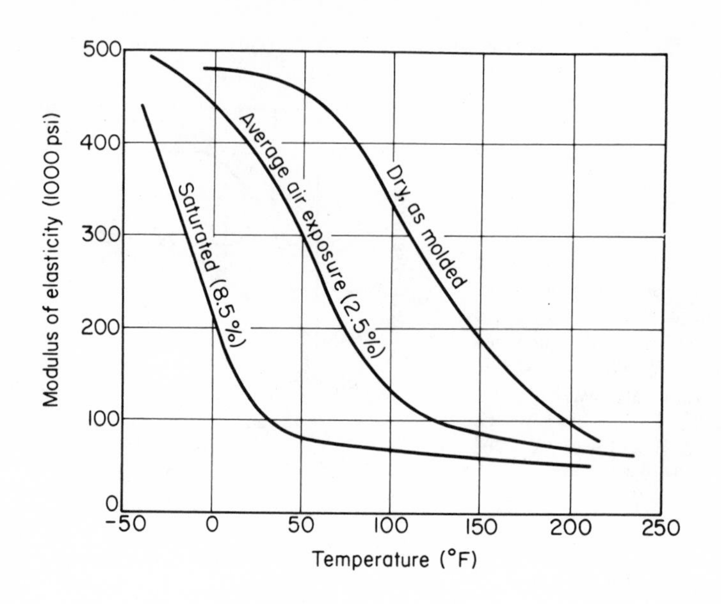

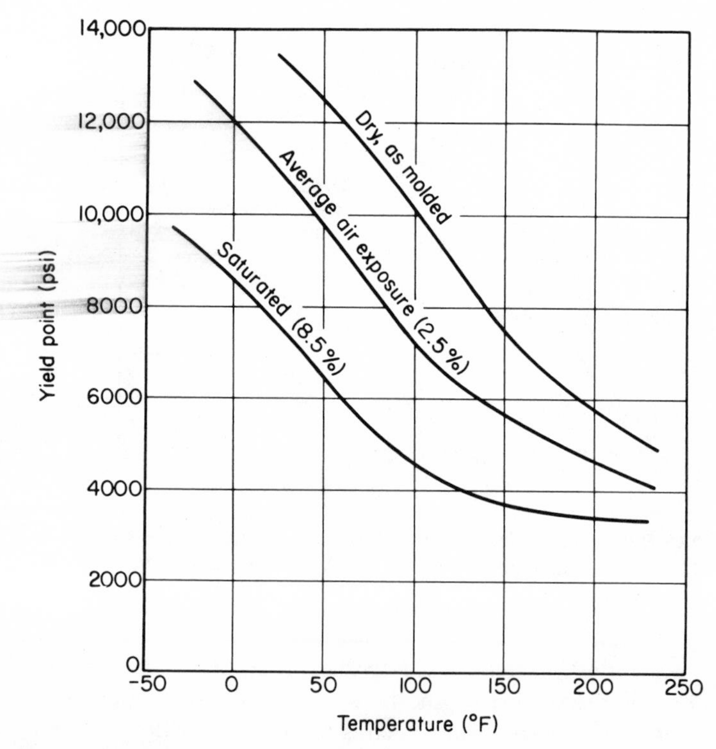

Fig. 10.16 (a) Modulus of elasticity and (b) yield point of "Zytel" 101 vs. temperature at various moisture contents. From "Designing with Du Pont Plastics," E. I. Du Pont de Nemours & Co., Inc., (1956).

acetal resins (e.g., "Delrin"),* are far less harmed by water or its vapor.

Cold Drawing—The flat portion of the stress–strain curve that resembles the lower yield jog in mild steels is known as *cold drawing*. The characteristics of this behavior have been more intensively studied than the rest of the curve since advantage is taken of this commercially—in uniaxial stretching to form fibers and in biaxial tension to produce films. In stretching a nylon or polyethylene rod, the deformation is reasonably uniform over the complete length of the specimen up to the yield point. During the next increment of strain, over which the applied load decreases, however, the rod develops a very deep constriction (Fig. 10.17a, b). The amount of area contraction during necking may be of the order of eighty per cent, whereas the thickness decrease in mild steel during the yield jog rarely exceeds five per cent. The ends of the constricted portion of the rod forms sharp shoulders, and as deformation proceeds these shoulders move apart (Figs. 10.17c, d, e, f, and g). As long as there is some length of rod available, allowing the

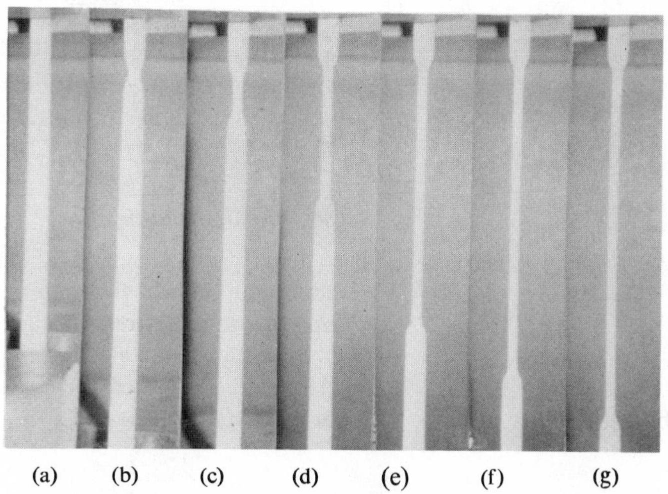

(a) (b) (c) (d) (e) (f) (g)

Fig. 10.17 Photograph of the formation of the neck during "cold drawing" of low density polyethylene. The cross-sectional area of the necked portion is about one-fifth that of the unnecked region.

shoulders to continue to separate, strain continues without a concurrent increase in nominal stress. After the rod is completely necked, the amount of tension required for further stretching again increases.

Cold drawing was first noticed in crystalline polymers and was found to coincide with the rearrangement of crystallites in the body from a random to an oriented condition. This orientation, as pointed out in Chapter 4, leads

* Du Pont trade name.

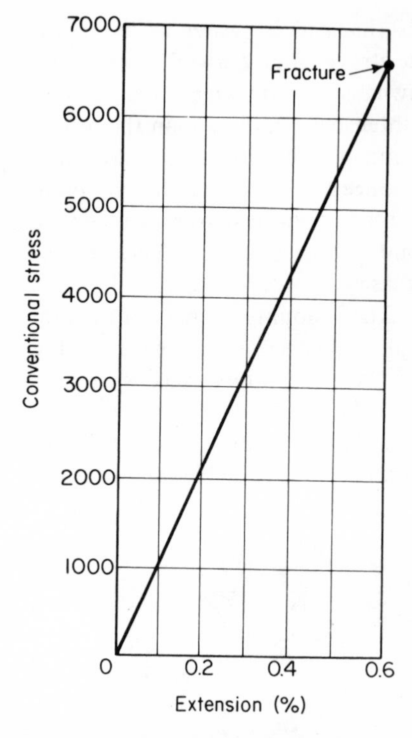

Fig. 10.18 Stress-strain curve of wood-flour filled BAKELITE*. (* Union Carbide Plastics Co. trade name.)

to far higher strengths in microcrystalline polymers, and for this reason stretching is always included as one step in the production of textile fibers or high strength films. Since cold drawing was first observed while the crystallites were being orientated, it was thought to result from this phenomenon. Subsequently it was also found in amorphous polymers, so that orientation must be considered simply a contributing factor. By measuring the temperature changes along a test bar while it was being stretched, a pronounced local heating in the vicinity of the shoulder was detected. This thermal wave front that results from the supplied mechanical energy has also been suggested as a cause of cold drawing, especially in amorphous polymers. No satisfactory single explanation for cold drawing, applicable to both crystalline and amorphous polymers, is available.

Thermosetting Plastics

Because of the very rigid binding of all the molecules in thermosetting plastics, their tensile stress–strain curve is quite simple. Only elastic deformation is possible as shown in Fig. 10.18. Hence the complete behavior is described by Eq. (10.12), and the value of α is small (≈ 0).

Rubber

The nonmetallic material whose tensile properties are best understood is rubber. Its deformation mode, and the mechanism that is thought to underlie it, were discussed in earlier chapters of this book. A typical stress–strain curve for a vulcanized rubber is shown in Fig. 10.19. This curve, like that of the plastics, generally consists of three sections. Its initial portion, over which the modulus is relatively low, involves elastic deformations of hundreds of per cent. The mechanism of atomic and molecular movement that accounts for this behavior was first proposed in the early 1930's and is not only valid for the natural rubbers, which were the only ones then known, but applies equally to the synthetics developed since. Briefly, rubber elasticity requires that one have a body made up of atoms that are tightly bound in one direction (along the backbone of the molecule) and loosely bound transverse to this direction (between molecules). In addition, it is necessary to have some single bonds such as —C—C— in order to allow for free rotation along the molecule length. Because it is easier for motion to occur transversely to the chain length than along it, molecules of this type "bunch up" and form statistically random shapes (see Fig. 1.6b). This leads to the maximum entropy. Motion is not stopped in the contracted molecules but it no longer has any preferred direction with respect to the external shape of the body. This motion is somewhat restrictive because the moving units are sections of molecules, and each segment must remain compatible to the part of the molecule to which it is joined. Nevertheless, there is sufficient freedom so that the motion is very much like that of the molecules of an ordinary liquid and is referred to as "micro-Brownian" motion. Although the entanglement of the individual molecules both with themselves and their neighbors would hinder flow in such a material, it would not be much different, mechanically, than an extremely viscous liquid. Hence, an additional requirement must be met; viz., an interlocking of the molecules at a few places along their length to form a three dimensional network. This interlocking, which prevents the individual molecules from flowing apart, is accomplished in rubbers by vulcanizing. In the light of this mechanism, it is not surprising to find that rubber becomes more difficult to extend (i.e., its modulus increases) as the degree of vulcanization, or the number of cross-links, is increased. A typical behavior is shown in Fig. 10.20.

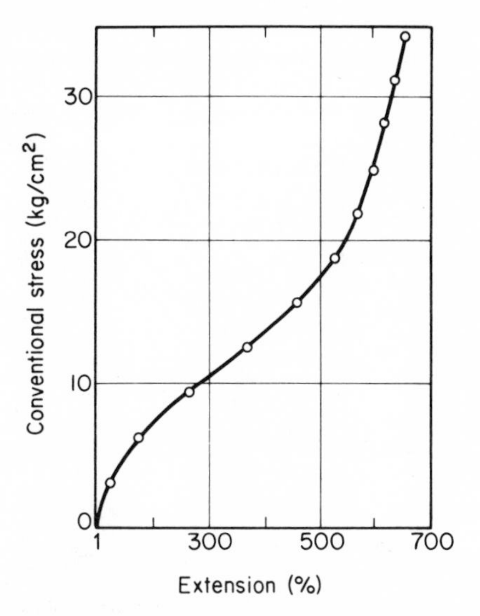

Fig. 10.19 Force extension curve for "pure gum" GR-S rubber at 2°C. From L. R. G. Treloar, *The Physics of Rubber Elasticity*, Clarendon Press, Oxford (1958).

Beyond the initial low slope portion of the tensile stress-strain curve, the crystallinity of the rubber becomes important. Although rubber, as described in the previous paragraph, is normally amorphous, it becomes crystalline at low temperatures. Crystal nuclei serve very much like cross linkages between molecules as far as mechanical properties are concerned. Hence, at low temperatures where the crystallinity becomes pronounced, both the

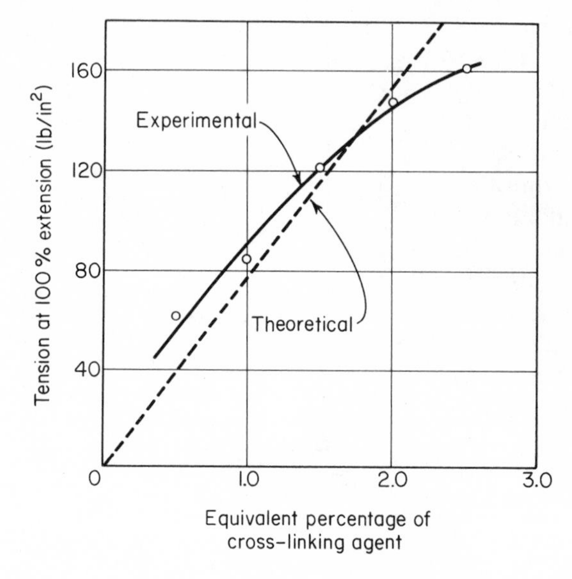

Fig. 10.20 Dependence of modulus on degree of cross-linking. GR-S rubber. From P. J. Flory, *Principles of Polymer Chemistry*, Cornell University Press, Ithica, N.Y. (1953).

modulus and tensile strength of rubber increase, and at the same time it becomes brittle. The existence of some small number of crystal sites in natural uncured rubber is probably one reason it has "rubbery" properties in spite of its lack of chemical cross-links.

Crystallinity is also developed in rubber during stretching. This is identical with low temperature crystallinity excepting that, in this case, the crystals are oriented along the stretching axis. When crystals form, some stability is given to the portion of the extended molecules that forms a part of the crystal so that, over a range of extensions, the slope of the stress-strain curve decreases from its initial values (around 300–400 per cent in Fig. 10.19). Increased straining causes more crystallization that in turn gives more junction points along the molecules length. This additional linking becomes the dominating feature when the extensions are large; the action of the crystal sites is now equivalent to increased vulcanization, and the slope of the stress-strain curve rises sharply.

Crystallization is probably not the sole cause for the upward curvature of the stress-strain curves. The statistical nature of the network (i.e., the average shape of a molecule) also predicts an increase in slope with extension.

GLASS

The factors that influence the strength of glass have been mentioned previously on numerous occasions in this book. For example, the formation of cracks by water vapor corrosion was discussed on page 94, and the influence of flaws on page 246. Because of the growing importance of plastic-reinforced fiberglass, the influence of some of the variables on the tensile properties of fibers are further discussed here.

The most conspicuous property of fibers, as compared with bulk glass, is their very high strength. Whereas bulk glass has a tensile strength that varies from 8,000 to 25,000 psi, depending largely on its composition, glass fibers have been produced in the laboratory with strengths in excess of 1,000,000 psi, and commercially produced fibers have strengths of 500,000 psi.

The most significant difference between bulk and fiber material, viz., the number of flaws encountered in the small volume of fibers as compared with the bulk material, has been discussed in Chapter 9. In addition, however, fiber structures are inherently stronger than the bulk material. This appears to result from the difference in thermal history encountered in producing the two forms. Fibers are cooled very rapidly from the melt (about 1300°C.) to room temperature so that the structure that is "frozen in" is an arrangement of atoms corresponding to a higher temperature equilibrium than that from which bulk glass attains its structure. Because the latter is cooled slower, there is more time for atomic rearrangements during chilling. Not only is the strength higher in fibers, but other physical properties such as density, viscosity, elastic modulus, and index of refraction, all agree with properties characteristic of glass at a higher temperature. Fibers have a density that is 10 to 15 per cent less than that of bulk glass, and this is reflected in a lower modulus. When glass fibers are heated, all of their physical properties revert toward, but do not reach, those of bulk glass; i.e., their structure approaches one characteristic of equilibrium at a lower temperature. This is evidenced by an increase in density and elastic modulus and is known as thermal compaction.

Two hypotheses have been proposed to explain the increased strength of the small filament. The first considers it an orientation effect. The smallest (covalent) bonds line up in the direction of drawing, causing the high strength in the direction of the fiber axis. The second is based on the assumption that the glass is partially covalent and partially ionically bonded. Fast quenching is thought to increase the proportion of ionic bonds.

TENSILE PROPERTIES OF ADHESIVE JOINTS

Adhesive joints are unique in their tensile behavior. The application of a uniaxial tensile load to an adhesive bond (Fig. 10.21) results in a triaxial stress state, and a stress concentration is developed in the joint at the edges of the adhesive–adherend interface. This complex stress system results because the adhesive wants to contract more than the adherends under the influence of an identical applied stress. In this respect, the system is not unlike a necked tensile bar. The amount of transverse strain in the adherend is $\varepsilon_a = \nu_a\sigma/E_a$, that in the adhesive $\varepsilon_b = \nu_b\sigma/E_b$. Hence, the triaxial stress

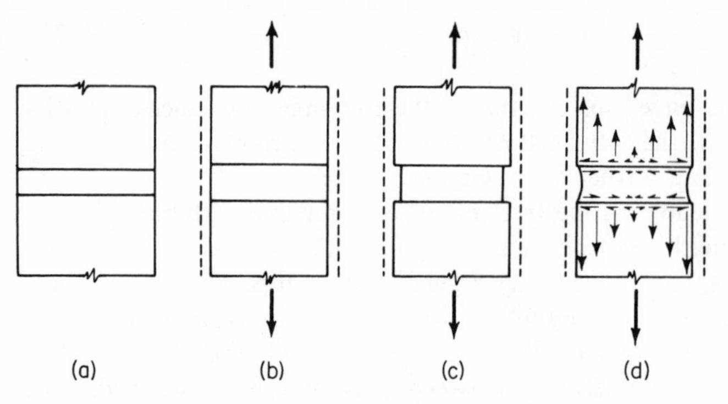

Fig. 10.21 Schematic behavior of adhesive bond in tension. (a) Adherend-adhesive system, unstressed. (b) Under tensile stress, adherend and adhesive have same ratio of modulus of elasticity E to Poisson's ratio ν. (c) Adhesive with larger ratio of E/ν than adherend. (d) Stresses set up at interface under condition (c). From G. H. Dietz, in *Symposium on Tension Testing on Non-Metallic Materials, ASTM Spec. Tech. Pub.*, No. 194 (1956.)

state and stress concentration occur, unless there is a unique relationship between the adhesive and the adherend in which

$$\frac{E_a}{\nu_a} = \frac{E_b}{\nu_b} \tag{10.13}$$

where E_a = Young's modulus of adherend
E_b = Young's modulus of adhesive
ν_a = Poisson's ratio of adherend
ν_b = Poisson's ratio of adhesive.

Since Young's modulus of metals is orders of magnitude higher than those of most polymers used for adhesives (see Table 5.1), (10.13) is seldom approached.

The bond strength, because of this complex stress, and possibly because of the orientation of the polymer near the metal surface, is seldom identical with that of the adhesive when tested in bulk form. The stress state and orientation are a function of the joint dimensions. For this reason the measured bond strength is also dependent on dimensions, as shown in Fig. 10.22.

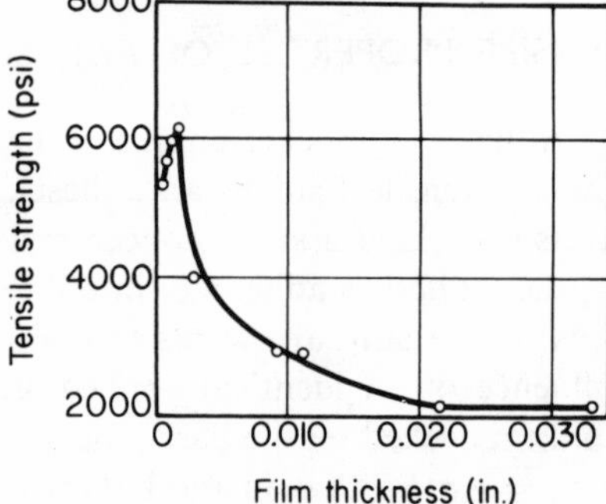

Fig. 10.22 Bond strength in tension vs. film thickness for Buna N-Phenolic resin on steel. From G. H. Dietz, in *Symposium on Tension Testing on Non-Metallic Materials, ASTM Spec. Tech. Pub.*, No. 194 (1956).

PROBLEMS

1. A tensile testing machine-specimen system resembles a hard and soft spring in series; the soft spring is the test piece, and the hard one the machine. Make a schematic drawing of a test setup to show why this is so.

2. A 0.505 inch diameter steel test bar is threaded into one inch diameter loading rods. The test piece is 2 inches long, and the top and bottom loading bars are each 1 foot long and $2\frac{1}{2}$ inch diameter. The elastic position of the stress–strain curve is measured between the cross-heads, i.e., over the 26 inch length. Assuming there is no slipping in the threads, what extension is measured when the specimen is loaded to a stress of 100,000 psi?

3. In some tests, specimen extension is assumed to be identical with cross-head travel. In Prob. (2), what is the percentage error if this is done to calculate E of the test sample?

4. An aluminum coated steel rod carries a 100,000 lb. weight. The rod is one foot long so that its own weight can be neglected. The diameter of the steel is 2 inches and aluminum coating is $\frac{1}{2}$ inch thick. How much of the load is carried by the steel and how much by the aluminum? Assume τ at the steel–aluminum interface is negligible. ($E_{\text{steel}} = 30 \times 10^6$; $E_{\text{aluminum}} = 10 \times 10^6$.)

5. Using the following data, sketch the conventional stress vs. strain and true stress vs. natural strain diagram as best you can.

$$E = 30 \times 10^6 \text{ psi} \qquad e_f = 18\%$$
$$\sigma_{o(0.2\%)} = 30{,}000 \text{ psi} \qquad e_n = 15\%$$
$$\sigma_f = 75{,}000 \text{ psi} \qquad q_f = 25\%$$
$$S_u = 50{,}000 \text{ psi}$$

6. What is the modulus of the resilience of the metal in problem (5)? What are the values of K and n?

7. Aluminum has a $\Delta H \approx 3.5 \times 10^5$ cal/mole, according to Fig. 8.7, and $R \approx 2$ cal/°K/mole. What strain rate intervals are equivalent to 20° → 30°C; 20° → 120°C?

8. Assuming E_o is independent of time, discuss Fig. 8.3 in terms of Eq. (10.11).

9. Show that the tangent in Fig. 10.8b intercepts the ordinate at the value of S_u.

10. A steel tie-rod holds the opposite walls of a 24 foot long bin together. At room temperature (75°F.) the stress on the rod is 2,000 psi. What is the stress at −20°F and at +120°F? The coefficient of thermal expansion of steel is 6.5×10^{-6} in/in/°F.

REFERENCES FOR FURTHER READING

(R10.1) Hollomon, J. H., and L. D. Jaffee, *Ferrous Metallurgical Design*. New York: John Wiley and Sons, Inc., 1947.

(R10.2) Nadai, A., *Theory of Flow and Fracture of Solids.* New York: McGraw-Hill, Inc., 1950.

(R10.3) Hoffman, O., and G. Sachs, *Introduction to the Theory of Plasticity for Engineers.* New York: McGraw-Hill, Inc., 1953.

(R10.4) "Symposium on Significance of the Tension Test of Metals in Relation to Design" *Proc. ASTM,* Vol. 40 (1940), 501.

(R10.5) "Symposium on Tension Testing of Non-Metallic Materials." *ASTM, Spec. Tech. Publication,* No. 194 (1957).

(R10.6) Timoshenko, S., *Strength of Materials, Part I, Elementary Theory and Problems.* 3rd Edition. Princeton, N.J.: D. Van Nostrand Co., Inc., 1955.

(R10.7) *ASTM Standards,* Parts 1 to 11, ASTM, Phila., Pa. (most recent edition).

11

PROPERTIES UNDER COMPRESSIVE LOADS

Compressive properties are important in three classes of materials:

1. Brittle structural materials—e.g., concrete, brick, or natural rocks such as granite or limestone—whose tensile strength is only a fraction of their compressive strength. In structural applications these are invariably loaded in compression.
2. Malleable metals, shaped by pressure into various commercial forms such as rails, tubing, forgings of all kinds, sheets, etc.
3. Soft metals, plastics, and certain other substances in such applications as bearing linings, washers, or gaskets.

COMPRESSION OF BRITTLE SOLIDS

The stress-strain curve of a brittle solid in uniaxial compression is similar to that of a glass rod in tension (Fig. 1.2) except that a much higher strength level is reached before fracture occurs. Some brittle substances—like glass, quartz, or diamond—show a nearly perfect Hookean relation between stress and strain, whereas in others a slight curvature is observed (Fig. 11.1). The amount of permanent strain prior to fracture is, however, very small and rarely exceeds a fraction of one per cent. In these conditions the compressive strength and the compressive fracture strength become synonymous, and it is meaningless to speak of a "yield stress."

The compressive or crushing strength of a selection of brittle materials are compared with their tensile strengths in Table 11.1.

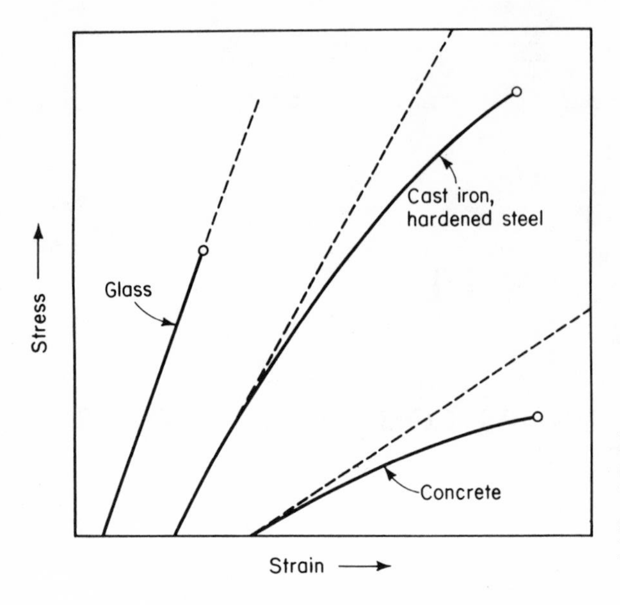

Fig. 11.1 Compression stress-strain diagrams of brittle materials (schematic, not to scale). Elastic lines are shown dashed.

TABLE 11.1

COMPARISON OF COMPRESSION AND TENSILE STRENGTH OF BRITTLE MATERIALS

Material	*Fracture Strength* $\times 10^3$ *psi* Compression	Tension
Granite	20	0.6
Marble	14	0.85
Brick	4.5	
Tungsten carbide	750	130
Cast iron	100	30
Plexiglas or Lucite	9	15
Polystyrene (molding compound)	7	14

Fracture Modes of Brittle Solids in Uniaxial Compression

Compression itself has no tendency to form or propagate internal cracks and thus to cause fracture. This is not true, however, of the accompanying shear and secondary tensile stresses; shear fractures inclined to the compression axis are among the most typical in brittle materials (Fig. 11.2).

Quite often cylindrical specimens fracture along a plane parallel to the compression axis rather than by shear (Fig. 11.3). Although the cause is not obvious at first sight, it was found by numerous photoelastic tests on Plexiglas that secondary tensile stresses, which develop spontaneously across

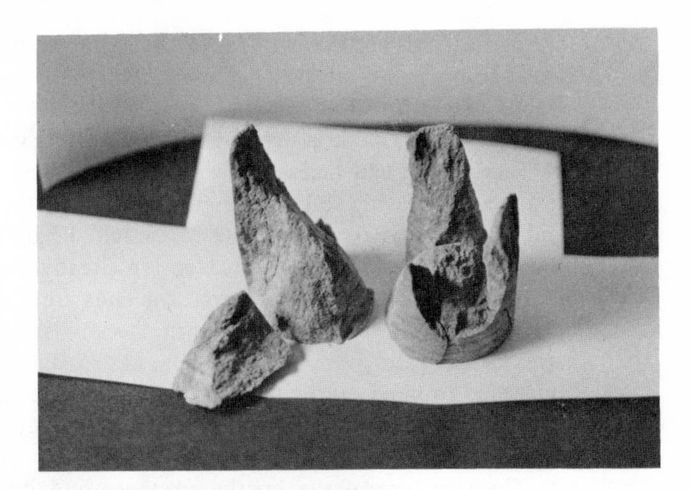

Fig. 11.2 Fracture of sandstone cylinders in compression. Path of fracture follows shear stresses but is unusually steep. Courtesy of Prof. T. R. Seldenrath, Mining Laboratory, Technical University, Delft, Holland.

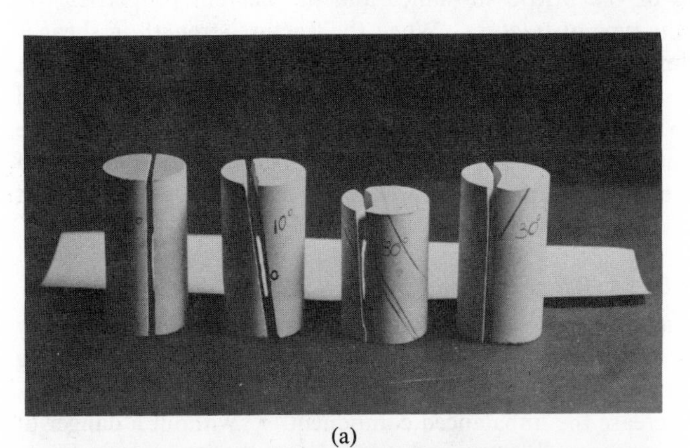

(a)

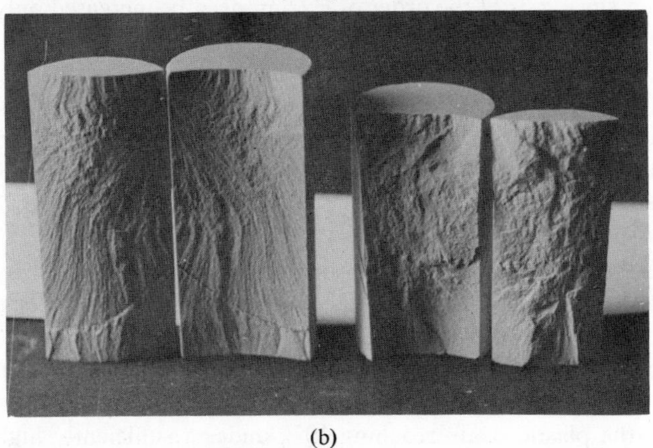

(b)

Fig. 11.3 Fracture of lithographic sandstone along an axial plane. From T. R. Seldenrath and J. Gramberg, (R11.5).

the vertical axis in the end thirds of the specimen, reach about one half of the axial compressive stress. These stresses may easily cause the observed fracture (R11.5).

The fracture mode is strongly affected by the ease with which each surface element of the end faces of the brittle specimen can move outward. For instance, by placing soft rubber gaskets between the specimen and anvil, the end faces tend to expand relative to the mid–height portion of the specimen, thereby generating the transverse tension already mentioned with consequent fracturing (Fig. 11.3). If, however, the ends are confined (e.g., by relatively rigid reinforcing rings or recesses in the anvils), the transverse tensile end stresses are largely suppressed. In this case a much higher axial compression is needed to build up the tensions to the fracture stress value in the bulging out central portion. As a result the apparent fracture strength rises, sometimes to twice its "free-end" value.

The structure of the brittle substance and its inherent properties will decisively affect the type of fracture. When the fracture strength in shear is low, oblique failures will occur no matter what the degree of constraint at the ends. Coarse grained materials like sandstone, marble, or cement will fail in this fashion, and the effect of end confinement is slight. In cylinders the typical fracture may propagate part way along the periphery of a cone, and the separated tapered section may then wedge the rest of the material apart (see Fig. 11.2).

Brittle Materials Under Triaxial Compression

In such cases as Fig. 11.3, where fracture results from the spontaneously arising transverse tensile stresses, σ_t, fracturing is avoided by superimposing upon the uniaxial compressive stress, σ_c, a hydrostatic component, $\sigma_h > \sigma_t$. One could now increase the unbalanced component, σ_c, without a danger of fracture. Moreover, since σ_t is of the order of $\sigma_c/2$, σ_h need be increased only at about half the rate at which σ_c increases to prevent cracks from spreading.

In practice a truly hydrostatic pressure is developed by submerging the entire test assembly in a pressurized liquid (Fig. 11.4a), and the three principal stresses become $\sigma_1 = \sigma_c + \sigma_h$, $\sigma_2 = \sigma_3 = \sigma_h$. A biaxial or lateral liquid pressure is often simpler to apply (Fig. 11.4b), resulting in principal stresses $\sigma_1 = \sigma_c$, $\sigma_2 = \sigma_3 = \sigma_h$. In the first case, $\tau_{max} = (\sigma_1 - \sigma_3)/2$ is simply $\sigma_c/2$ whereas, in the second, it is $= (\sigma_c - \sigma_h)/2$.

Note that in the above cases the conventional notation was reversed by calling σ_1 the largest compressive stress. This is generally done in problems involving mostly compressive stresses.

Kármán was the first to demonstrate that, with a sufficiently high $\sigma_3 (= \sigma_h)$, brittle rocks like marble and sandstone could deform plastically, without failure (Fig. 11.5); the plastic strain reaching 10% under a sufficiently high σ_h. Similar experiments were performed on brittle cast zinc by Böker who

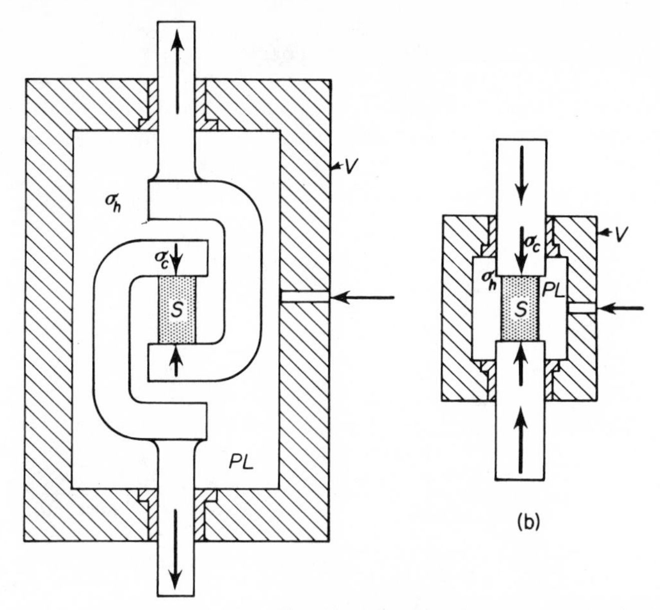

Fig. 11.4 Compression tests under hydrostatic pressure. (a) All-sided liquid pressure σ_h on specimen. (b) σ_h applied laterally only. *V*–high-pressure vessel. *PL*–pressurized liquid. *S*–specimen.

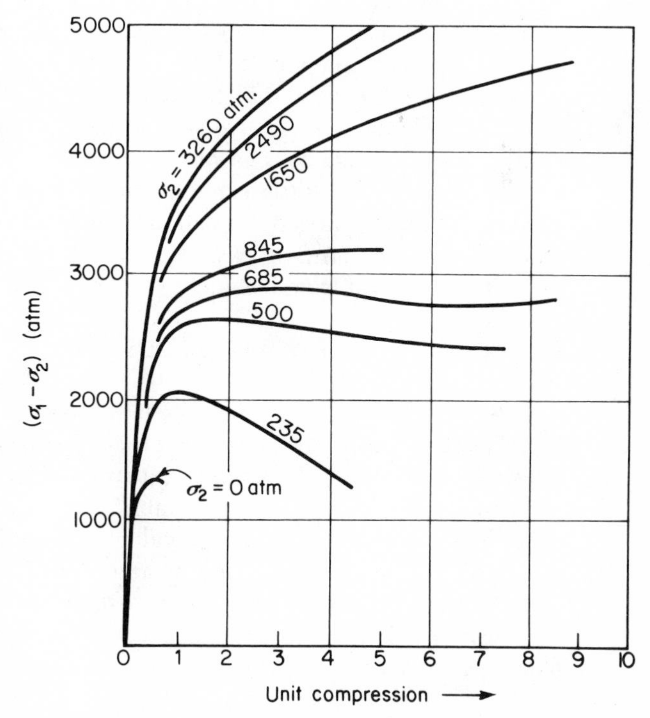

Fig. 11.5 Results of Kármán's tests on marble cylinders under combined axial (σ_1) and lateral hydraulic compression (σ_2). The lateral pressure σ_2 was kept constant in each test. After (R11.4).

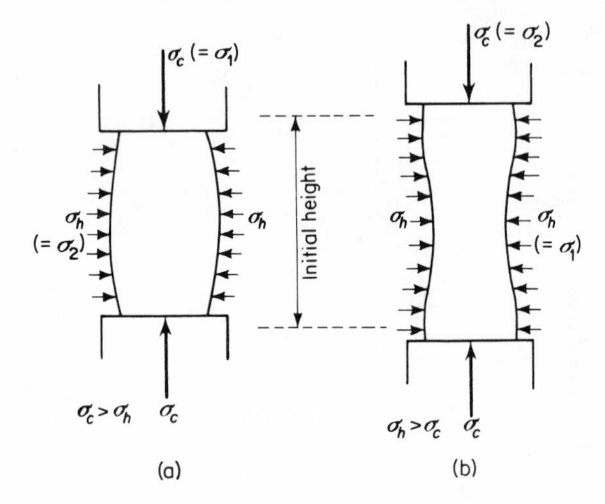

Fig. 11.6 (a) Plastic bulging of Kármán's cylinder. (b) Necking of cast zinc in Böker's experiments.

used a lateral pressure in excess of the axial pressure in some instances. This caused his specimens to contract laterally, forming a neck instead of the bulging observed by Kármán (Fig. 11.6). As a result of the hydraulic pressure a neck and an extension occurred in the axial direction.

Despite the fact that in these tests σ_h was applied laterally along the lines of Fig. 11.4b, the Kármán experiments could be regarded as uniaxial compression tests on which an all-sided hydrostatic pressure is superimposed, the net uniaxial component being $\sigma_c - \sigma_h = \sigma_1 - \sigma_2$. In Böker's tests, $\sigma_2 < \sigma_1$ and $\sigma_h > \sigma_c$ so that the difference, $\sigma_c - \sigma_h$, was equivalent to a uniaxial tension. The negative sign of the $\sigma_c - \sigma_h$ difference is again due to the "inverted" sign convention.

The plasticity of brittle materials under high hydrostatic pressures can account for the "crackless" formation of undulations and bends in geological strata of normally brittle rocks. The high pressures produced by the earth layers prevent them from breaking.

COMPRESSION OF DUCTILE SOLIDS

The Geometry of Homogeneous (Parallelepipedal) Compression

At least 90% of all the metals used today have undergone a plastic shaping operation like rolling, forging, or extrusion somewhere between its initial extraction from the ore and final fabrication. In these operations, all of which involve compression, the total plastic strain suffered by the metal often exceeds the tensile ductility by an order of magnitude. The ability to undergo large plastic deformations and their limits in a particular material can be imitated in metals by using compression tests on small cylinders, blocks, or strips, and the stress-strain relations thus determined can be employed to predict the pressures and forces arising in industrial forming operations. We shall now analyze the compression behavior of plastically deformable bodies.

The simplest conceivable type of compressive straining is a homogeneous deformation, defined in Chapter 3 as one in which originally plane and parallel surfaces remain plane and parallel. A homogeneously deformed cylinder grows shorter and increases its diameter while maintaining an exactly cylindrical form (Fig. 11.7a). A cube transforms into a square prism, whereas a rectangular prism gradually shortens while its length to width ratio remains constant (Figs. 11.7b and c). Since the deformation is plastic, the volume does not change, so that

$$A_o h_o = A_1 h_1 = A_2 h_2 = \cdots Ah$$

A_o and h_o being the initial cross-sectional area and height, and A_1, A_2, A and h_1, h_2 and h the deformed cross-sectional areas and heights. The mean compressive stress σ_1 at any stage is

$$\sigma_1 = F/A$$

Since

$$A = A_o \frac{h_o}{h}$$

$$\sigma_1 = \frac{Fh}{A_o h_o} \tag{11.1}$$

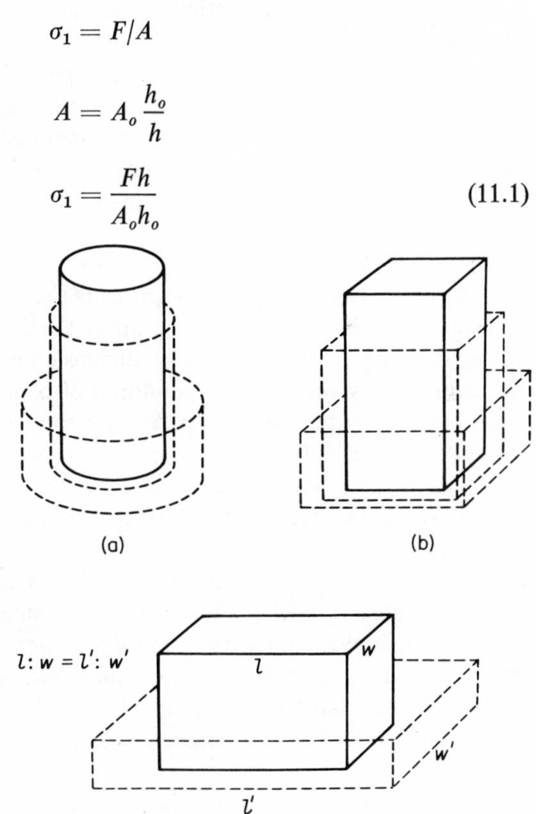

Fig. 11.7 Homogeneous compression of: (a) Cylinder. (b) Square prism. (c) Rectangular prism.

The compressive stress or pressure is thus determined from measurement of the initial dimensions of the cylinder or parallelepiped, its instantaneous height and applied force.

The same principles also hold for a prismatic body that is restrained from changing one of its principal dimensions (plane strain conditions). This particular case lends itself readily to quantitative treatment, and the formal analysis of various metal working processes is often patterned on the plane strain compression prototype.

Compression involves one important factor that is absent from stretching—the friction between the ends of the specimen and the pressure plates or anvils. As the material is upset, it must slide sidewise over the plates, against the frictional resistance acting at the tool and material interface. These frictional forces can never be completely eliminated and they may become large enough to bring the process to a standstill or to increase the apparent compressive resistance beyond the strength of the compression tools. Hence an analysis of plastic compression involves not only consideration of the inherent properties of the metal but also the effects of friction on the process.

A truly homogeneous or "parallelepipedal" compression is never completely realized because of the ever present friction. This condition can be approached, however, and it will be assumed here that the presence of friction does not disturb the homogeneity of flow, so that the following analysis is based on this simplified premise.

Homogeneous Compression of a Wide Slab

The upper half of Fig. 11.8 represents the side view of a rectangular slab of length l, height h, and infinite width in the direction normal to the surface of the drawing. Because the width dimension is so large, deformation in this direction is prevented; i.e., a condition of plane strain exists. (This is consistent with the statement on p. 56 that this strain state occurs when two dimensions of a body are small and one is large.) Each section, formed by two cuts parallel with the page and a unit distance apart, is acted on by identical stresses. Hence, it is sufficient to analyze the stress acting on only one such section.

Consider a length element, dx, at a distance, x, from the end, K', of the slab. Along dx, σ_1 is virtually constant and, since homogeneous deformation is assumed, it does not vary in the h direction either. Consequently, when yielding occurs, the horizontal component, σ_3, acting in the length direction will also be constant over the element face. To find the value of σ_1 necessary to start plastic flow, the influence of the coefficient of friction at the surface, μ, must also be considered.

Stress, σ_3, must be zero on the two free external faces, K' and K'', but it changes on moving toward the center plane, N-N. Accordingly, if elementary

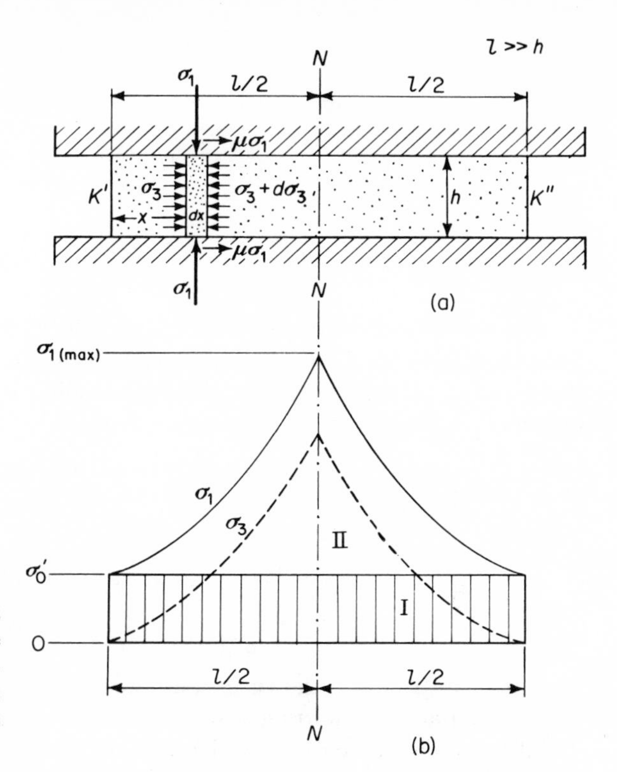

Fig. 11.8 Homogeneous compression of a very wide plastic slab between parallel plates. (a) Dimensions and stresses. (b) Pressure distribution.

slice, dx, is acted upon by σ_3 on the surface located at a distance x from K', a stress, $\sigma_3 + d\sigma_3$, will act on a surface located $x + dx$ from K'. The resultant horizontal force acting on the slice is $(\sigma_3 + d\sigma_3)h \cdot 1 - \sigma_3 h \cdot 1 = h\, d\sigma_3$. To maintain a static equilibrium, this force must be balanced by some other force of the same magnitude but of opposite sign. The balancing force is provided by the friction acting on the top and bottom faces of the element and it equals $2\mu\sigma_1\, dx$. The equilibrium equation is thus:

$$h\, d\sigma_3 = 2\mu\sigma_1\, dx,$$

or (11.2)

$$h\, d\sigma_3 - 2\mu\sigma_1\, dx = 0.$$

The two unknown stresses, σ_1 and σ_3, are interdependent because the maximum distortion energy criterion of yield under plane strain (p. 152, problem 1) is:

$$\sigma_1 - \sigma_3 \simeq 1.15\sigma_o = \sigma_0'. \tag{6.7}$$

Taking differentials of both sides,

$$d\sigma_1 - d\sigma_3 = 0; \quad d\sigma_1 = d\sigma_3$$

Equation (11.2) can thus be rewritten in the form

$$h\, d\sigma_1 = 2\mu\sigma_1\, dx$$

or

$$\frac{d\sigma_1}{\sigma_1} = \frac{2\mu\, dx}{h}$$

Integration yields

$$ln \sigma_1 = \frac{2\mu x}{h} + C \tag{11.3}$$

The constant C is found from the condition that for $x = 0$, $\sigma_3 = 0$. Since there is no horizontal pressure on the free surfaces K' and K'', it follows that $\sigma_1 = \sigma_0'$ at these locations and, substituting σ_0' for σ_1 in (11.3), one obtains $C = ln\, \sigma_0'$. Hence, the pressure formula is

$$ln\, (\sigma_1/\sigma_0') = 2\mu x/h$$

or

$$\sigma_1 = \sigma_0' e^{2\mu x/h} \tag{11.4}$$

The maximum pressure is reached at the center plane N-N where $x = l/2$ and

$$\sigma_1(\max) = \sigma_0' e^{\mu l/h} \tag{11.5}$$

The σ_1 distribution shown in the lower portion of Fig. 11.8 represents an exponential curve starting at $\sigma_1 = \sigma_0'$ for $x = 0$ and terminating at N-N. The distribution curve over the slab portion to the right of N-N is a mirror image of the former.

Assuming that the lower plate is stationary and the upper moves downward, any arbitrarily chosen point of the body will be gradually displaced down and in the direction away from N-N. The only points that will not move sidewise are those located on plane N–N which is usually designated as the *no-slip* or *neutral* plane.

In order for the yield criterion to be satisfied, the difference between σ_1 and σ_3 must be σ_0' along each vertical plane normal to the direction of material flow. The horizontal pressure diagram is thus identical with that of σ_1 except that it is lower by a distance σ_0' (dashed lines in Fig. 11.8b).

According to (11.5), the maximum pressure required to start plastic flow increases rapidly with μ and with the l/h ratio. If it were possible to eliminate friction completely, the vertical pressure would have been equal to the yield stress σ_0' at every point.

The pressure diagram of σ_1 can thus be regarded as consisting of two parts: a rectangular "yield stress base", *I*, of length l and height σ_0', and of a "friction hill," *II*, whose height is a function of μ.

Experience confirms the qualitative relationship between σ_1, μ, and l/h. However, the growth of σ_1 is often not as rapid as predicted by the formula (11.4), and the relation between pressure and $\mu l/h$ appears to be linear rather than exponential. This may arise because the deformation is never exactly homogeneous; also, μ is not necessarily constant over the contact faces.

Indeed we shall see in the following section that, if μ is sufficiently high, a linear relation between σ_1 and x is expected.

The total compressive force, F, computed from (11.4) on the basis of a slab of unit width, is

$$F_1 = 2\int_o^{l/2} \sigma_1\, dx = 2\sigma_o' \int_o^{l/2} e^{2\mu x/h}\, dx$$

$$= \frac{\sigma_o' h}{\mu} [e^{(\mu l/h)} - 1] \tag{11.6}$$

The expressions derived above refer to a stationary, instantaneous situation. If compression is continuous, both h and l change gradually, the first decreasing and the second increasing. However, since the volume and the width of the slab remain the same, the product, lh, is constant at all times so that either l or h can be eliminated from (11.6). This circumstance enables F_1 to be determined at any stage of deformation.

The work or energy expended in reducing the thickness of a slab from h_1 to h_2 is the product of work and displacement. Since F_1 is variable, the work of deformation per unit of slab width becomes

$$W_1 = \int_{h_1}^{h_2} F_1\, dh \tag{11.7}$$

Numerical integration of (11.7) leads to a cumbersome expression. Further difficulties are caused by the fact that F_1 depends on yield stress, σ_o', which usually varies with the degree and rate of deformation. These make an analytical computation of the deformation work impractical so that a graphical method of evaluating W_1 as the area beneath the force–height diagram is most commonly used.

Homogeneous Compression of Cylinders

An expression analogous with (11.4) can be developed for compression of a circular disc or short cylinder by using polar coordinates and equating the resultant of the radial components of σ_r and σ_θ (as in the derivation of (2.26)) with the product of segment surface and frictional drag $\mu\sigma_r$ (Fig. 11.9). Easy considerations show that $\varepsilon_r = \varepsilon_\theta$, and, therefore $\sigma_r = \sigma_\theta$. The resulting equation is

$$\sigma_1 = \sigma_o e^{2\mu(r_o - r)/h} \tag{11.8}$$

where r_o is the outside radius, and r any intermediate radius of the disc. Since no lateral constraint is involved, σ_o and not σ_o' is the yield stress.

The maximum pressure for $r = 0$ is

$$\sigma_1(\max) = \sigma_o e^{2\mu r_o/h} \tag{11.9}$$

Fig. 11.10 shows a family of compression stress–strain curves of copper cylinders with various h_o/d_o ratios. The data are in qualitative agreement with

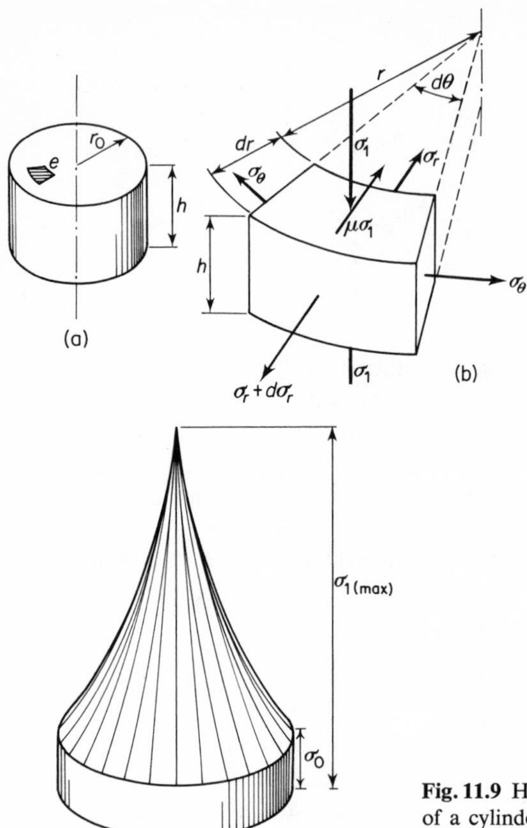

Fig. 11.9 Homogeneous compression with end friction of a cylinder. (a) Location of elementary segment e in cylinder. (b) Stresses acting on segment. (c) Friction hill over cylinder.

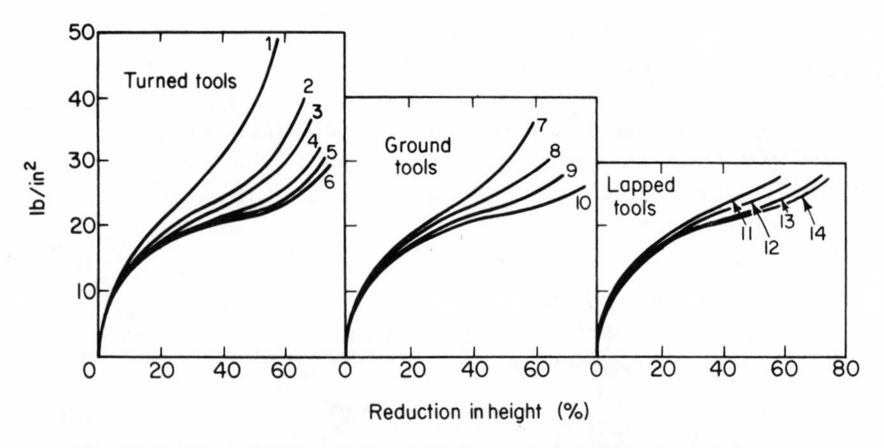

Fig. 11.10 Effect of h_o/d_o and of end friction on compressive stress-strain curves of annealed copper. Ratio h_o/d_o decreases from 2 (curves 1, 7, 11) to $\frac{1}{3}$ (curves 6, 10, 14). From M. Cook and E. C. Larke, *J. Inst. of Metals*, **71**, 371 (1945).

the analytical relations between σ_1, μ, and h/d, a decrease of h/d resulting in an upshift of the appropriate stress–strain curve. The shorter cylinder has a higher resistance to deformation and appears to be "harder."

Limiting Friction Conditions: "Sticking"

The foregoing considerations and mathematical treatment are based on the assumption of sliding friction (Coulomb friction) between the compressed metal and the anvils. The unit frictional force or drag opposing the lateral displacement of the material is then $\mu\sigma_1$ and physically represents a shear stress, τ_μ. At first sight there should be no upper limit to the value that can be reached by the frictional drag $\tau_\mu = \mu\sigma_1$. In actual fact τ_μ cannot exceed the yield strength in shear, $\tau_o = \sigma_o/2$ (or $\tau_o' = \sigma_o'/2 = \sigma_o/\sqrt{3}$ under plane strain conditions). Whenever the product of the coefficient of friction and normal pressure reaches or exceeds this value, the Coulomb friction mechanism will be replaced by plastic shear in the layer of the metal immediately beneath the anvil. Under these conditions the material "sticks" to the anvil instead of sliding over it. The term "sticking" does not imply, however, an absence of relative movement between the two bodies in contact. What actually happens is that the compressed metal "smears" over the anvils while moving across their faces.

The mathematical expression of "sticking" is obtained by replacing $\mu\sigma_1$ by τ_o' in (11.2). This leads to

$$h\,d\sigma_1 = 2\tau_o'\,dx$$

or

$$h\sigma_1 = 2\tau_o'x + C \qquad (11.10)$$

Assuming that "sticking" occurs throughout the contact surface, $\sigma_1 = \sigma_o'$ for $x = 0$. Hence $C = h\sigma_o'$ and

$$\sigma_1 = \sigma_o'\left(1 + \frac{x}{h}\right) \qquad (11.11)$$

If $\mu \geq 0.5$, $\tau_\mu \geq \sigma'/2 \geq \tau_o'$ even at $x = 0$. Sliding is thus possible over the entire contact faces as long as $\mu \leq 0.5$. Only on rough, dry, or hot surfaces can μ reach 0.5 or more.

Ordinarily μ may be of the order of 0.1, and sticking will occur only when the ratio σ_1/σ_o' becomes about 5. This point will be reached somewhere between the free edge and the neutral plane. From there on the exponential curves, Eq. (11.9), will be replaced by straight lines, according to Eq. (11.11). The constant, C, in (11.10) is to be computed for the σ_1 value at the boundary between the "slipping" and "sticking" zones, and it will be found that the straight line portions are tangent to the exponentials at the boundary. The general effect of sticking is to reduce the peak pressure by an amount dependent on $\mu l/h$. The higher this factor the more conspicuous will be the difference between the two pressure peaks (e.g., Fig. 11.11). There, a strip

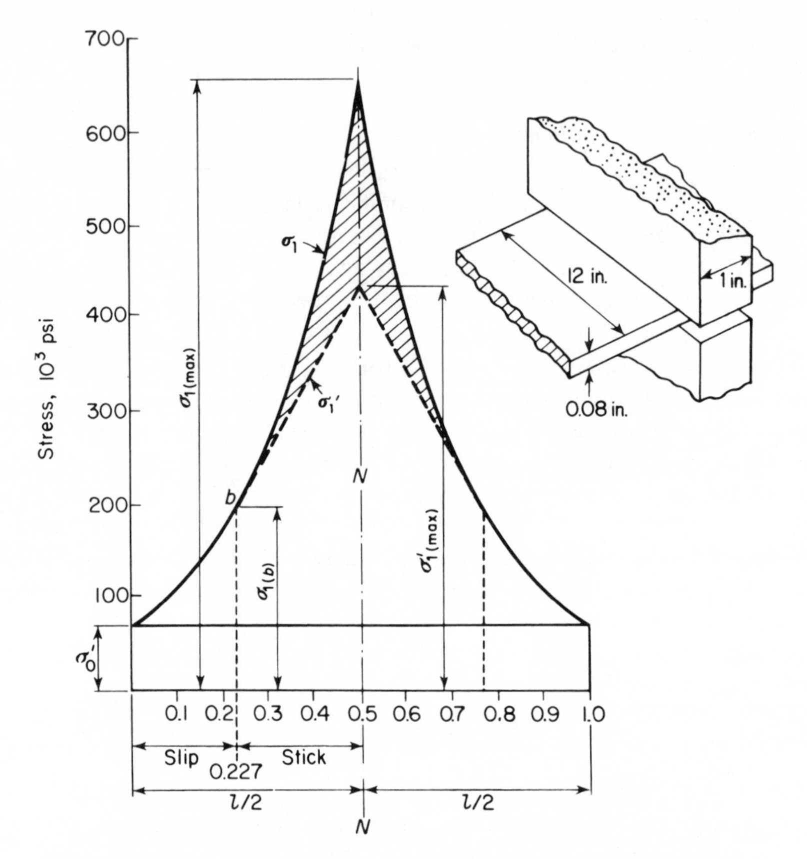

Fig. 11.11 Friction hill diagrams for conditions involving both Coulomb friction and "sticking." By taking "sticking" into consideration, the calculated force is reduced by the cross-hatched area.

0.080 in. thick and 12 in. wide is compressed between parallel anvils which are 1 in. in the long direction of the strip and overlap the strip on both sides. Let the coefficient of friction between the anvil and strip be 0.18, and the yield stress in tension of the strip 61,000 psi. In terms of the nomenclature already used (formulae 11.4, 5, 10, and 11) we have

$$h = 0.08 \text{ in.} \quad l = 1 \text{ in.} \quad w = 12 \text{ in.} \quad \mu = 0.18 \quad \sigma_o = 61{,}000 \text{ psi}$$

In view of the relatively large width (12 in.), it can be safely assumed that there will be no plastic flow in the width direction. Plane strain conditions will thus prevail, with $\sigma_o' = 1.157\sigma_o \approx 70{,}000$ psi. From (11.4) the pressure σ_1 required to produce plastic yielding of the strip is

$$\sigma_1 = \sigma_o' \, e^{2\mu x/h}$$

with a maximum at $x = l/2$:

$$\sigma_1(\max) = \sigma_o' e^{\mu l/h} = \sim 665{,}000 \text{ psi}$$

The shear stress (or frictional drag) at the interface between anvil and strip is $\tau_\mu = \mu\sigma_1$. However, as has already been explained, τ_μ cannot exceed the yield stress in shear; i.e., $\tau_o' = \sigma_o'/2$. Therefore, when σ_1 reaches a value such that $\mu\sigma_1 = \sigma_o'/2$, slipping changes to sticking, and the exponential relation between σ_1 and x breaks down. The boundary between the two zones is found from

$$\mu\sigma_1 = \mu\sigma_o' e^{2\mu x_b/h} = \sigma_o'/2 \tag{11.12}$$

It follows

$$e^{2\mu x_b/h} = \frac{1}{2\mu} \tag{11.13}$$

and

$$x_b = 0.227$$

The vertical pressure at this distance is

$$\sigma_{1(b)} = \sigma_o'/2\mu = 70{,}000/0.36 = \sim 195{,}000 \text{ psi.}$$

Since the frictional drag, τ_μ, in the range, $x_b < x < l/2$, is constant and equal to τ_o', the pressure must vary according to the linear law (11.11). Hence,

$$\sigma_1' = \sigma_{1(b)}\left(1 + \frac{2\mu x'}{h}\right)$$

in which x' is measured from boundary b. We find that σ_1 (max) $= 438{,}000$ psi, which is only two thirds of σ_1 (max) computed above on the Coulomb friction assumption.

By taking the first derivatives of (11.4) and (11.11), it is found that their values at $x = x_b$ are equal. The transition between the curvilinear and straight line positions of the diagram is, therefore, smooth without a break.

Nonhomogeneous Compression of Ductile Metals

The assumption of homogeneity is a convenient means for formulating basic relations between the various parameters of the compression process. In practice, however, it is not possible to maintain pure homogeneity on account of the friction between the metal and the anvils. Because of this factor, the material located at some distance from the contact surfaces flows more readily than that in their immediate vicinity, causing a cylindrical specimen to become barrel-shaped. The degree of barreling developed for a given reduction of height increases with increasing frictional resistance at the tool-to-material interface (Fig. 11.12). In extreme conditions (such as shown on the right) there was no slippage at all at the bases of the cylinder. The expansion of the ends occurred by an inversion of the cylindrical surface, a part of which was flattened out to become the peripheral ring of the circular base (see also Fig. 11.16, left). Since the metal adhering to the

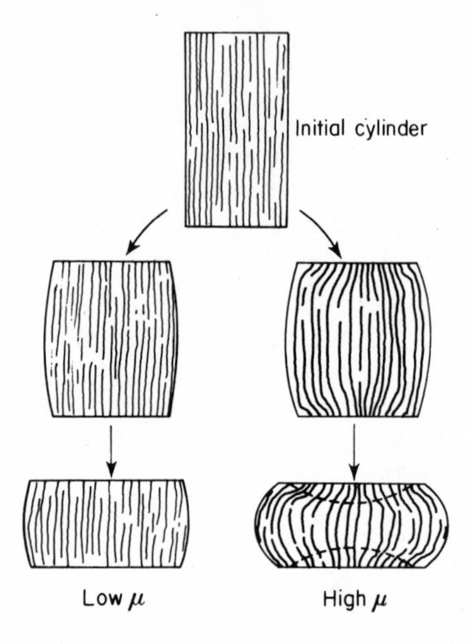

Fig. 11.12 Cross sections of steel cylinders compressed various amounts with low and high friction. Streaks represent gradual distortion of initially aligned impurities.

compression plate does not move, the plastic strain in it is zero, and it suffers no work-hardening.

If a cylinder, compressed under high friction conditions, is sectioned along an axial plane, and the degree of work-hardening examined from point to point, a picture of the type shown in Fig. 11.13 is found. The two end zones, A, are work-hardened least. Hardening increases from almost zero

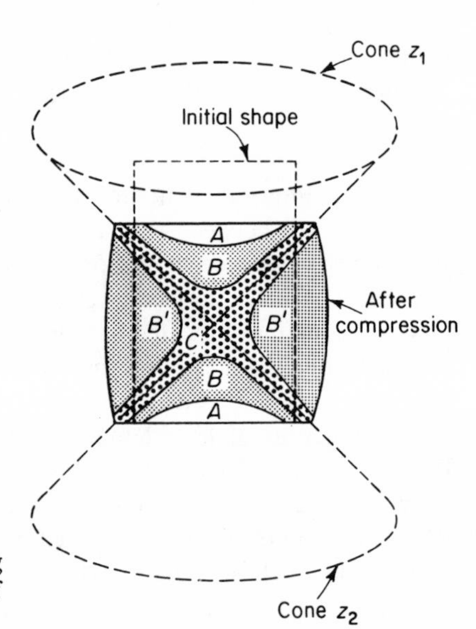

Fig. 11.13 Zones of variable work-hardening in a ductile metal after a moderate amount of plastic compression between rough anvils.

in the immediate vicinity of the contact surfaces through intermediate zones, *B*, to a maximum in *C*. The latter zone extends along the imaginary cone surfaces z_1 and z_2, which pass through the edges of the end faces and meet at the center of the cylinder. These cones approximately coincide with the surfaces of maximum shear, but their base angles are between 35° and 40° rather than 45°.

The differential degrees of work-hardening can be spectacularly exposed by annealing the specimen and etching a polished cross section to reveal the grain size. The light portions of the cones adjacent to the anvils show no change in grain size (Fig. 11.14a). Adjacent to the dark, not recrystallized areas are layers of very coarse crystals along which the small critical amount of cold work was reached. A light fully recrystallized zone, with a decreasing grain size indicative of a large deformation, covers the center sections.

When care was taken to reduce friction to a minimum, the deformation was practically homogeneous, and the differential strain–hardening was virtually absent. The resulting recrystallized grain size was very uniform (Fig. 11.14b).

Slip Cones

Since the end zones of a specimen suffer little distortion during nonhomogeneous compression, they may be regarded as essentially elastic, passive transmitters of pressure from the anvils to the plastic portions of the compressed body. The term, *slip cones*, which was coined to designate them

Fig. 11.14 Effect of compression on recrystallized grain size in a low carbon steel. Compressed between: (a) rough and (b) smooth (ground and polished) anvils.

(a)

(b)

is indicative not only of their shape but also implies that slip occurs principally along their peripheries.

Difficulties in certain industrial forming processes such as forging, coining, and rolling were attributed to the occurrence of these cones. In particular, the high resistance to deformation of thin discs or sheets had been ascribed to the meeting and intersection of two opposite slip cones (Figs. 11.15a, b). Direct observation does not confirm the idea of touching or interpenetrating slip cones, and, indeed, friction alone gives a satisfactory explanation of the increased resistance. Studies of work-hardening gradients performed on cross sections of compressed cylinders of decreasing initial heights indicated a progressive flattening of the cones as the specimens grew shorter. In very short cylinders the slip cones not only flatten but also begin to decrease in diameter when the h/d ratio falls below a certain minimum. These gradual changes are illustrated schematically in Figs. 11.15, c through f. The diagrams show the approximate shapes of the elastic end zones in a cylinder during progressive compression. A definite shape of the slip cones is characteristic of a particular current h/d ratio. It is fundamentally immaterial whether the specimen had these proportions initially (h_o/d_o) or the appropriate shape was reached by gradual compression of an initially tall cylinder.

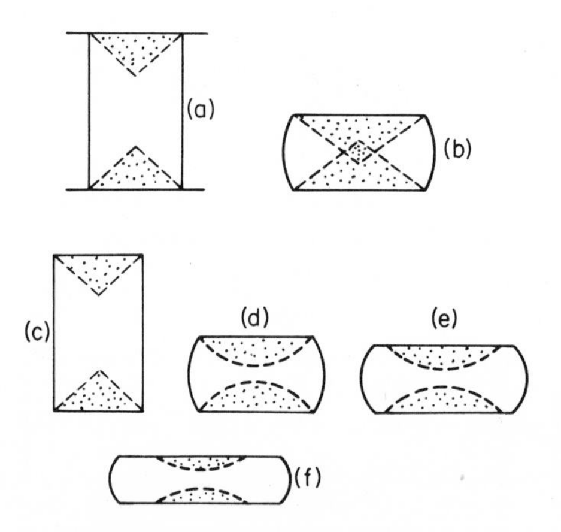

Fig. 11.15 Illustration of "interpenetrating" (a, b) and actually observed (c–f) forms of slip cones.

The schematics in Fig. 11.15 are based on experiments of the type seen in Fig. 11.16. The three copper cylinders were purposely tarnished and their ends polished prior to compression between degreased, dry anvils. The dark continuous ring at the edge of the base in (a), observed at 50% compression, represents a "folded over" or "inverted" portion of the original periphery. End slippage was absent due to the combined effects of the high h_o/d_o ratio and friction between the dry plate and specimen. Although the much shorter cylinder (b) behaved substantially like (a), the presence of sporadic radial scratches on the outer oxide ring is indicative of incipient radial sliding.

Fig. 11.16 Photographs of $\frac{3}{4}$ in. diam. copper specimens tarnished on periphery and compressed 50% between dry anvil. Initial height to diameter ratios 2, 1, and 0.4.

The appearance of the end surface of the short specimen (c) shows traces of extensive radial movement and the narrow oxide annulus at the edge was scraped off almost entirely by the interface shearing action.

The size of the slip zones goes through a maximum during compression of a tall cylinder ($h/d = 1.5$ to 2.2). This occurs according to the following sequence:

1. While the specimen is still tall, the initial slip zone grows by inversion (Figs. 11.12 right and 11.15c and d). This "added on" portion of the rigid zone is not soft since it underwent plastic deformation to achieve its new position.

2. As compression continues, the specimen begins to deform as a short rather than a tall cylinder. Now, according to the behavior of the right-hand cylinder in Fig. 11.16 and that in Figs. 11.15e and f, the corners of the cylinder slide over the anvil, and the slip cone shrinks.

This process starts at the periphery and gradually extends toward the axis of the specimen. When h/d is about 1:4 or less, the zones of stagnant metal represent only a small fraction of the total volume.

A mathematical interpretation of inhomogeneous compression of a cylinder is not available. However, Prandtl* developed one for the qualitatively related but simpler case of a wide rigid-plastic slab squeezed between

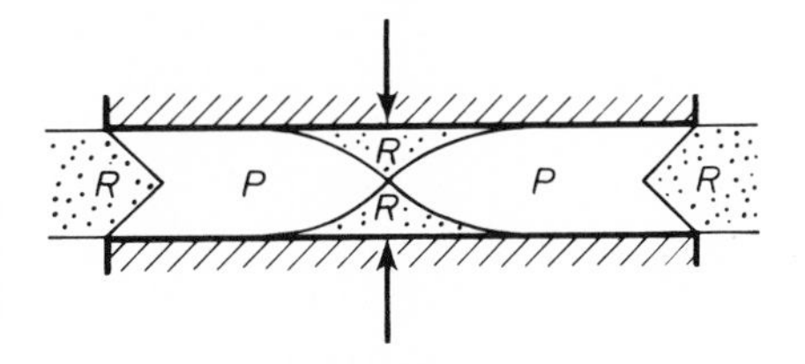

Fig. 11.17 Flow of plastic mass compressed between overlapping rough and parallel plates. *P*–plastic. *R*–rigid, non-yielding regions. After Prandtl.

rough plates under conditions of plane strain (Fig. 11.17). Prandtl's solution indicates that the *P* regions are plastic whereas those marked *R* are rigid. The two opposing rigid wedges at the center are apparently analogs of slip

---

* See (R11.4)

cones. The shear stress along their bases is less than τ_o', and relative motion is thereby prevented.

Shape Distortion Caused by Slip Cones

The existence of areas of stagnant and relatively rigid material during compression under high friction conditions has a characteristic effect upon the shape of the body being upset. In general, a prism or cylinder compressed between rough anvils will develop a single or double bulge, depending on whether the ratio of height to minimum transverse dimension is low or high (Fig. 11.18).

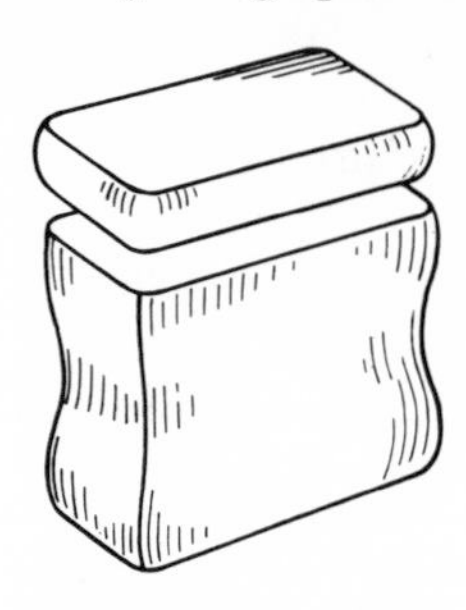

Fig. 11.18 Shape distortion of short (top) and tall (bottom) cylinders or prisms during compression with friction present.

The origin of these two types of distortion can be explained in the following manner: the fully developed stagnant end zones adjacent to the plates are approximately bounded by cones with 37° base angles (p. 311). When the distance between the anvils is such that the vertices of the cones are close to one another, the metal between the rigid cones is squeezed out, forming a single bulge. Experience shows that such a condition exists in short cylinders with $h_o/d_o < 1.3$. In tall cylinders, on the other hand, the cones occupy only a small proportion of the volume and their "wedging" action extends only over a limited distance from the end faces. Consequently, the mid-height section of the cylinder is sufficiently remote from the slip cones to deform essentially in a homogeneous fashion. Shapes of the type seen in Fig. 11.18 occur in upsetting of billets in forging operations under hammers or presses.

RESISTANCE OF WIDE SLABS TO INDENTATION

It has been repeatedly indicated that the elementary theory of plane homogeneous compression, and also the Prandtl inhomogeneous solution, apply only to conditions where the thickness or height, h, of the material is small relative to the compressed length, l. In many technical problems the reverse is true; i.e., $h \gg l$. For instance, in a process known as step-forging, a thick slab is reduced in thickness by a step-by-step compression between two relatively narrow dies. What is the pressure required to cause the material to yield? This and similar problems can be reduced to one of indenting a thick block with narrow dies (Fig. 11.19).

Observe that, when the slab face beneath the die just starts to yield, the dotted plastic area is very small and certainly does not penetrate to the center of the slab because the compressive stress decreases with the distance from the face of the die. However, since the "plasticized" material is forced out from under the die to the right and left (small curved arrows), it tends to pull or drag with it the still elastic material, M. The latter, therefore, is stretched elastically by the wedge-like action of the small plastic zones. We thus have not only to overcome the yield strength of the material directly under the die but also the elastic resistance of the bulk of the material in which the plastic zones are imbedded. As a result, the average pressure per unit of die surface rises above the inherent yield strength of the indented material. This pressure evidently increases with increasing h/l ratio.

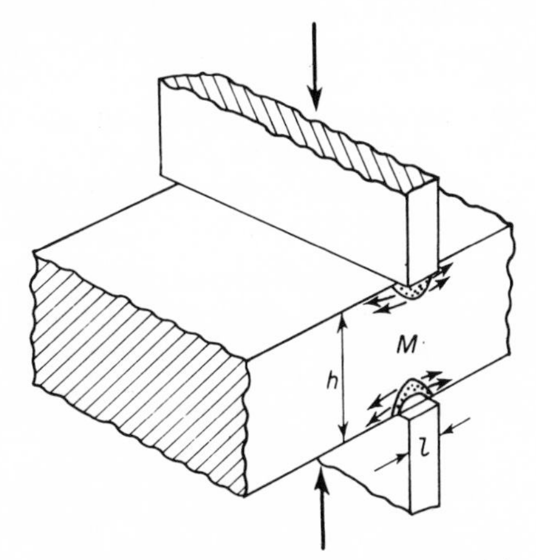

Fig. 11.19 Indentation of a thick slab with a narrow die. Dotted area is plastic.

Theoretical considerations by R. Hill and associates showed that the mean pressure, σ_1, should rise from σ_o' at $h/l = 1$ to $\sigma_o'(1 + \pi/2) = 2.57\sigma_o'$ when $h/l = 8.74$ (Fig. 11.20) but remain constant for still higher h/l ratios. In terms of unconstrained compressive (or tensile) yield strength, the mean pressure on the die face is $(1 + \pi/2)(2/\sqrt{3})\sigma_o = 2.95\sigma_o \approx 3\sigma_o$. Hence the effect of the unyielded elastic bulk on the actual yield pressure is very large.

In hardness testing (Chapter 12) a ball- or pyramid-tipped plunger is immersed into the material under a known load and the size of the indentation thus produced is measured. This problem, and the one just discussed, are related although the stress condition in hardness tests is triaxial rather than plane. It was found, however, that the mean pressure on the projected area of such indenter is also about 2.8 to $3\sigma_o$, a feature that enables the yield strength to be estimated from a hardness test result.

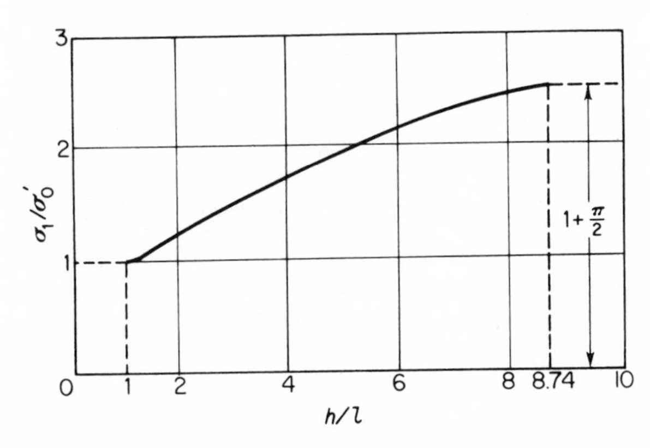

Fig. 11.20 Relation between the mean indentation pressure σ_1 and h/l for a slab of height h indented on opposite sides by a flat die of width l. From (R11.2).

PROBLEMS

1. A brass cylinder of 1 in. diam. and 1.5 in. tall was compressed 27% between smooth anvils by an axial force of 91,000 lb. At this point the cylinder fractured along a plane inclined under 45 degrees to the axis. Calculate the fracture stress.

2. Assume that the stress–strain relation for this brass is straight between $\sigma_0 =$ 25,000 psi and the fracture point when the abscissa is plotted as percent decrease of height, while ordinate is true stress. Draw the $\sigma - \bar{\varepsilon}$ curve for this material.

3. A metal disc 3 in. in diameter and 0.6 in. thick was compressed between smooth well lubricated plates ($\mu = 0$), while another identical disc was similarly compressed between rough plates. The forces at the yield point were 280,000 lb. in the first case and 350,000 lb. in the second. Determine the coefficient of friction.

4. A metal strip 2 in. wide and 0.1 in. thick is compressed between 0.5 in. wide parallel anvils which extend across the entire strip width. The ends of the strip are pulled apart with a force of 12,000 lb. Assuming a 60,000 psi yield stress and $\mu = 0.1$, find the force to be applied to the anvils to cause plastic yielding.

5. Determine the position of the neutral plane in Problem (4) when the tension is applied to one end of the strip only.

6. Determine the minimum coefficient of friction required to produce incipient sticking in the configuration described in Problem (4), but without any tensions applied.

7. Deduce a formula for true deformation rate at any stage of compression, taking the initial height as h_0 and assuming a uniform press ram speed v.

REFERENCES FOR FURTHER READING

(R11.1) Cook, M., and E. C. Larke, *Journal of the Institute of Metals*, 71, 371 (1945).

(R11.2) Hill, R., *Mathematical Theory of Plasticity*. London: Oxford University Press, 1951.

(R11.3) Johnson, W., and P. B. Mellor, *Plasticity for Mechanical Engineers*. London: D. Van Nostrand Co., Ltd., 1962.

(R11.4) Nadai, A., *Theory of Flow and Fracture of Solids*. New York: McGraw-Hill, Inc., 1950.

(R11.5) Walton, W. H. (ed.), *Mechanical Properties of Non-Metallic Brittle Materials*. New York: Interscience Publishers, Inc., 1958.

12

HARDNESS

The notions of "hardness" and "softness" are as old as they are ambiguous. When one speaks of a hard seat cushion or automobile spring, he has in mind primarily the elastic properties of the particular object, large deflections at low loads being synonymous with "soft." A soft rock or building material will often imply a heterogeneous substance showing a mechanical weakness of a single ingredient (binder) rather than the absolute "hardness" of some of the other, perhaps very strong, constituents. For instance, lime used in bricklayer's mortar is "soft" while the sand it contains is very hard. Likewise, rubber-bonded grinding wheels, while "soft" and flexible, contain a large proportion of very hard abrasive matter capable of grinding steel.

With reference to apparently homogeneous materials such as industrial metals, it is common knowledge that a steel file or even a nail will abrade, scratch, or dent a copper or aluminum plate, but not vice versa. However, a steel needle will not scratch glass although glass is certainly "weaker" than the needle.

These examples demonstrate that the term "hardness" encompasses a wide variety of properties but it generally can be reduced to resistance to elastic and plastic deformations and even to fracture. For instance, scratching of a glass plate with a diamond involves, first, a localized elastic strain, followed by a separation of fine glass particles from the plate's bulk (a multitude of small-scale brittle fractures). Indentation of a piece of aluminum by a steel punch or chisel entails a combination of elastic and permanent deformations, whereas measuring the "hardness" of rubber from its deflection under a specified load generally involves elastic deformation alone because the rubber will regain its original shape when the load is removed. With all this in mind, one can readily see that there is no such thing as an all embracing generic definition of hardness. It is, therefore, not surprising that various hardness definitions and measuring methods coexist, some better suited for specific categories of materials than others. However, the physical meaningfulness, ease of determination, coupled with reproducibility and a need for some kind of standardization resulted in a widespread acceptance of certain hardness measures in preference to others.

THE BRINELL (BALL) TEST

Realizing the cost, time, and destructive character of a tension test on a finished part, the Swedish engineer J. A. Brinell set out to devise a rapid and nondestructive method of measuring some property of metals which would yield information similar to that obtained from a tension test. He selected indentation hardness for this purpose and used a hard steel sphere as an indenter. He found that, by properly selecting the relation between ball diameter and the load with which it is forced into the metal, the average pressure in kilograms per square millimeter calculated on the spherical surface of the impression (and since known as the Brinell Hardness Number, *BHN*) was about three times the ultimate tensile strength of a variety of steels. At that time, as often today, S_u was regarded as the most important single strength characteristic; so the Brinell test almost immediately gained universal acceptance and came into widespread use around 1900.

The basic data associated with this test are shown in Fig. 12.1a, where D

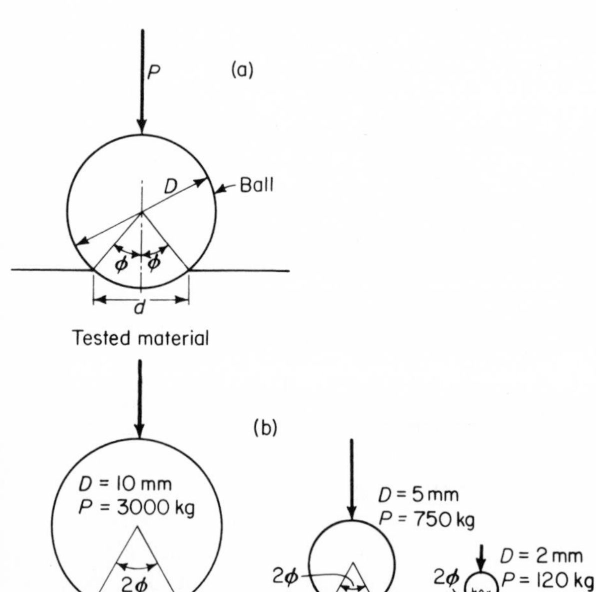

Fig. 12.1 The Brinell (ball) test. (a) Basic parameters. (b) Series of P-D combinations which produce geometrically similar impressions (2ϕ = constant).

is the diameter of the test ball, P the applied load, and d the diameter of the impression in the surface of the tested object.

From geometry we find the spherical surface of the impression,

$$A = \frac{\pi D}{2}[D - \sqrt{(D^2 - d^2)}]$$

and the average pressure

$$p = BHN = \frac{P}{\frac{\pi D}{2}[D - \sqrt{(D^2 - d^2)}]} \tag{12.1}$$

Since $d = D \sin \phi$, the Brinell hardness is also given by

$$BHN = \frac{P}{\frac{\pi}{2} D^2(1 - \cos \phi)} \tag{12.2}$$

Impressions obtained with balls of different diameters are geometrically similar when the included angle 2ϕ (or ϕ) is constant. Hence, to obtain identical Brinell numbers with different balls of diameters, D_1, D_2, etc., the applied load must be varied according to

$$\frac{P_1}{D_1^2} = \frac{P_2}{D_2^2} = \frac{P_3}{D_3^2} \text{ etc.} \tag{12.3}$$

or

$$\frac{P_1}{P_2} = \frac{D_1^2}{D_2^2}; \quad \frac{P_1}{P_3} = \frac{D_1^2}{D_3^2}; \text{ etc.} \tag{12.4}$$

The most common Brinell tester for steel has a 10 mm ball (D_1) and uses a 3,000 kg load, P_1. If a 2 mm ball (D_2) is employed, the appropriate load will be $P_2 = P_1(D_2^2/D_1^2) = 3{,}000{:}25 = 120$ kg, and similarly for other test loads (Fig. 12.1b).

The Brinell test is related to the compression test; the lesser the resistance to compression, the larger the indentation area and the lower the BHN.

The Mechanics of the Ball Test

The physical meaning of the ball test is rather complex. If the hard ball is brought in contact with the tested material and the load P is gradually increased, the deformation of both ball and material is at first elastic, and no impression is left when the load is removed. According to the theory of elastic contact between spheres (a plane is regarded as a sphere with $R = \infty$) developed by Hertz and Föppl, the maximum shear stress arises at some distance below the contact surface. Using either one of the yield criteria (Chapter 6), we find that plastic deformation begins in the softer metal when the pressure between the two surfaces reaches some value between σ_o and $1.15\sigma_o$. The impression formed at that moment is very shallow, and ϕ is quite small, so that the *BHN* is virtually identical with the initial yield strength of the metal being tested.

As the load is further increased, the impression deepens and the plastic region expands. However, this region is "encased" in the main mass, which is elastic and must be forcibly expanded to allow a displacement of the plastic material by the moving indenter (Fig. 12.2). Hence a condition analogous to the one discussed on p. 312 develops, where the plastic flow is opposed by the relatively rigid elastic mass.

It was found that this extra resistance quickly grew until a constant pressure of $3\sigma_o$ calculated on the projected area A_p of the impression was reached. This is almost identical with the value of 2.57×1.15 given on p. 313. The value of $3\sigma_o$ holds until the ball sinks into the specimen up to the equatorial circle. If the rate of strain-hardening is very low, the quotient P/A_p is nearly constant within wide limits of P. This is the case with metals having a nearly flat stress–strain curve such as is observed after cold working.

The P/A_p ratio is termed Meyer's Hardness Number (*MHN*) and is plotted in Fig. 12.3 for initially cold worked as well as soft metals (*M* curves).

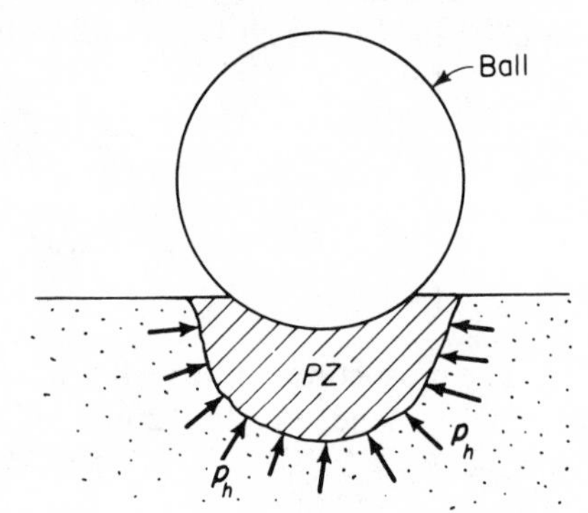

Fig. 12.2 Plastic zone *PZ* under ball and hydrostatic pressure p_h opposing flow.

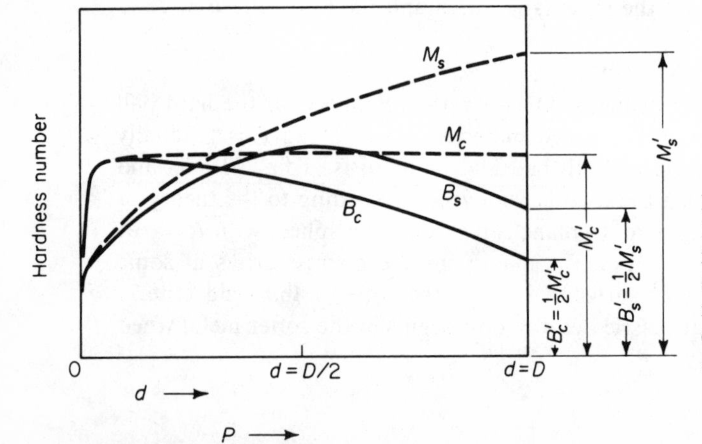

Fig. 12.3 Meyer (M) vs Brinell (B) hardnesses as functions of impression size. Subscripts (s) and (c) refer to soft (annealed) and cold worked metals, respectively.

It is seen that the cold worked curve, M_c, is nearly horizontal. However, when the metal is initially in the soft state so that its strain-hardening rate is high, progressive ball penetration encounters a growing resistance due to the rising yield strength. For this reason the Meyer hardness of an annealed metal is represented by a relatively steep curve, similar to the appropriate plastic σ–ε relation in compression or tension. It may be mentioned that Meyer's hardness scale, introduced after Brinell's, was never extensively used.

Contrary to the *MHN* curves, plots of load vs. *BHN* show maxima at certain P values (Fig. 12.3, B_s, B_c). Since $A > A_p$, it follows that $BHN < MHN$. The two are almost identical for very light loads when $\phi \approx 0$ and $A \approx A_p$. Later they separate, Meyer's hardness rapidly reaching a nearly constant value ($\approx 3\sigma_0$) or continuing to rise as has been explained. The Brinell hardness, on the other hand, begins to drop at a certain stage because, for deep impressions, the spherical surface A increases faster than the extra resistance to deformation caused by progressive work-hardening. As a result, the ratio P/A ($= BHN$), begins to fall, and, when the ball is immersed to its equator, $A = 2A_p$ and $BHN = \frac{1}{2}MHN$.

The steeper the σ–ε curve, the higher the load at which the maximum occurs. Conversely, the Brinell number of most metals that were cold worked about 15–20% or more, shows the drop at relatively low loads (curve B_c).

These differences in the location of the maximum on $BHN = f(P)$ curves lead to an interesting situation. At low loads a certain work-hardened metal may appear harder than another, annealed metal, whereas at a higher P the situation is reversed (e.g., curves B_c and B_s in Fig. 12.3). The B_s shape of the Brinell curve is advantageous from the practical point of view because most metal products are not cold worked and measurements over a range of loads will then yield almost identical hardness numbers.

If a particular material is very soft so that, for a given indenter load, the impressions are unusually large, a lower test load must be used. For this reason, when soft metals like annealed copper, aluminum, cadmium, tin, and lead alloys are tested with a 10 mm ball, a 500 kg load is applied. The optimum test range for any given load is often regarded as one which gives impression diameters, d, such that $0.25D < d < 0.5D$.

For a 3,000 kg load and a 10 mm diam. ball this corresponds to *BHN* between

$$3{,}000\Big/\left(\frac{\pi \times 2.5^2}{4} \times 1.016\right) = 590 \text{ kg/mm}^2 \text{ at } d = 2.5 \text{ mm}$$

and

$$3{,}000\Big/\left(\frac{\pi \times 5^2}{4} \times 1.072\right) = 143 \text{ kg/mm}^2 \text{ at } d = 5 \text{ mm}$$

whereas for a 500 kg load and the same ball, the range covered will be from $\sim$100 down to 24 *BHN*.

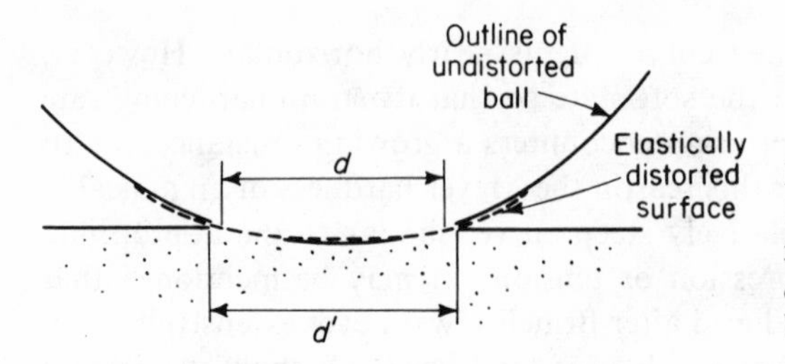

Fig. 12.4 Elastic deformation of ball surface in testing hard materials. Note the abnormally large d' caused by lateral expansion of ball.

On practical grounds it was necessary to extend the 3,000 kg range to include hardness from 90–100 *BHN* which are typical of very soft low carbon steels. Thus the maximum acceptable indentation size is about 0.6 to 0.65 D.

It is appropriate to point out that, at hardnesses above 400 *BHN*, the hardened steel ball commonly used as a Brinell indenter undergoes a significant and rapidly increasing elastic distortion. It is compressed along the load application axis and expands in the plane normal to it, leaving an oversize but shallow indentation (Fig. 12.4) and resulting in an unduly low *BHN* reading. For this reason tungsten carbide balls ($E = {\sim}90 \times 10^6$ psi) are advisable for Brinell tests on hard metals. Even diamond balls are occasionally employed in high precision laboratory work.

The Brinell test is affected not only by the elasticity of the ball but also by the friction between the ball and test piece. However, the influence of the latter is so small that it can be disregarded for most practical purposes.

Meyer's Power Relation for the Ball Hardness Test

From a large number of measurements Meyer found that the following relation holds for the ball test:

$$P = ad^m \tag{12.5}$$

where a and m are constants. This is similar to Eq. (10.6) which described the true tensile stress–natural strain curve. Again, taking logarithms of both sides, Eq. (12.5) becomes

$$\log P = \log a + m \log d \tag{12.6}$$

This linear relation can be conveniently used to determine a and m, the latter being known as *Meyer's exponent* or *Meyer's index*.

For this purpose several impressions, preferably not less than four, are made under increasing loads. The diameters of the imprints are measured, $P = f(d)$ is plotted on log-log coordinates, and m and $\log a$ are found (Fig. 12.5). The successive indentations are preferably made at a single spot, beginning with the lowest load to eliminate the effect of metal inhomogeneity.

For metals in a progressively cold worked condition, m was found to

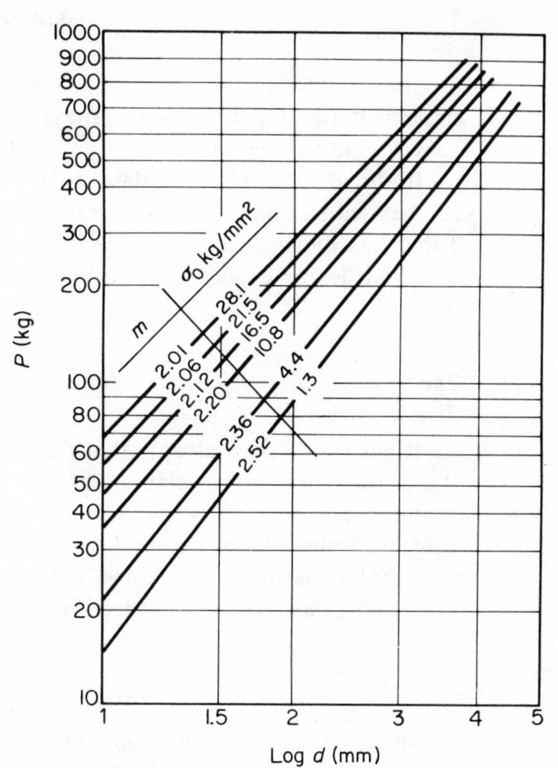

Fig. 12.5 Change of Meyer's index, *m*, in progressively cold worked copper. From N. A. Shaposhnikov, *Mechanical Testing of Metals*, Moscow: Mashgiz, 1954.

rapidly approach 2.0; for annealed (soft) metals, *m* varies from about 2.2 to 2.6, depending on the slope of the $\sigma - \varepsilon$ curve as illustrated by the following data:*

		m
Brass (62 % Cu, 38 % Zn),	annealed	2.61
Brass (62 % Cu, 38 % Zn),	cold-rolled 10 %	2.17
Nickel	annealed	2.50
Nickel	cold-rolled 10 %	2.14
Carbon steel 0.6 %C.	annealed	2.25
Aluminum	annealed	2.20

Note that, when $m = 2.0$, it follows from (12.5) that

$$a = P/d^2;$$

$$\frac{4a}{\pi} = \frac{P}{\pi d^2/4} = MHN \text{ or}$$

$$a = \frac{\pi}{4} \cdot MHN$$

* H. O'Neill, *Proc. Inst. Mech. Eng.*, 151 (1944), 115–30.

Since the Meyer hardness is the mean pressure on the projected area of the indentation ($\approx 3\sigma_o$), intercept a is about 2.3 times the yield strength of a heavily cold worked metal.

In the absence of tensile data, a *Meyer analysis* on the lines of Fig. 12.5 can be used to approximate the strain-hardening exponent of a metal or alloy. Tabor has shown that Meyer's exponent, m, is roughly related to the tensile strain-hardening exponent, n, by

$$n \approx m - 2 \tag{12.7}$$

The Relation Between Brinell Hardness and the Yield and Tensile Strengths

It has been stated already that the mean pressure on the projected area, A_p, of the ball indenter is about $3\sigma_o$ for a material displaying a nearly horizontal stress–strain curve with a Meyer's exponent $m \approx 2.0$. For normal Brinell hardness tests, the diameter of the impression, d, is from $0.25D$ to $0.5D$. The limiting A/A_p ratios are, therefore, 1.016 and 1.072, respectively, with 1.05 as an average. Hence the unit pressure on the spherical area, A (which is the BHN), is about $3:1.05 = 2.85\sigma_o$, or, equivalently,

$$\sigma_o(\text{kg/mm}^2) = 0.35\ BHN$$

Since the tensile strength, S_u, and the yield strength, σ_o, just about coincide in many metals cold worked 20% or more, the above ratio can be written as

$$\sigma_o \approx S_u \approx 0.35\ BHN \tag{12.8a}$$

This is valid for σ_o and S_u expressed in metric units (kg/mm²). For the U.S. system (σ in psi), the factor 0.35 is replaced by

$$0.35 \times 1422 = \sim 500$$

and

$$\sigma_o \approx S_u \approx 500\ BHN \tag{12.8b}$$

The appropriate formula, as applied to British units (σ in long tons per sq. in.), is

$$\sigma_o \approx S_u \approx 0.225\ BHN \tag{12.8c}$$

When the metal is soft, or when the amount of cold work was not high enough to raise σ_o to nearly the S_u level ($m > 2$), the multiplication factors used in (12.8) are too low. For example, for soft copper, brass, or aluminum, $k_B = BHN/S_u \approx 0.55$ (metric units); for annealed nickel, $k_B \approx 0.5$, while, for commercially cold drawn nickel (~20–25% cold work), $k_B \approx 0.4$. On the other hand, most steels used in structural and machine applications show a rather low strain-hardening rate ($m \approx 2.25$) compared with annealed copper ($m \approx 2.45$) or nickel ($m \approx 2.5$). This, combined with the relatively high σ_o/S_u ratio caused by the yield point anomaly in steel (see pp. 188–190), results in a k_B factor quite close to the 0.35 ratio calculated above for metric units. Typically, $k_B \approx 0.36$ in annealed or hot rolled steels of this category but about 0.32 to 0.33 in the cold worked or heat treated condition.

Advantages and Deficiencies of the Brinell Test

The Brinell test can be made with any ball size although the 10 mm ball is by far the most common. The relatively large impression it produces has the advantage of averaging out any internal heterogeneities. Its shape and size are also little affected by minor surface irregularities such as scratches, machining marks, or the inherent roughness of forged or cast surfaces. On the other hand, the large ball and indenting load cannot be safely applied to small or precision objects in view of the excessive distortion it causes. In these cases a smaller ball (5, 2.5, or 1 mm. diam) is used with a lower load according to Eq. (12.4).

The principal disadvantage of the ball test is that, for any ball diameter, the *BHN* is not a constant but varies with load (Fig. 12.3). It is not possible, therefore, to cover with a single load the entire range of hardnesses encountered in commercially used metals.

PYRAMID AND CONE INDENTERS: THE VICKERS TEST

The deficiencies inherent in the use of a ball indenter led to the development of other hardness testing methods. They employ a pointed indenter which causes fully developed plastic deformation in the material, even at very low loads, with the elimination of a gradual elastic–to–plastic transition when the load increases from zero upwards. To eliminate the variable A_p/A, it was necessary to use a tapered tool with rectilinear generators such as a cone or a multisided pyramid. In practice, only a right-circular cone and a four-sided pyramid were accepted.

Figure 12.6 shows a cone-tipped indenter. The mean pressure p' acting on the projected area of the impression,

$$p' = \frac{P}{\pi\, d^2/4}$$

is identical with Meyer's hardness, *MHN*, discussed in connection with the

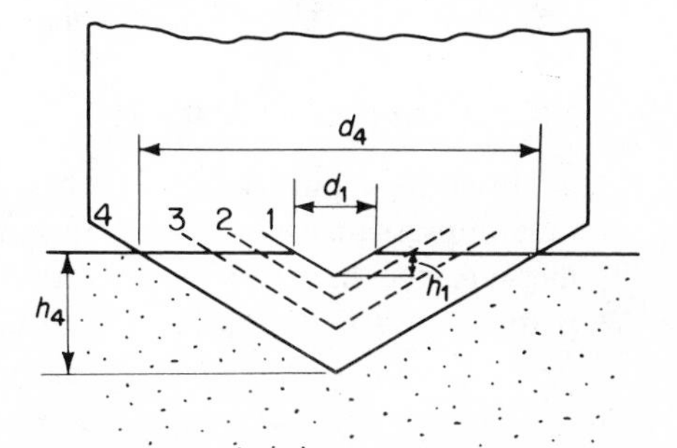

Fig. 12.6 Successive portions of cone indenter under increasing load.

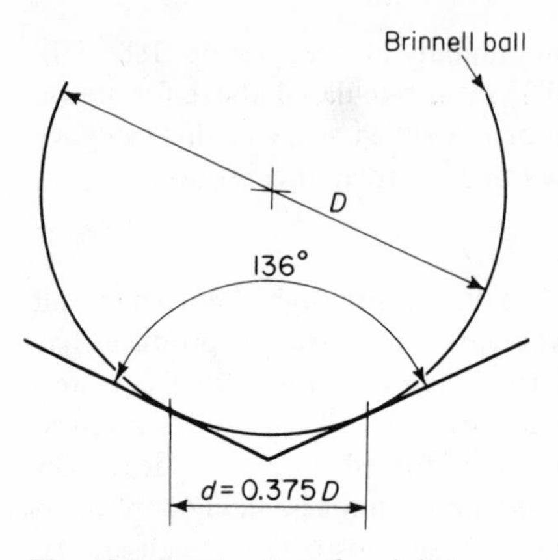

Fig. 12.7 Construction of angle between two opposite faces of diamond pyramid (Vickers) indentor.

ball test. The plastically deformed volumes beneath the indenter always have geometrically similar shapes, independent of the depth of the impression. Its size varies, of course, directly with d or h. There will be no elastic–plastic "incubation" region at low loads, and $p' = f(P)$ will be represented, therefore, by a horizontal line.

Instead of Brinell's spherical indentation surface A, we now have a conical surface of the indent, A'. However, this is a variable controlled by the cone semiangle, ϕ. To eliminate this variable, ϕ must be fixed. When Smith and Sandland first proposed, in 1922, the diamond pyramid or Vickers test, it was quite obvious that it could compete with Brinell's only if it could give very similar readings over a wide hardness range. This was achieved by making $2\phi = 136°$ so that opposite faces of the pyramid were tangent to a sphere along a circle in which $d = 0.375D$ (Fig. 12.7). Recalling that the diameter of a Brinell indentation is normally between 0.25 and $0.5D$, the factor 0.375 represents the median between the two. Moreover, the diamond pyramid number (DPN) was defined as P/A', where A' is the pyramidal surface of the indent. The relation between A' and the projected area A_p of the pyramid is

$$A' \approx 1.08A_p$$

which is within 2% of the A/A_p ratio in a Brinell test for $d = 0.375D$.

As a result, DPN (Vickers) and BHN (Brinell) hardnesses are just about identical as long as the Brinell impressions are of normal depth. At high hardnesses, when the steel ball deforms excessively (as already discussed), there is an increasing discrepancy between the two, as illustrated by the following comparison:

BHN	100	200	300	400	500	600
DPN	100	200	303	419	547	716

MICROHARDNESS AND THE KNOOP INDENTER

In many applications it is necessary to operate with extremely small hardness impressions to obtain a meaningful result. For instance, one cannot measure the hardness of very thin coatings, such as produced by nickel plating, with the tests heretofore described. Even the Vickers indentation

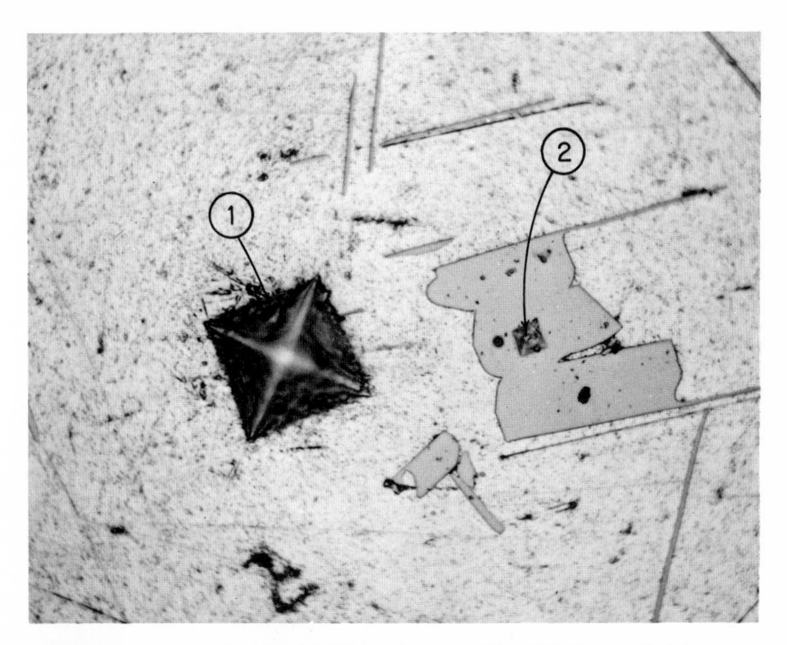

Fig. 12.8 Diamond pyramid indentations on Al +Hf alloy. Al-rich matrix—22 D.P.N., $HfAl_3$ phase—480 D.P.N.

has an edge dimension of the order of 0.010 in. while the surface layer may be only about 0.001 in. thick. This precludes the use of conventional hardness testing equipment because the pyramid breaks through the surface layer and penetrates to the base metal underneath giving an erroneous value for surface hardness. As another example, many alloys consist of a soft matrix in which hard particles of a second phase are imbedded. By using light-load indenters, the hardnesses of each component could be separately measured (Fig. 12.8).

Hardness measurements with loads of 1 kg or less are often considered as *microhardness tests* although opinions differ as to the value of the upper load and even the terminology. Bückle*, in his comprehensive review, classified all diamond pyramid hardness tests as follows:

1. *Microhardness;* Loads up to 200 g, indentation diagonals up to about 50 microns.

2. *Low-load hardness;* loads from 200 g to 3 kg, indentation size up to about 300 microns.

3. *Standard hardness*: loads of over 3 kg.

Testing may be carried out with Vickers type indenters, but the Knoop indenter (Fig. 12.9) became very popular for this application. Its long

* H. Bückle, "Progress in Micro-Indentation Hardness Testing," *Metallurgical Reviews*, **4,** 49 (1959).

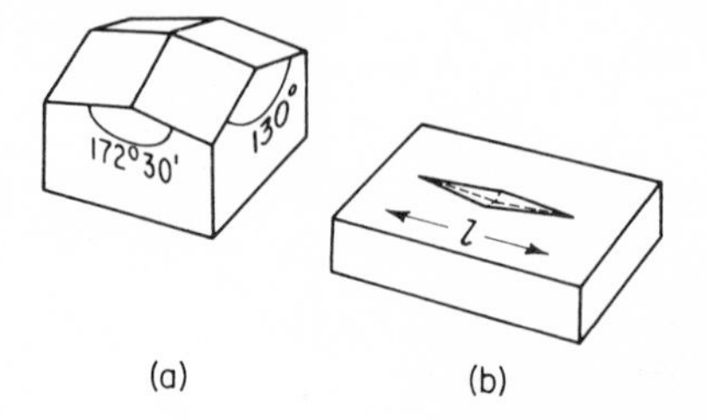

Fig. 12.9 (a) The diamond indenter used in Knoop hardness measurements. (b) The indentation. This has a length seven times its breadth.

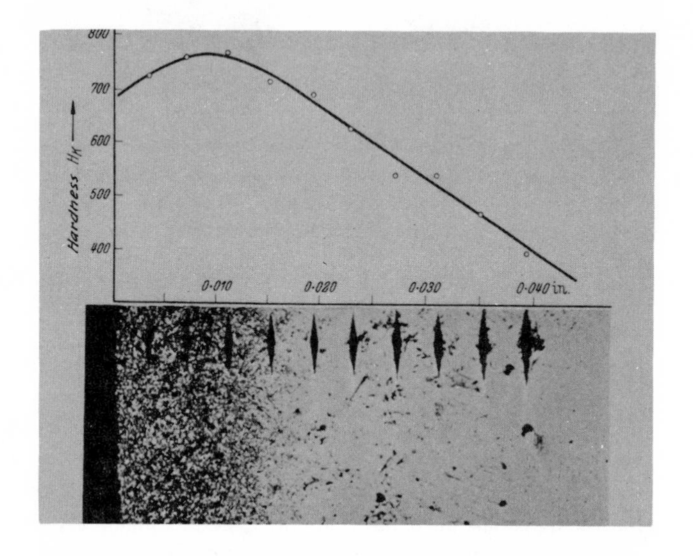

Fig. 12.10 Hardness gradients in surface of case-hardened steel. Knoop indenter, 500 g load. From V. E. Lysaght, *Indentation Hardness Testing*, New York: Reinhold Publishing Corp., 1949.

diagonal is about 7 times the short one, and, for the same load, the impression is one half as deep as for the Vickers. This makes the Knoop indenter particularly suitable for studying the properties of surface layers. The Knoop hardness numbers are quite close to the Vickers scale although they are based on a measurement of the long diagonal alone.

Figure 12.10 shows a Knoop hardness survey across the thickness of a carburized and quenched (case-hardened) surface of a steel part. The decreasing indentation size is indicative of the high hardness of the case.

Microhardness tests require care, and their reproducibility is not as good as that of standard tests. Sources of possible errors are many, including instrumental errors as well as those stemming from various properties of the material. Among the latter, surface preparation and anisotropy are perhaps the most important. The shallowness of the Knoop impressions makes them

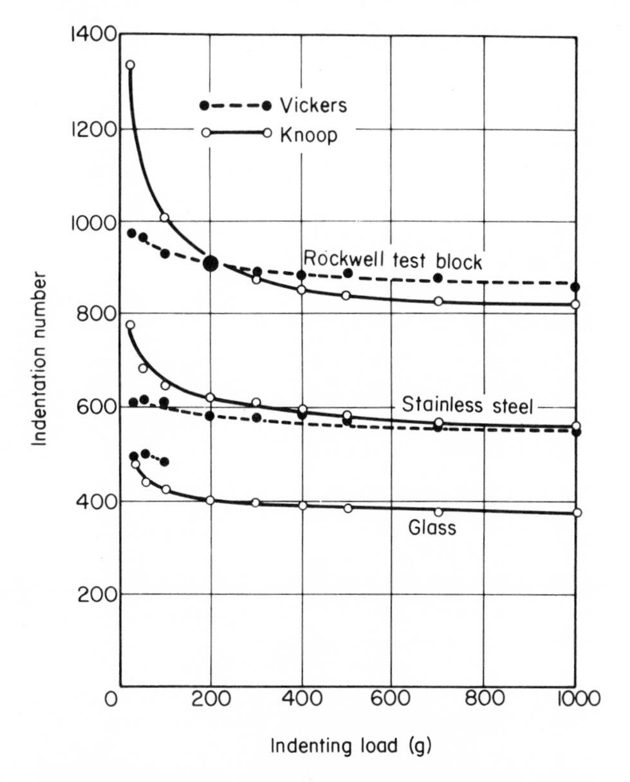

Fig. 12.11 Effect of test load on Knoop and Vickers hardness numbers of mechanically polished surfaces. From D. R. Tate, *Trans. Am. Soc. Metals*, **35**, 374 (1945)

more sensitive to the surface condition than the Vickers indentation. For example, abrasive polishing produces a shallow work-hardened layer, and, when such surfaces are tested with both the Vickers and Knoop indenters, the latter is found to be far more load-dependent than the former (Fig. 12.11). The Vickers test, by penetrating deeper, measures a larger proportion of the softer underlayer at low loads, thereby "diluting" the high surface hardness.

THE ROCKWELL CONE AND BALL TESTS

The Rockwell tester was developed principally for production testing of large numbers of parts. It requires no microscope and the results are recorded on a clock-type dial.

To dispose of the necessity of measuring the diameter or other lateral dimension of the indentation, the Rockwell machine measures its depth. To improve accuracy and to eliminate the effects of surface irregularities on the readings, a 10 kg ("minor") preload is applied first, and the resulting position of the indenter tip is taken as zero. An additional "major" load is then applied, the new depth measured and subtracted from a constant number. This difference is shown on the dial.

A conical 120° diamond indenter with a rounded tip is used on hard metals with a 140 kg major load (Rockwell *C* Scale). For softer materials, a $\frac{1}{16}$ in. ball is the indenter, and the major load is 90 kg (Rockwell *B* Scale). For special purposes, such as testing thin sheets, very soft or hard surface layers, other Rockwell scales exist with loads down to 15 kg. These low load machines are called "superficial."

The theory of the Rockwell cone test is related to the Vickers and similar devices that produce geometrically similar indentations at all loads.

By reference to Fig. 12.6, the pressure on the projected indentation area is again

$$p' = \frac{P}{\pi\, d^2/4}$$

However, if we measure h instead of d, we have

$$h = \frac{d}{2} \cot \phi, \quad \text{and} \quad d = 2h \tan \phi$$

where ϕ is the cone semiangle. Consequently,

$$p' = \frac{P}{\pi \tan^2 \phi} \cdot \frac{1}{h^2}$$

The load P is constant in a Rockwell test, so the fraction $P/(\pi \tan^2 \phi) = C$ and

$$p' = C/h^2$$

or

$$h = \sqrt{C/p'} = C_1/\sqrt{p'}$$

Since the depth measured is subtracted from a fixed value, H, the number indicated on the dial as a Rockwell C (R_c) hardness is

$$R_c = H - h = H - C_1/\sqrt{p'} \tag{12.9}$$

Since p' is proportional to *DPN*, the relation between R_c and *DPN* is parabolic.

$$R_c = H - C_2/\sqrt{DPN} \tag{12.10}$$

In view of various complicating factors, such as the elastic "spring-back" of the impression when the load is taken off and the rounded tip of the cone, constant C_2 must be fitted to secure good agreement.

Figure 12.12 shows an experimentally determined relation between R_c vs.

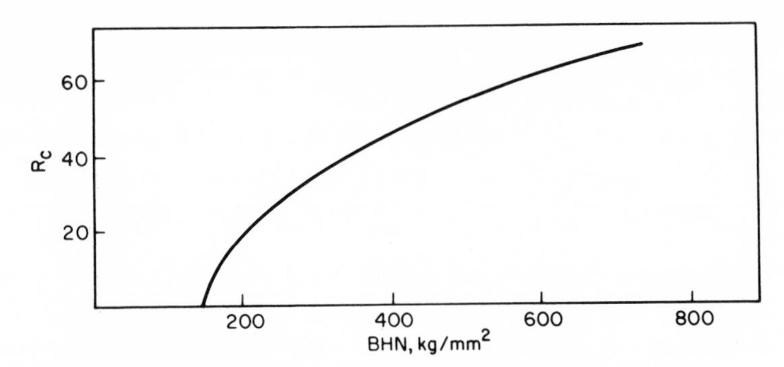

Fig. 12.12 Plot of Brinell vs. Rockwell "C" hardness. From D. Tabor, (R12.3).

BHN. The R_c scale is used in practice when $R_c > 18$–20 and the gently curved branch of the parabola can be approximated by the straight line $BHN \approx 10\ R_c$ for the range between 200 and 500 *BHN*.

The Rockwell hardness test has one serious drawback; namely, since the depth measurement includes both the plastic and elastic displacement, the type and shape of backing materials becomes very important. For example, if a sheet of rubber is placed between the specimen being tested and the machine anvil, the low modulus of the rubber would cause an erroneously low hardness value. Although this is an extreme example, there are practical occasions when the properties of the backing do become of decisive importance. To eliminate variability in the "anvil effect," a diamond anvil is employed as a backing in Rockwell testing of thin sheets.

THE REBOUND (SHORE) TEST

Contrary to the static slow-speed hardness tests described so far, the Shore test measures the loss of potential energy suffered by a steel weight dropped vertically onto a surface and rebounding from it. The difference between height of fall, which is constant, and the height of rebound represents a measure of hardness.

The difference between the two potential energies is used up mainly to produce an identation in the material. Hard materials with high elastic limits and yield stress values will absorb very little energy while soft ones will cause the falling weight (a tiny cylindrical hammer with a spherical tip) to lose much of its bounce.

The theory of this test is difficult, and various uncertain assumptions are needed for even an approximate analytical treatment* (not reproduced here). The use of the test is limited because of several hard-to-control factors that

* See (R12.3)

affect the test results. These include surface roughness and the difficulty of aligning the tester to assure that the weight drops normal to the surface on which hardness is being measured.

The commercially available instrument based on this principle is Shore's Scleroscope. It consists of a calibrated glass tube, a steel or diamond-tipped free falling piston (weight) with a pneumatic lifting, holding, and releasing device.

The lightness of this instrument contributed to its usefulness for checking the hardness of large objects such as castings or forgings *in situ* (e.g., rolls in rolling mills, dies in presses, etc.). Empirical tables for conversion of Shore units into Rockwell and Brinell units are available. A Shore hardness of about 95-100 is equivalent to 63 Rockwell *C* on steel. A universal hardness conversion is not possible. For instance, the high elasticity of rubber may result in Shore hardness numbers higher than steel.

SCRATCH HARDNESS

A scratch test can be used to estimate the relative hardness of two solids, the harder one being capable of scratching the other but not vice versa. The method has been used in a systematic manner by mineralogists after Mohs introduced a ten-grade scale, the hardness increasing with the number, as shown in the table below:

Mineral	*Mohs Number*
Talcum	1
Gypsum	2
Calcite	3
Fluorite	4
Apatite	5
Orthoclase	6
Quartz	7
Topaz	8
Corundum	9
Diamond	10

Various other minerals, artificial compounds, organic materials, and metals can be fitted into this scale which is, however, much too coarse for materials in which hardness variations are continuous and must be accurately specified.

In metals research, scratch hardness is occasionally measured by drawing a diamond tip across a surface under a constant load. The width of the scratch is the measure of hardness. This method is much too cumbersome and slow for practical application but it rather spectacularly reveals the different hardness of adjacent grains or phases in metallic alloys.

TIME EFFECTS IN HARDNESS MEASUREMENTS: HOT HARDNESS

In standard hardness tests on metals, the duration of load application is fixed by the design of the instrument (Rockwell, Shore) or it can be adjusted by a suitable release mechanism. At temperatures well below the recrystallization range, time has generally a minor effect on the size of the indentation. As the recrystallization temperature is approached, however, creep becomes pronounced, and a prolonged loading will produce oversize impression and result in a lower hardness reading.

High temperature hardness properties with their strong time–dependence are related to creep properties in compression. But long time indentation tests are beset with difficulties on account of the softening or other undesirable changes in the indenter. Diamond and corundum tips were used for research purposes.

HARDNESS OF PLASTICS

The methods developed for metals are suitable for measuring the hardness of plastics. But, in view of their viscoplastic behavior, the time effect here is as important as in hot hardness tests, and constant duration of load application must be maintained to secure comparable data.

The Brinell numbers for the harder plastics are up to 50 but may be as low as one tenth of that. Special Rockwell superficial scales are widely used: The *R* scale for soft varieties with a $\frac{1}{2}$ in. diam. ball, 10 and 60 Kg loads, and the *M* scale with a $\frac{1}{4}$ in. ball and 10/100 Kg. loads.

Scratch or mar resistance is an important requirement of the plastics used for facing (table tops) or optical applications where maximum transparency is prerequisite. Various imitative tests are used for this purpose such as impinging with abrasive particles.

HARDNESS OF RUBBER

The accepted hardness measurement methods for rubber are similar to indentation tests on metals, but this similarity is superficial inasmuch as rubber hardness conveys the meaning of elasticity rather than of resistance to permanent deformation.

The standard *ASTM* test (for hardness of rubber*) is based on the depth of an indentation produced by a hemispherically ended vertical plunger with a 0.0469 in. tip radius forced into the rubber specimen with a 3 lb. dead load. The specimen is pressed against the instrument table by a 0.625 in. diam. washer-like presser foot equipped with a 0.109 in. center hole through which the indenter passes. The load on the presser foot is 5 lb. In this manner a certain hydrostatic pressure is created in the specimen. The hardness is read

* *A.S.T.M. Standards*, 1961, Standard D314–58.

off a dial gauge measuring the indentation depth after 30 seconds application time.

The International Standard Hardness of Rubber (*ISHR*)* is also based on a penetration test, except that a 30 gm preload is used first. The major load is 580 gm for a 2.50 mm diam. ball, the loading duration is again 30 seconds, and the depth is measured with a dial gauge. The *ISHR*, a function of Young's modulus E, is 100 for $E = \infty$. The relation is selected so as to give large changes of *ISHR* in the practical E range encountered in various vulcanized rubbers (Fig. 12.13).

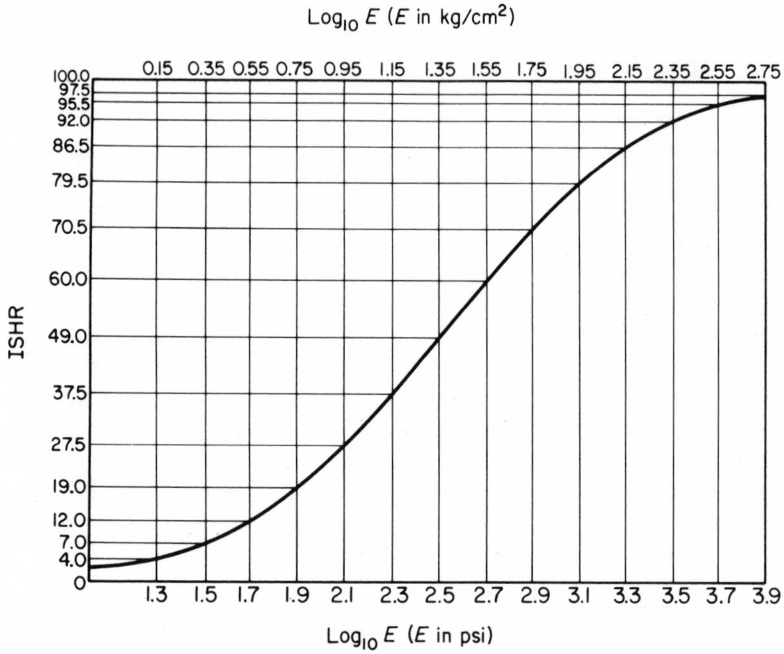

Fig. 12.13 Relation between modulus E and International Standard Hardness of Rubber.

Rockwell tests used on plastics can be applied to the harder rubbers. The "Durometer" uses a 30 degree tapered indenter with a radiused tip ($r = 0.004$ in.). A calibrated spring forces the indenter into the specimen, and a dial gauge indicates the depth of penetration in *Durometer numbers* which are directly proportional to the load on the spring**

* Ibid., Standard D1415–56T, p. 615.

** *A.S.T.M. Standards*, 1961, Standard D1484–59, pp. 674–76.

A number of other instruments, some of which use the rebound principle (pendulum), are available on the market but will not be described. Microhardness testers for light load testing of thin rubber sheets have also been developed.* Needless to add, a standard temperature must be maintained in all tests on rubber.

PROBLEMS

1. What effect on the shape of a Brinell and Vickers indentation will the anisotropy of the tested material have?

2. Calculate the actual depth of a Rockwell *C* indentation, under minor and major loads, in a steel with 150,000 yield strength. Assume the tip of the cone to be sharp and disregard work-hardening.

3. A 4.6 mm diam. impression was produced in a work-hardened alloy plate by a Brinell tester using 3000 lb. load and a 10 mm ball. What is the approximate tensile strength of this alloy?

4. Assuming that the relationship in Fig. 11.20 applies also to three-dimensional stress states, assess the minimum thickness of a 150 *DPN* hard strip at which a correct Vickers hardness test could still be performed with a 5 kg load.

REFERENCES FOR FURTHER READING

(R12.1) Mott, B. W., *Micro-Indentation Hardness Testing*. London: Butterworths & Co., 1950.

(R12.2) O'Neill, H., *The Hardness of Metals and its Measurement*. London: Chapman and Hall, Ltd., 1934.

(R12.3) Tabor, D., *The Hardness of Metals*. Oxford: Clarendon Press, 1951.

---

* J. R. Scott and A. L. Soden, *Trans. Institution Rubber Industry, 36* p. 1, (1960).

13

BENDING

Bending is the term applied to deformation in which a change of curvature is produced in relatively long thin members. It usually involves forces having at least one component normal to the long dimension of the element. Bending is probably the most common type of deformation encountered in engineering structures. Some examples are shown in Fig. 13.1. In (a) the lever is a beam deflected downward by force P; (b) is an aircraft wing lifted by uniformly distributed air pressure; (c), a piston ring in an engine or pump pressed diametrically by a force distributed along the perimeter, causing the ring to close by decreasing its diameter. In a closed ring such as a chain link (d), the ring portions at x would be bent to a larger, and those at y to a smaller, radius.

In some cases the shape change in bending is not immediately obvious. The frame in (e) is rigidly anchored at A and D and has rigid right angle corners at B and C. Application of force, P, causes a deflection of the frame so that members AB and DC are bent. For right angles B and C to be maintained, member BC is forced to assume the S-like shape, $B'C'$, even though the applied force, P, acts coaxially with this member.

Bending is also widely used in testing. Experimentally, it is about the easiest stress system to apply. For example, the effect of stress in a corrosive atmosphere requires holding large numbers of samples under stress for very long times, and these studies are almost invariably carried out by using elastically bowed sheet samples (Fig. 13.2). In addition, brittle materials, such as cast iron and rigid plastics, are tested in bending. Weld evaluations are also frequently made on bend specimens (Fig. 13.3).

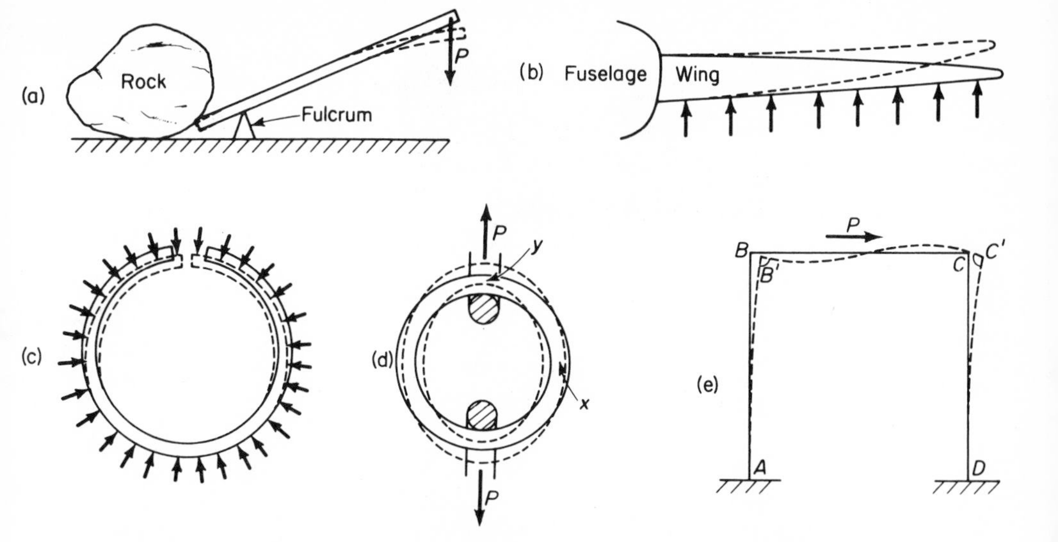

Fig. 13.1 Example of structural members subjected to bending: (a) Simple lever. (b) Airplane wing. (c) Piston ring. (d) Chain link. (e) Simple frame.

Fig. 13.2 Stress corrosion tests on elastically bent strips. From E. H. Phelps and A. W. Loginow, *Corrosion*, **16,** No. 7, 97 (1960).

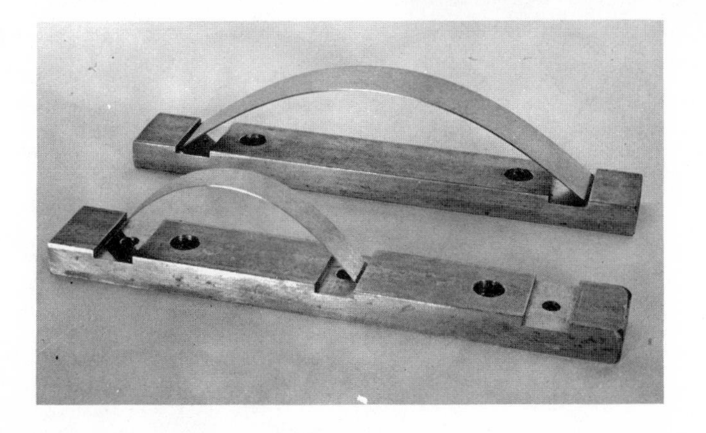

Fig. 13.3 Guided bend test on weld sample.

SUPPORT REACTIONS AND TYPES OF BEAMS

In order to analyze the stress in a beam, it is first necessary to determine all forces acting on it. In a practical beam problem the applied loads are generally specified while the support reactions are calculated by using the principles of statics. For static equilibrium, the sum of all forces acting in *any* direction, as well as the sum of the moments at any point, must be zero; i.e.,

$$\sum P + \sum R = 0 \tag{13.1a}$$

$$\sum M = 0 \tag{13.1b}$$

where P = applied (external) loads
R = reaction forces at supports
and M = bending moments

Several different types of supports are shown in Fig. 13.4. Roller support, A, can only transmit forces normal to its base, vertical in this case; pinned support, B, can also transmit oblique forces while still allowing free rotation of the beam at the support. If all external loads were vertical, the reaction forces at B would also be vertical. An obliquely applied force, such as P_1', will produce an additional horizontal component at B.

Beams are classified according to the manner in which they are supported. The *simply supported beam* rests on two supports which allow it free rotation and unhindered longitudinal expansion. Although roller supports of the type shown schematically in Fig. 13.4a are used in some service application (such as bridge spans), a beam lying freely on two fixed supports can be regarded as simply supported.

Beams, rigidly built in at one end (Fig. 13.4b) are known as *cantilever beams* and are held in equilibrium not only by the reaction force C but also by a reaction moment M_c in the fixed support. The support reaction is again calculated by Eq. (13.1a), and the moment by (13.1b). So long as the support reactions can be calculated from the static Eqs. (13.1), the beams are *statically determinate*.

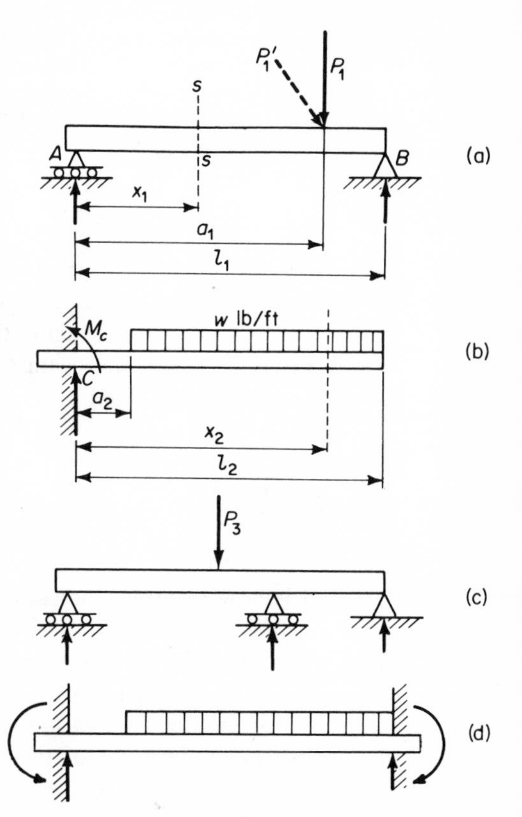

Fig. 13.4 Examples of statically determinate (a, d) beams with concentrated (a, c) and uniformly distributed (b, d) loads.

In Fig. 13.4, beam (c) has three supports and (d) is built in at both ends. Consequently, in (c) there are three unknown reaction loads at the three supports, and at (d) there are four unknowns (i.e., two support reactions and two support moments). Since only two static equilibrium equations are available, the additionally required equations must be supplied by methods other than those of statics. In cases such as these, the beams are said to be *statically indeterminate*. The calculation of these reactions is discussed in a later section since it requires information about the deflection of the beam.

BENDING MOMENTS AND SHEAR DIAGRAMS

The bending moment and shear force at any cross section along the beam length can be readily calculated by taking advantage of the fact that these internal forces and moments are in static equilibrium with their external counterparts. To demonstrate how these quantities are calculated, let us find the shear force and then the bending moment at one point in each of the top two beams in Fig. 13.4.

Starting with beam (a), to find reaction B, the equilibrium of moments taken about point A, according to Eq. (13.1b), is

$$P_1a_1 + Bl_1 = O$$

$$B = -P_1a_1/l_1$$

the negative sign indicating that B points in a direction opposite to P_1, i.e., up. Reaction A can be found by using Eq. (13.1a) or (13.1b). From (13.1a),

$$A + P_1 + B = O;\ A = -(P_1 + B) = -P_1 + \frac{P_1a_1}{l_1} = -P_1\left(1 - \frac{a_1}{l_1}\right)$$

All external loads acting on the beam are now known.

To find the shear force and bending moment at section s–s (a distance, x_1, from the left support) imagine the beam to be cut at this location and consider the portion of the beam to the left of the cut section. The shear force, V_1, on s–s is obtained by Eq. (13.1a),

$$A + V_1 = O; \qquad V_1 = -A$$

This internal shear force must, of course, also be in equilibrium with the right-hand section of the beam,

$$V_1 + P_1 + B = O; \qquad V_1 = -(P_1 + B) = A$$

The sign is now reversed, indicating that the shear force on two sides of a cut are oppositely directed. To avoid ambiguity in selecting the sign of a shear force, the following convention is used: The shear force is regarded as positive if the sum of all external forces acting on the portion of the beam to the left of the cut tend to move this portion upwards. Similarly, if the sum of forces to the right of the cut moves the right-hand portion down, the shear force is again positive.

The external forces acting on either side of a cut (such as s–s) exert a moment, M_x, on the cut surface. The magnitude of M_x can be calculated by adding the products of the individual forces and their distances from the cut (with due regard to their signs). By convention, an external force acting to the left of a section produces a positive moment when it tends to rotate the arm on which it acts in a clockwise direction; by the same token, the moment of a force to the right of the section will be positive when it acts counterclockwise. Another way to state this convention is: Positive moments cause the beam to bend concavely upward (collect water), and negative moments bend it concavely downward (shed water). Identical bending moments are obtained by using the right or left forces. To satisfy the equilibrium condition, $\Sigma M = O$, these external moments must be balanced by equal but oppositely directed internal moments developed within the beam.

Turning to the example in Fig. 13.4a, and considering cut-section, s–s, the only external force to its left is A. Hence, the bending moment on the section is

$$M_{x_1} = Ax = \left(1 - \frac{a_1}{l_1}\right)P_1x_1$$

Note that the sign of M_x would have been negative had the negative (as calculated) value of A been used. However, M_x is regarded as positive in view of the aforementioned convention.

The calculation of the shear force and bending moment in beam (b) is carried out as follows: First, reaction force, C, is found by summing the vertical forces,

$$C + w(l_2 - a_2) = O \qquad \text{so that} \qquad C = -w(l_2 - a_2)$$

The reaction moment, M_c, again balances the external applied moment,

$$M_c = \underbrace{-w(l_2 - a_2)}_{\text{load}}\underbrace{\left(a_2 + \frac{l_2 - a_2}{2}\right)}_{\text{lever arm}} = -\frac{w}{2}(l_2^2 - a_2^2)$$

with the sign of M_c being negative.

The shear force, V_2, at position x_2, considering the right-hand portion of the cut beam and using Eq. (13.1a), is

$$V_2 = w(l_2 - x_2)$$

The shear force is positive since the external force tends to move the cut section downwards.

The bending moment at x_2 is

$$M_{x_2} = -w(l_2 - x_2)\left(\frac{l_2 - x_2}{2}\right) = -\frac{w}{2}(l_2 - x_2)^2$$

The negatives indicate that the beam is bent down, i.e., sheds water.

In many problems it is desirable to have a continuous diagram of the shear force and bending moment along the entire beam. *Bending moment* and *shear force diagrams* are used for this purpose. It was shown above (Fig. 13.4a) that $V = A$ when x varied between O and a_1. Hence, the shear diagram to the left of load P_1 is shown in Fig. 13.5a as a horizontal line at A units above the abscissa. At position a_1 the shear force abruptly becomes negative and equal to B since $P = -(A + B)$.

The bending moment M_{x_1}, over length a_1, is Ax. Because P produces a negative moment, the total bending moment continuously decreases beyond a_1 and vanishes at B. There is always a singularity in the shear force and bending moment diagrams at the points where the loading conditions abruptly change.

The shear force diagrams and bending moment for beam (b) are obtained in a similar fashion. The reaction force, C, was found above to be $w(l_2 - a_2)$, so that the shear force is constant over length a_2. Over the portion of the beam that carries the uniformly distributed load, w, the shear force decreases linearly and vanishes at the free end of the beam, D.

The largest bending moment ($= M_c$) is at wall. It decreases linearly from C to E and parabolically from E to D.

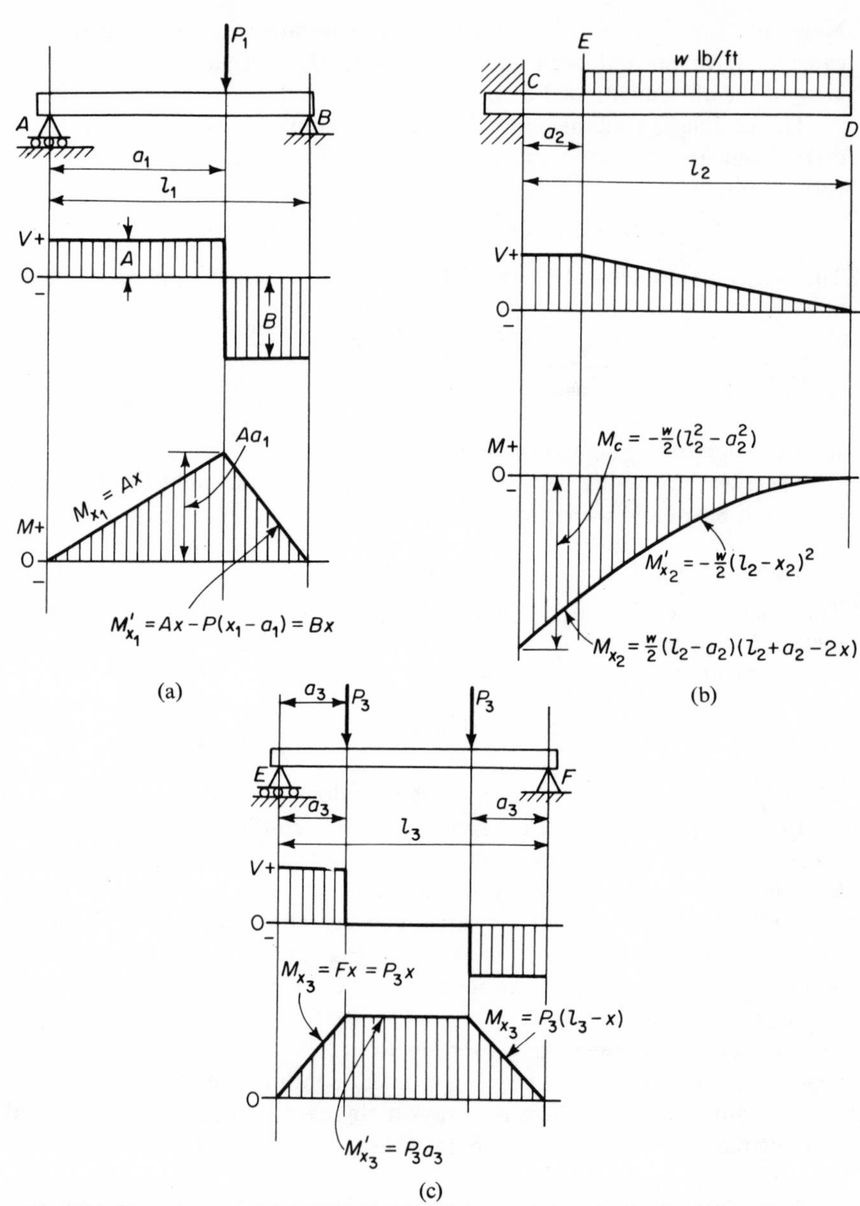

Fig. 13.5 Shear force and bending moment diagrams for statically determinate beams.

By reference to Fig. 13.6, it can be shown that there is a simple relationship between the shear force and bending moment in any beam. When a short section of a beam, dx long, between cuts s–s and s_1–s_1 is examined, the shear force and bending moment on the two sides generally differ. Since

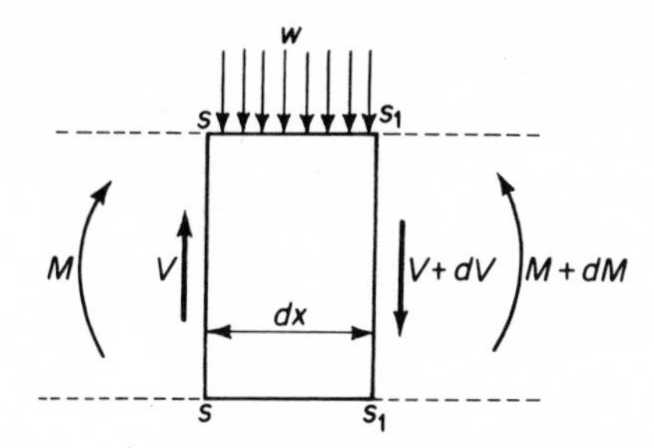

Fig. 13.6 Bending moments and shear forces acting on a beam element.

the section is in equilibrium, the sum of the moments about any point must be zero. Taking moments, for example, about s_1,

$$M + V\,dx - w\,dx\,\frac{dx}{2} - (M + dM) = 0$$

and

$$dM = V\,dx - \frac{w(dx)^2}{2} \approx V\,dx$$

Hence,

$$V = \frac{dM}{dx} \tag{13.2}$$

If there were no external loads between s and s_1, $dV = 0$, and Eq. (13.2) is again valid. This is apparent from the shear force and bending moment diagrams in Fig. 13.5, the former being the derivative of the latter.

One case of special interest in bending is *four point loading* (Fig. 13.5c). When identical loads, P_3, are applied at equal distances from the two supports, F and G, the beam length between the loads carries a constant moment and, in accordance with (13.2), no shear force.

STRESSES AND STRAINS CAUSED BY BENDING MOMENTS IN ELASTIC BEAMS

The relations between bending moments and the stresses and strains they produce in beams will now be determined. For this purpose a short length of the beam, confined between two planes, s–s and s_1–s_1, normal to the beam's long dimension and l' apart, will be considered (Fig. 13.7).

The bending moment, M, in this location (assumed positive) is equivalent to one produced by a pair of equal but oppositely directed forces, F_M, spaced m apart so that $M = F_M \cdot m$. The lower outwardly directed force, F_M, tends to elongate the lower part of the element, while the upper part is compressed and shortened. This is the "inner" mechanism by which the beam bows downward.

The elementary theory of elastic bending discussed below is based on two assumptions:

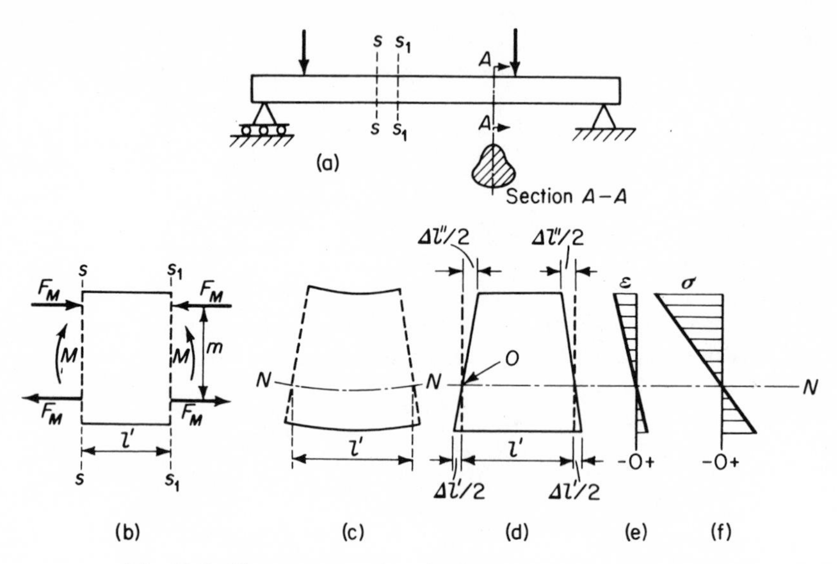

Fig. 13.7 Fiber stresses and strains in an elastically bent beam.

(1) Initially plane cross sections remain plane and normal to the beam axis under load.

(2) The materials obey Hooke's laws (5.1 and 5.2).

The first assumption requires element, s–s_1, to take on the shape shown in Fig. 13.7c or, disregarding the curvature which is usually slight, the form in Fig. 13.7d. The length change suffered by the element varies from maximum compression $\Delta l''$ on top to extension $\Delta l'$ on the bottom.

The longitudinal strains, $\varepsilon = \Delta l''/l'$, along the height of the element are shown in Diagram (e). In view of Hooke's law, the stresses, σ, are proportional to the appropriate strains, ε, Diagram (f). The length changes Δl, the strains ε, and stresses σ, all vanish at the *neutral surface*, N–N. Stress σ is the *bending stress*, and the question of whether it is a principal stress or not will be determined later.

It is often convenient to refer to a *neutral axis* of a particular cross section of a beam. This axis is defined as the intersection, O–O, of N–N and the cut plane, s (Fig. 13.8a). In Fig. 13.7 the trace of the neutral axis appears as O.

POSITION OF NEUTRAL AXIS

The distance between the neutral axis and the upper or lower extremity of the beam's section is found from Fig. 13.8. To locate the neutral axis, assume that the stresses, σ_x' and σ_x'', in the outer fibers are known. Since the cross-sectional dimensions of the beam are specified, the distances, y_1 and

y_2, can be found directly from

$$\sigma_x' : \sigma_x'' = y_1 : y_2$$

and

$$y_1 + y_2 = h$$

Stress σ at a distance y from N–N (or O–O) is

$$\sigma = \sigma_x' \frac{y}{y_1} \qquad \text{or} \qquad \sigma = \sigma_x'' \frac{y}{y_2} \tag{13.3}$$

Viewing the cross section (Fig. 13.8c), it is seen that a surface element of width b_y and height dy is acted upon by force $\sigma b_y\, dy$. All such forces above O–O must add up to F_M (Fig. 13.7b), which in turn must be balanced by a numerically equal but oppositely directed force below O–O. With reference to Fig 13.8b, this results in the equality,

$$\int_0^{y_1} \sigma b_y\, dy = \int_0^{y_2} \sigma b_y\, dy \tag{13.4a}$$

Fig. 13.8 Diagram used for determining the position of the neutral axis.

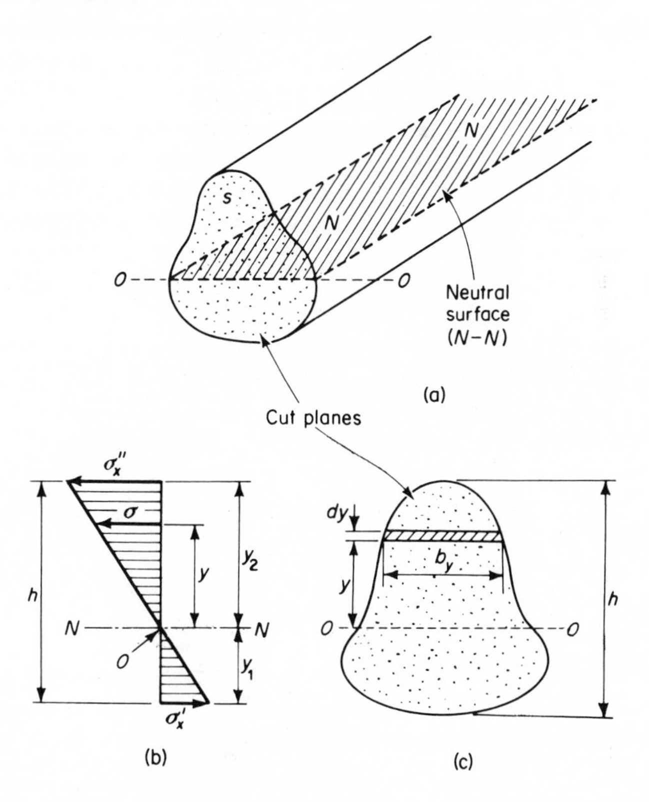

or, using (13.3),

$$\int_0^{y_1} \sigma'_x \frac{y}{y_1} b_y \, dy = \int_0^{y_2} \sigma'_x \frac{y}{y_1} b_y \, dy \tag{13.4b}$$

Since σ'_x and y_1 are not variables,

$$\int_0^{y_1} y b_y \, dy = \int_0^{y_2} y b_y \, dy \tag{13.5a}$$

or

$$Q = \int_0^{y_1} y \, dA = \int_0^{y_2} y \, dA \tag{13.5b}$$

Both equations state that the statical moments Q of the cross-sectional areas above and below the neutral axis are equal. Hence, finding the neutral axis of a given cross section is synonymous with determining its centroid.

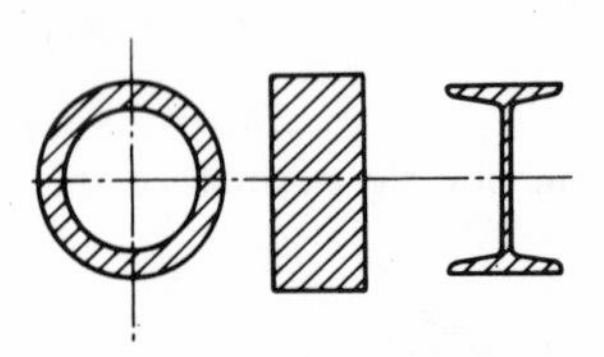

Fig. 13.9 Examples of cross section symmetrical with respect to the neutral axis.

If a beam is loaded in the vertical direction and its cross section is symmetrical about a horizontal plane (as shown by the examples in Fig. 13.9), the axis of symmetry is identical with the neutral axis. Otherwise the position of O–O must be found by integration. As an example, let us locate the neutral plane of a triangular cross section (shown in Fig. 13.10), assuming that the external forces are acting in the vertical direction. Assuming that the neutral axis O–O is at a distance, y_1, from A, the static moment of an area element $2z\,dy$, with respect to O–O, is $2zy\,dy$. Now $z:(y_1 - y) = a:2h$, and

$$z = \frac{a(y_1 - y)}{2h}$$

The total static moment of the area below O–O is

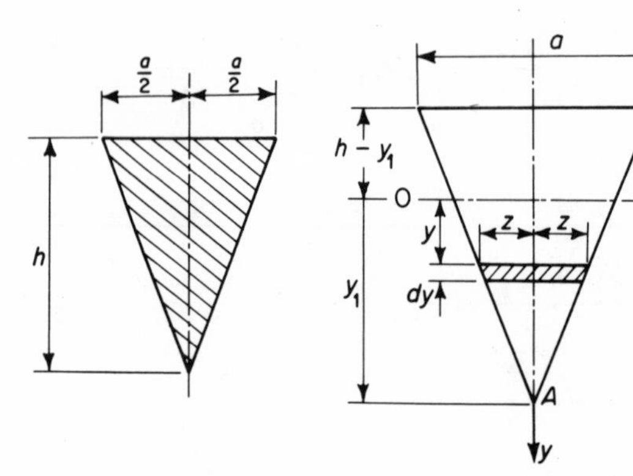

Fig. 13.10 Diagram for determining the neutral axis in an assymetrical (triangular) beam.

$$\int_0^{y_1} \frac{a}{h}(y_1 - y)y\,dy \tag{a}$$

For the trapezoidal area above O–O, we have

$$z:(y_1 + y) = a:2h$$

and

$$z = \frac{a(y_1 + y)}{2h}$$

The static moment is here

$$\int_0^{h-y_1} \frac{a}{h}(y_1 + y)y\,dy \tag{b}$$

after executing (a) and (b), they are equated

$$\frac{y_1}{2}(h - y_1)^2 + \frac{(h - y_1)^3}{3} = \frac{y_1^3}{2} - \frac{y_1^3}{3} = \frac{y_1^3}{6}$$

which leads to

$$y_1 = \tfrac{2}{3}h$$

The positions of the neutral axis for other asymmetric shapes are computed in substantially the same manner. Many of them are tabulated in various engineering handbooks.

It must be stated that the cases of bending treated in this text are restricted to beams loaded in their longitudinal plane of symmetry.

RELATION BETWEEN BENDING MOMENT, GEOMETRY OF CROSS SECTION, AND MAXIMUM LONGITUDINAL STRESS

From the standpoint of the material, the most important stress is σ_{max}. This stress acts in the longitudinal fiber most distant from the neutral surface, or axis (Fig. 13.7f). It is desirable, therefore, to directly relate moment M to σ_{max} and the cross-sectional dimensions.

The equilibrium of the section is secured by the fact that the total moment M_i of the internal forces taken over the entire surface A, is equal to, and balanced (or opposed) by, the externally applied bending moment M. Consequently,

$$M = M_i = \int_A dM_i \tag{13.6}$$

The problem is now narrowed down to calculating the value of the above integral. The integration limits are y_1 and y_2 (Fig. 13.8), the elementary force is $\sigma\,dA$, and its variable moment arm, y. Substitution of these into (13.6) yields

$$M = \int_{y_2}^{y_1} \sigma_x y\,dA \tag{13.7}$$

Denoting the y value pertaining to $\sigma_{\max}$ by c, and inserting these terms in (13.3),

$$\sigma_x = \sigma_{\max} \frac{y}{c} \tag{13.8}$$

When this is substituted in (13.7),

$$M = \int_{y_2}^{c} \frac{\sigma_{\max}}{c} y^2 \, dA = \frac{\sigma_{\max}}{c} \int_A y^2 \, dA = \frac{\sigma_{\max} I}{c} \tag{13.9}$$

The integral represents the moment of inertia, I, of the cross section with respect to the neutral axis, O–O, and it can be readily computed when $dA = f(y)$ is specified. For instance, in a rectangular cross section of height h ($= 2c$) and width b, and where O–O bisects h, $dA = b\,dy$. The moment of inertia is thus,

$$I = \int_A y^2 \, dA = \int_{-h/2}^{+h/2} y^2 b \, dy = \frac{by^3}{3}\Bigg|_{-h/2}^{+h/2} = \frac{bh^3}{12} \tag{13.10}$$

By substituting this into (13.9), with $c = h/2$, the result becomes

$$M = \frac{\sigma_{\max} I}{c} = \sigma_{\max} Z = \sigma_{\max} \frac{bh^2}{6} \tag{13.11}$$

where the quotient, $I/c = Z$, is termed *section modulus.* In a rectangular beam, the section modulus is

$$\frac{bh^3}{12} \Big/ \frac{h}{2} = \frac{bh^2}{6}$$

Equation (13.11) allows one to directly find the longitudinal stress, σ_x, at any distance, y, from the neutral axis. For this purpose $\sigma_{\max}/c$ is replaced by σ/y in accordance with (13.8). The result is

$$\sigma_x = \frac{My}{I} \tag{13.12}$$

To determine the stresses in a symmetrical I-beam (Fig. 13.11), one would have to calculate first its cross-sectional moment of inertia, using Eq. (13.10) considering the center web and the flange separately. The

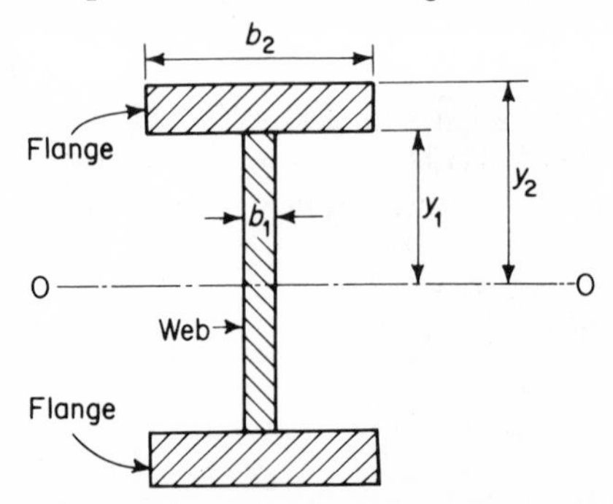

Fig. 13.11 Division of symmetrical I-beam into simple sections for calculating moment of inertia.

moment of inertia is first computed for integration limits $O - y_1$ and width b_1, and then for limits $y_1 - y_2$ and width b_2. The two results are added and the sum multiplied by two. The longitudinal stress at any distance y from O–O is found from (13.12) for the specified M.

The I and Z values for a variety of other common sections can be found in many engineering manuals. For special sections, both the position of O–O and hence c, as well as I or Z, must be calculated by solving the integral.

SHEARING STRESSES GENERATED BY BENDING

It was shown earlier (13.2) that the transverse force in a beam subjected to bending is $V = dM/dx$. The shear force, V, disappears over the portion of the beam along which the moment does not change (Fig. 13.5c).

The total shear force, V, can be regarded as a sum of unit shear stresses, τ, acting at each point of the particular cross section. Hence,

$$V = \int_A \tau \, dA \tag{13.13}$$

To determine the way in which τ varies over the beam height, an indirect approach is used, taking advantage of the fact that coplanar shear stresses on adjacent sides of an elementary cube are equal; i.e., $\tau_{xy} = \tau_{yx}$. By

Fig. 13.12 Diagram for finding distribution of shear stress across beam.

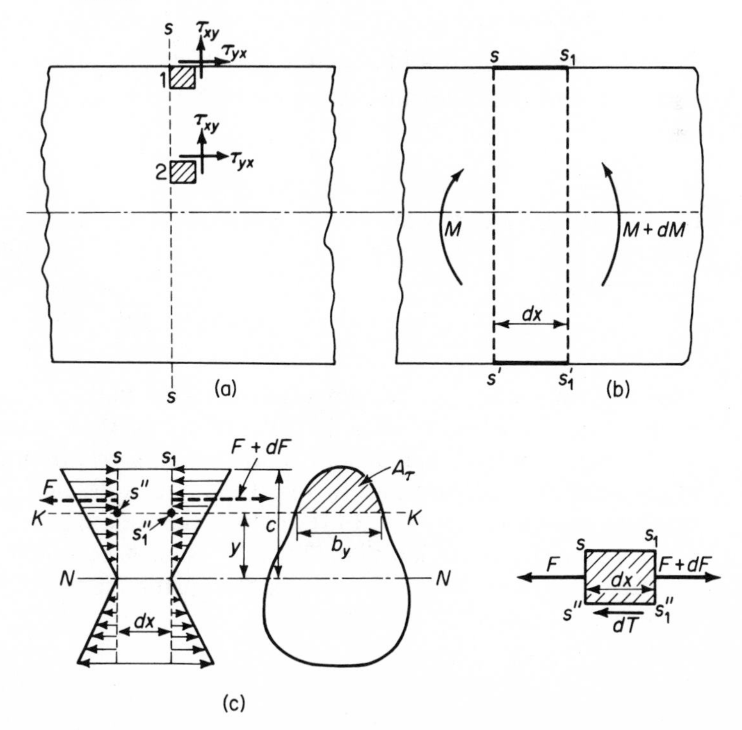

reference to Fig. 13.12a, it is seen that component $\tau_{yx} = 0$ at location 1 since no shear stress can act on a free external surface. Consequently, τ_{xy} at 1 is also zero. The values of τ_{xy} elsewhere, e.g., at position 2, is deduced from the variation of bending stress, σ_x.

It will be assumed that the bending moment varies along the beam, its value being M at some cross section, s–s' and $M + dM$ at another cross section, s_1–s_1', a distance dx away (Fig. 13.12b). The resulting $\sigma_{x(\text{left})}$ and $\sigma_{x(\text{right})}$ distributions are computed from (13.12) for the appropriate I, y, and left and right moment values.

The shear stress acting on plane K–K, located at an arbitrary distance, y, from the neutral plane, N–N, can now be determined.

The horizontal forces, F and $F + dF$ (Fig. 13.12c and d), acting over areas, s–s'' and s_1–s_1'', of the portion of the beam element above K–K (area A_τ) are not equal. Their difference, dF, must then be balanced by a shear force, dT, along s''–s_1'' to provide equilibrium.

With the aid of (13.12) one finds, on the surface above K–K,

$$F = \int_{A_\tau} \sigma_x \, dA = \int_y^c \frac{M}{I} y \, dA = \frac{M}{I} \int_y^c y \, dA \tag{c}$$

on the left A_τ face (s–s'')
and

$$F + dF = \int_{A_\tau} \sigma_x' \, dA = \int_y^c \frac{M + dM}{I} y \, dA = \frac{M + dM}{I} \int_y^c y \, dA \tag{d}$$

on the right-hand face (s_1–s_1'').
Noting that $Q_\tau = \int_{A_\tau} y \, dA$ is the statical moment of A_τ about O–O (identical with N–N in Fig 13.12c), and subtracting (c) from (d),

$$dF = \frac{Q_\tau \, dM}{I} = dT$$

Shear force, dT (Fig. 13.12d), acts on a surface dx long and b_y wide, b_y being measured in the direction normal to the surface of the page (see Fig. 13.12c right). Hence,

$$\tau_{xy} = \frac{dF}{b_y \, dx} = \frac{Q_\tau \, dM}{Ib_y \, dx} = \frac{Q_\tau V}{Ib_y} \tag{13.14}$$

The sought stress, τ_{xy}, is thus proportional to the static moment about the neutral axis of the "outer" area, (A_τ), cut off by the shear plane.

An important limitation of (13.14) becomes apparent from consideration of the inner face of the flange in Fig. 13.11. At a distance y_1 from O–O, the shear stress according to (13.14) would be the same at the exposed surface as across the flange-web junction. This is impossible because the shear stress at a free surface must be zero. Hence, the equation is valid for rectangular cross sections only.

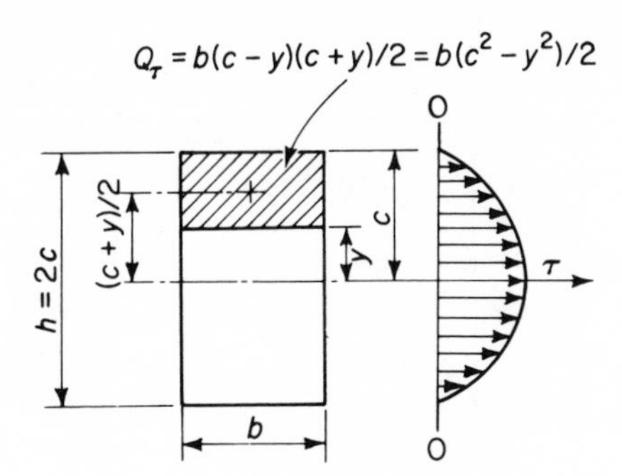

Fig. 13.13 Shear stress distribution in a rectangular beam.

SHEAR STRESS DISTRIBUTION

In the simplest case, the beam has a rectangular cross section of height, h, and width, b, and from (13.10), $I = bh^3/12$. The static moment of the area between y and $c = h/2$ (Fig. 13.13) is $Q_\tau = b(c^2 - y^2)/2$. After substitution, the shear stress is obtained from (13.14) as

$$\tau_{xy} = \frac{6(c^2 - y^2)}{bh^3} V$$

or, since $c = h/2$,

$$\tau_{xy} = \frac{3(h^2 - 4y^2)}{2bh^3} V \tag{13.15}$$

The shear stress is at a maximum when $y = 0$, i.e., along the neutral plane. Its value is then,

$$\tau_{\text{max}} = \frac{3V}{2bh} = \frac{3V}{2A} \tag{13.16}$$

which is 50 per cent higher than would be obtained on the trivial and obviously incorrect assumption that $\tau = V/A$. Similar calculations will yield the shear stress function (13.14) for other cross sections.

PRINCIPAL STRESSES IN ELASTICALLY BENT BEAMS

Since the shear stresses vanish at the extreme points (faces) of the beam farthest from the neutral plane, the longitudinal stress σ_x is the principal stress at this location. The other principal stress is normal to σ_x and has a value of zero on the free surface.

At the neutral plane $\sigma_x = 0$, while τ_{xy} is at a maximum. Consequently, the conditions here correspond to simple shear (page 39) with the principal stresses, $\sigma_1 = \sigma_2 = \tau_{xy}$, at 45 degrees or 135 degrees to the shear direction. At intermediate points, the angle between one principal stress and the neutral plane varies between 0 and 45 degrees, with the other principal stress at right angles to the first.

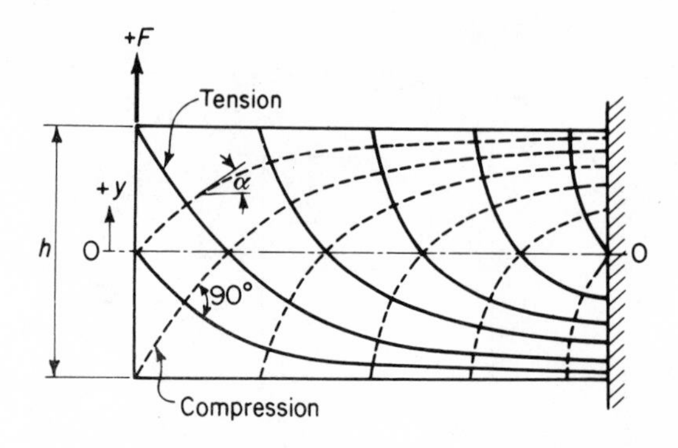

Fig. 13.14 Stress trajectories in a beam subjected to bending and shear. The relative contribution of shear decreases from left to right because of increasing bending moment.

Hence, the principal stresses and their directions at various distances from the neutral axis can be found by drawing Mohr's circles, or else by using the appropriate plane stress formulae (2.22, 23). The needed values of σ_x and τ_{xy} are taken from (13.12) and (13.14), respectively.

The direction of the principal stresses at any point are tangent to the two families of orthogonal curves known as stress trajectories (see p. 130) in Fig. 13.14.

INCONSISTENCIES OF THE ELEMENTARY THEORY OF ELASTIC BENDING

The relations between the internal bending and shear stresses, σ and τ, and the bending moment, M, and transverse force, V, were developed by assuming explicitly that plane cross sections remain plane in the strained condition (Fig 13.7). However, the above assumption can be satisfied only if the shearing stresses were either absent or uniformly distributed. Shearing distorts an element parallel to N–N in the manner indicated in Fig. 13.15. The degree of "warping" varies from zero in the layers carrying the highest longitudinal stress, and where $\tau = 0$, to a maximum at the neutral plane. Hence, the initially plane cross sections cannot, in reality, remain plane. Moreover, since there is generally a variable amount of shear distortion along the beam length, significant normal stresses, (σ_y), arise between adjacent layers.

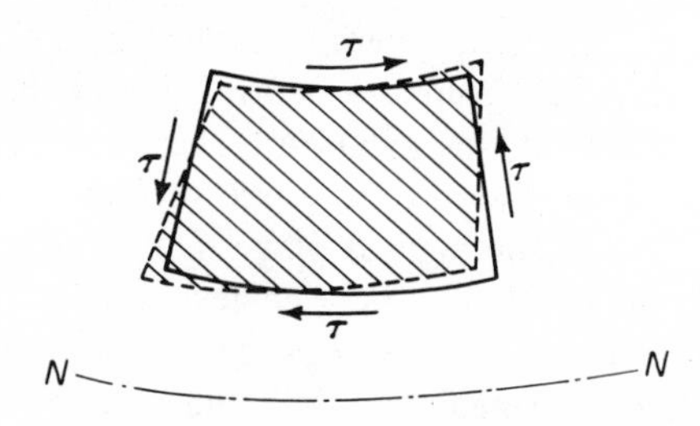

Fig. 13.15 Distortion of bent element by shear stresses.

These complicating factors were disregarded for the sake of simplicity. The exact solutions can be developed by methods of the theory of elasticity (see Chapter 5). They are much more difficult to arrive at and are more elaborate mathematically. Luckily, when they are applied to long beams, their results differ little from those obtained with the aid of the approximate equations.

It is also appropriate to point out that the assumption of $\tau_{xz} = \tau_{yz} = 0$ is valid only for beams of rectangular cross-section. Reasoning similar to that used in connection with Fig. 5.14 will show that such assumption is not usually true. As an exercise the reader may convince himself that, in the absence of stress components other than σ_x, σ_y, and τ_{xy}, an equilibrium of the triangular "end tips" of the elementary layer in Fig. 13.10b is impossible without shear components in the z-direction. Again, the errors in the vast majority of practical strength calculations caused by disregarding these refinements are insignificant.

CURVATURE OF ELASTICALLY BENT BEAMS

It is apparent from Fig. 13.7e that the increase of linear strain with distance from the neutral surface, N–N, causes the beam to bend. The resulting curvature of the neutral surface is defined in terms of its radius, R, or degree, $1/R$.

Let the angle α in Fig. 13.16 be so chosen that the length of arc, N_1–N_2, is exactly unity. Since beams are designed to deflect very little under foreseeable loads, radius R is normally very large compared with the beam's cross-sectional dimensions. Consequently, angle α is very small, usually less than one degree.

An arbitrary layer, DH, at a distance, y, from N–N is now selected, and line N_2F drawn parallel to radius CA. Since α is small, $N_1N_2 \approx DF$. Therefore, FH represents, the extension of a layer originally one unit long and, hence, strain ε. Noting that triangles CDH and N_2FH are similar,

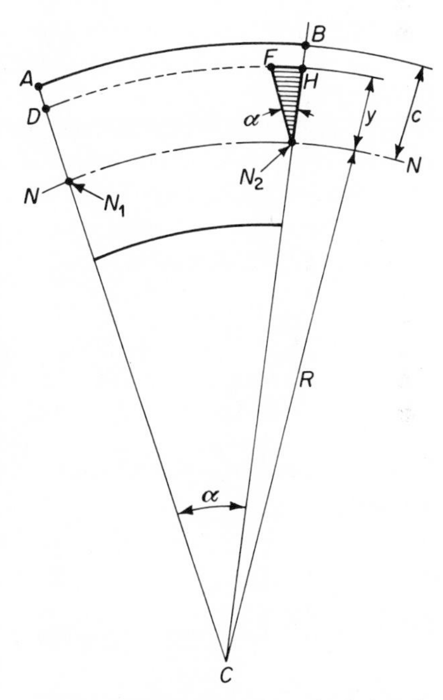

Fig. 13.16 Diagram for calculating the curvature of a bent beam.

$$CN_2 : N_1N_2 = N_2H : FH,$$

or

$$R : 1 = y : \varepsilon$$

or

$$R = y/\varepsilon$$

or
$$1/R = \varepsilon/y \tag{13.17a}$$

Using the outer layer at the maximum distance c from the neutral surface, and recalling that $\sigma = E\varepsilon$, the following terms are equivalent:

$$R = c/\varepsilon_{\max} = Ec/\sigma_{\max}$$

$$R = y/\varepsilon \quad = Ey/\sigma \tag{13.17b}$$

or their reciprocals for $1/R$. Since from (13.9), $\sigma_{\max} = Mc/I$ and $\sigma_{\max}/c = M/I$, Eq. (13.17b) can be replaced by

$$\frac{1}{R} = \frac{|M|}{EI} \tag{13.18}$$

An absolute value of M is indicated since the sign of the curvature has not yet been defined. Even with a constant beam cross section, the curvature generally varies from point to point and, to make the analysis more useful, it is advantageous to express the change of curvature along the beam in terms of coordinates x and y rather than R and then directly "translate" the curvature values into deflection data. The latter operation is important because many beams are designed on a maximum permissible deflection rather than maximum stress.

It is known from differential geometry that the curvature of a curve expressed as $y = f(x)$ is

$$\frac{1}{R} = \frac{\left|\dfrac{d^2y}{dx^2}\right|}{\left[1 + \left(\dfrac{dy}{dx}\right)^2\right]^{3/2}}. \tag{13.19}$$

Since dy/dx is the tangent of a very small angle, the squared term is negligible, leaving

$$\frac{1}{R} \approx \left|\frac{d^2y}{dx^2}\right| \tag{13.20}$$

Combining (13.20) and (13.18) yields the differential equation,

$$\frac{d^2y}{dx^2} = -\frac{M}{EI} \tag{13.21}$$

which is readily integrated for any $M = f(x)$ and constant EI.

The reason for the negative sign at the right-hand side of (13.21) is as follows:

If the deflection y caused by a positive bending moment (see p. 340 for convention) is also regarded as positive, as it is here, the positive x and

y directions will be as shown in Fig. 13.17. Since the slope of the deflected centroidal axis decreases with increasing x, the sign of the derivative d^2y/dx^2 must be negative. Hence, a negative sign is necessary in (13.21) to bring the two sides of the equation into agreement.

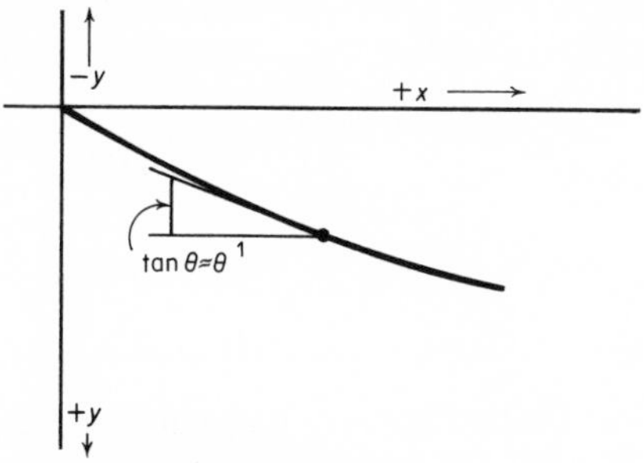

Fig. 13.17 Coordinate signs used to derive the equation for deflection (13-21).

ELASTIC DEFLECTION OF BEAMS

It is evident that a single integration of Eq. (13.21) will yield the slope, dy/dx, of the deflected element, while a second integration will result in the deflection y itself. However, each integration will introduce a constant that must be found from "boundary conditions." These define certain relationships between y and x which must be satisfied for any combination of loads and moments. Using Fig. 13.4 as an example, one sees that in (a) and (b) the deflections must be zero at the support points A, B, and C; in (b) both deflection y and slope dy/dx are zero at C. Moreover, since adjacent beam "zones"

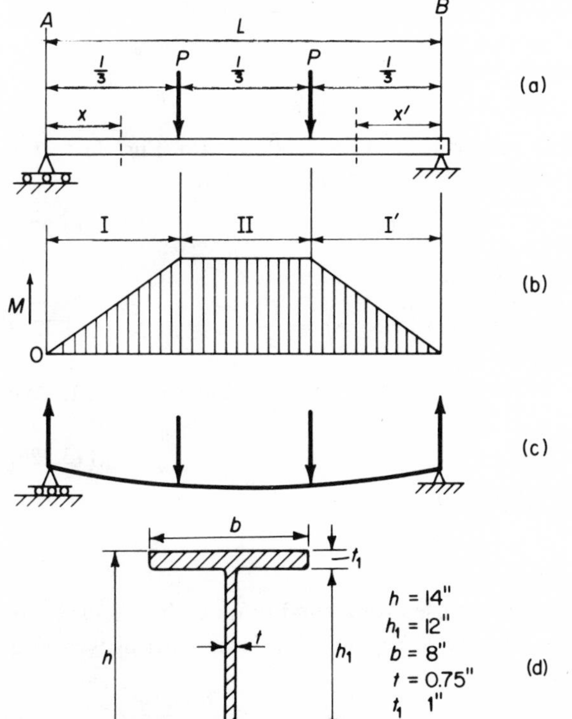

Fig. 13.18 Beam used for sample calculations of stresses and deflections.

I, II, III, etc. [i.e., lengths to which a single moment equation applies (see Fig. 13.5)] must merge smoothly and without sharp breaks, the conditions,

$$\frac{dy}{dx}(\mathrm{I}) = \frac{dy}{dx}(\mathrm{II}); \qquad \frac{dy}{dx}(\mathrm{II}) = \frac{dy}{dx}(\mathrm{III}); \quad \text{etc.}$$

must be satisfied at the "junctions." In addition, at these same locations

$$y(\mathrm{I}) = y(\mathrm{II}); \qquad y(\mathrm{II}) = y(\mathrm{III}); \text{ etc.}$$

One exemplary calculation of deflections for the steel *I*-beam shown in Fig. 13.18 will now be performed.

(a) *Support Reactions A and B*

From (13.1b), reaction B is

$$B = -\left(\frac{Pl}{3} + \frac{2Pl}{3}\right)\Big/ l = -P \tag{13.22a}$$

From (13.1a),

$$A = -(-P + 2P) = -P = B. \tag{13.22b}$$

A and B are thus equal, as could be expected from the symmetrical arrangement of the loads and distances relative to the center of the beam. The negative signs indicate that the reactions are directed oppositely to the acting loads.

(b) *Bending Moments*

Following the procedure described in connection with Fig. 13.5, the moments must be calculated separately for zones, I and II.

In zone I,
$$M_I = Ax \tag{13.22c}$$

counting x from A.

In zone II
$$M_{II} = Ax - P\left(x - \frac{l}{3}\right) = \frac{Pl}{3} \tag{13.22d}$$

The moments in I′ are obviously a mirror image of those in zone I. We thus have

$$M_I' = Px' \tag{13.22e}$$

where x' is measured from support B.

(c) *Moment of Inertia,* I

This calculation is based on that used for a solid beam with a rectangular cross section. It is done in two steps, separately for the vertical web and horizontal flanges (Fig. 13.18d).

WEB. The moment of inertia, I_W, is found directly from (13.10), using height h_1, and thickness t.
Hence: $I_W = th_1^3/12$
FLANGES. Here it is convenient to take twice the moment, of a single flange, $2I_F$, using the integral in (13.10) with $h/2$ and $h_1/2$ as limits. Accordingly,

$$2I_F = 2\int_{h_1/2}^{h/2} y^2\,dA = 2\int_{h_1/2}^{h/2} by^2\,dy = \frac{b(h^3 - h_1^3)}{12}$$

The total moment, I, is now

$$I = \frac{th_1^3 + bh^3 - bh_1^3}{12} = \frac{bh^3 - (b - t)h_1^3}{12} \qquad (13.22f)$$

After substitution,

$$I = \frac{8 \cdot 14^3 - (8 - 0.75) \cdot 12^3}{12} = 783 \text{ in.}^4$$

(d) *Deflection Equation for Zone* I(I′)

By inserting M_I into (13.21), the differential equation becomes

$$-EI\frac{d^2y}{dx^2} = Px$$

First integration,

$$-EI\frac{dy}{dx} = \frac{1}{2}Px^2 + C_1 \qquad (13.22g)$$

Second integration,

$$-EIy = \tfrac{1}{6}Px^3 + C_1x + C_2 \qquad (13.22h)$$

Integration Constants C_1 and C_2

Since $y = 0$ for $x = 0$, a substitution of these values into (13.22h) produces $C_2 = 0$.

To calculate C_1, the deflection equation for zone II must first be integrated.

(e) *Deflection Equation for Zone* II

Here M_{II} is constant and equal to $Pl/3$, so that

$$-EI\frac{d^2y}{dx^2} = \frac{Pl}{3} \qquad (13.22i)$$

Integrating once,

$$-EI\frac{dy}{dx} = \frac{Plx}{3} + C_3 \qquad (13.22j)$$

Second integration,

$$-EIy = \frac{Plx^2}{6} + C_3x + C_4 \tag{13.22k}$$

Integration Constant C_3

Note that at the mid-length of the beam, ($x = l/2$), the tangent to the deflection line is horizontal; i.e., $dy/dx = 0$
Using Eq. (13.22j),

$$0 = \frac{Pl^2}{6} + C_3; \qquad C_3 = -\frac{Pl^2}{6}$$

Integration Constants C_1 and C_4

It is seen from Fig. 13.17c that at $x = l/3$, the same values must be obtained for y and slope dy/dx, whether computed for zones I or II. This will provide two equations from which C_1 and C_4 can be found

With $C_3 = -Pl^2/6$, (13.22j) becomes

$$EI\frac{dy}{dx} = \frac{Pl}{6}(l - 2x);$$

For $x = l/3$, this gives

$$EI\frac{dy}{dx} = \frac{Pl^2}{18} \tag{13.22m}$$

On the other hand, (13.22g), for $x = l/3$, becomes

$$EI\frac{dy}{dx} = -\frac{Pl^2}{18} - C_1 \tag{13.22n}$$

Equating the right sides of (13.22m) and (13.22n)

$$\frac{Pl^2}{18} = -\frac{Pl^2}{18} - C_1;$$

$$C_1 = -\frac{Pl^2}{9}$$

Finally, the deflections calculated from (13.22h) and (13.22k) are equal for $x = l/3$. Substituting the constants already known, the following is obtained:

$$\frac{Pl^3}{162} - \frac{Pl^3}{27} = \frac{Pl^3}{54} - \frac{Pl^3}{18} + C_4, \tag{13.22o}$$

$$C_4 = \frac{Pl^3}{162}$$

All the constants are now substituted into the appropriate equations which acquire the following final forms:

	Zone I	*Zone* II
Slope	$EI\dfrac{dy}{dx} = \dfrac{P}{18}(2l^2 - 9x^2)$	$EI\dfrac{dy}{dx} = \dfrac{Pl}{6}(l - 2x)$
Deflection	$EIy = \dfrac{Px}{18}(2l^2 - 3x^2)$	$EIy = \dfrac{Pl}{162}(-27x^2 + 27lx - l^2)$

The final deflection line equation for Zone II can be checked by noting that identical deflection must be obtained for $x = l/3$ or $x = 2l/3$. The slope, dy/dx, is positive for $x < l/2$ and negative for $x > l/2$.

DEFLECTION CURVES FOR COMPLEX LOADING. PRINCIPLE OF SUPERPOSITION

Solution of differential equation (13.21) is simple but generally quite laborious, especially when a number of concentrated and distributed loads act on a beam (e.g., Fig. 13.19a). In this particular example there are four zones and a total of eight constants to compute.

In such cases it is helpful to use the superposition method (see Chapter 5), whereby the problem is broken down into a number of elemental cases (Fig. 13.19 b–d). The "component" curves are determined separately and added. A great number of deflection problems have been solved, and the resulting equations can be readily found in many applied mechanics and general engineering manuals. By adding the appropriate equations, the stresses and deflections for complex loaded beams are readily found. Some of the more common beam deflection formulae are contained in Table 13.1. The total deflection and other parameters pertaining to Fig. 13.19 can be obtained by the use of this Table.

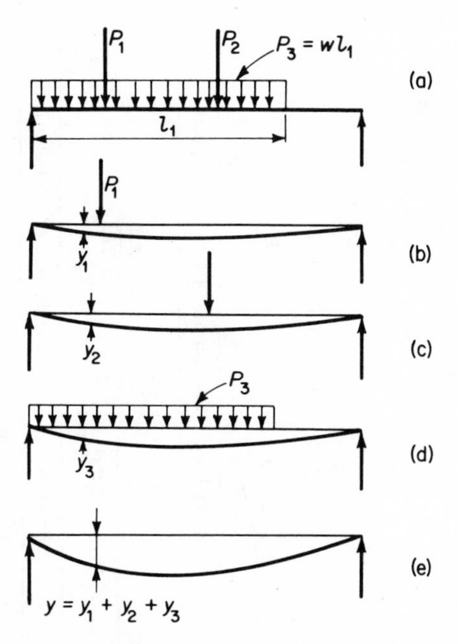

Fig. 13.19 Use of principle of superposition to simplify solution of a complex loaded beam. (y_1, y_2, etc. are deflections)

TABLE 13.1

BEAM DEFLECTION FORMULAS*

Beam Type	Slope at Free End	Deflection at any Section in Terms of x: y is Positive Downward	Maximum Deflection
1. Cantilever Beam—Concentrated load P at the free end			
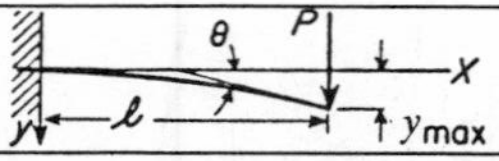	$\theta = \frac{Pl^2}{2EI}$	$y = \frac{Px^2}{6EI}(3l - x)$	$y_{max} = \frac{Pl^3}{3EI}$
2. Cantilever Beam—Concentrated load P at any point			
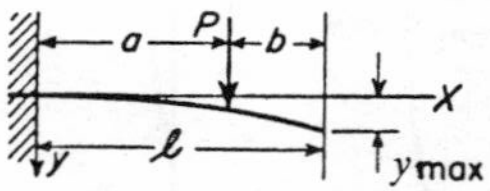	$\theta = \frac{Pa^2}{2EI}$	$y = \frac{Px^2}{6EI}(3a - x)$ for $0 < x < a$ $y = \frac{Pa^2}{6EI}(3x - a)$ for $a < x < l$	$y_{max} = \frac{Pa^2}{6EI}(3l - a)$
3. Cantilever Beam—Uniformly distributed load of w lb. per unit length			
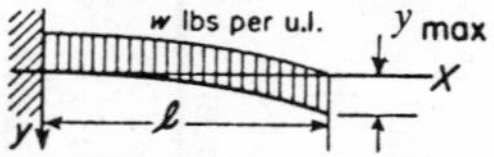	$\theta = \frac{wl^3}{6EI}$	$y = \frac{wx^2}{24EI}(x^2 + 6l^2 - 4lx)$	$y_{max} = \frac{wl^4}{8EI}$
4. Cantilever Beam—Uniformly varying load; maximum intensity w lb. per unit length			
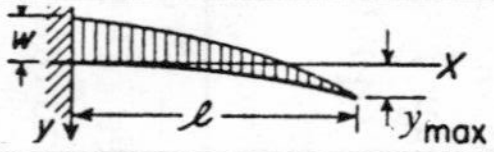	$\theta = \frac{wl^3}{24EI}$	$y = \frac{wx^2}{120lEI}(10l^3 - 10l^2x + 5lx^2 - x^3)$	$y_{max} = \frac{wl^4}{30EI}$
5. Cantilever Beam—Couple M applied at the free end			
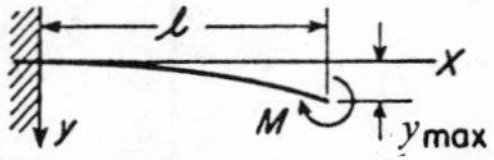	$\theta = \frac{Ml}{EI}$	$y = \frac{Mx^2}{2EI}$	$y_{max} = \frac{Ml^2}{2EI}$
6. Beam Freely Supported at Ends—Concentrated load P at the center			
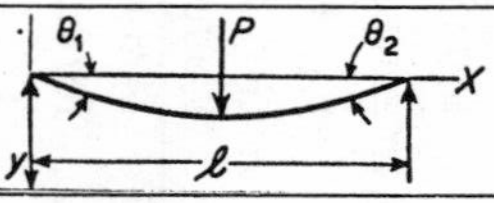	$\theta_1 = \theta_2 = \frac{Pl^2}{16EI}$	$y = \frac{Px}{12EI}\left(\frac{3l^2}{4} - x^2\right)$ for $0 < x < \frac{l}{2}$	$y_{max} = \frac{Pl^3}{48EI}$

TABLE 13.1 (continued)

Beam Type	Slope at Ends	Deflection at any Section in Terms of x: y is Positive Downward	Maximum and Center Deflection
7. Beam Freely Supported at Ends—Concentrated load at any point			
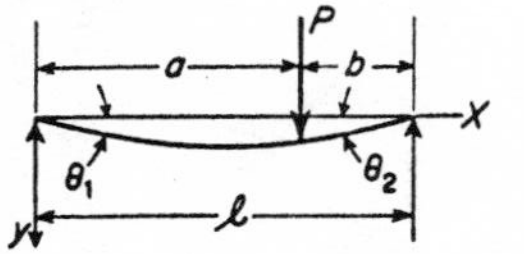	Left End. $\theta_1 = \frac{Pb(l^2 - b^2)}{6lEI}$ Right End. $\theta_2 = \frac{Pab(2l - b)}{6lEI}$	$y = \frac{Pbx}{6lEI}(l^2 - x^2 - b^2)[0 < x < a]$ $y = \frac{Pb}{6lEI}\left[\frac{l}{b}(x - a)^3 + (l^2 - b^2)x - x^3\right] [a < x < l]$	$y_{\max} = \frac{Pb(l^2 - b^2)c}{9\sqrt{3}lEI}$ at $x = \sqrt{\frac{l^2 - b^2}{3}}$ At center, if $a\,.\,.\,b$ $y = \frac{Pb}{48EI}(3l^2 - 4b^2)$
8. Beam Freely Supported at Ends—Uniformly distributed load of w lb. per unit length			
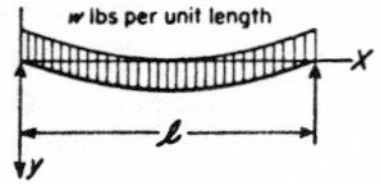	$\theta_1 = \theta_2 = \frac{wl^3}{24EI}$	$y = \frac{wx}{24EI}(l^3 - 2lx^2 + x^3)$	$y_{\max} = \frac{5wl^4}{384EI}$
9. Beam Freely Supported at Ends—Couple M at the right end			
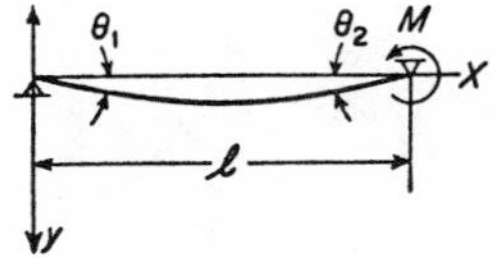	$\theta_1 = \frac{Ml}{6EI}$ $\theta_2 = \frac{Ml}{3EI}$	$y = \frac{Mlx}{6EI}\left(1 - \frac{x^2}{l^2}\right)$	$y_{\max} = \frac{Ml^2}{9\sqrt{3}EI}$ at $x = l/\sqrt{3}$ At center $y = \frac{Ml^2}{16EI}$
10. Beam Freely Supported at Ends—Uniformly varying load: max. intensity w			
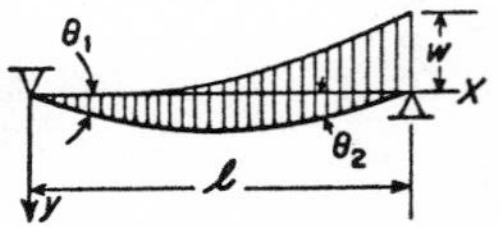	$\theta_1 = \frac{7wl^3}{360EI}$ $\theta_2 = \frac{wl^3}{45EI}$	$y = \frac{wx}{360lEI}(7l^4 - 10l^2x^2 + 3x^4)$	$y_{\max} = 0.00652\,\frac{wl^4}{EI}$ at $x = 0.519l$ At center $y = 0.00651\,\frac{wl^4}{EI}$

* From S. Timoshenko and G. H. MacCullogh. Elements of Strength of Materials, 4th ed., copyright 1962, D. Van Nostrand Co., Princeton, N.J.

As an alternative, various graphical methods were developed and are described in specialized textbooks on strength of materials. They also reduce the labor involved in solving beam problems.

SOLUTION OF BEAM PROBLEMS WITH THE AID OF SINGULARITY FUNCTIONS

The method just described constitutes a straight forward application of the conditions imposed upon the deflection by the requirements of continuity of the beam. However, while it always yields the correct solution, the computational labor it requires increases sharply with the number of load discontinuities, to the point of making it impracticable.

There are other methods that satisfy automatically the conditions of continuity and yet are capable of solving the bending problems, without increasing essentially the volume of calculations with the number of load discontinuities. The most elementary, and probably the most efficient of them will be briefly outlined below.

Let $x = a_i$ define a point along the beam at which a load of intensity

$$\langle x - a_i \rangle^n = \begin{cases} 0 & \text{for} \quad x < a_i \\ (x - a_i)^n & \text{for} \quad x > a_i \end{cases} \qquad (n = 0, 1, 2, \cdots) \tag{13.23a}$$

is applied (Fig. 13.20c, d, e, f). In addition, let the following symbolic designations be introduced:

$\langle x - a_i \rangle^{-1}$ for a unit concentrated force at $x = a_i$ (Fig. 13.20b) (13.23b)

$\langle x - a_i \rangle^{-2}$ for a unit moment at $x = a_i$ (Fig. 13.20a) (13.23c)

The class of functions

$$\langle x - a_i \rangle^n \qquad (n = -2, -1, 0, 1, 2, \cdots) \tag{13.23d}$$

is referred to in the literature as *singularity functions.* Their importance in the bending problem arises from the fact that virtually all loads and reactions encountered in practice are multiples or linear combinations of singularity functions.

The bending problem, as formulated in this chapter, may be described by the four differential relations

$$\frac{dV}{dx} = -w; \qquad \frac{dM}{dx} = V; \qquad \frac{d\theta}{dx} = -\frac{M}{EI}; \qquad \frac{dy}{dx} = \theta. \tag{13.23e}$$

Its solution requires a successive integration of these relations, each integration representing, respectively, a transition from the load to the shear force diagram V, to the bending moment diagram M, to the slope of the deflection line $\theta \approx \tan\theta$ and, finally, to the deflection y.

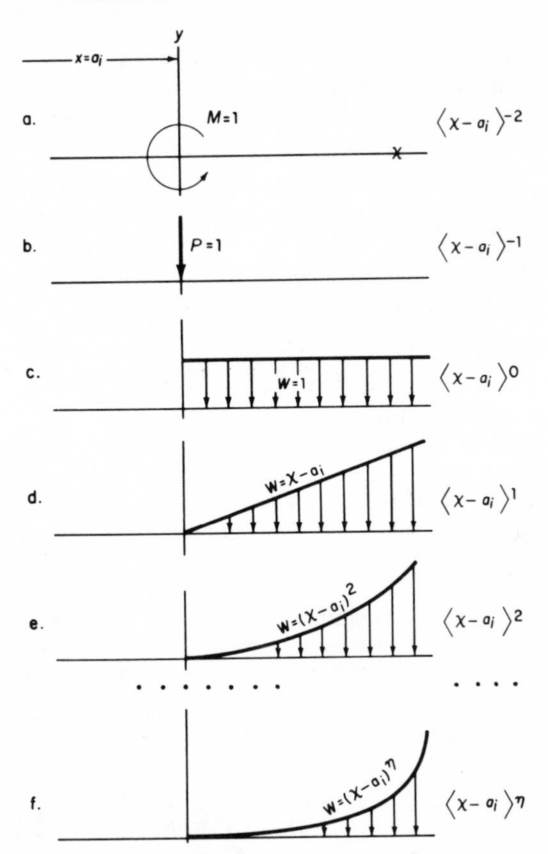

Fig. 13.20 Types of loading and the appropriate singularity functions.

By the virtue of the principle of superposition, these integrations may be executed separately for each load, provided of course, that eventually all loads and reactions are taken into account so that overall equilibrium is assured (or that the beam is temporarily clamped at some distant point to the right, i.e., beyond the section of interest).

It is not difficult to check that the integration of any of the four differential relations (13.23e) essentially amounts to moving one step down the column of singularity functions in Fig. 13.20. For instance, if the load under consideration is $\langle x - a_i \rangle^0$ (Fig. 13.20c), the shear force diagram it generates is $-\langle x - a_i \rangle^1$, the bending moment diagram $-\frac{1}{2}\langle x - a_i \rangle^2$, the slope

$$\frac{1}{2 \cdot 3} \frac{\langle x - a_i \rangle^3}{EI} + C_1$$

and the deflection

$$\frac{1}{2 \cdot 3 \cdot 4} \frac{\langle x - a_i \rangle^4}{EI} + C_1 x + C_o$$

The integration constants C_o and C_1 are necessary to adjust the derived slope and deflection so as to satisfy the conditions at the supports. No such constants were needed in the first two integrations because the conditions of equilibrium were met automatically.

It is now obvious that for each individual load belonging to the class of singularity functions $\langle x - a_i \rangle^n$, the integration of all four relations (e), and thus the calculation of the deflection, reduces to raising the exponent n by four units, multiplying by the appropriate factor, and adding the required constants of integration; in symbols

$$y = \frac{1}{EI}\frac{n!}{(n+4)!}\langle x - a_i \rangle^{n+4} + C_1 x + C_o \tag{13-23f}$$

if it is agreed that

$$n! = 1 \quad \text{for} \quad n \leq 0$$

It follows that if all the loads and reactions are expressible in terms of singularity functions, as is almost always the case, say,

$$\sum w_i \langle x - a_i \rangle^{n_i} \tag{13-23g}$$

the deflection is simply

$$y = C_o + C_1 x + \frac{1}{EI}\sum_i \frac{n_i!}{(n_i + 4)!}\langle x - a_i \rangle^{n_i+4} \tag{13-23h}$$

By the same token the expression for the slope

$$y' = \frac{dy}{dx} = C_1 + \frac{1}{EI}\sum_i \frac{n_i!}{(n_i + 3)!}\langle x - a_i \rangle^{n_i+3} \tag{13-23i}$$

Equations (h) and (i) show that the constants C_o and C_1 may be given a clear physical meaning. If the origin $x = 0$ is placed at *the left end* of the beam, C_o measures the deflection and C_1 the slope at this end of the beam.

We can now apply the method of singularity functions to the previous example (13.22). Noting that the loading is $P\langle x - 0 \rangle^{-1}$ for $0 < x < l/3$ and is $P\langle x - 0 \rangle^{-1} - P\langle x - (l/3) \rangle^{-1}$ for $l/3 < x < l/2$, the general expression for deflection is found, with the aid of (13.23), by raising the exponent (-1) of the singularity functions of load by four units to yield

$$y = \frac{P}{6EI}\left[-\langle x - 0 \rangle^3 + \left\langle x - \frac{l}{3} \right\rangle^3\right] + C_1 x + C_o$$

Since the second term in carets does not apply to the left end, we find that $C_0 = 0$ for $y = x = 0$ at this point.

The slope is

$$y' = \frac{P}{2EI}\left[-\langle x - 0 \rangle^2 + \left\langle x - \frac{l}{3} \right\rangle^2\right] + C_1$$

Since

$$y' = 0 \quad \text{at} \quad x = \frac{l}{2},$$

$$\frac{P}{2EI}\left[-\frac{l^2}{4} + \frac{l^2}{36}\right] + C_1 = 0$$

$$C_1 = \frac{Pl^2}{9EI}$$

and

$$y = \frac{P}{EI}\left[-\frac{\langle x - 0\rangle^3}{6} + \frac{\left\langle x - \frac{l}{3}\right\rangle^3}{6} + \frac{l^2}{9}x\right]$$

For $0 \le x \le \frac{l}{3}$, $\quad y = \frac{P}{EI}\left(-\frac{x^3}{6} + \frac{l^2x}{9}\right) = \frac{Px}{18EI}(2l^2 - 3x^2)$

For $\frac{l}{3} \le x \le \frac{l}{2}$, $\quad y = \frac{P}{EI}\left(-\frac{x^3}{6} + \frac{\left(x - \frac{l}{3}\right)^3}{6} + \frac{l^2}{9}x\right)$

$$= \frac{Pl}{162EI}(-27x^2 + 27lx - l^2)$$

These are identical with the tabulated values on p. 357. Application to more complex problems and a more detailed discussion can be found in (R13.1) and (R13.3).

STATICALLY INDETERMINATE BEAMS

The two static equilibrium equations (13.1) are sufficient to determine the support reactions and bending moments when the problem involves only two unknowns. This is the case for simply supported or cantilever beams (see Figs. 13.4a and b). Such problems were said to be statically determinate. The beam represented in Fig. 13.21, however, has three supports and, hence, three reactions at A, B, and C, all of which are unknown. In this and similar cases, it is necessary to provide an additional equation by making use of the properties of the deflection curve.

For this purpose it is at first assumed that the "redundant" third support does not exist. This assumption can be applied equally to any one of the three supports (e.g., to C). In its absence (Fig. 13.21b), the beam becomes statically determinate; the "provisional" reactions, A' and B', the moment equations, and the deflection curve are successively found, the imaginary deflection at C being y_c.

By referring again to Fig. 13.21a, it will be noted that the deflection at support C must be 0. Now force C can be calculated to provide a deflection

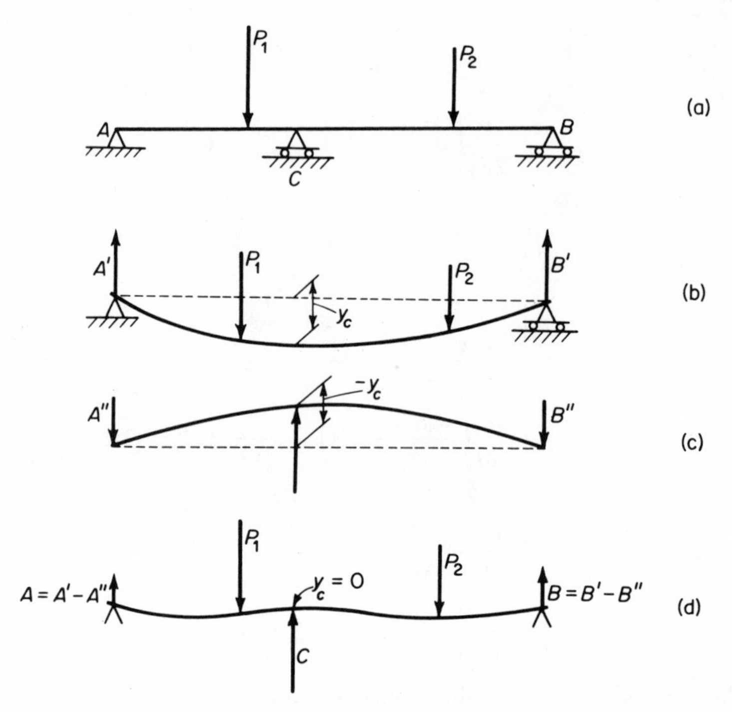

Fig. 13.21 Application of superposition to solution of a statically indeterminate beam of the first degree.

of $-y_c$ (diagram (c)) in the direction opposite to that in diagram (b). In this way the condition of zero deflection at support C will be satisfied, reaction C offsetting (or compensating for) the imaginary deflection caused by P_1 and P_2 in the absence of support C.

Reaction C acting by itself produces partial reactions, A'' and B'', at the beam's ends. By simply adding the data of the two diagrams (b) and (c), the final results are obtained in Fig. 13.21d.

The beam just discussed involved only one redundant reaction, and it represented a case of *indeterminacy of the first degree*. Another example of the same type is shown in Fig. 13.22. The beam in (a) carries only a distributed load, w, per unit length. A, B, and M_B are the unknown reactions. Diagrams (b) and (c) illustrate one method of breaking down (a) into two statically determinate "component" situations. It is evident that $A''(= A)$ must be chosen so as to offset deflection y'_A with an equal but opposite deflection y''_A.

The same result can be arrived at by assuming that the beam is not built in, but lies on two simple (or point) supports, and then superimposing upon

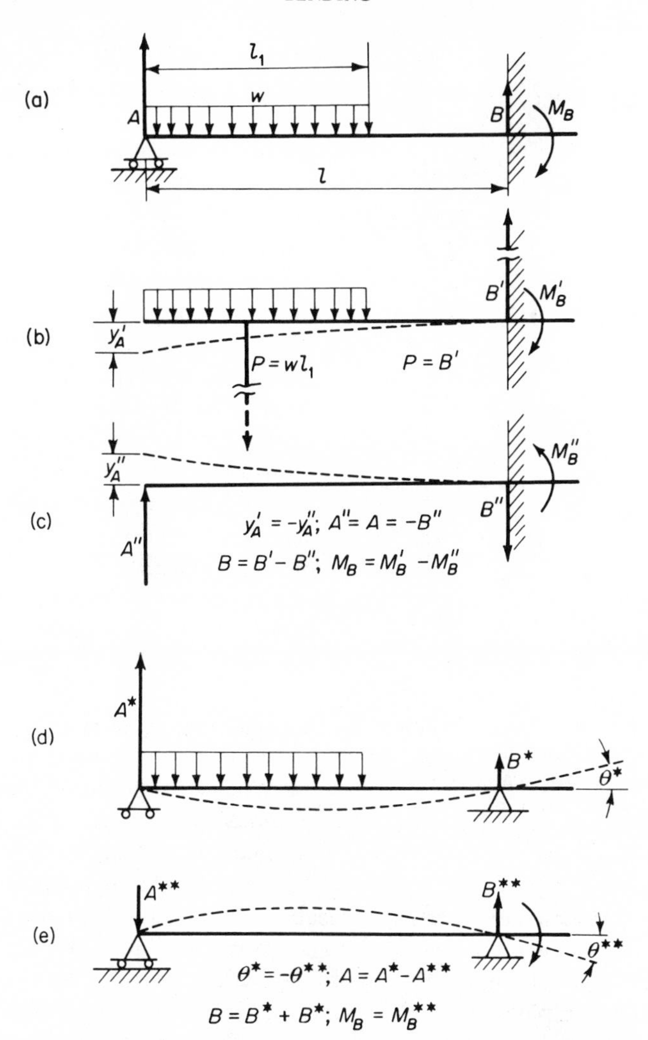

Fig. 13.22 Two ways of using the principle of superposition to determine bending moment in statically indeterminate beam. (a): Method I is represented by (b) and (c) while Method II is represented by (d) and (e).

it the loading-deflection conditions diagramed in (e). Since the beam must project horizontally from the wall at B, angles θ^* and θ^{**} must be equal but opposite in sign so as to cancel each other out. This is equivalent to satisfying the condition, $dy/dx = 0$, at B.

The complete bending moment diagram is obtained by adding two "fractional" diagrams corresponding to simplified situations (b) and (c) or (d) and (e) (Fig. 13.22). The procedure for superimposing the

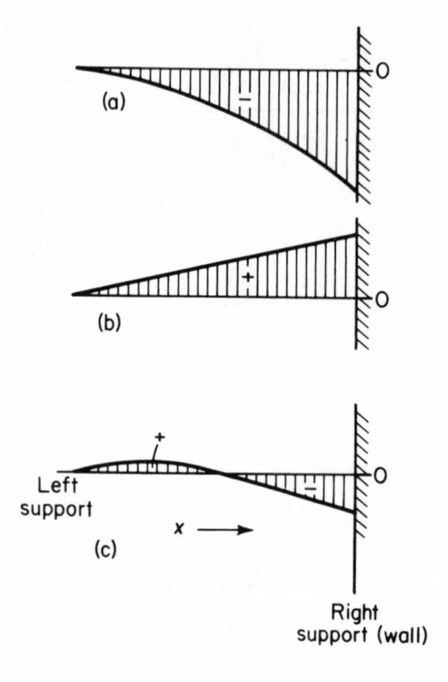

Fig. 13.23 Partial and resultant bending moment diagrams obtained by using Method I on the beam in Fig. 13.22a. Moment diagram (a) corresponds to beam (b), diagram (b) to beam (c), and the resultant diagram (c) to fully loaded beam in (a), Fig. 13.22.

"component" bending moments for the first pair is illustrated in Figs. 13.23a and b. These moment diagrams are added in (c) (which shows the resultant moment curve actually developed in the beam). It is the maximum moment on this diagram that determines the maximum stress in the beam.

If the same beam were built in at both ends, an additional moment M_A would have appeared at end A. This would increase the number of unknown reactions to four (A, B, M_A, and M_B), thereby resulting in a *second degree indeterminacy*. The additional equation required for a solution is found from the condition that the beam must also remain horizontal (i.e., $dy/dx = 0$) at A.

BENDING IN THE PLASTIC RANGE

When stress, σ, in the outer layer of a beam reaches the yield stress, σ_o, the linear stress distribution shown in Fig. 13.7 is altered because stress is no longer proportional to strain. Because the rate at which the stress increases with strain in the plastic range (plastic modulus) is only a small fraction of the elastic modulus, small load increments will cause the plastic deformation to rapidly propagate inwards and approach the neutral surface.

The plastic extension of the convex, and compression of the concave side, will result in a permanently bent beam. If such situation develops within a structure by accident, it may render it unsuitable for further use.

The assumption of plane transverse surfaces remaining plane after bending is retained in the plastic range. Since this will cause the strain to increase linearly with distance from the neutral surface (Fig. 13.24a), the fiber stress, σ_x, will be determined by the shape of the stress–strain curve, $\sigma_x = f(\varepsilon)$, shown in (b).

In most cases the stress–strain curves in tension and compression are about identical. Consequently, as long as the curvature is not excessive, and the beam has a symmetrical cross-section, the neutral surface will remain at the center of the beam for all practical purposes. However, the actual

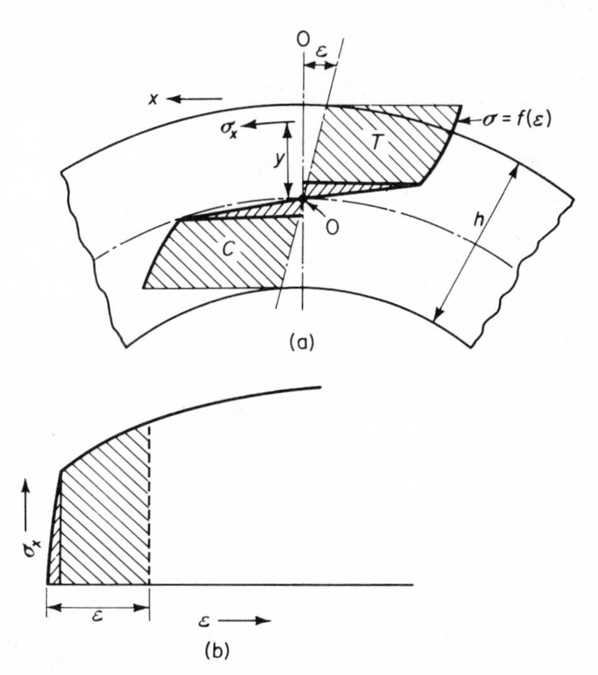

Fig. 13.24 Stress-strain relations across the thickness of plastically bent beam.

radius of curvature of the outer and inner surface layers differ and, assuming that the neutral plane did not shift, the former is $R + h/2$ while the latter $R - h/2$. As a consequence, the actual strain suffered by the concave surface is greater than that of the tension face, especially when R/h is not very large. This causes more work-hardening on the compressive side, and, to secure equilibrium, the neutral surface must be displaced toward the concave side of the beam. It may be added that in curved beams, the neutral axis is displaced in this direction even in the elastic range.

Disregarding this, and limiting the strains to a few per cent only, the equilibrium equation is

$$M = \int_{-h/2}^{+h/2} \sigma_x b \, dy \cdot y = b \int_{-h/2}^{+h/2} \sigma_x y \, dy \tag{13.24}$$

From (13.17a), $y = \varepsilon R$ or, $dy = R \, d\varepsilon$. Substitution in (13.24) now gives

$$M = b \int_{-h/(2R)}^{+h/(2R)} \sigma_x \varepsilon R \cdot R \, d\varepsilon = bR^2 \int_{-h/(2R)}^{+h/(2R)} \sigma_x \varepsilon \, d\varepsilon \tag{13.25}$$

Note that the integration limits were converted to read for outer fiber strains, ε, instead of distances, y, by using (13.17),

$$\varepsilon_{\max} = \pm h/2 : R = \pm h/(2R)$$

Equation (13.25) permits the bending moment to be calculated provided the stress–strain characteristic of the material is known.

Another useful relation is obtained when expression (13.25) is divided by R^2, differentiated with respect to $1/R$, and solved for σ_{xc}, the stress in the outer fiber. The result is

$$\sigma_{xc} = \frac{4}{bh^2}\left[\frac{1}{2R}\frac{dM}{d(1/R)} + M\right] \tag{13.26}$$

which makes it possible to construct the uniaxial stress–strain curve, $\sigma_{xc} = f(\varepsilon)$, from bending moment and curvature radius measurements.

For loading conditions which involve elastic and plastic strains of comparable magnitude, the elastic (13.7) and plastic (13.25) equations can be combined.

BENDING TESTS AND THEIR APPLICATION

Bending tests are used frequently to evaluate strength properties of brittle or "embrittled" materials because of the ease of preparing samples and the simplicity of carrying out the tests (see Chapter 15). Cast iron, hardened tool steel, ceramics, and plastics are tested in this way by using standard size specimens and distances between supports and a predetermined shape of the loading ram tip and supports.

Slow bend tests are rarely used on homogeneous ductile metals but quite frequently on welded specimens, with the weld located beneath the ram and parallel to it. The purpose of such a test is to compare the strength of the weldment with that of the base metal as well as to serve as a general quality control on welding procedures.

Various impact tests, like the V-notch, keyhole, and pin (Schnaadt) tests are all varieties of bend tests.

ELASTIC INSTABILITY OF AXIALLY COMPRESSED ELEMENTS

A straight rod subjected to an axial compressive force is the prototype for a slender tall column (Fig. 13.25a). In principle, such a rod carries only a longitudinal compressive stress but no bending moment. Assuming that the ends of the rod have pin tips, and that a small lateral force, F_h, is applied at mid-length while F is acting, the column will deflect by a small amount, y. Removal of force F_h will cause the rod to straighten out.

If F_h is kept constant while gradually increasing the axial force F, a condition will eventually be reached when moment $F \cdot y$ is large enough to keep the column partially deflected after F_h is removed. An even slight further increase of F by a fixed amount, ΔF, will cause the already existing deflection, y, to increase, thereby increasing in turn bending moment, $F \cdot y$, with a consequent increase of y, and so forth. As long as the rod remains elastic for any given F, the deflection reaches an equilibrium value. However, the bowing may be very large and exceed the permissible distortion by far.

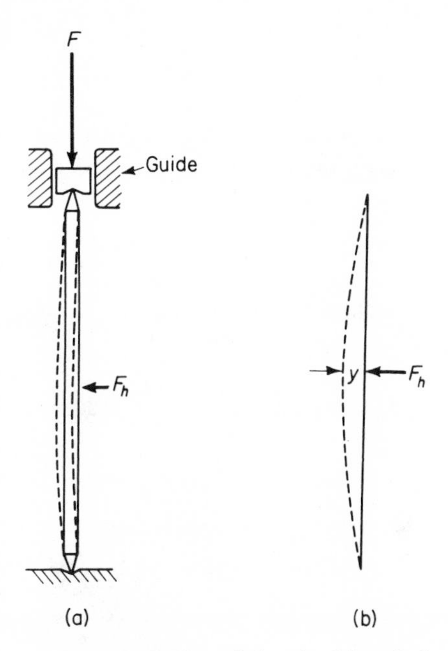

Fig. 13.25 Pin-ended rod of length l under axial load F: (a) Setup. (b) Deflection.

The force, F, at which the column first shows a deflection when the lateral force, F_h, is removed is called the *critical* or *Euler force*, F_c. For $F > F_c$, the deflections increase precipitously and the rod is said to *buckle*. Thus, the relation between F and y is not linear and buckling takes place suddenly, essentially without warning. It is important to realize that the axial stress in a column at the buckling stage, ($F = F_c$), may well be only a fraction of the compressive yield stress. The value of the latter is of importance only in very short columns where plastic yield can occur before F_c is reached.

The question may be now asked why a perfectly straight, axially compressed column should ever buckle. This requirement for perfection is never ideally satisfied; a minute lateral deflection may be caused even by "accidental" forces such as a wind acting on a television tower or a man leaning against a column. Once started, the deflection will spontaneously increase when $F > F_c$, until an equilibrium between moment, $F \cdot y$, and the stresses in the member is reestablished. However, the equilibrium may not occur until the deflection is excessive, say, 10 to 15 per cent of the column length.

DETERMINATION OF THE EULER LOAD FOR A PIN-ENDED COLUMN

According to (13.21), the relation between moment, M, and deflection, y, at any point is $d^2y/dx^2 = M/(EI)$. From Fig. 13.25, $M = -F \cdot y$ so that

$$\frac{d^2y}{dx^2} = -\frac{F_c y}{EI} \tag{13.27a}$$

at the critical load. Now putting $F_c/(EI) = k^2$, Eq. (13.27a) becomes

$$\frac{d^2y}{dx^2} + k^2 y = 0 \tag{13.27b}$$

The solution of this differential equation is

$$y = C_1 \sin kx + C_2 \cos kx \tag{13.27c}$$

Constants C_1 and C_2 are determined from the boundary conditions which, for the case in Fig. 13.25, are

$$\left.\begin{aligned} y &= 0 \text{ for } x = 0 \\ y &= 0 \text{ for } x = l \end{aligned}\right\}$$

The first condition yields $C_2 = 0$ so that (13.27c) is reduced to

$$y = -C_1 \sin kx$$

which, with the second boundary condition, becomes

$$C_1 \sin kl = 0$$

A nontrivial solution is obtained for $\sin kl = 0$, which is satisfied when $kl = n\pi$, where n is an integer.

From the definition of k as $\sqrt{F_c/(EI)}$, we then obtain

$$k^2 = \frac{n^2\pi^2}{l^2} = \frac{F_c}{EI}$$

and

$$F_c = \frac{n^2\pi^2 EI}{l^2} \tag{13.28a}$$

which is Euler's formula. The practical solution which results in the lowest F_c is clearly for $n = 1$ and $k = \pi/l$. Eq. (13.17) then becomes

$$F_c = \frac{\pi^2 EI}{l^2} \tag{13.28b}$$

For higher values of n (= 2, 3, etc.), the column will buckle into a sinusoidal curve, with the number of half waves equal to n (Fig. 13.26).

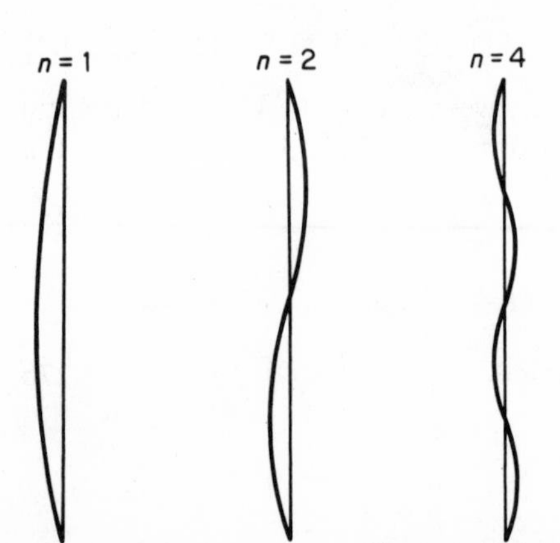

Fig. 13.26 Conceivable deflection curves of axially loaded slender columns.

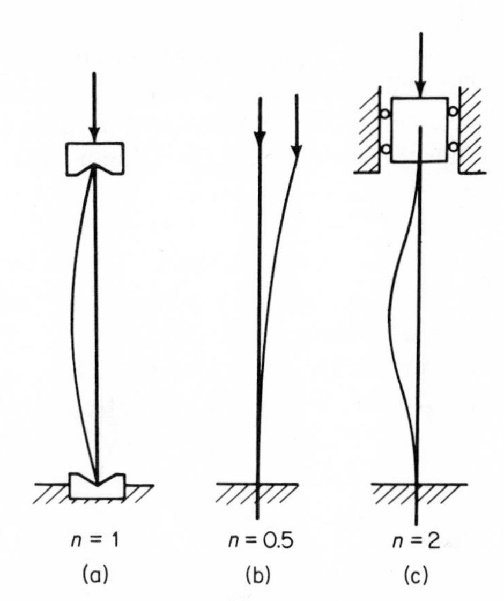

Fig. 13.27 The number of deflection half-waves (n) involved in columns with different end conditions.

It is important to observe that the value of I to be used in Eq. (13.28b) is the minimum equatorial moment of the column cross section. For example, for a U section, it will be relative to the axis parallel to the web. Hence, it is generally uneconomical to use single, highly asymmetrical structural shapes for columns. Circular or square sections are the most effective, especially when hollow (tubular), in view of their uniformly high sectional moments of inertia.

EULER'S EQUATIONS FOR OTHER COLUMN END CONDITIONS

A pin-ended column is shown schematically in Fig. 13.27a. Diagram (b) represents a column fixed at one end and free at the other; diagram (c), one fixed at both ends. Euler equations for these two cases can be written down directly from consideration of the geometrical relations between them and that in (a).

A beam, built in at one end and free at the other, has, for the same end forces, F, a deflection curve identical with one half of that in diagram (a). Hence, column length l_1 in diagram (b) is effectively twice that in (a), so that $2l_1$ must be substituted for l in Eq. (13.28b). The critical force is then:

$$F_c = \frac{\pi^2 EI}{(2l_1)^2} = \frac{\pi^2 EI}{4l_1^2} \tag{13.29}$$

which is only one quarter that of an equally long but pinned-at-both-ends column.

On the other hand, it is evident that the lateral deflection at the same F and l will be much smaller in a column that is rigidly guided at both ends, according to Fig. 13.27c, since its ends cannot rotate or be laterally displaced. The deflection line here adds up, essentially, to two half-waves [$n = 2$ in Eq. (13.28a)] each of which has the same deflection as the pin-ended column. Thus the critical force for column (c) is

$$F_c = \frac{n^2\pi^2 EI}{l_2^2} = \frac{4\pi^2 EI}{l_2^2} \tag{13.30}$$

By the same token, one could have used $n = 0.5$ to derive (13.29) from (13.28a) directly.

These three analogous equations lead to the convenient notion of an *effective length of a column* l_e. In general,

$$F_c = \frac{\pi^2 EI}{l_e^2}$$

where $l_e = l/n$

or $\quad l_e = l/\sqrt{C} \quad$ for $\quad 0.5 < n < 2$

where C is the *coefficient of fixity* or of end constraint, and it varies from 0.25 to 4 for the three cases discussed.

Instability problems are very important in the design of thin-walled shells subjected to outside pressures. Examples are submarine structures, airplane and rocket shells (which are exposed to large pressures during acceleration and slow-down periods), oil well tubing, etc. A photograph of a collapsed shell is shown in Fig. 13.28. An analysis of shells is, however, beyond the scope of this book.

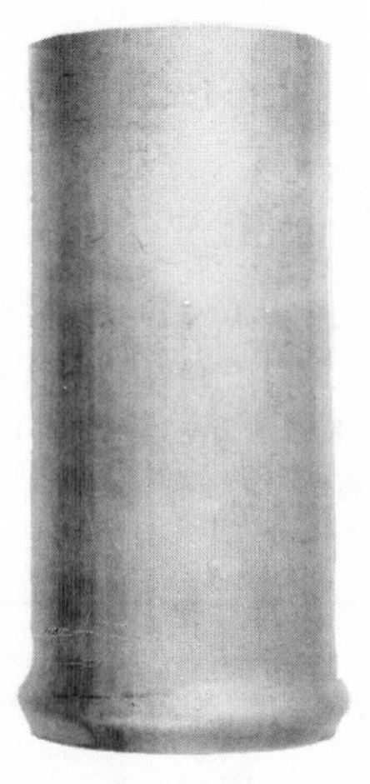

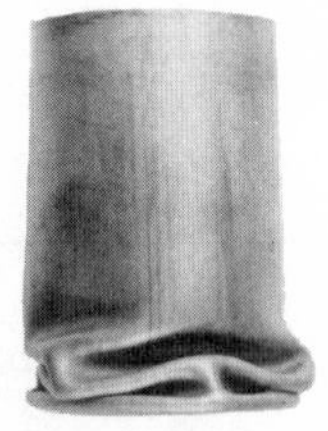

Fig. 13.28 Partial (plastic) buckling and folding of an axially compressed brass tube (two stages).

PROBLEMS

1. Determine the position of the neutral plane, maximum bending stress and shear stress distribution over the U-shaped section sketched below:

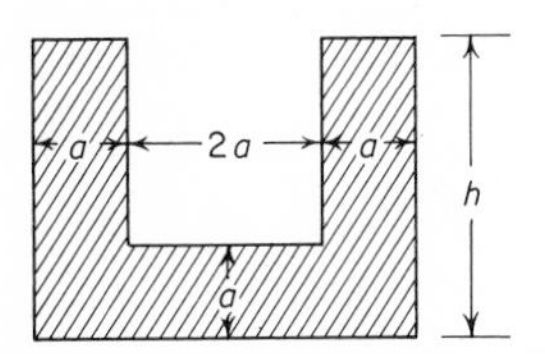

2. Calculate the height of 3 in. wide rectangular section in a 10 ft. long cantilever beam assuming that a 2,000 lbs. normal load acts at the far end and the allowable bending stress is 15,000 psi.

3. Using the result of Problem (2), calculate the length of the beam at which a bending stress of 15,000 psi is produced by the beam's own weight (no other load applied).

4. Calculate the deflection at the end of the beam in Problem 3.

5. A long tube of 3.5 in. O.D. and 2.8 in. I.D. is used as a beam lying on three equidistantly spaced supports and loaded as per sketch. (P = 6,000 lbs., a = 5 ft.) Calculate the maximum stress in the beam in the configuration shown as well as after the center support is eliminated.

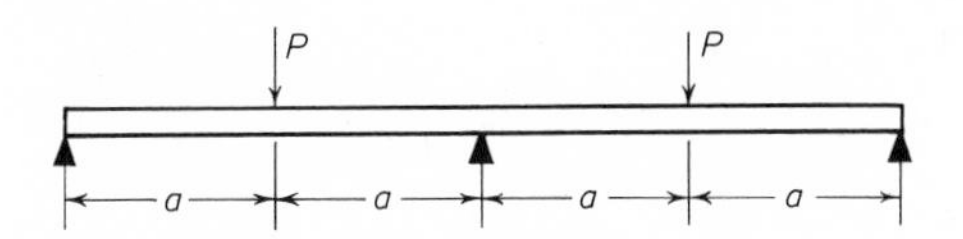

6. Calculate the bending moment in a round bronze rod ($E = 15 \times 10^6$ psi) of 0.5 in. diameter and 3 ft. long bent to a 15 ft. radius.

7. Calculate the load applied at mid-length of a simply supported beam at the onset of plastic yield. Data: square cross-section 1 in. $\times$ 1 in., 4 ft. between supports, $E = 26 \times 10^6$ psi, σ_0 = 40,000 psi.

8. The bar in Problem 7 was bent plastically over a 40 in. radius template. Calculate the approximate radius of curvature after the external force is released.

9. A steel column with a hollow square cross-section of 6 in. side and 0.5 in. thick wall carries a 40,000 lbs. axial load. Find its maximum permissible length assuming that both ends are clamped and the safety factor is 1.5.

10. Three steel tubes with O.D. = 2 in. and I.D. = 1.8 in. are arranged according to the drawing. ($l = 5$ ft., $a = 2$ ft.). Find load P at which buckling will occur.

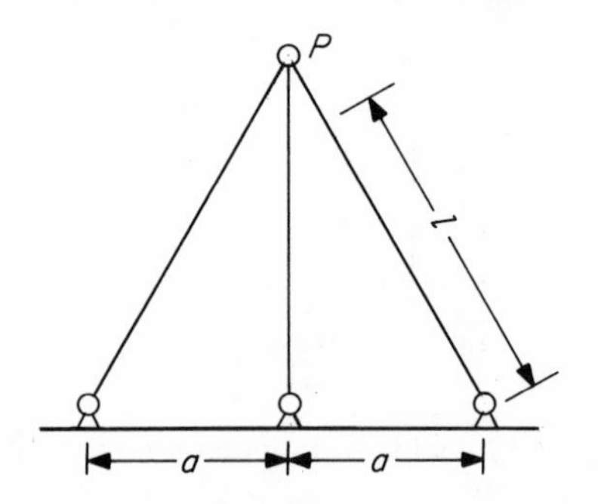

REFERENCES FOR FURTHER READING

(R13.1) Crandall, S. H. and Dahl, N. C., eds., *An Introduction to the Mechanics of Solids*. New York: McGraw-Hill Book Co., 1959.

(R13.2) Nadai, A., *Theory of Flow and Fracture of Solids*. Vol. I. New York: McGraw-Hill Book Co., 1956.

(R13.3) Panlilio, F., *Elementary Theory of Structural Strength*. New York: John Wiley & Sons, Inc., 1963.

(R13.4) Shanley, F. R., *Strength of Materials*. New York: McGraw-Hill Book Co., 1957.

14

TORSION AND SHEAR

SHEAR STRESSES IN ELASTICALLY TWISTED, CIRCULAR SECTIONS

When force couples act on planes normal to a member's axis rather than in an axial plane, as was the case with bending, *torsional moments* or *torques* arise. Typical examples are demonstrated in Fig. 14.1. The source of the torque in cases (a) and (b) is obvious; in (c), the torque in the spring coils is produced by axial load, P, acting on radius (or lever arm), R, so that $T = PR$.

Torque T, applied to the ends of an elastic cylindrical rod of radius a (Fig. 14.2a), causes it to deform (twist), and an initially straight generator, AB, becomes a steep helix, AB'. The shear in the surface layer is, therefore, $\gamma_a = \tan \phi$, where ϕ is angle BAB'. For small deformations, characteristic of metals (which are generally used in these applications), $\gamma_a \approx \phi$.

For any given length, l, of the rod, the shear strain, γ, is uniquely related to the angle of twist $BOB' = \theta'$. Since the arcuate distance, $BB' = a\theta'$ and $\gamma_a = BB'/l$, the shear strain in the peripheral layer is

$$\gamma_a = a\,\frac{\theta'}{l} \tag{14.1a}$$

or, when twist angle θ is referred to a unit of length of the cylinder,

$$\theta = \theta'/l$$

then

$$\gamma_a = a\theta \tag{14.1b}$$

$$\tau_a = G\gamma_a = G\theta a \tag{14.2}$$

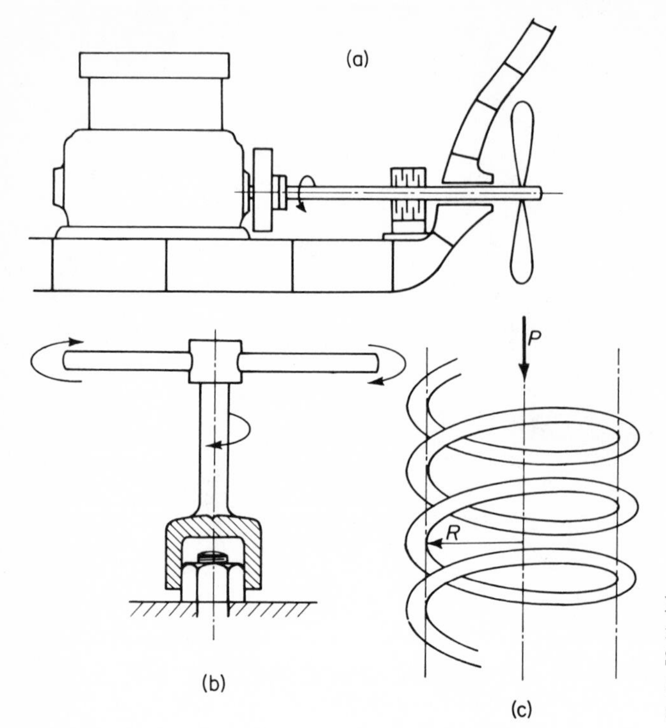

Fig. 14.1 Examples of elements loaded in torsion: (a) Ship propeller shaft. (b) Socket wrench. (c) Coil spring.

Fig. 14.2 Torsion of a cylindrical rod held rigidly at far end.

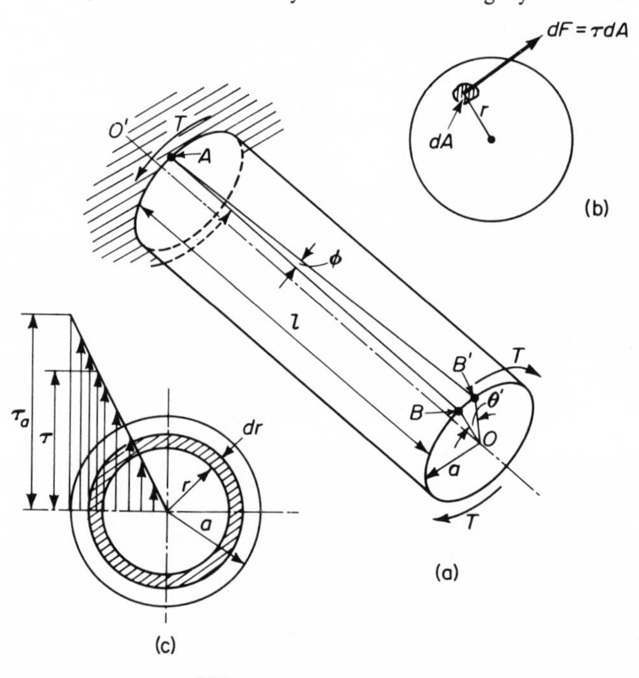

If a cylindrical surface, coaxial with the rod periphery has a radius r such that $0 < r < a$, the shear strain it undergoes, and the corresponding shear stress, will be

$$\gamma = \theta r \tag{14.3a}$$

and

$$\tau = G\theta r \tag{14.3b}$$

respectively.

It follows that the shear strains and stresses induced by elastic torsion in a cylindrical shaft vary linearly with the radial distance from the axis. If the maximum γ and τ in the peripheral layer are γ_a and τ_a, their values at intermediate points along a radius are

$$\frac{\gamma}{\gamma_a} = \frac{r}{a} \quad \text{or} \quad \gamma = \gamma_a r/a \tag{14.4a}$$

and

$$\frac{\tau}{\tau_a} = \frac{r}{a} \quad \text{or} \quad \tau = \tau_a r/a \tag{14.4b}$$

Shear stress, τ, over a surface element, dA, of a cross section represents a force, $dF = \tau\, dA$, with a vector normal to the radius along which dA is located (Fig. 14.2b).

The product, $(r)(dF) = \tau r\, dA = dT_i$, is therefore, an elementary internal torsional moment. The sum of dT_i over the entire cross section must balance the applied external moment, T. The summation is particularly simple for a cylinder since, by reference to Fig. 14.2c, the shear stress must be constant over a dr wide ring of radius r, according to (14.4b). The elemental torque then is

$$dT_i = \tau r\, dA = \tau r(2\pi r\, dr) = 2\pi\tau r^2\, dr$$

and, with (14.4b),

$$dT_i = 2\pi \frac{\tau_a}{a} r^3\, dr$$

The total integrated internal moment is, therefore,

$$T_i = \frac{2\pi\tau_a}{a}\int_o^a r^3\, dr = \tfrac{1}{2}\pi\tau_a a^3 \tag{14.5a}$$

and it is equal to the applied torque T.

For practical calculations, the bar diameter $D = 2a$ is often used, Eq. (14.5a) then becoming

$$T(= T_i) = \frac{\pi\tau_a D^3}{16} \approx 0.2\tau_a D^3 \tag{14.5b}$$

Observe that $2\pi r\, dr \cdot r^2$ is an elemental *polar moment of inertia*, dI_p, with respect to the bar axis. The total moment is

$$I_p = \int_0^a 2\pi r^3\, dr = \frac{\pi a^4}{2} = \frac{\pi D^4}{32} \approx 0.1 D^4 \tag{14.6}$$

Substituting (14.6) into (14.5) yields

$$T = \tau_a \frac{I_p}{a} = \tau_a \frac{I_p}{D/2}$$

or

$$T = \tau_a Z_p \tag{14.7}$$

where, by analogy with bending, Z_p is the *torsional section modulus.*

A change of the lower integration limit in (14.5a) from 0 to b, where $0 < b < a$, will result in a formula valid for hollow cylinders or tubes with inside diameter $2b$.

TORSION IN THE PLASTIC RANGE

When the yield stress in shear, τ_o, is exceeded in the outer layer, the elastic straight line relation between r and τ breaks down. On continued twisting, the "plasticized" annular zone rapidly penetrates toward the interior, eventually leaving only a very thin elastic core (Fig. 14.3). Just as was the case with plastic bending, the stress–strain relations along the radius are now determined by the shape of the stress–strain curve (which usually involves strain-hardening) in τ–γ coordinates.

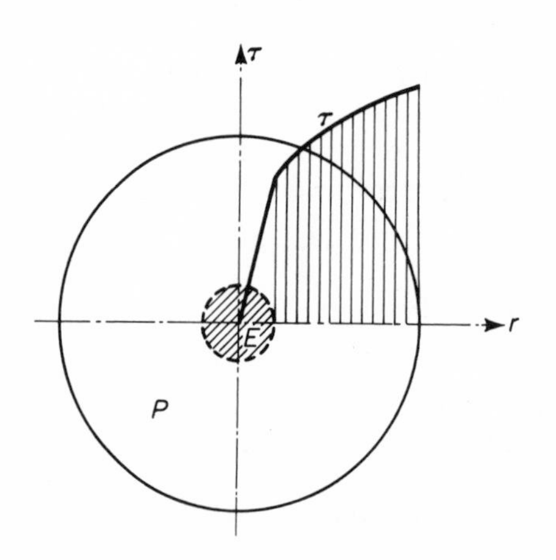

Fig. 14.3 Theoretical stress-strain relations in a plastically twisted bar. *E*–elastic core. *P*–plastic periphery.

The integration of the elementary internal torsional moments will be done in the same manner as for deriving (14.5).

$$T_i (= T) = 2\pi \int_o^a \tau r^2\, dr \tag{14.8a}$$

Inasmuch as the purely geometrical relations, (14.3a) and (14.4a), hold for the plastic as well as elastic strains, Eq. (14.3a) can be substituted into (14.8a), yielding

$$T = \frac{2\pi}{\theta^3} \int_o^{\gamma_a} \tau \gamma^2\, d\gamma \tag{14.8b}$$

where τ is given in terms of γ in the form of the τ vs. γ stress–strain curve.

This enables the twisting moment to be calculated from shear stress–strain data.

On the other hand, if the $T = f(\theta)$ curve is determined experimentally from a torsion test, the shear stress, τ, as function of γ for this material, can be derived from the curve. This is done by rewriting (14.8b) in the form,

$$T\theta^3 = 2\pi \int_0^{\gamma_a} \tau\gamma^2 \, d\gamma$$

or, since $\gamma_a = a\theta$, and expressing γ as a function of $(a\theta)$,

$$T\theta^3 = 2\pi \int_o^{a\theta} \tau a^2\theta^2 \, d(a\theta)$$

Differentiating the latter with respect to θ yields

$$\frac{d(T\theta^3)}{d\theta} = 2\pi\tau_a a^3\theta^2$$

or

$$\tau_a = \frac{1}{2\pi a^3\theta^2}\frac{d(T\theta^3)}{d\theta} \tag{14.9}$$

where τ_a is the shear stress in the outer layer. Hence, τ_a is proportional to

$$\frac{1}{\theta^2}\frac{d(T\theta^3)}{d\theta} = \theta\frac{dT}{d\theta} + 3T \tag{14.10}$$

which can be readily determined from an experimental torque, T, vs. angle of twist, θ, curve. The differentiation, $dT/d\theta$, can be done graphically.

Equations (14.8) to (14.10) are legitimately applicable to small plastic shear strains (of the order of 0.01 of 0.1) because of the use of conventional strain units. For larger deformation, the logarithmic strain scale is more appropriate (see next Section).

It should be stated in connection with Fig. 14.3 that in reality an elastic "core" can be observed only at very small plastic shear strains, γ_a. The central part of the cylinder generally work-hardens also, mainly because of the length increase of the peripheral portion observed in torsion tests. The extending periphery "drags" the central part with it, causing plastic flow in the core.

CONVERSION OF LARGE TORSIONAL STRAINS INTO EFFECTIVE STRAIN UNITS

It is sometimes necessary to compare the deformation in torsion (also termed simple shear) with linear strain induced by extension, drawing, or rolling. Since the principal normal stresses in torsion always act at 45° to the maximum shear stress, both can be readily expressed by effective stress

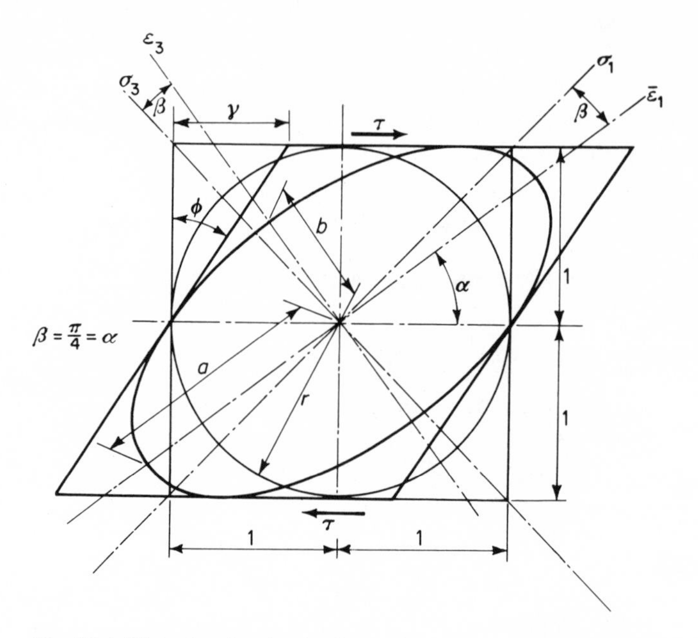

Fig. 14.4 Distortion by simple shear of a square with side length equal two units.

formulae. Using (5.17b) and (5.20a) with a single τ component, one obtains

$$\sigma_e = \tau\sqrt{3}; \qquad \tau = \sigma_e/\sqrt{3} = \sigma_1/\sqrt{3} \tag{14.11}$$

where σ_1 is the stress applied in a uniaxial tension test.

Stresses σ_1 and σ_3 act along the diagonals of a square Fig. 14.4 (compare with Fig. 2.13). Shear strain, γ, distorts this square into a parallelogram, and the maximum extension occurs along the large axis of an ellipse that was initially a circle of unit radius r inscribed in the original square.

Assuming that the strain in the third direction (normal to the page) is zero, the volume constancy, $\Delta V = O$, requires that $r^2 = 1 = (a)(b)$, a condition satisfied by the construction in Fig. 14.4.

It might seem that the effective strain should be similarly computed from (5.19) and (5.20b) for a single γ component and $\nu = 0.5$, by using natural units of normal strain. This would result in

$$\bar{\varepsilon}_e = \gamma\sqrt{3} \qquad \text{or} \qquad \gamma = \bar{\varepsilon}_e/\sqrt{3} = \bar{\varepsilon}_1/\sqrt{3} \tag{14.12}$$

However, a consideration of Fig. 14.4 will show that the directions of the maximum normal stress and strain in shear do not coincide. While this discrepancy is less than 2% at $\gamma = 0.6$ and is negligible in the elastic range of all metals, it becomes important after severe twisting, such as is encountered in torsion tests on ductile metals.

The proper equation for large strains is developed by using the following three relations which can be proved to hold.*

$$\cot 2\alpha = \tfrac{1}{2}\tan\phi = \tfrac{1}{2}\gamma, \qquad \gamma = a - b,$$

and further

$$a = \frac{1}{b} = \sqrt{1 + \frac{\gamma^2}{4}} + \frac{\gamma}{2}$$
$$b = \frac{1}{a} = \sqrt{1 + \frac{\gamma^2}{4}} - \frac{\gamma}{2} \qquad (14.13)$$

With $\bar{\varepsilon}_3 = -\bar{\varepsilon}_1 = ln\, a$

$$\bar{\varepsilon}_e = \frac{2}{\sqrt{3}}\, ln\, a \qquad (14.14)$$

A comparison of results obtained by (14.12) and (14.14), respectively, is given in the table below. The values from (14.14) must be used for large strains.

γ	0.2	0.4	0.6	1.0	2.0	3.0	5.0
$\bar{\varepsilon}_e$ (14.12)	0.116	0.231	0.347	0.578	1.16	1.73	2.89
$\bar{\varepsilon}_e$ (14.14)	0.115	0.230	0.342	0.556	1.017	1.380	1.65

TORSION TESTING

The torsion testing is much less common than the tensile test. It is employed mostly for metals expected to operate under severe torsional stresses. For instance, spring wire or torsion bars for automobiles and tractors, as well as certain shafting may be tested in this fashion, primarily to determine the exact value of the yield stress in torsion. Occasionally, hard tool steels, with a relatively low brittle strength, which fracture in tension practically without elongation, are tested in torsion. Advantage is taken of the fact that in torsion, rather high shear stresses can be applied before the normal fracture strength is reached. It will be recalled that in torsion, $\tau_{12} = \tau_{max} = \sigma_1$ (see p. 39), while in tension, $\tau_{max} = \sigma_1/2$. Since it is the shear stress that produces plastic yield, twice as high a shear stress can be applied by torsion than by tension for any specified maximum σ in the system. Accordingly, the shear yield stress, τ_o, may be reached in torsion before σ increases to its (brittle) fracture value. In this manner, it is possible to determine the true

* E. Siebel, ed., *Handbuch der Werkstoffprüfung*, Vol. 2. (Berlin: Springer-Verlag, p. 715 (1955).

yield stress of brittle materials despite the fact that their tensile ductility is nearly nil.

Ductile metals can withstand a considerable amount of twisting prior to fracture, which eventually occurs on a plane normal to the specimen axis. Although the true fracture stress may be calculated from (14.9) it is quite common instead to specify a so-called *modulus of rupture*, τ_u, obtained from (14.7) as

$$\tau_u = T_f/Z_p \tag{14.15}$$

where T_f is the torque at fracture. The rupture modulus is a number useful for comparison of various materials but has no physical significance. It is based on an assumed linear stress increase, from zero at the axis to a maximum at the surface, which is patently untrue. Because of this assumed linearity, τ_u is usually much higher than the actual $\tau_{\max}$ at fracture.

PLASTIC INHOMOGENEITY IN TORSION

Ductile metal rods in the soft condition twist uniformly along their entire length, and, except for the areas immediately adjoining the shoulders, the shear strain is constant (Fig. 14.5). One familiar exception to this rule is mild steel which produces a discontinuous yield point not only in tension (see Figs. 7.34, and 7.35) but also during twisting.

A second and much more pronounced strain inhomogeneity is encountered in twisting of initially cold worked metals. These materials display

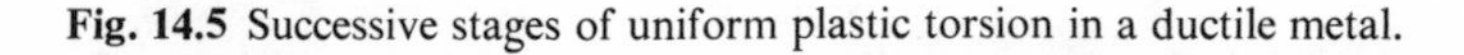

Fig. 14.5 Successive stages of uniform plastic torsion in a ductile metal.

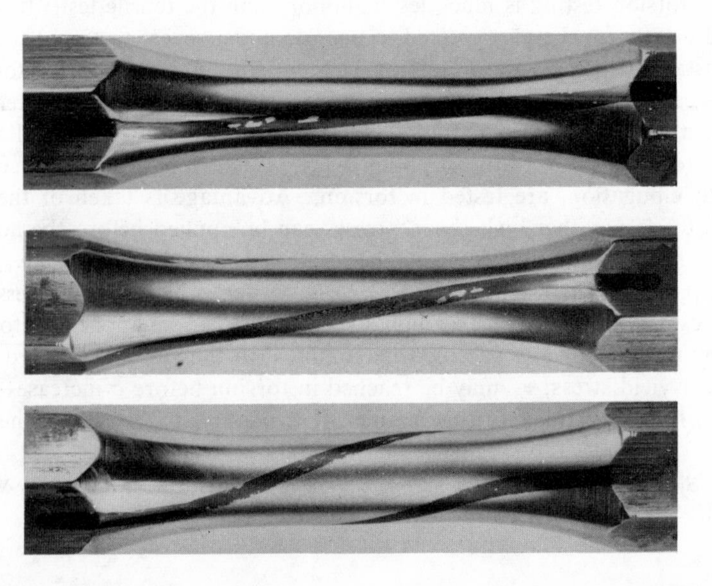

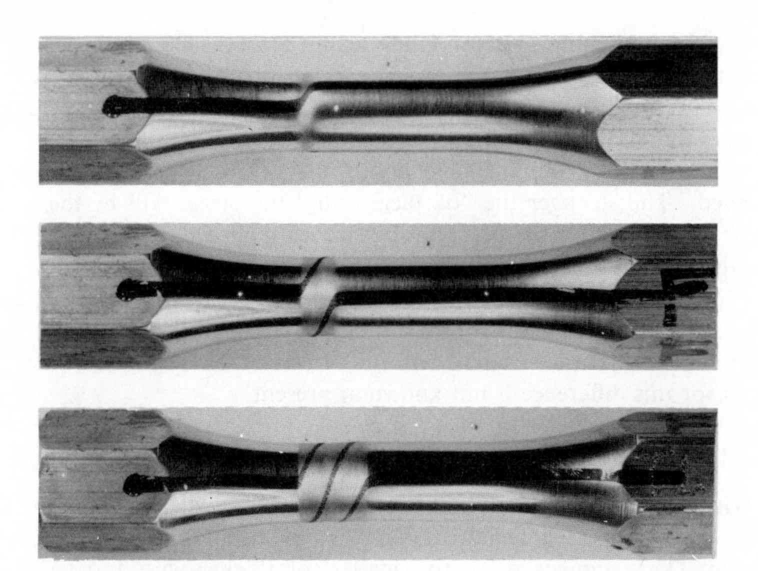

Fig. 14.6 Inhomogeneous plastic torsion: spreading of a localized shear band in heavily cold worked low carbon steel.

uniform deformation when first twisted after extension or drawing. However, this uniform strain is small and, as twisting is continued, it is replaced by more or less severe shearing strain confined to a narrow annular zone which progressively spreads over the length of the specimen (Fig. 14.6). When this stage is completed, and the entire length has thus been consumed, further twisting is uniform and it remains so until fracture.

The larger the initial extension, the more severe is the strain, γ_b, in the shear band during subsequent torsion. In fact, γ_b may be so large as to border on the fracture strain, and fracture may occur within the localized band even before the latter traverses the entire specimen (Fig. 14.7).

Fig. 14.7 Shear fracture within localized area of band. From N. H. Polakowski and S. Mostovoy, *Trans. ASM*, **54**, 567 (1961).

The probable cause for this spectacular effect is as follows: When the metal is first cold worked by, say, drawing, an extensive defect structure consisting of an entanglement of dislocations arises in it. This structure must have a definite orientation, which is a function of the type and amount of the pretorsional strain.

When a torsional stress is now applied, other dislocations begin to move on different slip planes than those that were active initially. However, they soon run into "walls," built up of the originally active dislocations, which are held together by a complex system of interacting forces. The applied torsional stress must be raised to a substantial level to force the newly activated dislocations through these built in obstacles. As soon as such breakthrough is effected, the stress required for further propagation is reduced (just as in the case of a discontinuous yield point), and a large local shear strain is observed. The stronger the "obstacle wall," the larger will be the surplus torque available after the breakthrough is accomplished, the farther will slip propagate, and the steeper will be the shear band.

The feature just described is conspicuously absent, even after extensive cold working, from two-phase metals, such as pearlitic steels, which consist of a uniform mixture of ferrite (iron) and cementite (Fe_3C) lamellae. The reason for this difference is not known at present.

TWISTING TESTS OF WIRE

Wires are sometimes subjected to twisting tests, wherein the number of twists that a specified length of wire can withstand prior to fracture is counted. This number is clearly smaller for thicker wire and vice versa. The appearance of localized twist zones in these tests on heavily drawn wire were formerly regarded as a manifestation of an internal weakness or flaw* although their presence is quite natural in the light of the foregoing discussion.

One repercussion of these facts is worthy of mention. In view of the tendency of some cold drawn wire to form these local "knots" of severe twist, one will observe a drastic fall of the number of twists to fracture

* E. Siebel, ed., *loc. sit.*, pp. 124–25.

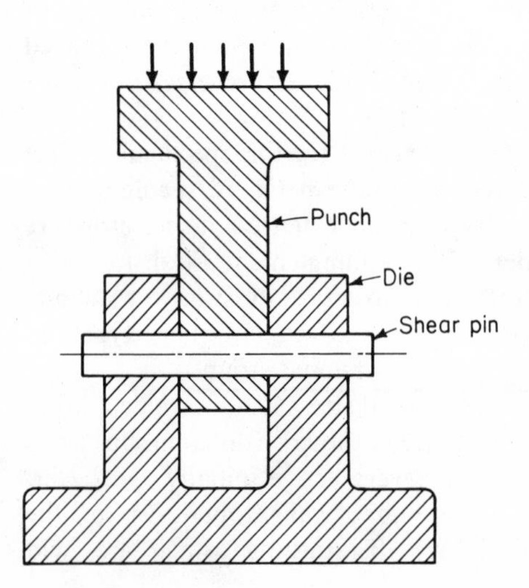

Fig. 14.8 Setup for double-shear test on a pin.

in heavily cold drawn copper, brass, mild steel and similar, essentially single-phased, metals. However, since the already mentioned pearlitic steels, often with as much as 0.8 to 0.9 per cent carbon, are free from this effect, they will twist uniformly even after 80–90 per cent of reduction by cold drawing. Wire of this sort, usually subjected to a so-called "patenting" heat treatment, is used for making steel cables for bridges, cranes, and elevators and, despite its high strength and hardness, it may exhibit more twists before fracture than would a cold drawn and soft (by comparison) low carbon fence wire.

The effect is, of course, spurious since the local shear strain in the low carbon steel wire is much greater than in the hard cable wire. However, the twisting is uniform in the latter case and it takes place all along the wire instead of being concentrated at one location.

DIRECT APPLICATIONS OF SHEAR

Elements such as rivets, various pins, or short axles represent beams where the length between supports is quite small compared with the lateral dimensions (e.g. diameter). A typical double shear test on a pin is shown schematically in Fig. 14.8. Because in tests of this type, the bending moment and longitudinal stresses are sufficiently small to be neglected, the major working stress originates from the transverse force, V (Chapter 13). In such cases the average shear stress over the section is calculated as

$$\tau = V/A \tag{14.16}$$

The general use of shear tests for structural materials is to determine the maximum shear stresses that can be applied without causing a catastrophic

Fig. 14.9 Microsections showing deformed zone at various penetrations when using a sharp-edged die. (a) (left) Penetration 23 per cent metal thickness; (b) (center) Penetration 56 per cent metal thickness; (c) (right) Penetration: 75 per cent metal thickness. From F. Howard, *Sheet Metal Ind.*, **37,** 342 (1960).

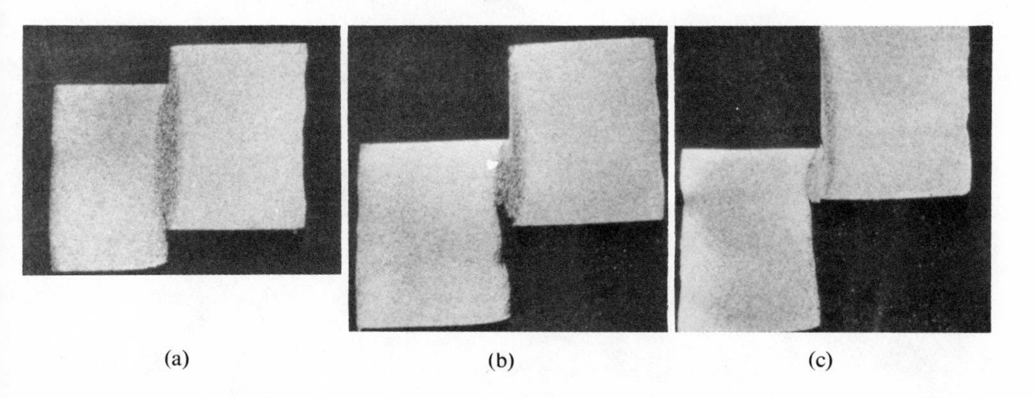

failure. In fabrication processes, such as shearing, or punching holes in sheets, on the other hand, one strives to minimize the force required to form the new fracture surfaces by suitable tool design. With sharp tool edges, the shear zone, within which separation takes place, is highly localized (Fig. 14.9), and the applied force is relatively low. Similar considerations are made in machining operations such as lathe turning (Fig. 14.10).

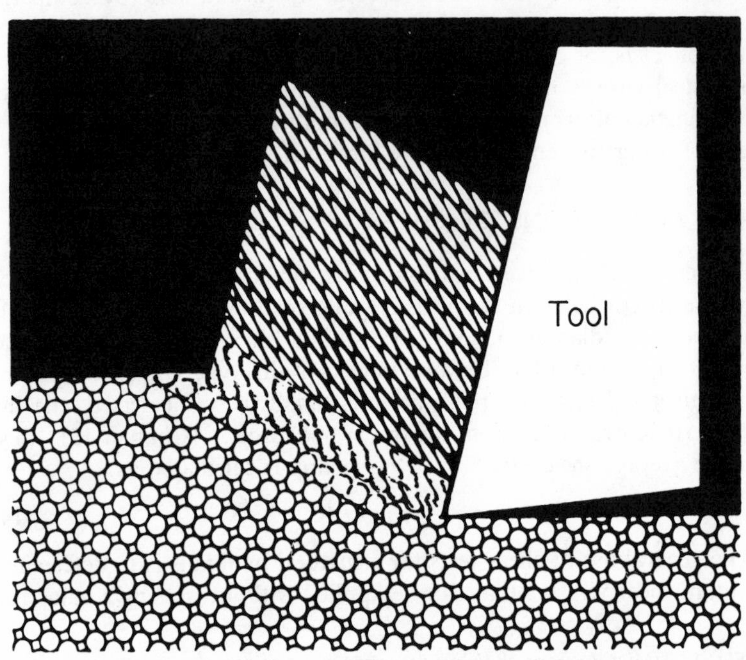

Fig. 14.10 Schematic diagram of shearing under the action of a cutting tool. From M. E. Merchant, *Jl. Appl. Phys.*, **16**, 267 (1945), by permission.

15

PROPERTIES THAT MEASURE RESISTANCE TO FRACTURE

A stressed member may fail either by breaking or by distortion to an extent that makes it useless. In the previous few chapters the properties that are relied on for evaluating deformation resistance were discussed. In this chapter the means used to measure fracture resistance on application of a continuously increasing load are described.

BRITTLE VS. "FRANGIBLE" MATERIALS

Fracturing can be preceded by a large amount of permanent deformation or by a small or even negligible amount. As long as it is certain that a structural member will distort excessively before breaking, there may be no concern with its fracturing characteristics because failure in service will be by yielding. If, on the other hand, the body breaks after a slight amount of deformation, its fracture behavior becomes all-important, and, as a consequence, properties that measure fracture resistance are primarily concerned with brittle fracture. This type of failure occurs in either of two classes of materials. There are those, such as marble or cast iron, that are brittle under almost any service or testing condition. The use of these materials is restricted to applications where the service tensile stresses, if any, are relatively small; e.g., marble is almost always loaded in compression. The fracturing characteristics of these were discussed in Chapter 11. Safe service stresses are given as some fraction of the laboratory measured stress for breaking, with the fraction being determined by the required safety factor.

Other structural materials, of which steel is the most important, perform in a far more complex fashion. These may be ductile under some conditions and brittle under others, making their behavior in service difficult to predict. They may be referred to as "frangible." Because of their dual nature, which was considered in developing Eq. (9.19) and their great practical importance, these are the materials on which most brittle fracture studies are conducted.

The property of particular interest in these frangible materials is toughness. This term had been defined in Chapter 9 as the amount of energy that is irreversibly absorbed in the process of fracturing.

Some values for toughness are easily obtained. If a tensile bar of, say, mild steel were stretched, its toughness would be found by integrating the area under its true stress–strain curve. Indeed, this particular integration was shown in Eq. (9.7). Toughness is expressed in units of energy. If the stress–strain curve were plotted in pounds per square inch and dimensionless strain, the units of toughness would be pounds per square inch. This, like the case for resilience, is a reduction of inch-pounds per cubic inch.

Unfortunately, toughness measured by such a simple procedure, especially at room temperature, does not correlate with fracture resistance in service. Frangible materials in uniaxial tension break in shear (i.e., show a fibrous surface) while brittle fractures that occur in service are found to have granular surfaces, indicating a cleavage mode of failure. Apparently, this lack of correlation between the laboratory tests and field behaviors is due to different mechanisms of failure. Much of the fracture or toughness testing, especially on steels, is aimed at measuring the relative ease with which cleavage can be produced. By comparing the "cleavage tendency" of one steel with that of another on which service experience is available, it is possible to rate the new material for a particular application.

A second shortcoming of measuring toughness by using uniaxial tensile specimens is that the two energy components (that required to initiate a crack and that needed to make it grow) are not separated, and, usually, the largest portion of the total measured energy is that required for crack initiation. Unlike ideally prepared laboratory specimens, structural elements in service contain cracks and other flaws of a wide variety of sizes. Breaking occurs in these when sufficient energy is supplied to propagate an already existing crack. Since the cracks are present even before the load is applied, all the supplied energy that is stored in the region that contains the cross-section where fracture is to occur is available for crack propagation. Again, much of the toughness testing that has been done consists of introducing an artificial crack into the test specimen and then seeing if it propagates in a brittle (cleavage) fashion to form a *transverse tensile fracture*, absorbing little energy, or in a more ductile (fibrous) manner to form an *oblique shear fracture*. Not only do notches lower the energy needed for fracture but they also cause high triaxial tensile stress concentrations and, as a result, raise the transition temperature (as in Fig. 10.10).

CHARPY AND IZOD TESTING

Notches of almost every conceivable shape have been cut into tension or bending specimens to simulate either abrupt section size changes (e.g., at the hatch corners of ships) or accidentally formed cracks. One of the most common of these notched test bars has a square cross section and is known as a *Charpy specimen* (Fig. 15.1). Of the two types of notches in use, the "V" and keyhole (Figs. 15.1a and b), the former is the more generally used since it is the more severe. These specimens are broken as simple beams in impact. The Charpy test is one of the most deeply rooted toughness tests and is included in the specifications for many products. It is performed on an impact machine equipped with a heavy pendulum hammer pivoted about a horizontal axis, which is raised to a fixed height prior to the test. As shown schematically in Fig. 15.1d, the pendulum is initially raised to height h_1, at

Fig. 15.1 (a) "V" notch Charpy bar. (b) Keyhole notched Charpy bar. (c) Izod bar. (d) Schematic diagram of impact machine.

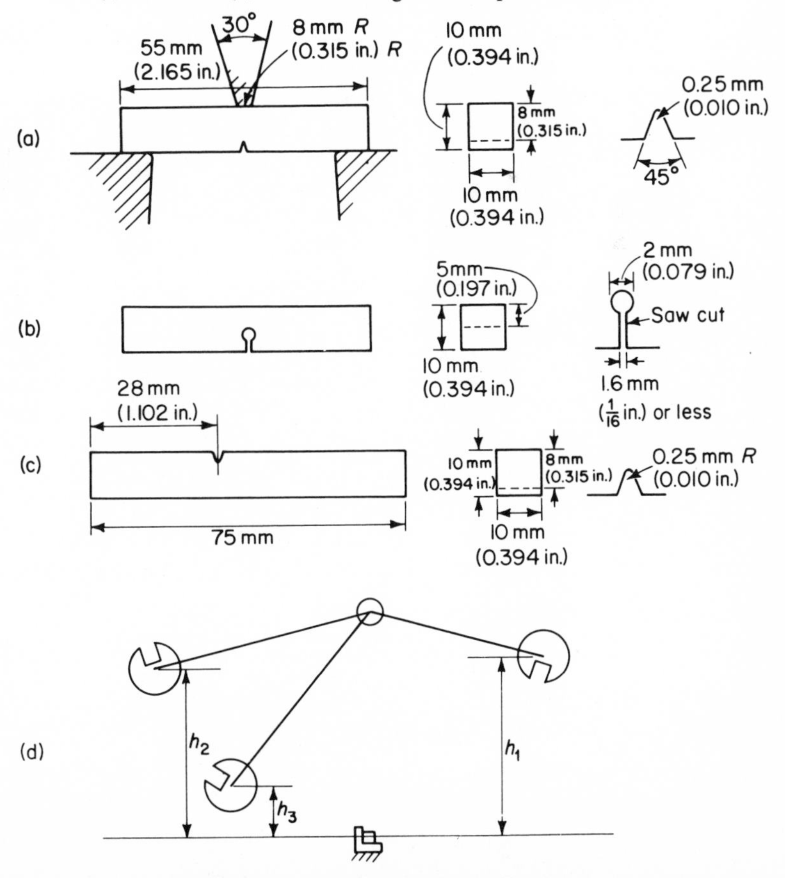

which position it has a potential energy U_1. When the pendulum is released, it swings through an arc to height h_2. The difference between h_1 and h_2 represents the wind losses, friction, etc. This new height, h_2, is taken as the zero position on the test equipment. In running impact tests, a specimen is placed in the path of the pendulum, and the energy required to break the sample is measured as the energy difference between h_2 and the height h_3 to which the pendulum rises after fracturing the sample. This absorbed energy, expressed in foot-pounds, is a measure of the material's toughness. The hammer is pointed at its lead end and is made to strike the Charpy bar behind the notch. Since the specimen is supported on its two ends, it breaks as a simple beam with the notch on the tension side.

Another notch bending bar, used more commonly in Great Britain than in the USA, is the *Izod* (Fig. 15.1c). In this test the specimen is clamped vertically just below the notch line so that it is tested as a cantilever beam. The testing machine may be identical with the Charpy machine except for a different type of specimen holder and striking edge on the hammer.

Impact tests yield their most significant data when applied to materials that exhibit a transition temperature (see p. 280). To locate this change from tough and ductile to brittle, tests must be run over a range of temperatures.

There are three methods for evaluating the degree of brittleness (or its absence) at each testing temperature: from the energy required to produce fracture; from the fracture appearance, i.e., the relative amounts of shear and cleavage; and from the ductility measured as lateral contraction at the notch bottom. Transition temperature values based on these are referred to as *energy transition*, *fracture appearance*, or *fracture transition* and *ductility transition*, respectively. Each of these methods was used to define the transition temperatures for keyhole and "V" notched Charpy bars in Fig. 15.2.

The absorbed energy and the per cent fibrous fracture are shown in Fig. 15.3 along with photographs of the fractured surfaces for a heat-treated steel tested at temperatures between −40 and −196°C. Notice in these photographs that even in the specimens tested at the lowest temperature, at which the fracture is almost completely brittle, there is some shear surface on the two sides of the fractured piece. These are known as *shear lips* and are found in many otherwise brittle fractures. It will be shown below in discussing fracture mechanics that, as the leading edge of a brittle fracture approaches an outside surface, a change to a ductile failure is expected.

Not only are there several methods for measuring the embrittling effect of temperature, but, in general, the transition based on a single type of measurement is not sufficiently abrupt to clearly define a specific temperature. For this reason a number of arbitrary means have been proposed to assign a definite value to the transition temperature, either by energy absorption or appearance. These include:

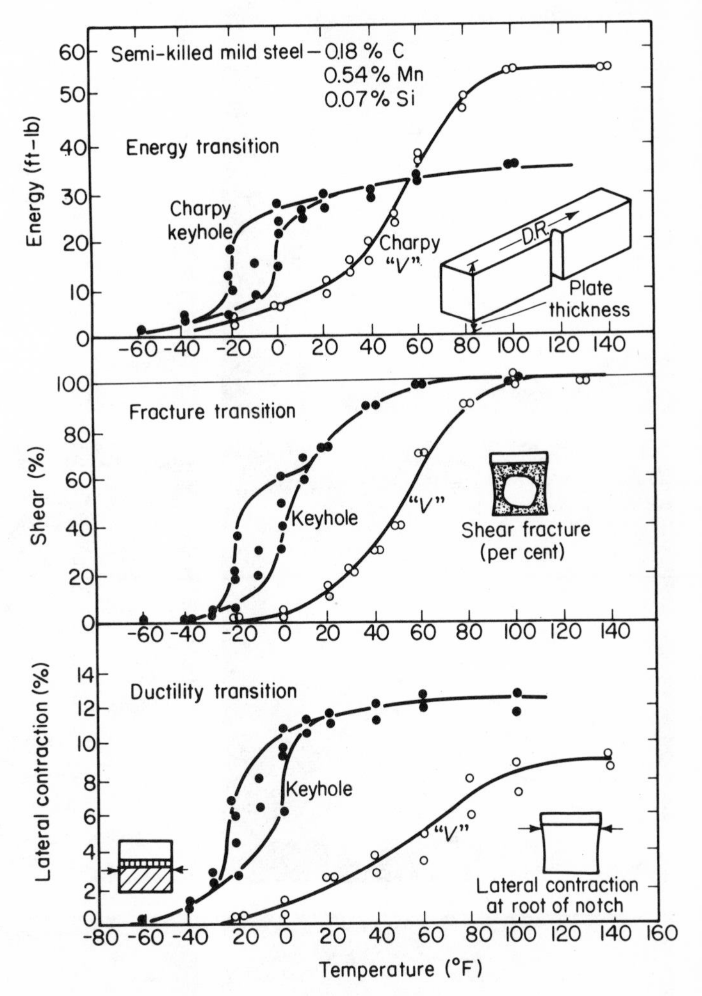

Fig. 15.2 Three methods for measuring the transition temperatures of Charpy "V" and Keyhole notched specimens. From W. S. Pellini in "*Symposium on Effect of Temperature on Brittle Behavior of Metals*," *ASTM, Spec. Tech. Pub. No. 158* (1954).

I. Energy Transition

a. The temperature at which the energy-temperature curve intersects the 15 ft.-lb. level. For example, a minimum impact energy of 15 ft.-lb. at −40°F. is frequently specified in acceptance tests.

b. The temperature at which the energy-temperature curve intersects the 40 ft.-lb. level.

c. "Average energy transition," the temperature at which the impact energy is midway between its values at high and low temperatures, i.e., upper and lower shelf heights, see Fig. 15.2.

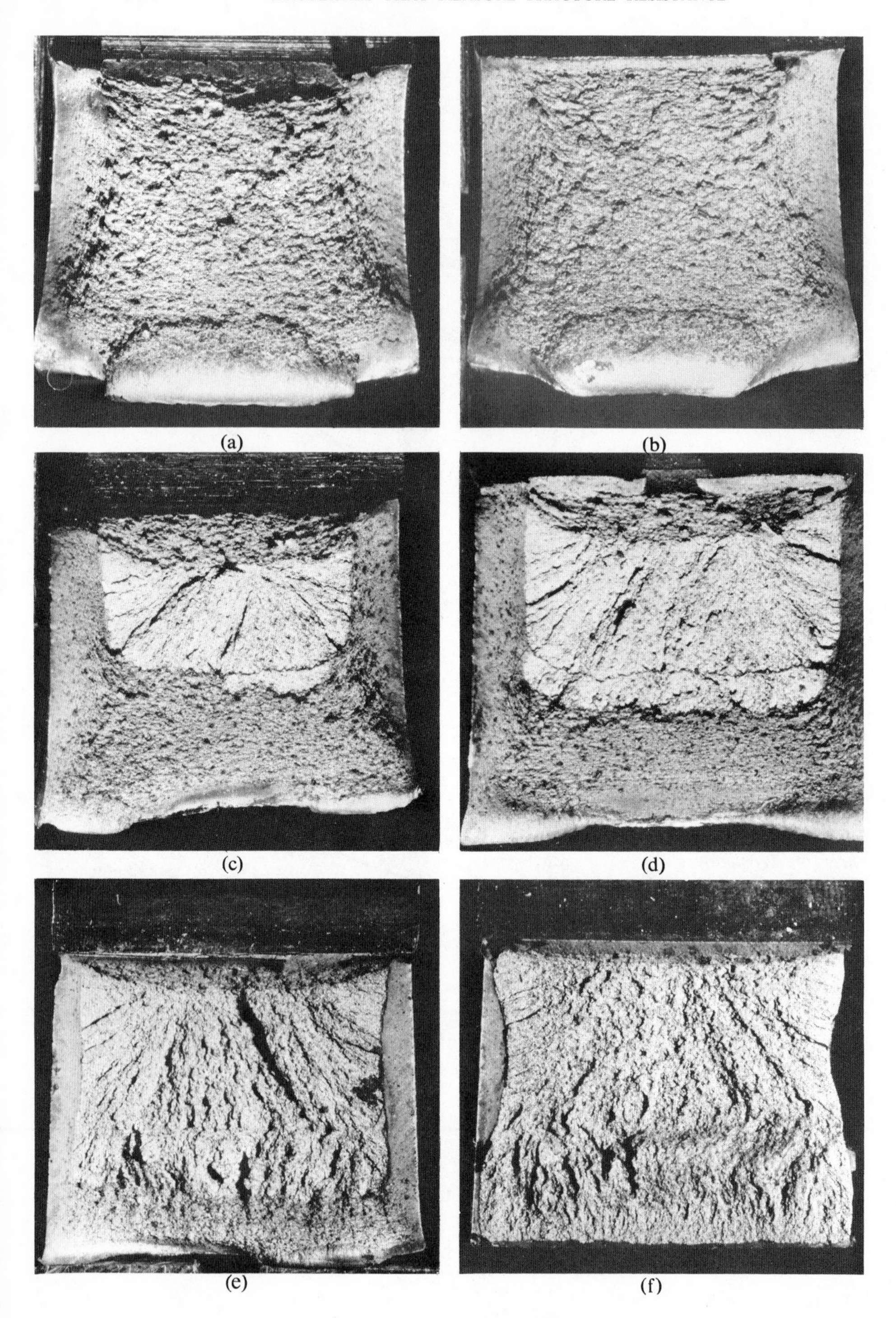
(a)
(b)
(c)
(d)
(e)
(f)

(g)

Note notch on top
in each photograph

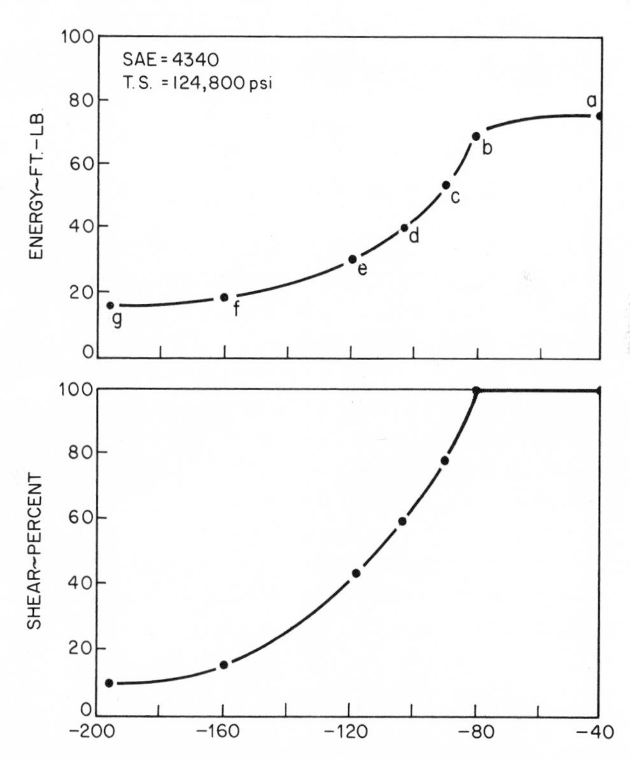

Fig. 15.3 Transition temperature and fracture appearance of a heat treated steel. Courtesy F. Larson, U.S. Army Materials Research Agency. Letters on photographs correspond with data points on energy chart.

II. Fracture Transition

a. The lowest temperature at which a 100% fibrous fracture is obtained. Notice how well defined this value is in both Figs. 15.2 and 15.3.

b. The highest temperature at which a 0% fibrous fracture is obtained.

c. The temperature at which 50% fibrous fracture is obtained.

The temperature determined by any of these procedures can be compared only with data obtained on identically dimensioned specimens since the measured values cannot be reduced to common units, such as psi used to measure uniaxial yield strength on variously shaped bars.

In certain applications, and especially in heavy military equipment, special significance is given to the highest temperature of the lower horizontal shelf of the transition curve. This particular value appears to be less dependent on the testing procedure (V-notch, keyhole) than the other toughness criteria. Since it is thought to be the temperature below which the steel cannot deform plastically in the presence of a notch, it is referred to as the *nil-ductility transition temperature* (*NDT*).

Although the Charpy impact test has received widespread acceptance, it has a number of inherent drawbacks. For one, it is greatly influenced by the test section size. This section size effect is found for both geometrically similar bars and for bars in which only one dimension is altered. As the absolute size of the specimen is increased, the impact energy generally does not increase proportionately. This is shown for the slow notch bend tests (Fig. 15.4a). If one dimension of the specimen, for example, its width, is changed, the toughness may increase or decrease. One would expect that as the specimen width is increased, the transverse restraint on the specimen would also increase, as described on p. 102. The additional restraint could cause tensile stresses and, as a consequence, a reduction of toughness. This is not always the case, however, as shown by the curves in Fig. 15.4b. The use of specific energy in this figure reduces the data to a constant width; i.e., the measured impact energy for the $2X$ specimen is divided by 2 while the $\frac{1}{2}X$ specimen would be multiplied by 2.

Another limitation of the Charpy test is that it does not separate the energy that is absorbed for initiating the crack and that required to propagate it. A technique that was sometimes used for overcoming the latter difficulty is based on the *low-blow transition temperature*. This test procedure consists of first striking the specimen in the impact machine by a light blow sufficient to initiate a crack in the specimen notch portion but not enough to allow the crack to propagate. The specimen is then tested in a normal fashion, and the measured energy is that required only for propagating the fracture.

Natural cracks are also put into Charpy bars by using fatigue loading (see Chapter 18). Both of these precracking procedures put the specimen into

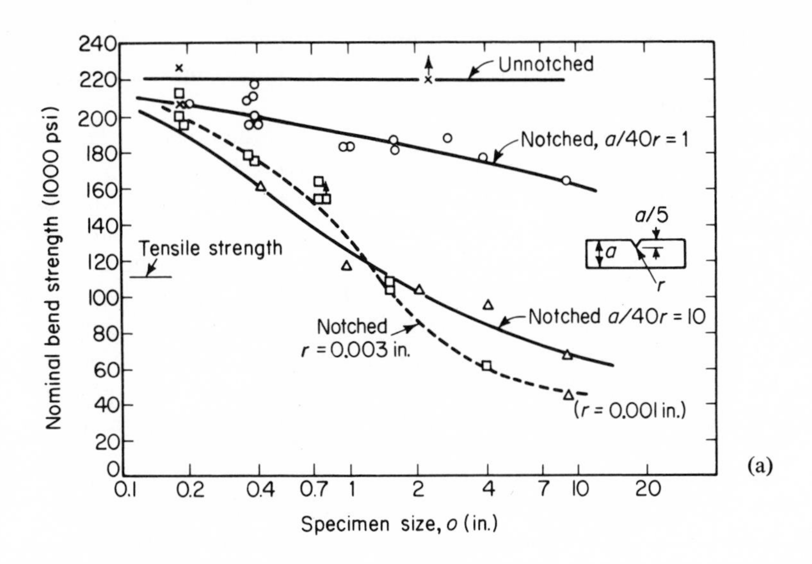

(a)

Fig. 15.4 (a) Effect of size and notch acuity on the bend strength of a Ni-Mo-V steel at room temperature. From J. D. Lubahn and S. Yukawa, *Proc. ASTM*, **58,** 661 (1958). (b) Effect of specimen width on energy absorption for five different steels. After C. E. Jackson, M. A. Pugacz, and F. S. McKenna, *Trans. AIME*, **158,** 263 (1944).

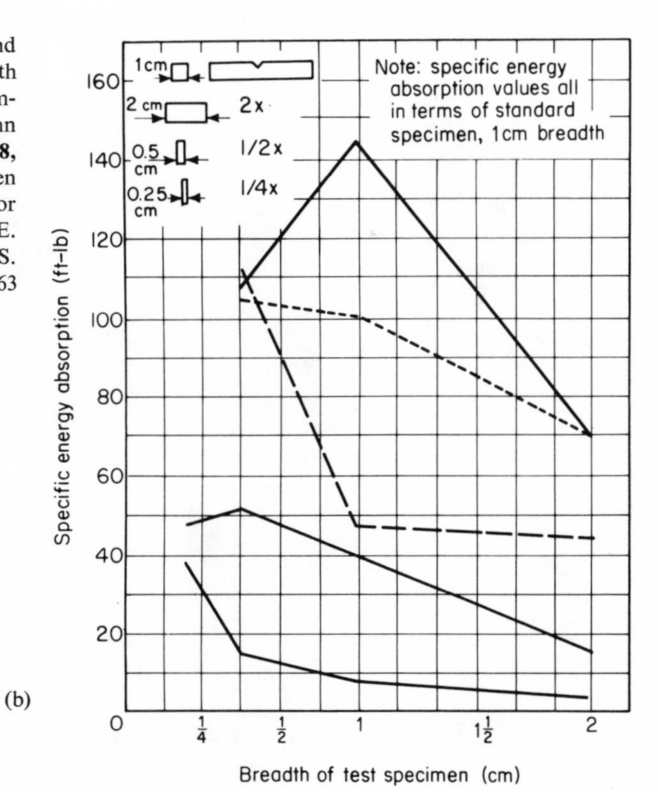

(b)

a condition that is more closely akin to members actually used in service. To analyze the data obtained on precracked samples requires some knowledge of fracture mechanics (discussed on p. 406).

The *NDT* described above has broader applications than just in Charpy impact testing. It is used as a generic term in any test to define the temperature at which a metal loses its capacity for noticeable plastic deformation in the presence of a crack. Two of the tests on which it is applied are the *explosion bulge test* and the *drop weight test*. Both of these are used for plate evaluation. In the first, a bead-on-plate brittle weld (hard surfacing type) is deposited near the center of a 14 by 14 inch plate to serve as a crack starter. The plate is then exposed to an explosive force sufficient to cause it to bulge at a "high" temperature. As the temperature is lowered, the plate cracks at a temperature again identified as the *NDT*, with no noticeable deformation (Fig. 15.5).

Fig. 15.5 Explosion crack-starter test series in 20°F. steps illustrating the dramatic increase in fracture toughness of steels above the NDT temperature. From W. S. Pellini, L. E. Steele, and J. R. Hawthorne, Naval Res. Lab. Report 5780, April 17, 1962.

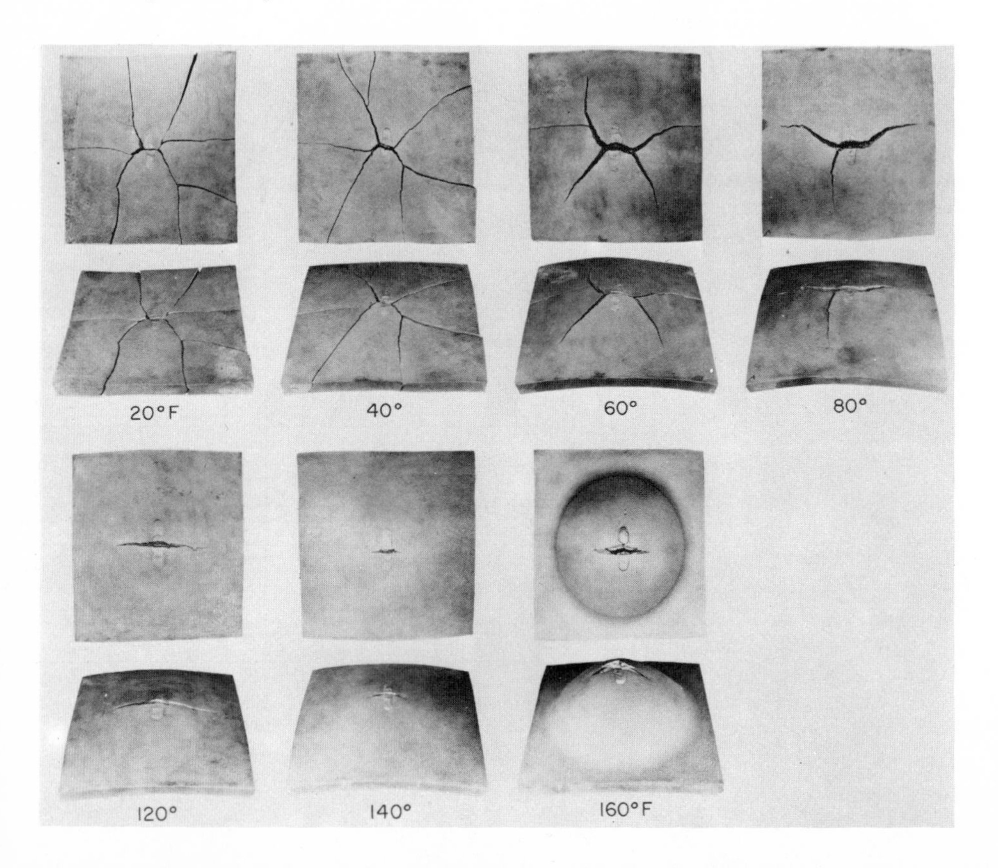

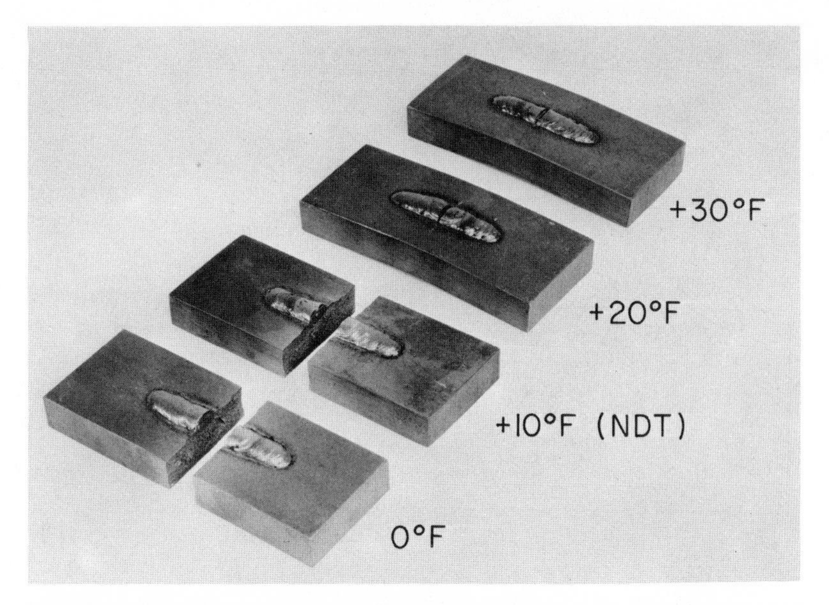

Fig. 15.6 Typical drop-weight test series, illustrating NDT at 10°F. From W. S. Pellini, L. E. Steele, and J. R. Hawthorne, Naval Res. Lab. Report 5780, April 17, 1962.

The drop weight test is similar in that a brittle weld is also used to initiate the crack. In this case a weight is dropped on the mid-span of a $3\frac{1}{2}$ by 14 inch plate with a brittle weld on its tension side. The *NDT* is the highest temperature at which fracture occurs within a five degree bend. Since it had been found that it takes three of the five degrees to crack the weld, it is seen that little, if any, plastic flow can occur in the base plate (Fig. 15.6).

STRESS ANALYSIS IN THE VICINITY OF NOTCHES

Because of its great practical importance, a considerable effort has been applied to analyzing the stresses in the vicinity of a notch or natural crack. This met with significant success so long as loading was primarily elastic. H. Neuber presented a systematic analysis of the stresses in the vicinity of a number of differently shaped notches in his book, *Kerbspannungslehre*, originally published in 1937 and, in a revised form, in 1958.* R. E. Peterson (R15.4) provided a practical guide for notch strength calculations by utilizing both Neuber's analytical procedures, as well as other techniques such as

* This is also available as a translation *Theory of Notch Stresses*, AEC-tr-4547 (1961), OTS, Dept. of Commerce, Washington, D.C.

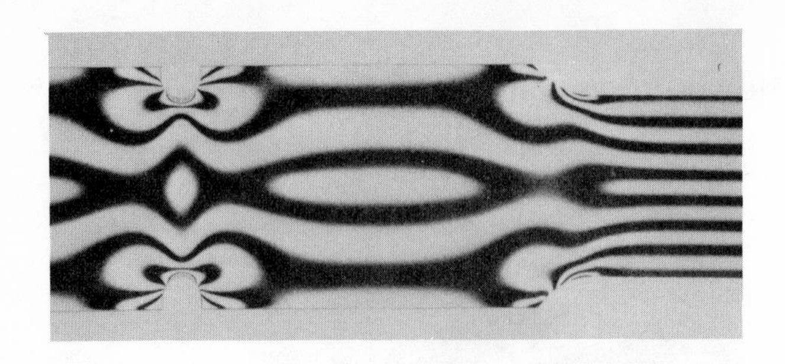

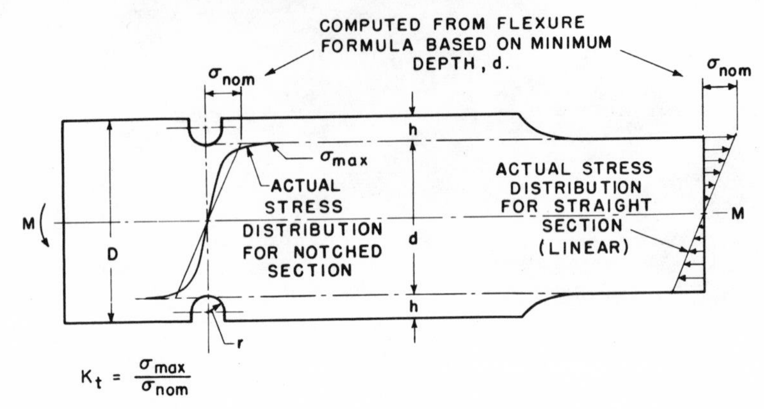

$$K_t = \frac{\sigma_{max}}{\sigma_{nom}}$$

Fig. 15.7 Stress concentrations due to an abrupt and gradual section size change. Relative density of fringe lines is a measure of stress concentration. From R. E. Peterson, (R15.4).

Fig. 15.8 Variation of stress concentration as an elliptical hole changes from one loaded parallel to its minor axis to one parallel to its major axis. From R. E. Peterson, (R15.4).

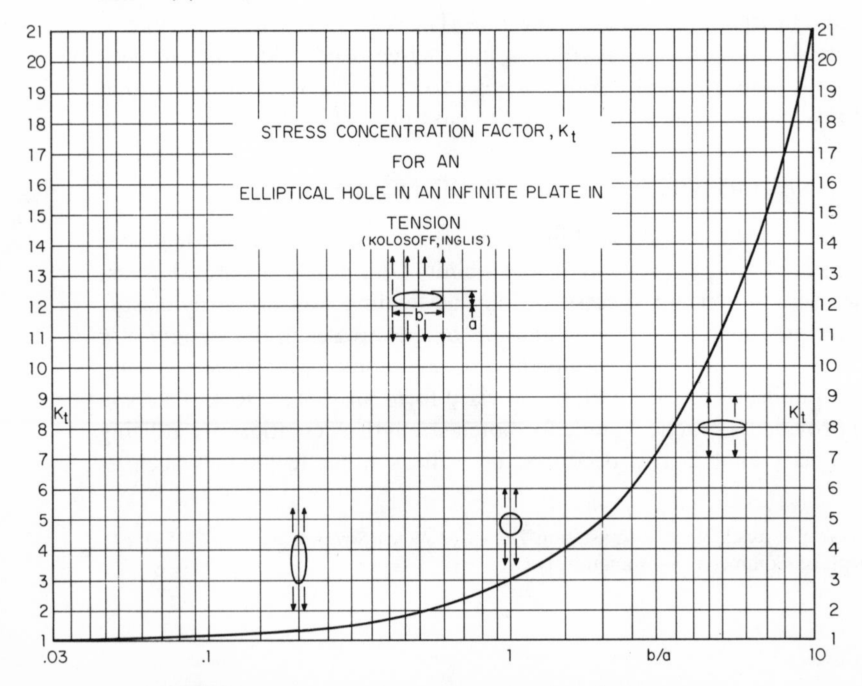

photoelasticity. Until recently this was the most useful data available and it was applied both to notches of varying sharpness and natural cracks. Later, equations describing the stresses near a crack tip have been developed. These older principles of notch analysis are presented first, followed by the more recently developed analysis of stresses in the vicinity of a "true" crack.

Notches cause high localized stresses by creating *stress concentrations*, the effects of which are defined by means of a *stress concentration factor*, K_t, where

$$K_t = \frac{\sigma_{max}}{\sigma_{nom}} \quad \text{for tension or bending}$$

$$\text{and} \qquad K_t = \frac{\tau_{max}}{\tau_{nom}} \quad \text{for torsion} \tag{15.1}$$

The values for σ_{nom} or τ_{nom} are the nominal (or mean) stresses calculated by the elementary formulas of Chapters 10 and 14. The subscript t in Eq. 15.1 indicates that K_t is always a theoretical factor. Experimental test factors, K_f, are generally lower than K_t; i.e., notches are less damaging than theoretical analysis would indicate.

The magnitude of stress concentration depends on how abruptly the section of a member varies. The stress gradients caused by a *U*-shaped notch, and those introduced by a fillet, in a bending member are compared in Fig. 15.7. This dependence of stress concentration factors on geometry has been thoroughly studied, and K_t has been calculated for many forms. For example, stress concentration that results from the introduction of an elliptical hole into an infinitely wide (in practice, very wide) plate is

$$K_t = 1 + 2a/b \tag{15.2}$$

where a and b are the major and minor axes, respectively. As the hole shape changes from an ellipse loaded parallel with its minor axis, to one loaded in the direction of its major axis (Fig. 15.8), the stress concentration factor decreases rapidly. When the hole is circular, $a/b = 1$ and $K_t = 3$.

The stresses in the vicinity of a discontinuity drop off rapidly with distance from its edge. This is shown for a round hole in Fig. 15.9.

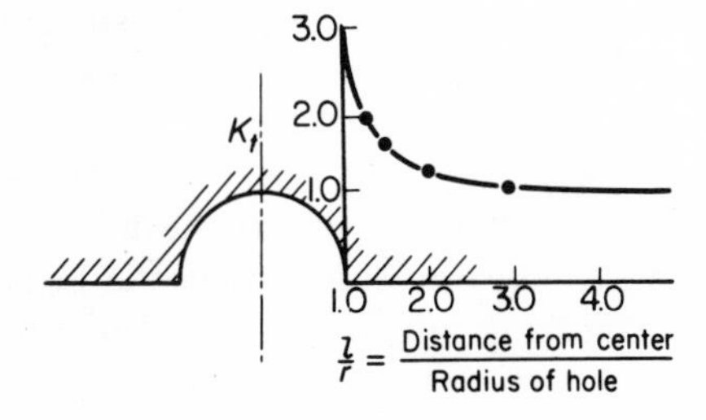

Fig. 15.9 Stress gradient at edge of a circular hole. (After E. G. Coker and N. G. Filon.) From J. M. Lessells, *Strength and Resistance of Metals*. New York: John Wiley and Sons, Inc. 1954.

Specimens with a center hole are frequently used in toughness evaluation by tensile loading because of the importance of holes in riveted or bolted structures. In other cases the severity of the discontinuity is increased by adding a fine saw cut and/or a natural crack in the direction of the expected fracture surface.

Another common tension test specimen is made with edge notches. An analysis for this shape for both round and rectangular cross-sectioned specimens, assuming the notch to be a hyperbolic groove, was given by Neuber. If the specimen is an edge notched plate, the stress concentration is

$$K_t = \sqrt{0.8d/r + 1.2} - 0.1 \tag{15.3a}$$

If the specimen is a circular cylinder with a hyperbolic notch, the stress concentration factor is

$$K_t = \sqrt{0.5d/r + 0.85} + 0.08 \tag{15.3b}$$

where $d =$ the specimen diameter or width at the notch bottom,
and $r =$ the minimum contour radius at the notch bottom.

These equations are valid only if the notches are relatively deep.

The combination of the methods of theory of elasticity and photostress techniques has lead to a reasonably complete understanding of the effects of notching so long as the stressed bar remains elastic. The stress analysis beyond the elastic range is far more difficult. Nevertheless, by the use of a more advanced procedure, these stresses can be reasonably well ascertained. (Neuber, see footnote p. 399.)

The distribution of stresses in the early portion of the plastic range is not much different from that in the elastic range. For example, Fried and Sachs showed the hardness contour lines in a fractured notch bar to be similar to the photoelastic pattern in a notched bakelite specimen (Fig. 15.10).

By starting with the known stress distribution in an elastic member, the progressive changes in stress can be inferred by taking account of the plastic flow criteria given in Chapter 6. Hence, we know that, in the elastic range, a notched tensile bar has stress peaks at the edges of the discontinuity (Fig. 15.11a), and, according to Fig. 15.9, these peaks occur close to the notch bottom. Because of this high local stress, plastic flow begins at the notch bottom (Fig. 15.11b) when this stress reaches the yield strength of the material. Since plastic flow relieves the stress, the yield strength limits the maximum stress that can be obtained.

Not only does the discontinuity produce a stress concentration, but because the material at the notch bottom is prevented from contracting, transverse (radial and tangential) stresses are also developed (see discussion of necking in Chapter 10). These transverse stresses, like the longitudinal, also vary along the specimen radius. Hence, for the condition of plasticity

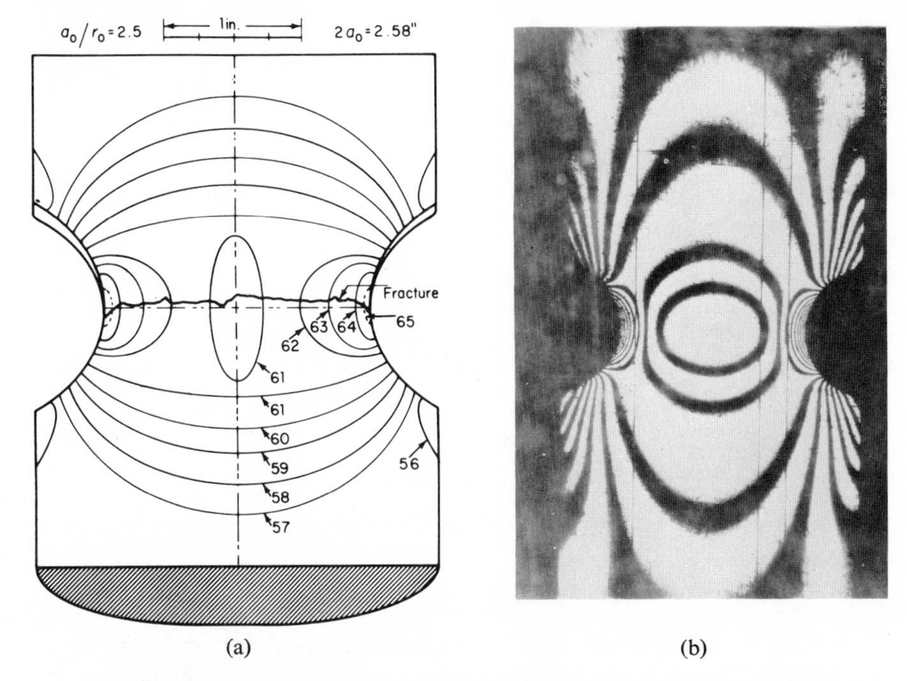

(a) (b)

Fig. 15.10 (a) Strain distribution on the longitudinal section of a fractured steel notched tension specimen revealed by hardness contour lines. (b) Photoelastic pattern in a flat notched Bakelite specimen under tension. From M. L. Fried and G. Sachs, in *ASTM Spec. Tech. Publication No. 87*, p. 85 (1949).

to be satisfied (e.g., $\sigma_o = \sigma_1 - \sigma_3$), the longitudinal stress required for plastic flow, and, consequently, the value of longitudinal stress attained, will also vary along a radius. Thus, a further rationalization of the longitudinal stress pattern necessitates an examination of the transverse stress.

The principal directions in the plane of the notch bottom are the longitudinal, radial, and tangential. Most is known about the values of these directions on the notch surface and at a point where the specimen axis intersects the notch plane. The radial stresses, of course, must be zero at the surface. Hence, even if the tangential stress is positive at this position, the plasticity condition is satisfied when the longitudinal stress, σ_1, reaches the yield stress, σ_o, since the smallest principal stress, the radial one, equals zero (Fig. 15.11b). At the specimen axis, the radial and tangential directions are indistinguishable so that the two transverse stresses must be equal, and, because of the restraint, they are both positive. Hence, as the applied load increases, the distribution of the three principal stresses varies as shown in Figs. 15.11a, b, c, and d.

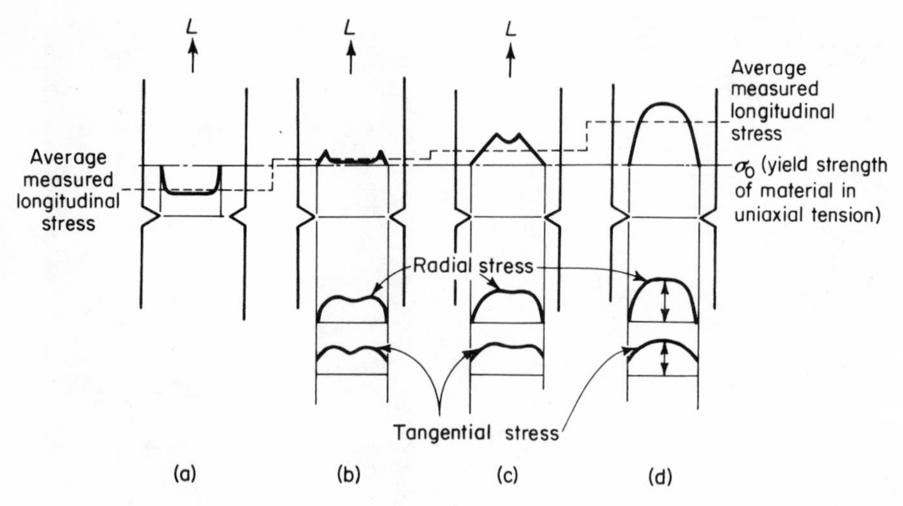

Fig. 15.11 Stress distribution in a circular notched tension bar.

USE OF NOTCHED TENSILE TEST FOR MATERIAL EVALUATION

In carrying out notch tensile tests, the material is evaluated by a quantity, *notch strength*, which is similar to the ultimate tensile strength and defined as

$$S_n = \frac{\text{maximum load}}{\text{original cross-sectional area at notch bottom}} = \frac{F_m}{A_{o(n)}} \quad (15.4a)$$

For cylindrical bars, a second quantity, *notch ductility*, much like the contraction in area, is also used,

$$q_n = \frac{A_{o(n)} - A_{f(n)}}{A_{o(n)}} = 1 - \frac{A_{f(n)}}{A_{o(n)}} \quad (15.4b)$$

where $A_{o(n)}$ = cross-sectional area at notch bottom before deformation;
$A_{f(n)}$ = cross-sectional area at notch bottom after fracture.

If a notch tension bar is made of a metal sufficiently tough not to fail at the notch bottom because of the initial stress concentration, its entire stress–strain curve is raised over the value it would have in an unnotched, smooth specimen (Fig. 15.12). This, of course, results from the plastic constraint at the notch bottom, so that the amount of the stress–strain curve elevation depends on the notch shape. G. Sachs and his associates have obtained many data on cylindrical specimens, using a 60 degree-sharp bottomed V notch (radius < 0.001 in.) that removes 50 per cent of the metal's cross-sectional area. This configuration with tough metals resulted in a notch strength of 1.5 times the tensile strength.

As an example of how notch strength is used in material evaluation, the method for establishing the maximum safe strength for steels will be

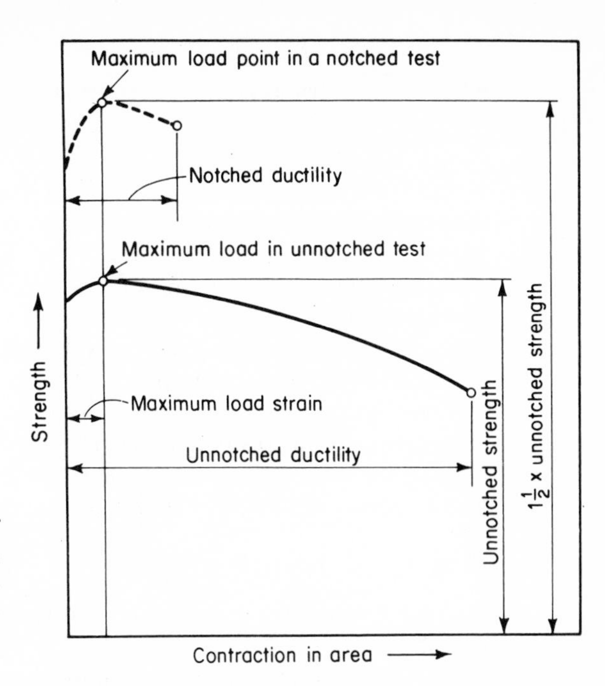

Fig. 15.12 Schematic diagram of notched and unnotched stress–strain curve of the same material. From E. J. Ripling, *Materials and Methods*, **34,** No. 3, 81, (1951).

considered. Steels can have their tensile strength increased by heat treating. For these, therefore, the notch strength S_n, can be plotted as a function of tensile strength, S_u (see upper right hand curve of Fig. 15.13). The dependence of the shape of this curve on notched and unnotched stress–strain curves

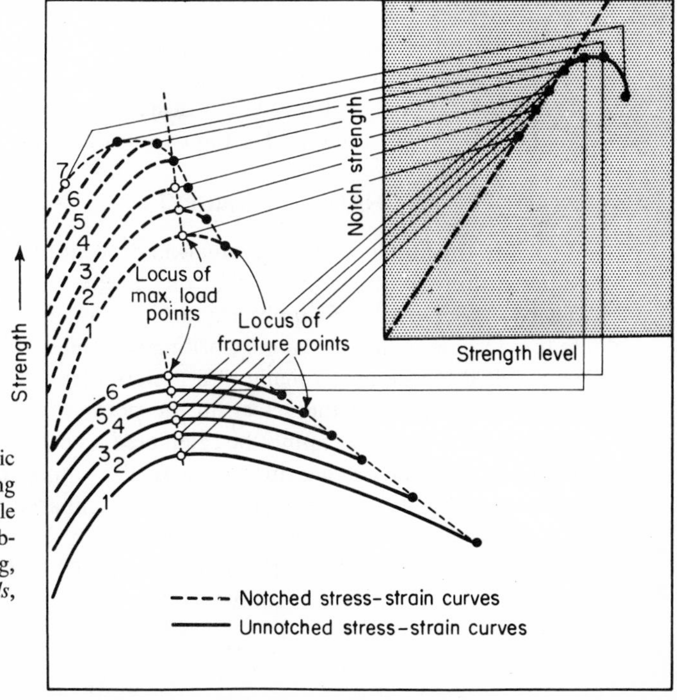

Fig. 15.13 Pairs of schematic stress–strain curves showing how notch strength-tensile strength curves are established. From E. J. Ripling, in *Materials and Methods*, **34,** No. 3, 81, (1951).

is also shown in this Figure. For the notch shape discussed above, the S_n vs. S_u curve rises at a slope of 1.5, so long as the strength levels are in the range over which the steel remains tough. Additional hardening of the metal beyond this range causes a reduction in S_n, and Sachs has used a *notch strength ratio*, $NSR = S_n/S_u$, of unity to separate notch insensitive from notch sensitive metals. If, instead of S_n/S_u, the notch ratio is defined as S_n/σ_o, and a value of one is still used to separate safe and unsafe behaviors, a less conservative requirement of notch properties results. A ratio of unity was selected because structures are designed on the basis of their tensile (or, more properly, yield) strength. Hence, if the machined sharp notch is equivalent to the most serious service flaw possible, and the service loads do not lead to plastic failure, they should also not cause failures due to the notch.

EFFECTS OF NOTCH CONTOUR

The three notch configuration variables, radius, depth, and angle, obviously affect notch tensile strength. The influence of each of these is shown separately (Fig. 15.14), and notch radius is seen to be the most important. The notch strength decreases with decreasing radii at a rate that depends on the material. Since the notch strength is continuously decreasing, even for radii less than one mil for the two steels, the minimum radius necessary to represent a crack ($\rho \approx 0$) is not established.

The notch depth (Fig. 15.14b) is shown in terms of the ratio d/D for theoretical reasons. According to Neuber's analysis, the stress concentration is a maximum for $d/D = 0.707$. This is found to be the case for the steel in the most brittle condition, which, of course, approaches an elastic failure.

The notch angle exerts the least influence on notch strength (Fig. 15.14c). The notch strength is expected to decrease as the angle decreases, but this appears to be the case only for the two lower stress levels.

FRACTURE MECHANICS CONCEPTS

All of the notch property evaluation procedures discussed up to this point suffer from the same failing in that they are not directly applicable to design. That is, none of these supply a quantity similar to, say, the yield strength, that a designer could use in his working formulae. If, for example, a Charpy or notched tensile test is made of a material from which a pressure vessel is built, the strength of the vessel, in the presence of a flaw that the notch is supposed to duplicate, cannot be calculated. The use of Charpy bars requires that one have experience in the particular application so that the energy absorption ability of one material can be compared with that of another on which service experience is available. In the use of notch tension bars, the separation of safe and unsafe strengths by $NSR = 1$ is not completely satisfactory because it does not allow for optimum high strength design. In many

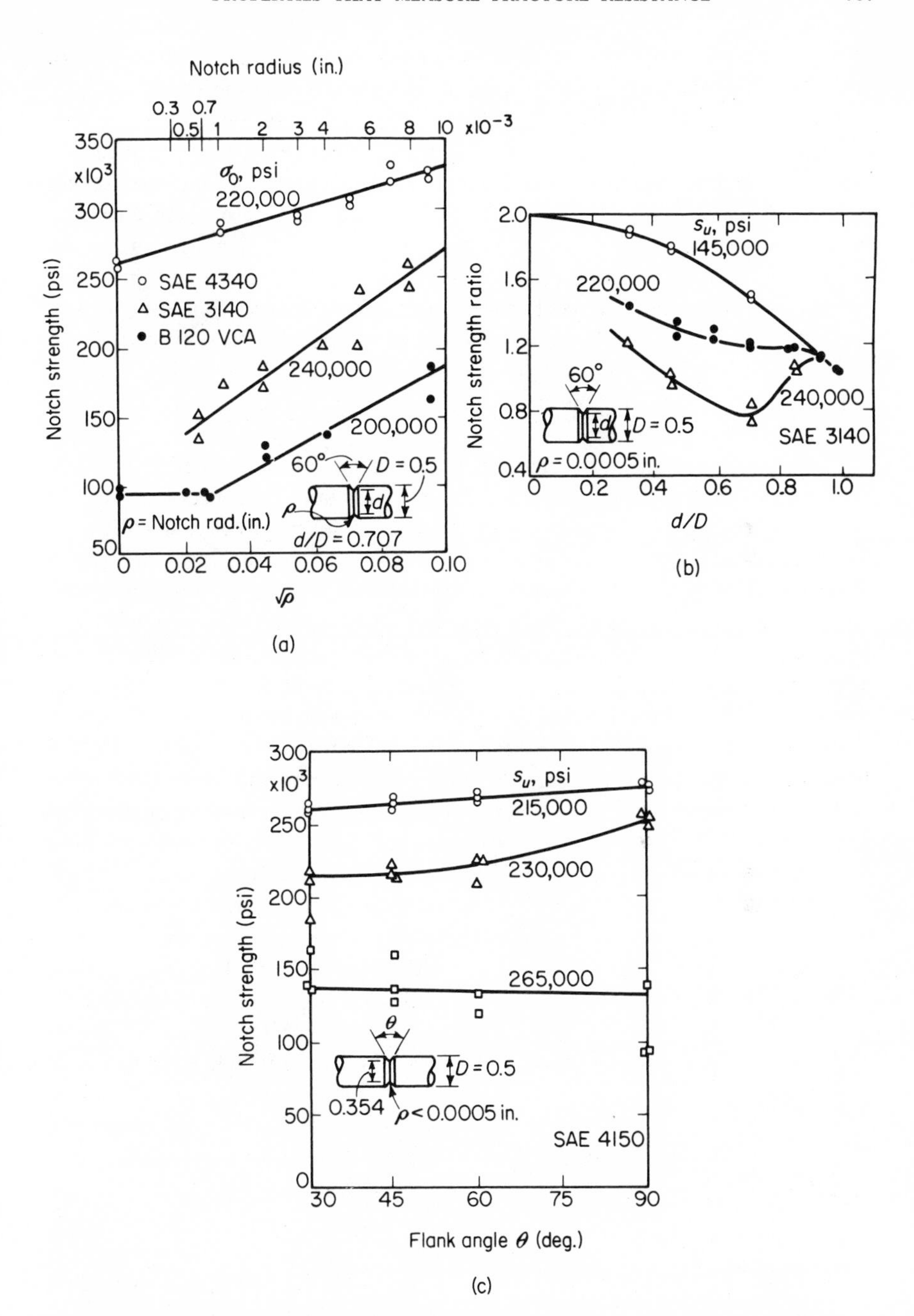

Fig. 15.14 Effect of three notch configuration variables: (a) Radius. (b) Depth. (c) Angle. From Special ASTM Committee, *Materials Research and Standards*, **2,** 196, (1962).

structures S_n/S_u is allowed to go to less than unity with satisfactory service life. This, for example, is the case in aluminum aircraft skin materials.

G. Irwin and his associates have developed a procedure known as *fracture mechanics* that is unique for evaluating brittle fracture because it measures a material constant which is directly applicable in design. It has the further advantage over Charpy testing in that it is readily usable on thin sheets of which many critical elements such as missile cases, aircraft skins, etc. are made. Earlier methods of testing thin sheets in impact required laminated specimens made up of many sheet blanks to produce an assembly that would not buckle. Indeed the methods of fracture mechanics are applicable to both brittle and frangible materials as well as to composite bodies like adhesive joints. Unlike standard impact testing, the methods of fracture mechanics also take into account section-size effects.

This concept, like the fracturing studies of Griffith described in Chapter 10, assumes that all structures are afflicted with flaws or cracks that vary in size from coalesced dislocations up to ones that may be visible by eye. Toughness, or fracture resistance, is then governed by the ease with which these cracks grow. In a low carbon structural steel, when the crack propagates slowly, there is evidence of plastic flow near the fracture; the new surface is fibrous, and the material, at least under the existing conditions of temperature, strain rate, initial crack size, etc., is considered to be tough. Under other conditions (e.g., a larger initial crack or a lower test temperature) the crack will grow catastrophically with little or no evidence of deformation. It will have a cleavage surface appearance, and the material will behave brittlely. Indeed, under the latter case, crack growth is usually an instability phenomenon, and propagation continues even with a decreasing load. The use of Griffith-Irwin fracture mechanics makes it possible to define the conditions of applied load, crack size, etc., that separate the slow ductile crack growth from the fast brittle type, at least in high strength steels ($S_u > 200{,}000$ psi) and aluminum alloys ($S_u > 60{,}000$ psi) where the effects of strain rate are small.

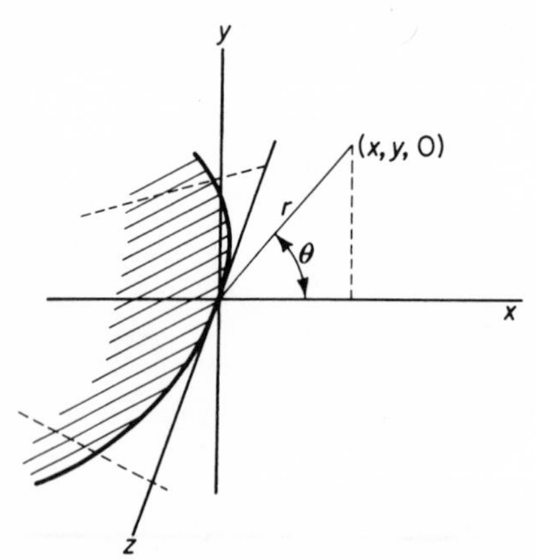

Fig. 15.15 Cartesian coordinates x, y, and z and r, θ at the tip of an interior crack. (Shaded portion is crack surface and curve is crack front. Crack lies in z-x plane.)

When a crack is present in a body, it can spread in any one of three ways. In selecting the coordinate system in Fig. 15.15, its origin was placed at the crack tip, and the z direction was taken as normal to the page. The crack which lies in the $x - z$ plane can grow in the x direction by an "opening" mode due to a tensile load in the y direction, or by either a "forward-" or "edge-sliding" (shearing) motion of one of its faces relative to the other. In the latter two cases, the crack expands much as a dislocation moves through a crystalline material, i.e., by an edge-gliding mode or a screw-gliding mode. In practice it is found that most of the fracture surface in reasonably homogeneous materials, such as a piece of steel, is perpendicular to the largest tensile stress in the element, indicating that the opening mode is of primary importance. In a heterogeneous system, such as an adhesive joint of aluminum-epoxy-aluminum, fracturing also frequently occurs by a forward-sliding mode.

Equations for the normal and shear stresses near the crack edge, for the opening fracture mode, have been developed by G. Irwin (R15.3). In polar coordinates, these are:

$$\begin{aligned}
\sigma_y &= \mathscr{K}\frac{\cos\theta/2}{\sqrt{2r}}\{1+\sin\theta/2\sin 3\theta/2\}\\
\sigma_x &= \mathscr{K}\frac{\cos\theta/2}{\sqrt{2r}}\{1-\sin\theta/2\sin 3\theta/2\}-\sigma_{ax}\\
\tau_{xy} &= \mathscr{K}\frac{\sin\theta/2}{\sqrt{2r}}\cos\theta/2\cos 3\theta/2\\
\tau_{yz} &= \tau_{xz} = 0\\
\sigma_z &= \nu(\sigma_y+\sigma_x)\text{ for plane strain, see Eq. (5.2b)}\\
\sigma_z &= 0\text{ for plane stress}
\end{aligned}\tag{15.5}$$

where σ_{ax} = a uniform stress parallel with the fracture surface;
$\mathscr{K}$ = *stress intensity factor*. Its value is a function of the applied load, the crack size, and the dimensions and shape of the body containing the crack. It has the units of psi$\sqrt{\text{in}}$.

The two parameters, $\mathscr{K}$ and σ_{ax}, completely define the stress field near the leading edge of the crack. The uniform stress, σ_{ax}, parallel to the fracture plane has been found to be unimportant, however, so that the stress field is related to the applied stress essentially by $\mathscr{K}$ alone. Since the intensity of the stress field determines whether or not a crack will run, the stress intensity factor has acquired great importance in high strength design, much like the yield strength in ductile design.

Factor $\mathscr{K}$ is closely related to the Griffith crack theory, in which cracking is controlled by the rate at which the stored strain energy can be supplied to the propagating crack. The relationship between $\mathscr{K}$ and the rate at which strain energy is released as the crack propagates is

$$\mathscr{G} = \frac{\pi \mathscr{K}^2}{E}(1 - \nu^2) \qquad \text{(plane strain)} \tag{15.6a}$$

and

$$\mathscr{G} = \frac{\pi \mathscr{K}^2}{E} \qquad \text{(plane stress)} \tag{15.6b}$$

where $\mathscr{G}$ is the *strain-energy-release rate* of a modified Griffith theory, with units in the U.S. system of in.-lb. per sq. in. or lb. per inch.*

The modified Griffith theory is used in (15.6) since it is recalled (from Chapter 9) that Griffith set the strain-energy-release rate equal to the surface energy required for forming a new crack (9.10, 9.11) since his fractures were completely brittle. In the present case no such restriction is placed on the type of fracture, and $\mathscr{G}_c$ is simply a measure of the force necessary to cause the crack to propagate at the rate at which it occurs. Parts of this released energy may be used for forming the new surface, for plastic flow, as kinetic energy, etc. This interpretation, suggested by Irwin, is akin to that given by Orowan (9.13).

The factor, $\mathscr{K}$, in (15.5, 15.6) is not to be confused with the stress concentration factor, K_t, of Eq. (15.1) or its related value, K_f. The latter two are nondimensional numbers representing the ratio of two stresses, while $\mathscr{K}$, the stress intensity factor, has the dimensions of a stress multiplied by the square root of a distance.

It was stated above that $\mathscr{K}$ or $\mathscr{G}$ was all that was required to define the stress field at the crack tip. Let us now see how these parameters are used to describe fracture toughness. When the applied load, F, on a cracked member is continuously increased, $\mathscr{K}$ also increases because it is a function of F. During this time the crack may or may not extend slowly. Slow crack extension can occur while the load increases only if the size of the plastic strain zone associated with the crack increases, so that a plastic area remains in front of the growing crack. This increase in volume of plastically deformed material partially offsets the intensification of the local stress field caused by the increased load and crack length. However, $\mathscr{K}$ increases faster than the plastic "cushion", and its value rises rapidly during the period of slow crack growth. At some combination of F and a, represented by a critical value of $\mathscr{K}$, the stress intensity at the crack edge becomes so large that the crack can propagate without additional energy being supplied. At this point F goes through a maximum, the crack becomes unstable, and its propagation changes from slow to very rapid. The critical value of $\mathscr{K}$, or the corresponding $\mathscr{G}$, is called the *fracture toughness* of the material and is designated by the symbol $\mathscr{K}_c$ or $\mathscr{G}_c$. This value separating stable from unstable crack

* The symbol $W_{\mathscr{G}}$ was used for the strain energy release in Eq. (9.10). Irwin's definition for $\mathscr{G}$ is $dW_{\mathscr{G}}/da$ where a has the same meaning as c = half the crack length.

propagation is found to be a material property. Like yield strength, it depends on the testing conditions such as temperature, strain-rate, etc.

It is apparent from the above discussion that the Irwin treatment is only strictly applicable to transverse, tensile fractures. (Techniques have been developed for handling a small amount of plastic flow ahead of the crack, however, as explained in a later section.) If the crack propagates through the loaded member by forming an oblique shear fracture, the elastic stress analysis used to find $\mathscr{K}_c$ or $\mathscr{G}_c$ becomes inaccurate. Nevertheless, even in this case, the computed values of toughness have some significance since $\mathscr{K}_c$ and $\mathscr{G}_c$ are underestimated, resulting in conservative, i.e., safe, design. Most materials become increasingly brittle, however, as their strength is increased. This would be true, for example, in a series of steels, heat treated to continuously higher yield strengths, or in polymers whose numbers of cross bonds is continuously increased. This suggests one very important use for fracture mechanics: it enables one to select the highest possible yield strength in a structure that still does not produce brittle fracture, even though cracks of some known size are present.

In addition to being a stress field parameter, $\mathscr{G}_c$ has been shown to be the amount of strain energy used to extend a crack of unit width by one unit of length. To derive an equation for $\mathscr{G}_c$ in terms of energy loss in the system, use is made of the fact that the crack and plastically strained area in its vicinity are surrounded by an elastic stress field. The total energy stored in the field is found by the methods discussed in Chapter 5, and the change in energy is equal to $\mathscr{G}_c$ when a crack of unit width moves one unit of length.

If a load F is applied to a specimen with spring constant M, a displacement δ results. According to (8.10),

$$F = M\delta$$

The elastic energy from F [see Eq. (5.12)] is

$$W_o = \frac{1}{2}F\delta \quad \text{or} \quad \frac{F^2}{2M}$$

If δ is held constant, i.e., a fixed-grip condition is assumed, then, although F and M vary as the crack grows, the ratio F/M is constant. Hence,

$$\left(\frac{\partial W_o}{\partial a}\right)_\delta = \frac{1}{2}\frac{F}{M}\left(\frac{\partial F}{\partial a}\right)_\delta$$

Also differentiating

$$\frac{F}{M} = \delta \quad \text{where} \quad \delta = \text{constant}$$

$$\frac{1}{M}\frac{\partial F}{\partial a} = -F\frac{\partial\left(\dfrac{1}{M}\right)}{\partial a}$$

so that

$$\frac{\partial W_o}{\partial a} = -\frac{1}{2}F^2 \frac{\partial\left(\frac{1}{M}\right)}{\partial a}$$

or

$$\mathscr{G}_c = \frac{1}{2}F^2 \frac{\partial\left(\frac{1}{M}\right)}{\partial a} \tag{15.7}$$

The strain-energy-release rate is, therefore, a simple function of the load and the slope of the compliance $(1/M)$ vs. a curve.

Equation (15.7) makes it possible to measure fracture toughness even in systems so complex that the stresses at the crack tip cannot be calculated. For example, high strength adhesive joints and other composite bodies exhibit brittle fractures. Because they are heterogeneous systems, neither $\mathscr{K}$ nor $\mathscr{G}$ can be found by using Eqs. (15.5, 15.6). If it is possible to put a starter crack into the material, however, $\mathscr{G}$ can be obtained experimentally by using the *calibration bar* technique. This consists of first finding M vs. a for the system. In adhesive joints, for example, this can be done by saw cutting cracks into the adhesive so that each length of cut represents a different value of a. The spring constant or elastic modulus of the sample, M, is then measured vs. crack length, so that $1/M$ can be plotted as $f(a)$ to find $\partial(1/M)/\partial a$. For this step, the crack front need not be sharp. The second step is one of finding the load, F_c, at which a starter crack runs, so that F_c and $\partial(1/M)/\partial a$ for the corresponding crack length can be substituted into (15.7). This value of $\mathscr{G}_c$ can, of course, be converted to $\mathscr{K}_c$ by using (15.6) if E is known.

In homogeneous systems, $\mathscr{K}_c$, is more commonly found directly from a stress analysis of the specific stress distribution function in the vicinity of the crack edge. Fortunately, analyses are available for most of the common test bars and for many simple structures. Irwin lists equations for the stress intensity factor under a number of different loading conditions. The most common of these are given below without derivation, and the configurations to which they apply are shown in Fig. 15.16:

1. A normal tensile stress acting at an appreciable distance from a through crack of length $2a$ in an infinitely wide plate:

$$\mathscr{K} = \sigma\sqrt{a} \tag{15.8a}$$

$$\mathscr{G} = \frac{\pi\sigma^2 a}{E} \tag{15.8b}$$

This equation would be applicable to a large element e.g., a cylindrical vessel such as a rocket motor casing, containing a relatively small crack.

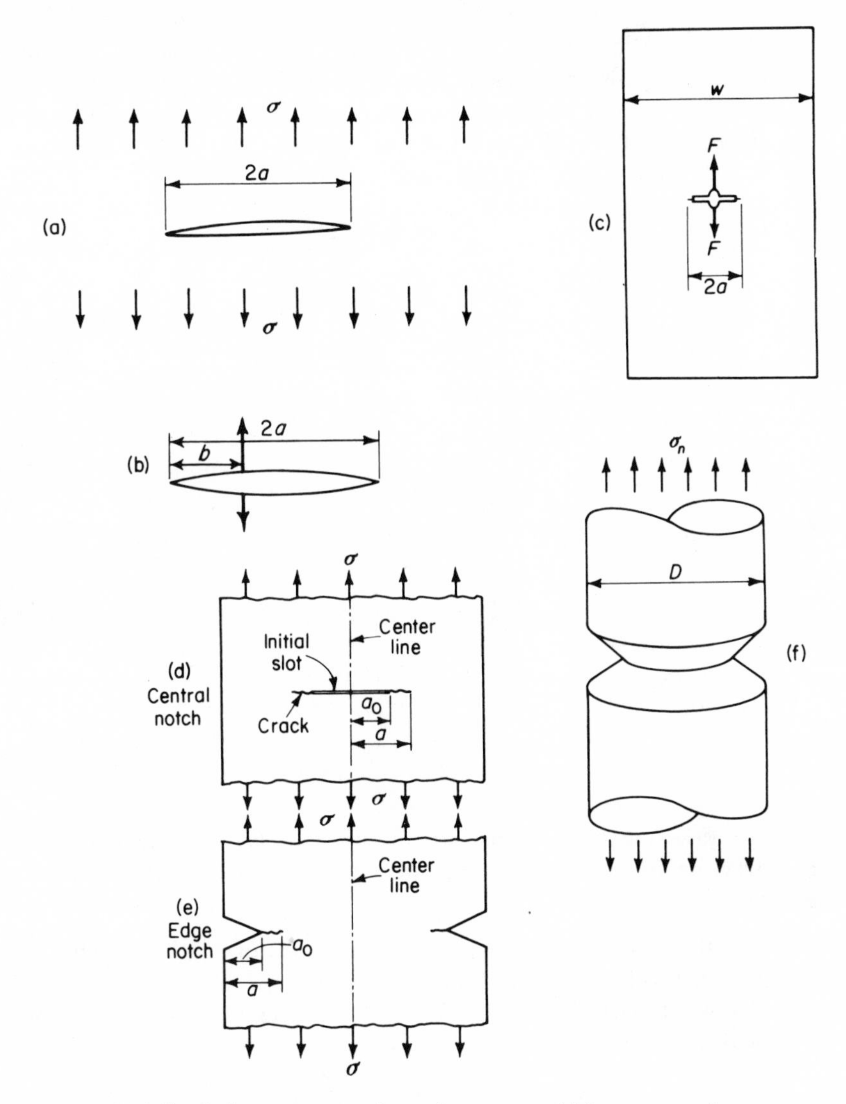

Fig. 15.16 Cracked structures and specimens on which stress analyses are available. (a) Eq. (15.8). (b) Eq. (15.9). (c) Eq. (15.10). (d) Eq. (15.11). (e) Eq. (15.12). (f) Eq. (15.13).

2. Again, a very large plate with a through crack, and with applied load F acting at the plane of the crack and normal to it at a distance b from the crack edge:

$$\mathscr{K} = \frac{F}{\pi}\sqrt{\frac{2a-b}{ab}} \tag{15.9}$$

It is interesting to note in this case that as F increases, and the crack expands, b also increases. Hence, with this type of loading, it is possible to stop and restart brittle cracks. If the plate width, w, is finite, and the load applied at the center of the crack surface, i.e., $a = b$, the equation for this loading condition is

$$\mathscr{K} = \sqrt{\frac{2F^2}{\pi w \sin \dfrac{2\pi a}{w}}} \tag{15.10a}$$

$$\mathscr{G} = \frac{2F^2}{Ew \sin \dfrac{2\pi a}{w}} \tag{15.10b}$$

3. A two dimensional crack, again $2a$ long, in the middle of a long plate of width w, with the applied load producing a tensile stress, σ, normal to the crack plane at some distance from it: (*Irwin tangent formula*)

$$\mathscr{K} = \sigma\sqrt{a\left(\frac{w}{\pi a} \tan \frac{\pi a}{w}\right)} \tag{15.11a}$$

This expression is accurate in the range $4a < w$. The same equation in a somewhat different form results from substituting 15.6b in 15.11a,

$$\mathscr{G} = \frac{\sigma^2 w}{E} \tan \frac{\pi a}{w} \tag{15.11b}$$

Because one of the most commonly used test specimens evaluates fracture toughness by means of a central notch, this equation has found wide usage.

4. Another commonly used specimen is a sheet of width w, deeply side-notched to have a central unbroken width of $2a_1$. If the stress on the full width part of the specimen (or *gross stress*) is σ, and at the notch width (i.e., av. *net section stress*) is σ_n,

$$w\sigma = 2a_1\sigma_n$$

By substituting $w - 2a_1$ for $2a$ and $2a_1\sigma_n$ for $w\sigma$ in Eq. (15.11a), the stress intensity factor becomes

$$\mathscr{K} = 2\frac{\sigma_n}{\pi}\sqrt{a_1 \frac{\pi a_1}{\left(w \tan \dfrac{\pi a_1}{w}\right)}} \tag{15.12}$$

5. A circumferentially cracked or sharply notched round tension bar of diameter D, in which one half the cross-sectional area is removed by

the crack or notch

$$\mathscr{K} = 0.23\sigma_n\sqrt{D} \tag{15.13}$$

where again σ_n is the average net section stress. The members described in (4) and (5) are those on which a considerable amount of experience has been gained by measuring notch strength and notch strength ratios.

The equations given above are used for evaluating fracture toughness. If one uses the appropriate stress for initiation of the brittle crack for σ or σ_n, and the length of the crack when it becomes unstable for $2a$ or $2a_1$ in (15.11) or (15.12) respectively, $\mathscr{K}$ and $\mathscr{G}$ acquire their critical values, $\mathscr{K}_c$ or $\mathscr{G}_c$. It is $\mathscr{K}_c$ (or $\mathscr{G}_c$) that is a material property.

The procedure for finding σ at the onset of instability is quite simple. As stated above, the load increases during slow crack propagation (Fig. 15.17); when instability occurs the stress drops abruptly and the maximum stress reached is substituted in Eqs. (15.8) to (15.13). Finding the crack length at instability is somewhat more difficult and a number of techniques have been used. Early in the development of these test procedures, a drop of India ink was placed at the notch edges or crack bottom of specimens of the type shown in Figs. 15.16c and 15.16d. As the crack grew slowly, the ink was drawn into the opening, coating it. The ink could not follow the rapid propagation so that, by examining the broken specimen, the crack size at instability could be ascertained. This method is no longer recommended, however, since the staining limit tends to be a function of the quantity and viscosity of the staining fluid, and it is impractical at temperatures that are far from ambient. Since the slow and rapid crack surfaces have distinctly different appearances in strain rate sensitive materials such as plexiglas, the length of the crack at instability can be determined visually.

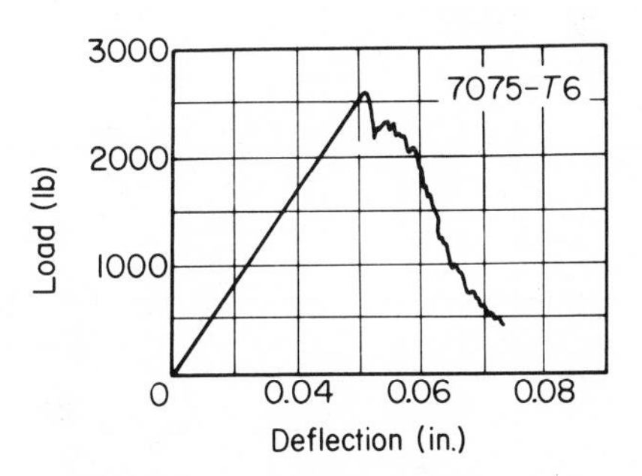

Fig. 15.17 Load deflection curve for a crack test bar. From G. R. Irwin, J. A. Kies, and H. L. Smith, *Proc. ASTM*, **58,** 640 (1958).

With the large number of methods for measuring fracture toughness, almost any material can be tested although it may be difficult for strain-rate sensitive materials. Values are available on materials as different as glass and heat treated steels (see Table 15.1).

In discussing Fig. 15.15, we made the point that a crack can expand by an opening mode, by forward shear, or by edge-sliding. Eqs. (15.8 to 15.13)

TABLE 15.1

FRACTURE TOUGHNESS OF A VARIETY OF MATERIALS. FROM G. IRWIN, (R15.3).

Material	*Yield stress* $10^3(\mathrm{lb/in^2})$	$\mathcal{G}_c$ (lb/in)	*Thickness* (in)	*Temperature* (°C)	*Method* (see Fig. 15.16)	*Fracture*
Glass, lantern slide cover, moist	>100	0.04	0.05	Room	C	Plane-strain
Glass, lantern slide covers, 2% r.h.	>100	0.08	0.05	Room	C	Plane-strain
Polymethylmethacrylate, cast	~10	4	0.25	Room	C	Plane-strain
Polymethylmethacrylate, hot-stretched	~10	25–40	0.25	Room	C	Plane-strain
Aluminum-epoxy-aluminum	—	1–4	0.25–1.0	Room	***	Plane-strain
Ship steel (108 in. wide C steel)	36	>1070	0.75	0°	C	Plane-stress shear
Ship steel	105*	~100	0.75	−80°	Residual stress zone tests	Plane-strain cleavage
Rotor steel Ni-Mo-Va quenched and drawn	101	960	6	Room	SD**	Plane-strain
Rotor steel Ni-Mo-Va pearlitic and upper bainitic	89	135–180	6	Room	SD**	Plane-strain
Steel SAE 4340	230	130–150	0.250	Room	C and NBT***	Plane-strain
Steel SAE 6150	260	275	0.075	Room	C	50% shear
Al-alloy 2024-T3	47	~1500	0.032	Room	C	Plane-stress
	47	340	4	Room	NBT***	Plane-strain
Al-alloy 7075-T6	68	790	0.032	Room	C	Plane-stress
	68	134	0.750	Room	NBT***	Plane-strain
	67	115	2(bar)	Room	C	Plane-strain

*Dynamic values. ** Centrally notched spinning disk. *** Using Eq. (15-7).

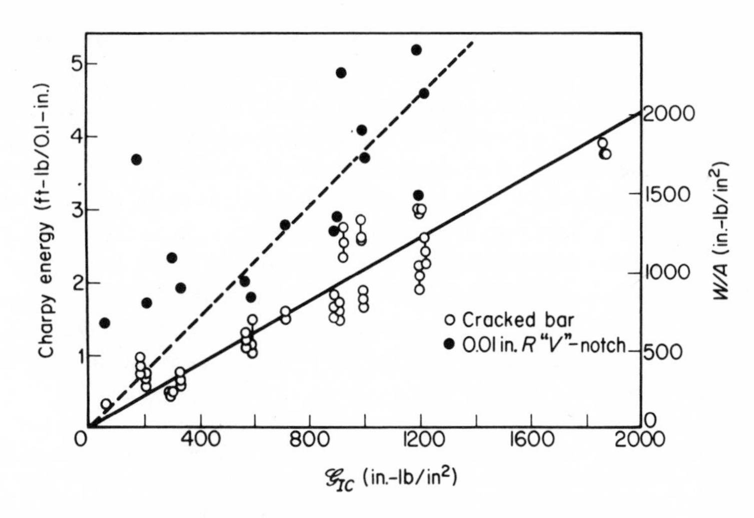

Fig. 15.18 Relation between W/A in a Charpy text and $\mathcal{G}_{Ic}$ in tension for a variety of sheet steels. From Orner and Hartbower, informal communication to ASTM Committee on Fracture Testing, 1962.

were derived by using the stress-field parameter for an opening mode. To restrict the fracturing toughness symbol to this type of cracking, it is designated $\mathcal{K}_{Ic}$ or $\mathcal{G}_{Ic}$. Since most testing is done for opening mode toughness, $\mathcal{K}_c$ or $\mathcal{G}_c$ implies $\mathcal{K}_{Ic}$ or $\mathcal{G}_{Ic}$. Equation (15.7) defines the strain-energy-release rate in a manner that does not require knowledge of the stress field so it can be used for any of the fracturing modes. Toughness for a forward-shearing mode is referred to as $\mathcal{K}_{IIc}$ or $\mathcal{G}_{IIc}$, and for an edge-sliding mode as $\mathcal{K}_{IIIc}$ or $\mathcal{G}_{IIIc}$. To use (15.7) to calculate the latter two, a shearing force, P, is applied perpendicular to the crack front for the IIc case, and parallel with it for $IIIc$. The change in compliance, $\partial(1/M)/\partial a$, must also be obtained for the appropriate loading condition.

On p. 394, the use of precracked Charpy bars was mentioned. Since $\mathcal{G}_{Ic}$ is the energy required to propagate a pre-existing crack, these are sometimes used for evaluating fracture toughness. An exemplary correlation of energy per unit of uncracked area, W/A, in a Charpy test and $\mathcal{G}_c$, measured by Eq. (15.11), is shown in Fig. 15.18. The use of cracked impact bars for finding toughness has two advantages: first, the test is simple and, hence, inexpensive, and, second, the toughness is measured at a higher strain rate which often approaches cracking rates in certain service conditions more closely than do ordinary tensile testing rates.

Equations (15.8) through (15.13) were derived by the methods of the theory of elasticity; in all cases it was assumed that no plastic flow occurred during brittle crack extension. If the test specimens used for evaluating $\mathcal{K}_c$ and $\mathcal{G}_c$ are sufficiently wide, the amount of plastic flow accompanying brittle cracking is small enough to be neglected. However, in the region of practical specimen sizes, an appreciable amount of flow frequently precedes the crack.

This nonlinear deformation partially relieves the stress at the crack tip. In this regard plastic flow is equivalent to having a larger crack than is actually the case, so that the measured $\mathscr{K}_c$ values are lower than they should be. Consequently, tests made on narrow sheets in the laboratory do not yield the proper values for analyses of fracture in wide sheets which are most generally encountered in service. Correcting the stress equation to compensate for plastic flow requires that one use a corrected value for crack length, $2a'$. The value for a' is approximated by substituting the yield strength, σ_o, for σ_y, and zero for θ in (15.5), in which case the first of these equations is written:

$$\sigma_o = \frac{\mathscr{K}}{\sqrt{2r_o}} \tag{15.14a}$$

or

$$r_o = \frac{\mathscr{K}^2}{2\sigma_o^2} \tag{15.14b}$$

where r_o = radius of plastic zone. The local stress relaxation due to plastic flow at the crack edge is then equivalent to an added crack length of r_o, and a' becomes

$$a' = a + \frac{\mathscr{K}^2}{2\sigma_o^2}$$

Replacing this value for a in (15.11a), and squaring both sides, yields for the centrally notched specimen,

$$\mathscr{K}^2 = \sigma^2 \frac{w}{\pi} \tan\left(\frac{\pi}{w} a + \frac{\pi \mathscr{K}^2}{2w\sigma_o^2}\right) \tag{15.15}$$

For the edge notched specimen,

$$\mathscr{K}^2 = \sigma^2 \frac{w}{\pi} q(\mu) \tag{15.16}$$

where $q(\mu) = \tan \mu + 0.1 \sin 2\mu$
and

$$\mu = \frac{\pi a}{w} + \frac{\pi \mathscr{K}^2}{2w\sigma_o^2}$$

It is apparent that the fracture toughness in (15.15, 16) is not explicitly defined so that the equations are difficult to use. For this reason, a graphical method for finding fracture toughness has been suggested by a special ASTM committee.* The simplified equation used in this procedure is

$$K_c{}^{**} = \sigma_m \sqrt{q_c w} \tag{15.17}$$

* Ref. (R15.2).

** The block symbol K is frequently used in test programs rather than the script $\mathscr{K}$. The relationship between these two is: $K = \mathscr{K}\sqrt{\pi}$.

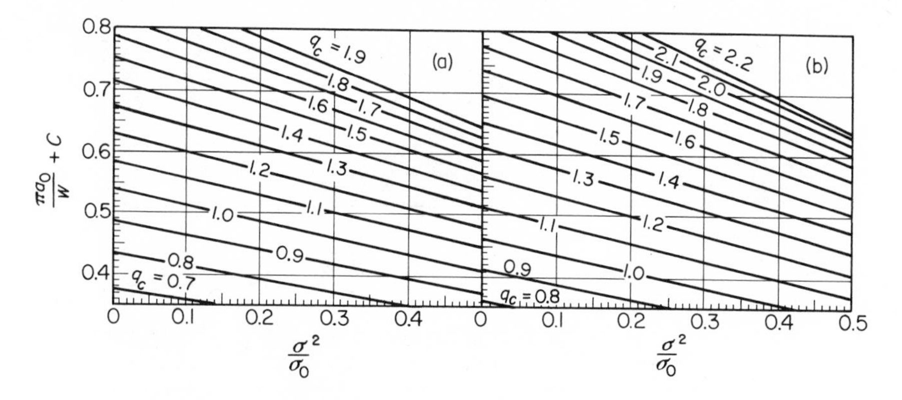

Fig. 15.19 Charts used for finding q_c in Eq. (15.17). From (R15.2).

where σ_m = maximum gross stress, i.e., σ at instability at a position removed from the crack or notch,

and q_c = a number of the order of unity, found graphically by means of Fig. 15.19.

As an example of the use of Eq. (15.17) and Fig. 15.19, tabulated data used in calculating K_c for a steel tempered at a variety of temperatures is shown in Table 15.2. Values for two steels and a titanium alloy at different plate thicknesses and testing directions are shown in Table 15.3.

Effect of Thickness (Constraint)

The stress required to cause plastic flow in a member depends on the stress state. Indeed, Chapter 6 was primarily concerned with a discussion of the criteria used to relate yield stress with stress state. Similarly the fracture toughness depends on stress state. Although the yield criteria were capable of considering stresses in all three directions, the stress analysis in the vicinity of a crack was limited to two dimensional models, with the extremes being plane stress and plane strain.

The measured value of fracture toughness is very much dependent on which of these extremes exists. Whether fracturing occurs under a condition of plane stress or plane strain depends on the plate thickness. If the plate is thin, it is not able to develop large stresses through its thickness so that the stress normal to the surface is close to zero, and cracking occurs under a condition of plane stress. If the plate is thick, on the other hand, contraction in thickness is restrained by the surrounding mass of elastic material, large stresses are developed through the thickness, and a condition of plane strain arises.

This change from plane stress to plane strain occurs abruptly as the plate

TABLE 15.2

TYPICAL DATA FOR CALCULATION OF $K_c = \sigma\sqrt{q_c w}$. FROM (R15.2).

Material AISI 4340, air melt; specimen: longitudinal, $t = 0.080$ in., $w = 3.00$ in., $w:t = 37.5$, center notch.

Tempering Temperature, deg. F.	σ_o 1000 psi	σ, 1000 psi	a_0, in.	P	$\frac{\pi a_0}{w}$	c	$\frac{\pi a_0}{w} + c$	$\left(\frac{\sigma}{\sigma_0}\right)^2$	q_c*	$q_c w$, in.	$\sqrt{q_c w}$, $\sqrt{\text{in.}}$	K_c 1000 psi $\sqrt{\text{in.}}$	Avg K_c 1000 psi $\sqrt{\text{in.}}$
350	208.3	93.6	0.39	0.62	0.412	0.024	0.436	0.202	0.89	2.67	1.63	153	...
	...	97.7	...	0.87	...	0.055	0.468	0.220	0.99	2.97	1.73	169	...
	...	113.9	...	0.87	...	0.055	0.468	0.299	1.03	3.10	1.76	174	...
	...	96.9	...	0.75	...	0.040	0.452	0.216	0.94	2.82	1.68	163	165
425	203.9	108.1	0.39	0.95	0.412	0.065	0.478	0.281	1.04	3.12	1.77	191	...
	...	107.1	...	0.80	...	0.046	0.459	0.276	0.98	2.94	1.72	184	...
	...	114.4	...	1.00	...	0.071	0.484	0.315	1.08	3.24	1.80	189	...
	...	103.9	...	0.95	...	0.065	0.478	0.259	1.02	3.06	1.75	182	187
500	197.9	98.4	0.39	0.70	0.412	0.034	0.446	0.247	0.93	2.80	1.67	164	...
	...	96.1	...	1.00	...	0.071	0.484	0.236	1.02	3.06	1.75	168	...
	...	106.0	...	0.60	...	0.024	0.436	0.286	0.94	2.82	1.68	179	...
	...	95	...	0.80	...	0.046	0.453	0.230	0.94	2.82	1.68	160	168
700	181.6	107.5	0.39	1.00	0.412	0.071	0.484	0.351	1.10	3.30	1.82	195	...
	...	120.6	...	1.00	...	0.071	0.484	0.441	1.15	3.45	1.86	225	...
	...	58.2	...	0.50	...	0.009	0.421	0.102	0.82	2.46	1.57	91.2	...
	...	120.6	...	1.00	...	0.071	0.484	0.440	1.15	3.45	1.86	225	186.5

* See Fig. 15.19a; read on interpolated line through point $\left(\frac{\sigma}{\sigma_0}\right)^2, \left(\frac{\pi a_0}{w} + c\right)$.

where a_o = initial crack length
w = specimen width
σ_m = maximum gross stress
σ_o = yield strength
P = shear lip fraction

TABLE 15.3

TYPICAL DATA FOR VARIOUS MATERIALS. FROM (R15.2)

Specimens	t, in.	σ_0 1000 psi	σ, 1000 psi	a, in.	K_c 1000 psi $\sqrt{\text{in.}}$
Steel 6434					
Air melt					
Longitudinal	0.07	190	102	0.67	205
	0.22	190	107	0.76	251
	0.07	210	107	0.66	218
	0.22	210	88.5	0.82	185
Vacuum melt					
Longitudinal	0.07	190	100	0.72	200
	0.22	190	88	0.77	184
	0.07	210	97.4	0.69	191
	0.22	210	83.5	0.78	165
Transverse	0.07	190	95.7	0.73	191
	0.22	190	83.6	0.83	181
	0.07	210	100.4	0.83	197
	0.22	210	74.6	0.78	140
Steel "Tricent"					
Vacuum melt					
Longitudinal	0.07	190	93	0.76	183
	0.22	190	69.4	0.61	108.5
	0.07	230	109	0.75	214
	0.22	230	77.9	0.74	132
Transverse	0.07	190	95	0.77	187
	0.22	190	70	0.56	108.8
	0.07	230	108.5	0.76	214
	0.22	230	77.4	0.78	152
Ti alloy Ti-6Al-4V					
Longitudinal	0.072	159	68.7	0.63	123
Transverse	0.072	164	72.8	0.68	132

thickness is increased (Fig. 15.20), leading to the concept of a thickness transition. Within this transition range, the material toughness, i.e., $\mathscr{K}_c$, undergoes a sudden change similar to that observed in certain metals with a change of temperature. This transition is controlled in both cases by the ratio of plastic zone size to section thickness.

Because $\mathscr{K}_c$ depends so much on elastic constraint, a crack traveling along a plate does not have a plane front perpendicular to the plate surface. The crack travels fastest at the mid-thickness of the plate where the elastic restraint is greatest and, therefore, the plate is most brittle. This center crack

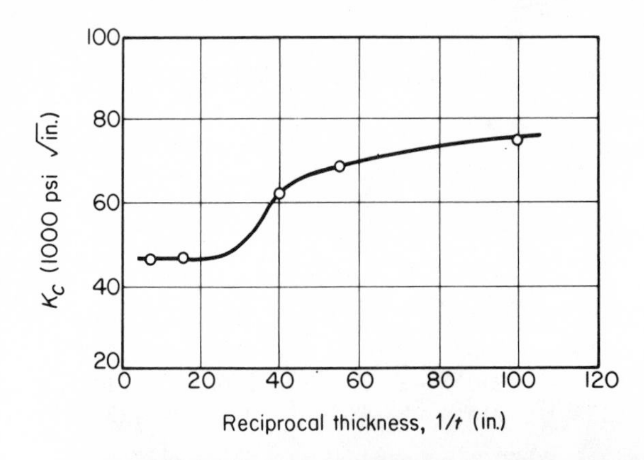

Fig. 15.20 Thickness transition in a titanium alloy. After A. J. Repko, M. H. Jones, and W. F. Brown, Jr.

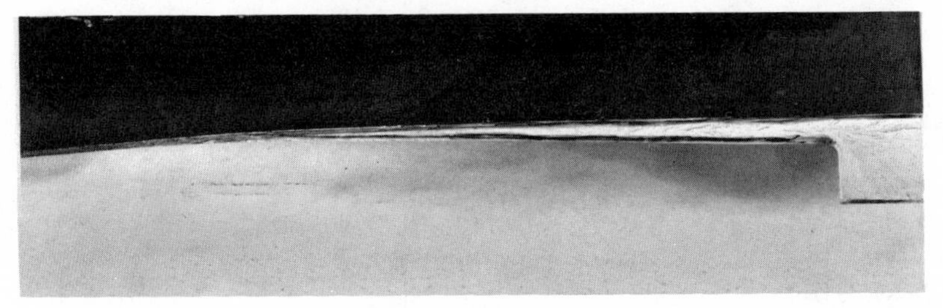

Fig. 15.21 Fracture in pressure chamber of high-strength steel. It started at left, outside the view, and ran to the right. The mode of fracture changes from all shear to 55% shear as the wall thickness increases from $\frac{3}{16}$ to $\frac{1}{4}$ in. From G. R. Irwin and J. A. Kies, *Metal Progress*, **78,** No. 2, 73 (1960).

is generally of a cleavage type but, as the crack approaches the plate surfaces, restraint is greatly reduced because the distance from the crack edge to the plate surface is small, so that fracturing changes from plane strain to plane stress, accounting for the formation of a shear lip near the plate surfaces. The influence of wall thickness on fracture toughness, and the shear lip appearance, can be seen in Fig. 15.21. Cracking in this pressure chamber started at the far left beyond the field of the photograph; there, the plate was thin and the fracture toughness high. As the crack ran to the right, the wall was thicker and so less tough. The fracture changed, during travel, from all shear to just slightly more than half shear, when the wall thickness increased from $\frac{3}{16}$ to $\frac{1}{4}$ inch. If a meaningful value of $\mathscr{K}_c$ is to be measured in a laboratory test for application to a service structure, the test bar should obviously have the same thickness as the structure being analyzed.

When a crack is small and does not extend through the entire plate thickness, the value of $\mathscr{K}$ developed in a plate is low because of the crack size, but $\mathscr{K}_c$ is also small because of the elastic constraint. Consider, for example, the plate sketched in Fig. 15.22, with an initial crack shown shaded. The crack grows slowly from position 1 to 2. At position 2 it reaches $\mathscr{K}_c$ under plane strain so that the crack abruptly expands to position 3. When it becomes a through crack, however, the fracture toughness for rapid propagation again changes from plane strain to plane stress, and fracturing is arrested if the applied load is not too large.

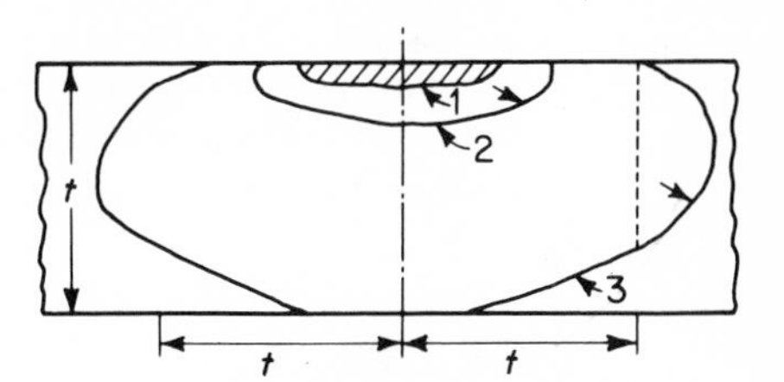

Fig. 15.22 Schematic diagram showing how a crack propagates through a plate of thickness t. Initial crack shown shaded, slow growth from 1 to 2, followed by fast growth to 3, where the crack is arrested. From (R15.2).

The lowest value that $\mathscr{K}_c$ or $\mathscr{G}_c$ can attain, when the elastic restraint is at a maximum, is found by extrapolating a plot of toughness vs. thickness to infinite thickness. This is most easily done by plotting the property as a function of reciprocal thickness, and extrapolating to zero (as in Fig. 15.23 for two aluminum alloys and one steel). In this limiting case of purely plane strain, the influence of the shear lips at the free side surfaces is negligible. This critical value for toughness is designated by $\mathscr{K}_{ic}$ or $\mathscr{G}_{ic}$.

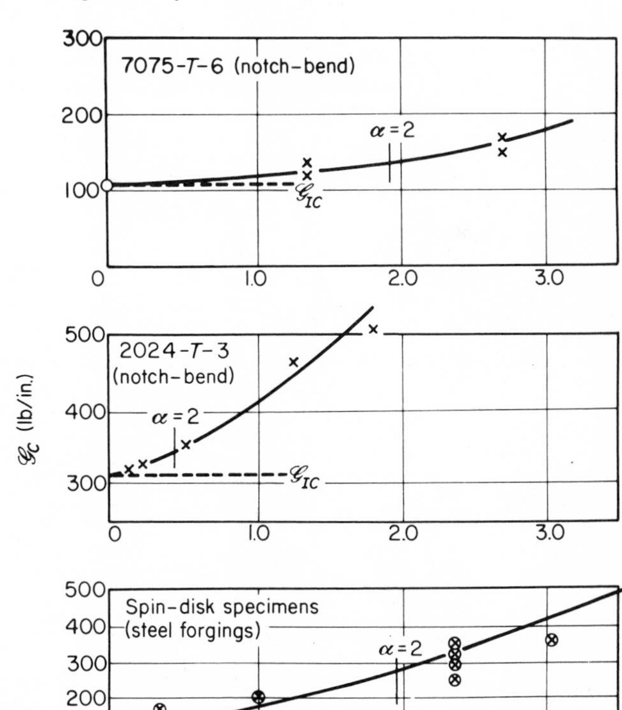

Fig. 15.23 Fracture toughness of two aluminum alloys and a steel as a function of reciprocal plate thickness. From G. R. Irwin, J. A. Kies, and H. L. Smith, *Proc. ASTM*, **58**, 645 (1958).

The plate thickness, that at least roughly separates plane stress from plane strain, can be found from Eqs. (15.6), relating $\mathscr{K}$ and $\mathscr{G}$ for plane strain, and (15.14b) which gives the length of the plastically deformed material preceding the crack. Hence,

$$r_0 = \frac{\mathscr{K}_{ic}^2}{2\sigma_o^2} = \frac{E\mathscr{G}_{ic}}{(1-\nu^2)2\pi\sigma_o^2} \tag{15.18}$$

To suppress plastic flow, the plate thickness, t, must greatly exceed r_o, so that

$$t \gg \frac{E\mathscr{G}_{ic}}{(1-\nu^2)2\pi\sigma_o^2} \tag{15.19}$$

if a plane-strain fracture is to occur.

Because of the importance of the plastic zone size relative to section thickness in fracture mode transition, a ratio, β_{ic}, is sometimes used as defined by,

$$\beta_{ic} = \frac{E\mathscr{G}_{ic}}{t\sigma_o^2} \tag{15.20}$$

A β_{ic} value of unity is a mid-range condition relative to the change from conditions of plane stress ($\beta_{ic} > 1.5$) to those of plane strain ($\beta_{ic} > 0.5$), (Fig. 15.24).

Fig. 15.24 Critical crack-extension-force values as a function of the reciprocal of the relative thickness parameter α. From G. R. Irwin, (R15.3).

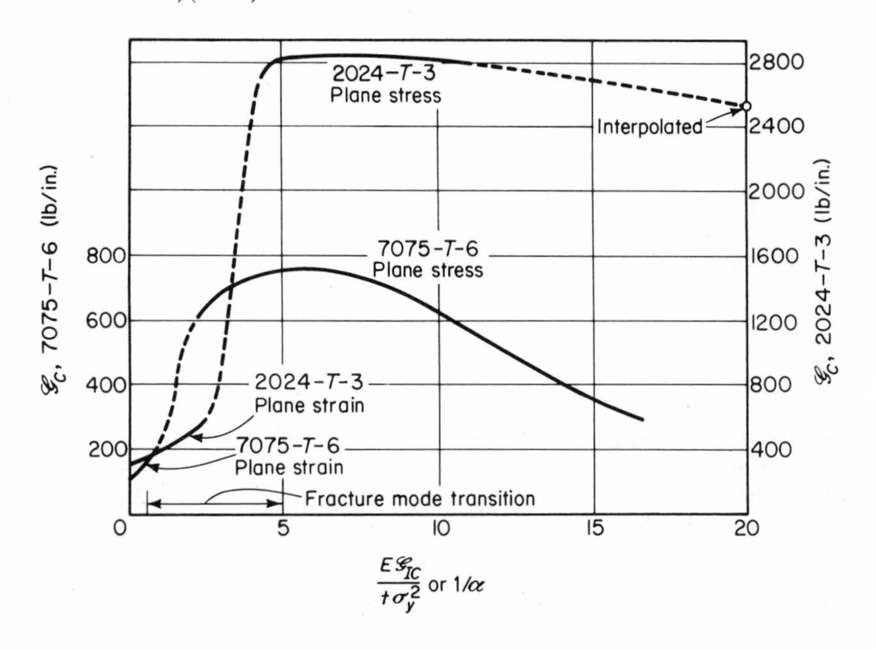

Example of Application of Fracture Mechanics

Some applications of fracture mechanics to pressure vessel failures has been discussed by H. Bernstein.* One of his examples was the cracked motor case shown in Fig. 15.21. This failed in the weld at a pressure of 750 psig which corresponded to a hoop stress of 92,000 psi, although the steel from which it was made had a yield strength of 200,000 psi. Examination of the fracture showed an initial discolored crack one inch long and 0.06 inches deep. The markings on the fracture surface showed this to be the origin, and an examination of the microstructure showed the crack to be present at the time the chamber was heat treated.

The original 1 × 0.06 in. crack expanded slowly during pressure to a 1.5 inch crack through the thickness and then became unstable, causing fast fracture of the vessel. The slow and fast crack sizes were deduced from fracture appearance. Eq. (15.8b) yields

$$\mathcal{G}_c = \frac{\pi \times (92{,}000)^2 \times 0.75}{30 \times 10^6} = 665 \text{ lbs./in.}$$

Fracture toughness measures of the intact chamber weld gave $\mathcal{G}_c$ values between 687 and 875 lbs./in.

PROBLEMS

1. In Fig. 15.2, what is the value of the Charpy "V" notch transition temperature based on each of the following criteria?

a. 15 ft/lb
b. 40 ft/lb
c. average energy
d. 100% fibrous
e. *NDT*

2. Plot σ_y as $f(\theta)$, and r = const., for the coordinate system in Fig. 15.15. (Hint: use units of $\mathcal{K}\sqrt{2r}$ for the ordinate.)

3. The strain-energy-release rate, $\mathcal{G}_c$, is the amount of stored strain energy absorbed by a crack of unit width extending one unit of length. How is the bulk of this energy dissipated in glass and in metals?

4. Certain steels, at their *NDT* temperature, are thought to have $\mathcal{G}_c \approx 25$. If $\sigma_o = 36{,}000$ psi for the steel at *NDT*, what is the "starter–crack" size if a very large steel plate fractures at one-half the yield strength?

5. If the aluminum used in an aircraft skin has a $\sigma_o = 68 \times 10^3$ psi, what size crack could be tolerated if the working stresses $= \sigma_o/2$ in a sheet 0.032 inch thick and in structural member 0.750 inch thick (see Table 1)?

* H. Bernstein in *High Strength Steels for the Missile Industry*, edited by H. T. Sumson, *A.S.M.*, 1961.

6. The following compliance ($1/M$), crack length a, and P_c data were collected on a one inch wide aluminum-epoxy-aluminum joint as the crack grew:

a	$\frac{1}{M}$	P_c
2.0 in.	12.5×10^{-6} in./lb.	533 lb.
3.0	37	336
4.0	80	264
5.0	145	220
6.0	240	188
7.0	360	166

What is the average value of $\mathscr{G}_c$?

REFERENCES FOR FURTHER READING

(R15.1) Biggs, W. D., *The Brittle Fracture of Steel.* New York: Pitman Publishing Corp., 1961.

(R15.2) Fracture Testing of High-Strength Sheet Materials: A Report of a Special ASTM Committee. *ASTM Bulletin* No. 243, January 1960, No. 244, February 1960, and continued in *Materials Research and Standards*, ASTM Vol. 1, No. 11, November 1961.

(R15.3) Irwin, G. R. in *Structural Mechanics*, ed. by J. N. Goodier and N. J. Hoff. New York: Pergamon Press, 1960.

(R15.4) Peterson, R. E., *Stress Concentration Design Factors.* New York: John Wiley and Sons, Inc., 1953.

(R15.5) Szczepanski, M., *The Brittleness of Steel.* New York: John Wiley and Sons, 1963.

16

CREEP AND STRESS RUPTURE

In Chapter 8 it was shown that the deformation experienced by a loaded member depends not only on the magnitude of the applied stress but on the time over which the stress operates as well. In many practical cases the time scale exerts such a modest effect that it can be ignored, e.g., in steel structures operating at ambient temperature. In other cases time becomes most important, as shown by its influence in establishing the stress–strain relationships for one type of polystyrene in Fig. 16.1. In the study of time-dependent deformation, either the load or the stress acting on the member, and its temperature, are generally held constant so that strains are solely a function of time. Deformation of this type is termed *creep*, and the mechanism by which it occurs was discussed in Chapter 8. In this chapter the phenomenological aspects of creep are discussed.

Creep is most commonly evaluated by applying a constant load to a tensile specimen whose extension is measured as a function of time. This type of deformation is not restricted to uniaxial tension however, and occurs whenever a body is loaded. Consequently compression or bending creep is often studied. Although in most cases the test period is long, compared with that of a tensile test, the range of testing time is very broad but, whenever practical, the testing times corresponds to the expected service life of the structure for which data is being collected. Creep properties have been reported for times as long as 100,000 hours on one hand, and complete tests have been conducted in just a fraction of a minute at the other extreme. Even at these short times, the tests are classified as creep since the load (or

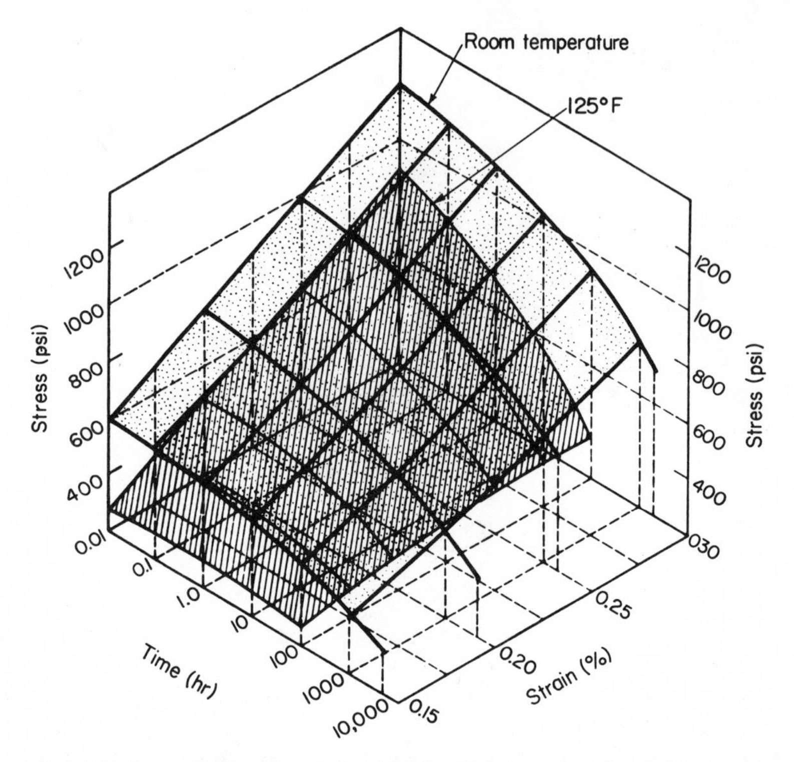

Fig. 16.1 Stress–strain-time relationship for high impact polystyrene at room temperature and 125°F. From G. A. Patten, *Materials in Design Engineering*, **55**, No. 5, 116 (1962).

stress) is held constant during the test, rather than continuously increasing as is the case for ordinary tensile testing.

Structural metals creep appreciably only at elevated temperatures while other materials, such as lead, rubber, or plastics, may do so even at ambient temperatures. The material characteristics that are of interest in creep testing are similar to those in short time testing; e.g., how much permanent deformation can be expected under a specific loading requirement, and under what conditions will the member fracture? Another related property that is of interest is relaxation as discussed in Chapter 8. This becomes important in applications like metal fasteners (e.g. bolts) used at elevated temperatures, in shrink fit or pressed assemblies, or in nonmetallic compression parts such as gaskets.

THE SHAPE OF THE CREEP CURVE

When a constant tensile load is held on a specimen, its time–extension curve, typically, has the shape shown in Fig. 16.2a. Experimental curves of

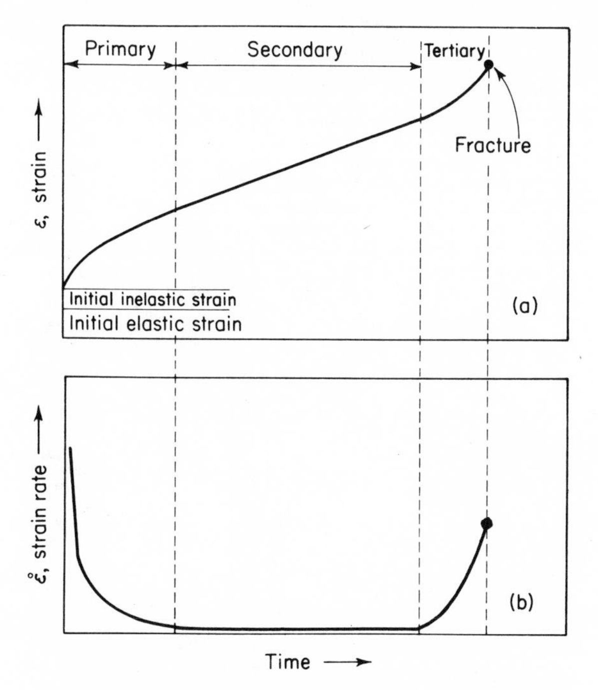

Fig. 16.2 Typical creep and creep rate curves.

this type obtained on a steel and a polymer were discussed briefly in connection with Fig. 8.5. For analytical convenience creep curves are divided into three regions even though this separation does not imply three physically different types of flow. The first section is referred to as *primary creep* and encompasses the portion of the curve over which the *creep rate*, i.e. $d\varepsilon/dt$, is continuously decreasing as shown in Fig. 16.2b. The second section is a period of approximately constant creep rate and is termed *secondary creep*; the creep rate is at a minimum over this region. The third stage or *tertiary creep* represents a continuously increasing creep rate and terminates in fracture of the specimen.

The initial instantaneous extension, prior to time-dependent flow is the result of both elastic and plastic strain. Although it is not ordinarily considered creep, since, by definition, it is not time-dependent, it may contribute significantly to the total allowable strain in the member. In British terminology, however, this initial strain is classified as a part of the creep curve so that the entire curve is said to have four parts. Following the customary practice, the initial time–independent strains are not considered here.

Even though the division of the creep curve into distinct regions does not imply three completely different deformation mechanisms, specific types of flow are major contributants to each region. A large portion of primary

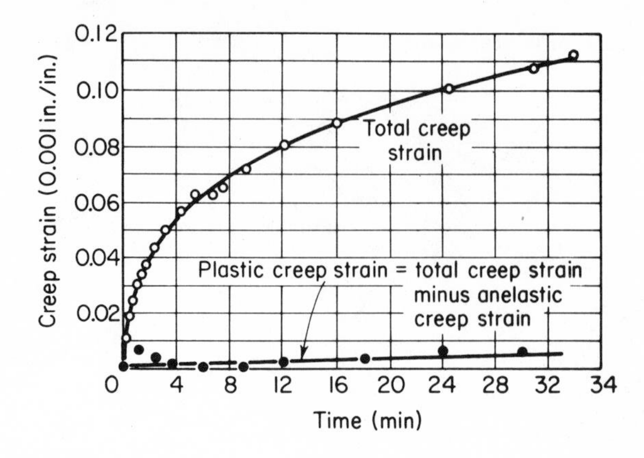

Fig. 16.3 Total creep strain and plastic creep strain as a function of time showing the large component of anelastic strain during primary creep. From J. D. Lubahn, *Trans. Am. Soc. Metals*, **45**, 787 (1953).

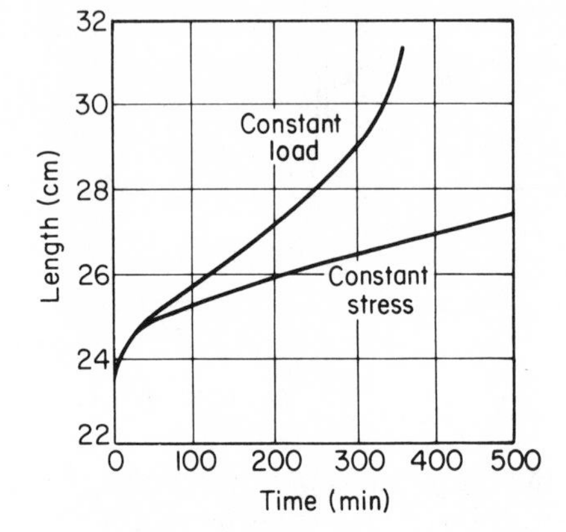

Fig. 16.4 Comparison of constant load and constant stress creep curves for lead. After E. N. da C. Andrade.

creep is anelastic and for this reason is sometimes referred to as *transient* or *anelastic strain*. Lubahn has separated a creep curve obtained on steel into anelastic and plastic components (Fig. 16.3). It is apparent that, at these short times, most of the deformation is anelastic. As time is increased, less of the primary creep is anelastic, however, so that the frequent use of transient or anelastic strain as a synonym for primary creep is not quite proper, especially in metals.

Beyond very small strains, the creep curve is complicated by two factors: first, with a constant applied load the true stress increases during the test because the cross section of the specimen continuously decreases; further, large creep strains imply long times, and, since creep is of most interest at elevated testing temperatures, the material may not be stable during the test (e.g., metals might recrystallize or age). If the material were stable, and a constant stress rather than constant load were applied to it, the creep rate would have continuously decreased during straining; i.e., only primary creep would have occurred. This is seen from the extension vs. time curves for constant stress and constant load for lead (Fig. 16.4). The reduced creep rate with time is thought to result from strain-hardening, much as was the case for tensile testing (Chapter 10). When a constant load rather than a constant stress is used on a structurally stable material, the inflection point on the creep curve occurs at about the same strain at which necking occurs in an ordinary tensile test.

Even with constant stress, commercial materials generally exhibit secondary creep because they are not stable at the times and temperatures of interest. This softening with time at elevated temperature is seen when specimens are tested in tension at room temperature after they have been held for various periods at elevated temperatures (Fig. 16.5). If the metal normally softens at the test temperatures and times, even in the absence of stress, its creep curve is the result of superimposing extensions attributable to two sources—first, an extension caused by the constant load applied to the progressively softening material; and, in addition, a creep contribution that would continuously decrease because of strain-hardening. The interaction of these two produces the observed secondary creep curve. Obviously then, if an unstable material is tested at constant load, the inflection point on the creep curve will occur at a smaller extension than that at which instability (necking) would have occurred in a tensile test. Constant stress-creep curves of stable materials are sometimes thought to exhibit secondary creep simply because of the manner in which the data are presented. This occurs most commonly when the creep test is not carried to failure. During primary creep, the deformation rate continuously decreases, so that, if a very broad

Fig. 16.5 Loss in room temperature tensile strength by holding samples at an elevated temperature for various times. From W. F. Brown, Jr., G. B. Espey, and M. H. Jones, *Trans. SAE*, **70**, 583 (1962).

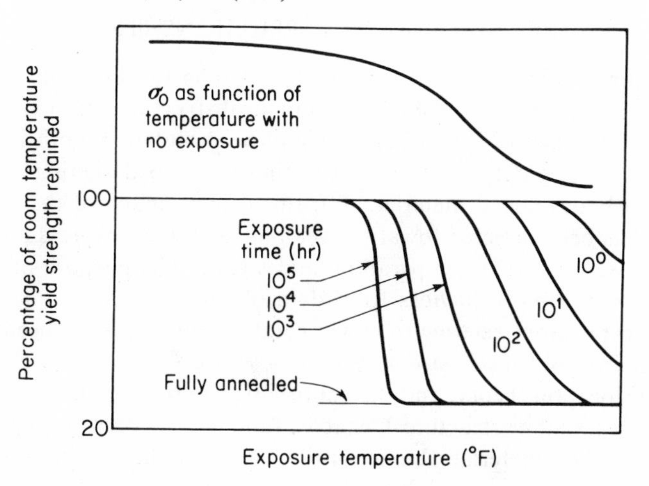

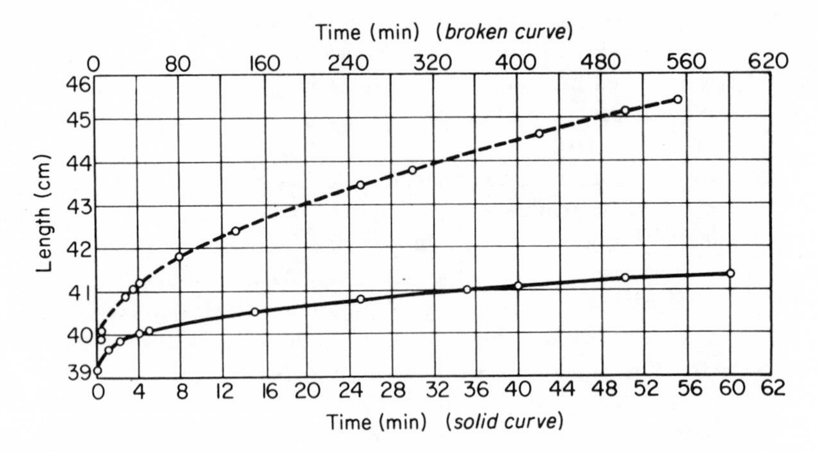

Fig. 16.6 Creep curve on an expanded and a condensed time scale. The solid curve appears to be exhibiting secondary creep after about five minutes. This is shown to be primary creep on the more condensed (dashed) curve. From (R16.3).

time base is selected (as in Fig. 16.6), the extension time curve appears linear; this apparent linearity is lost on a more condensed time scale.

Tertiary creep, like necking in a tensile test, is an instability phenomenon. Its occurrence is not completely understood but it is at least partially the result of having a strain-hardening rate that is not high enough to compensate for the specimen's loss of cross-sectional area during straining. Another contributant to tertiary creep is the formation of fine cracks, especially during low-stress, long-time tests. These again cause an accelerated creep rate by lowering the load carrying area of the specimen.

CREEP–TIME–STRESS–TEMPERATURE RELATIONSHIPS

A major problem in creep is that of obtaining the very long time data often required to simulate the life of a practical structure. In many cases an experimental attack of this problem would be prohibitive, both in terms of cost and the time delay involved in testing prior to designing a structure from a newly developed material. For this reason, much of the work done on creep has been directed toward developing analytical expressions relating creep-extension to time, temperature, and stress. The purpose of this is to determine long time behaviors from data obtained in short time tests. In addition, expressions relating time, temperature and stress are needed to use creep data in problems of stress analysis.

The dislocation theory, and the appreciation that creep involves diffusion (at least in crystalline structures), enabled fundamental relationships between creep and other material characteristics, such as the coefficients for self diffusion, to be developed for pure metals and dilute alloys. However, in spite of progress in relating the shape of the creep curve to the atomic mechanisms by which creep occurs, present understanding of the process is not so complete as to eliminate the need for empirical equations. Most creep–resistant materials used in practice have very complex microscopic and submicroscopic structures. Hence, while it is on these materials that most practical interest centers, these are also the least amenable to a mechanistic understanding and interpretation.

So many creep expressions have been developed over the years that only a few can be discussed in the scope of this text. One of the first significant equations was proposed by Andrade.

$$l = l_o(1 + \beta t^{\frac{1}{3}})\, e^{\kappa t} \tag{16.1}$$

where l = length at time t

l_o, β and κ = empirically determined constants

The constant, l_o, is approximately equal to the length of the specimen after applying the instantaneous load. When β is made equal to zero, and Eq. (16.1) is differentiated with respect to time, the constant, κ, is found to represent a constant extension rate per unit length; i.e.,

$$\kappa = \frac{1}{l}\frac{dl}{dt} \tag{a}$$

Flow at a uniform strain rate, according to (a), is termed κ-flow, and at a constant stress would be identical with viscous flow. By definition, strain rate is proportional to stress during viscous flow (see Eqs. 8.5). Since κ-flow is not proportional to stress, it is sometimes referred to as *quasi-viscous flow*. Special significance is also attached to β by Andrade. When $\kappa = 0$,

$$l = l_o(1 + \beta t^{\frac{1}{3}}) \tag{b}$$

and

$$\frac{dl}{dt} = \frac{d\varepsilon}{dt} = \tfrac{1}{3} l_o \beta t^{-\frac{2}{3}} \tag{c}$$

Thus the term β-flow is used to describe deformation whose rate decreases with time. The similarity between β-flow and primary creep, on the one hand, and κ-flow and secondary creep, on the other, is obvious.

By using Andrade's concept of two types of flow, the general shape of the creep curve can be described by the ratio of κ to β. When the ratio is high, creep approaches a steady state or quasi-viscous-condition; if it is low, the creep curve approaches a $t^{\frac{1}{3}}$ form. The ratio, κ/β, is similar for metals when they are tested at equal *homologous temperatures*, θ. The latter is defined as the testing temperature divided by the melting point of the metal, both on

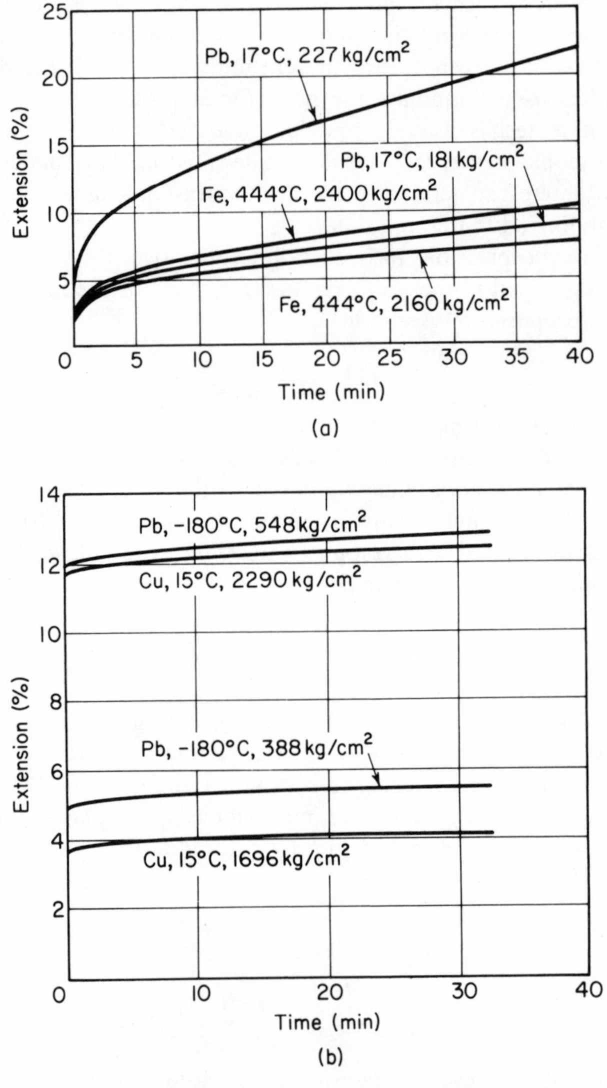

Fig. 16.7 (a) Creep of lead at 17°C. and iron at 444°C. (b) Creep of lead at −180°C. and copper at 15°C. From Andrade in (R16.1).

the absolute scale. This shape similarity is shown in Fig. 16.7. At 17°C., θ for lead is 0.48 and, for iron at 444°C., it is 0.40 (Fig. 16.7a). Lead at −180°C. and copper at 15°C. have θ values of 0.16 and 0.21, respectively (Fig. 16.7b).

Equation (16.1) is an adequate description of the behavior of pure metals and some alloys and plastics during the first two stages of creep. It is too restrictive, however, to represent the great variety of shapes of creep-time curves that are found, so a great number of other laws have also been proposed.

The general shape of the curves suggests that they are close to linear when plotted as log strain vs. log time (illustrated in Fig. 16.8). The equation describing this behavior is

$$\varepsilon = At^{a} \tag{16.2}$$

where A and a are constants.

Differentiation of Eq. (16.2) yields

$$\dot{\varepsilon} = aAt^{(a-1)} \tag{16.3a}$$

Fig. 16.8 Creep curves for three materials plotted on log-log coordinates. In each case curves (b) refer to a higher applied stress than (a). From R. G. Sturm, C. Dumont, and F. M. Howell, *Trans. ASME*, **58**, A62 (1936).

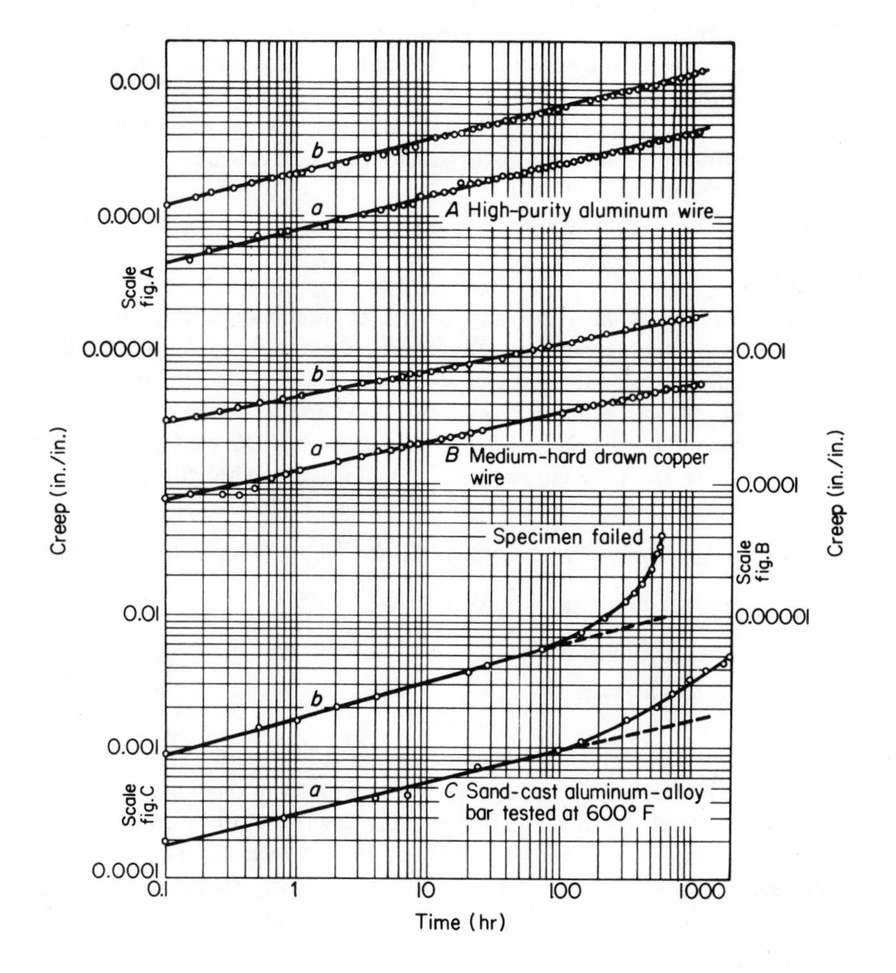

Equation (16.3a) has the same form as the one proposed by Cottrell on the basis of examining many of the various time laws.

$$\dot{\varepsilon} = bt^{-n} \tag{16.3b}$$
$$\text{where } b = aA \text{ and } n = 1 - a$$

A large number of differently shaped creep vs. time curves are possible with Eq. (16.3b) depending on the value of n.

When $n = 0$, the creep rate is constant, representing secondary (or κ) creep. This behavior is most commonly found at high temperatures. Assigning, in turn, a value of unity to n, and integrating Eq. (16.3b), gives

$$\varepsilon = a' \ln t + b \tag{16.4}$$

This is termed *logarithmic creep*, a type displayed by rubber, glass and some concretes. Equation (16.4) suggests that creep is an exhaustion process. This also occurs in metals at low temperatures as would be expected on the basis of the mechanism discussed in Chapter 8. Upon initial load application, dislocations move easily but, as deformation progresses, their interactions make additional movement more difficult. When the temperatures are low, diffusion and consequent climb are limited, and the trapped dislocation becomes immobile.

When $0 < n < 1$, integration of (16.3b) yields

$$\varepsilon = a'' t^m + c \tag{16.5}$$

This is termed *parabolic creep* and occurs at intermediate and high temperatures; a'' increases exponentially with stress and temperature, while m decreases with stress and increases with temperature. When $m = \frac{1}{3}$, Eq. (16.5) is identical with Andrade's expression in the absence of κ flow.

Expressions (16.4, 16.5) were first proposed in 1936 by Gentner, who divided creep vs. log time curves into: (1) concave downward, (2) straight, or (3) convex downward. He defined these three, respectively, as

$$\text{parabolic:} \quad \varepsilon = a_1 t^{m_1} + t b_1 \; (0 < m_1 < 1, a_1 > 0) \tag{16.6}$$

$$\text{exponential:} \quad \varepsilon = a_2 \log t + b_2 \; (a > 0) \tag{16.7}$$

$$\text{hyperbolic:} \quad \varepsilon = a_3 t^{m_3} + b_3 \; (m_3 < 0, b < 0) \tag{16.8}$$

The last, like logarithmic creep, is useful in describing the behavior of concrete.

All the equations given above were designed to describe the creep-time curve for the first two stages of creep. For any material these are, of course, affected by the applied stress and temperature as shown in Fig. 16.9. Experience in incorporating the effects of stress and temperature into these equations is limited. Because the problem is so complex, most attempts to add the influence of stress on the creep curve are restricted to its effect on the

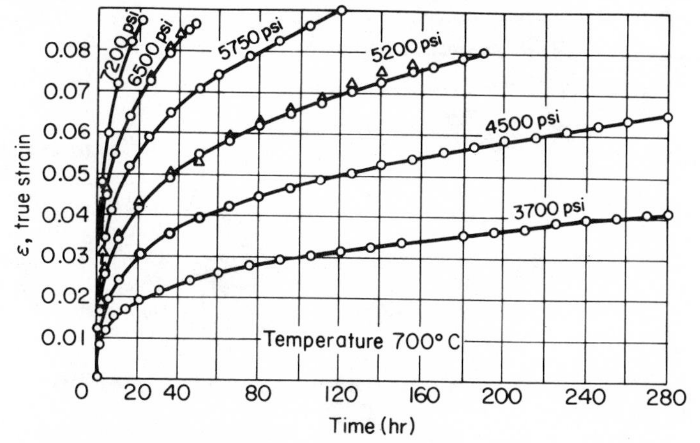

Fig. 16.9 Creep curves with stress as a parameter. From E. R. Parker, *Trans. Am. Soc. Metals*, **50,** 52 (1958).

minimum creep rate; e.g.,

$$\dot{\varepsilon} = B\sigma^n \ (n < 1) \tag{16.9}$$

Assuming that σ is independent of time, Eq. (16.9) can be integrated to yield

$$\varepsilon = Bt\sigma^n + C \tag{16.10}$$

If the constant C is small compared with $Bt\sigma^n$, and can be neglected, the so-called "log-log" creep stress–time relationship is obtained. An equivalent but more convenient form of the above expression is

$$\varepsilon = B't\left(\frac{\sigma}{\sigma_1}\right)^n \tag{16.11}$$

where σ_1 is some arbitrary value of stress. The term "log-log" results from the fact that the minimum creep rate and stress are linearly related on a log-log plot. So long as the instantaneous deformation and the amount of primary creep are small compared with secondary creep (e.g., long time applications where modest amounts of creep do not interfere with the operation of the equipment) Eq. (16.11) has appreciable use. Commercial data are frequently presented as $\dot{\varepsilon}$ vs. stress curves, with temperature as the parameter (Fig. 16.10). With these curves, designers calculate the stress, at a particular temperature, that will not cause more than a tolerable amount of creep over the expected life of the member.

A variety of other expressions have been proposed, including a modification of Eq. (16.11) by Marin and Hu, to take account of the instantaneous

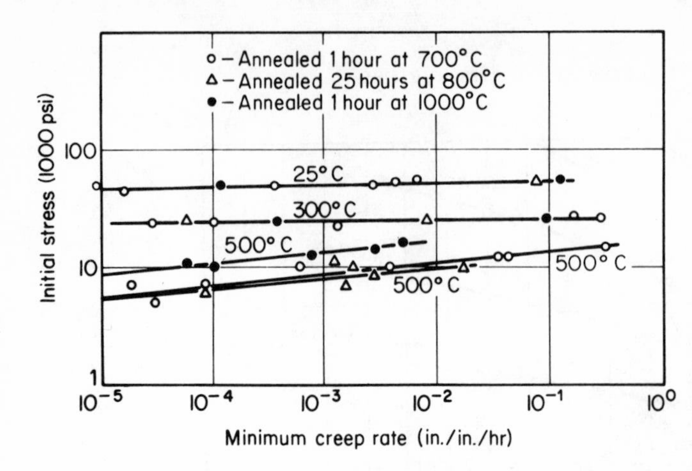

Fig. 16.10 Initial stress vs. minimum creep rate on a log-log plot for unalloyed zirconium. From R. W. Guard and J. H. Heller, *Trans. Am. Soc. Metals*, **49**, 449 (1957).

extension. This is done by adding a term, ε_o, to the expression for initial elastic and plastic flow:

$$\varepsilon = \varepsilon_o + B't\left(\frac{\sigma}{\sigma_1}\right)^n$$

or letting

$$\varepsilon_o = B''\left(\frac{\sigma}{\sigma_1}\right)^{n'},$$

$$\varepsilon = B''\left(\frac{\sigma}{\sigma_1}\right)^{n'} + B't\left(\frac{\sigma}{\sigma_1}\right)^n \tag{16.12}$$

If the instantaneous, primary, and secondary creep must all be considered in a creep expression, the total strain is most easily found by adding the various deformation components (as was done in the rheological models in Chapter 8). Hence, by superposition, the most general expression for creep is

$$\varepsilon = \frac{\sigma}{E} + k_1\sigma^m + k_2(1 - e^{-qt})\sigma^n + k_3t\sigma^p \tag{16.13}$$

If n and p are both unity, the terms in Eq. (16.13) are recognizable as

$\varepsilon_1 = \dfrac{\sigma}{E}$ —elastic (Hookean) (see Eq. (5.1)) (16.13a)

$\varepsilon_2 = k_1\sigma^m$ —plastic (see Eq. (10.6)) (16.13b)

$\varepsilon_3 = k_2(1 - e^{-qt})\sigma$ —anelastic* (see Eq. (8.13d)) (16.13c)

and

$\varepsilon_4 = k_3t\sigma$ —viscous (Newtonian) (see Eqs. (8.2) and (8.11)) (16.13d)

* In Table 8.3, the Kelvin body that is represented by this equation is said to represent "retarded elasticity" and anelasticity is represented by the more general Poynting-Thomson body. The former, however, is simply a special case of the latter.

The need for exponents n and p points out a shortcoming of the rheological models described in Chapter 8; i.e., that their response to load was always linear. According to Eq. (16.13), neither the anelastic nor the viscous components of strain are linearly related to stress. In many practical problems the stress is not high enough to cause plastic flow so that this component is neglected in Eq. (16.13).

The equations discussed above were developed empirically although they are generally consistent with the proposed mechanisms for creep. Investigators who attempted to include the influence of temperature in these creep-time–stress expressions have generally examined the fundamental aspects of creep rather than attempting to develop purely empirical equations. This approach has the obvious virtue that it is not only descriptive but, more important, explains creep in terms of its mechanism and, hence, should require no arbitrary constants. Its disadvantage is that it is only applicable to simple systems like pure metals and dilute alloys.

Dorn* has made a notable contribution to understanding creep by taking this approach. He emphasized, as did others, that creep only occurs in the presence of some thermally activated deformation process; otherwise strain at any temperature would depend only on stress. If a single thermally activated process were involved, the strain rate would be given by

$$\dot{\varepsilon} = Ae^{-\Delta F/RT} = Ae^{\Delta S/R}(e^{-\Delta H/RT}) \tag{16.14a}$$

where $\dot{\varepsilon}$ = creep rate

$\Delta F = \Delta H - T\Delta S$ = free energy of activation

ΔH = energy of activation

ΔS = entropy of activation

R = gas constant

T = absolute temperature

A = strain per second per activation

He then showed that, over a range of high temperatures, a constant activation energy for creep is obtained in some metals (Fig. 16.11), indicating that only one type of deformation is controlling. In discussing creep in Chapter 8, it was shown (Fig. 8.7) that the activation energy for high temperature creep and for self-diffusion is the same, indicating that dislocation climb is the controlling factor.

Since only a single deformation mechanism is active, it is apparent that ΔS and ΔH in (16.14) are constants. Consequently, the creep rate at any instant depends on the applied stress, σ, and the amount by which the material has already been strained. The latter can be treated as an instantaneous structure parameter of the material, s. Hence, (16.14a) can be rewritten as

$$\dot{\varepsilon}e^{\Delta H/RT} = Z = f(\sigma, s) \tag{16.14b}$$

---

* See for example, pp. 225–283 in (R16.1) and pp. 215–228 in (R18.8).

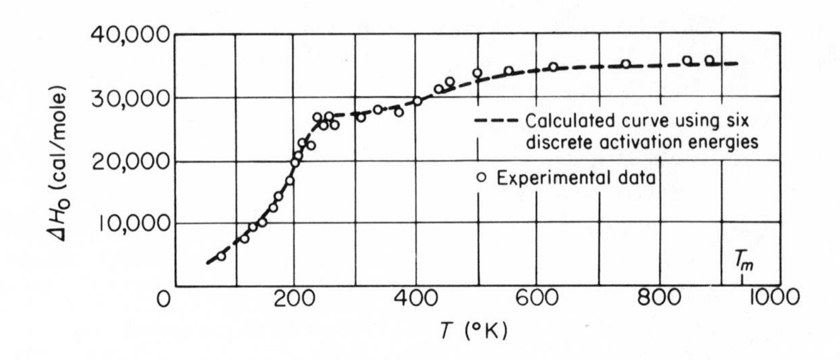

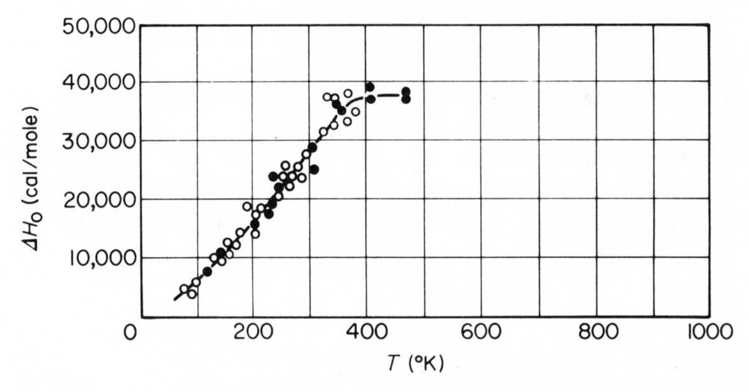

Fig. 16.11 Activation energies for: (a) Pure aluminum; (b) Pure copper as a function of absolute temperature. From J. E. Dorn in (R16.1).

Factor s, since it depends on complete strain history, will, of course, vary with the time over which creep occurred. To account for this dependence of s on both time and temperature, a temperature-compensated time was introduced, and defined as

$$\theta = te^{-\Delta H/RT} \tag{16.15}$$

where t = duration of test.

Consequently, at a constant stress,

$$\dot{\varepsilon}e^{\Delta H/RT} = f_1(\theta) \tag{16.16}$$

At constant temperature, $d\theta = dte^{-\Delta H/RT}$ from (16.15). Thus, $e^{\Delta H/RT} = dt/d\theta$ and after inserting the latter in (16.16), one obtains

$$d\varepsilon = f_1(\theta)\, d\theta$$

On integrating,

$$\varepsilon = f_2(\theta) \tag{16.17}$$

The validity of (16.17) is supported by the data of Fig. 16.12.

Since all of the points in Fig. 16.12 fall on the same curve, secondary creep also occurs at a constant value of θ, termed θ_s, which again is only a function of stress.

$$\dot{\varepsilon}_s e^{\Delta H/RT} = F(\sigma) \tag{16.18}$$

This relationship for high purity aluminum is shown in Fig. 16.13.

Equations (16.14 to 16.18) do not specify the relationships between creep

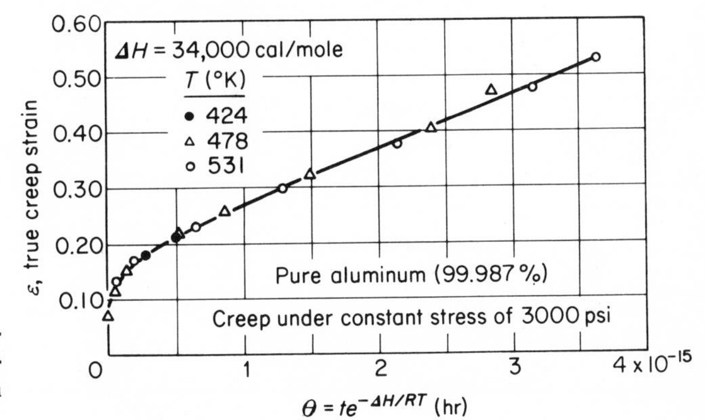

Fig. 16.12 Strain as a function of temperature compensated time. From J. E. Dorn in (R16.1).

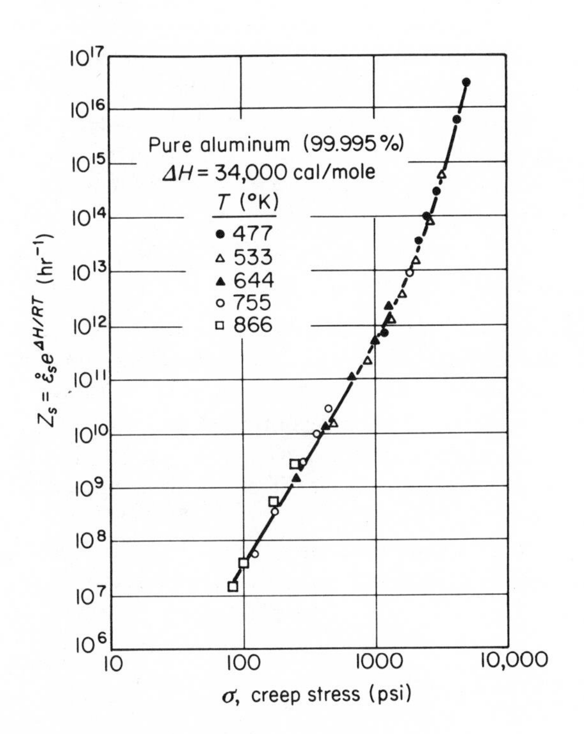

Fig. 16.13 Correlation of secondary creep rate data by means of Z. From J. E. Dorn in (R16.1).

deformation and stress. Dorn found that two different expressions were required to take into account this additional parameter.

$$\dot{\varepsilon} = s' e^{-\Delta H/RT} e^{\beta\sigma} \quad (16.19a)$$

was proposed to describe high temperature creep at high stresses, and

$$\dot{\varepsilon} = s'' e^{-\Delta H/RT} \sigma^n \quad (16.19b)$$

for low stresses,
where s' and s'' are structure parameters,
β = a material constant independent of temperature and precreep for annealed metals,
n = material constant.
According to Eq. (16.19), the structure factors are multiplicative and are not dependent on stress; actually, experimental studies indicate that this is not quite the case.

RELAXATION

It was stated at the beginning of this chapter that creep is most commonly evaluated by applying a constant load or stress to a specimen and measuring its extension with time. With this type of loading, relatively large amounts of creep strain are considered since they are generally tolerable in service, and usually the minimum creep rate is of most interest. When elements such as bolts or gaskets are used in certain applications, their usefulness is lost even with the occurrence of small amounts of time-dependent strain. For these, primary creep, which was shown to be largely anelastic, is decisive. This is evaluated by studying the relaxation characteristics of the material.

Relaxation was discussed in Chapter 8, but a few of its practical aspects remain to be considered. The Kelvin body was used to describe the manner in which stress (or applied load) vanishes in members extended or compressed to some constant size. Extension with time was given as

$$\delta = F/B(1 - e^{-t/T_k}) \quad (8.13d)$$

The property that determined relaxation was the retardation time, T_k. A T_k value obtained for a single parallel spring and dash-pot arrangement generally does not duplicate the behavior of real materials as well as groups of parallel elements do (see Chapter 8). To demonstrate the advantage of using four Kelvin elements in series as compared with using just one, Lubahn (1953) compared the creep recovery (anelasticity or dimensional change on unloading) of a steel at 800 degrees F. with the two models (Fig. 16.14).

In general, the accuracy of fit of a calculated curve to the experimental data can be improved as the number of terms is increased. This leads to the concept of a relaxation spectrum consisting of an infinite number of terms; i.e.,

$$\delta = \sum_{i=0}^{\infty} a_i(1 - e^{-t/T_{ki}}) \quad (16.20)$$

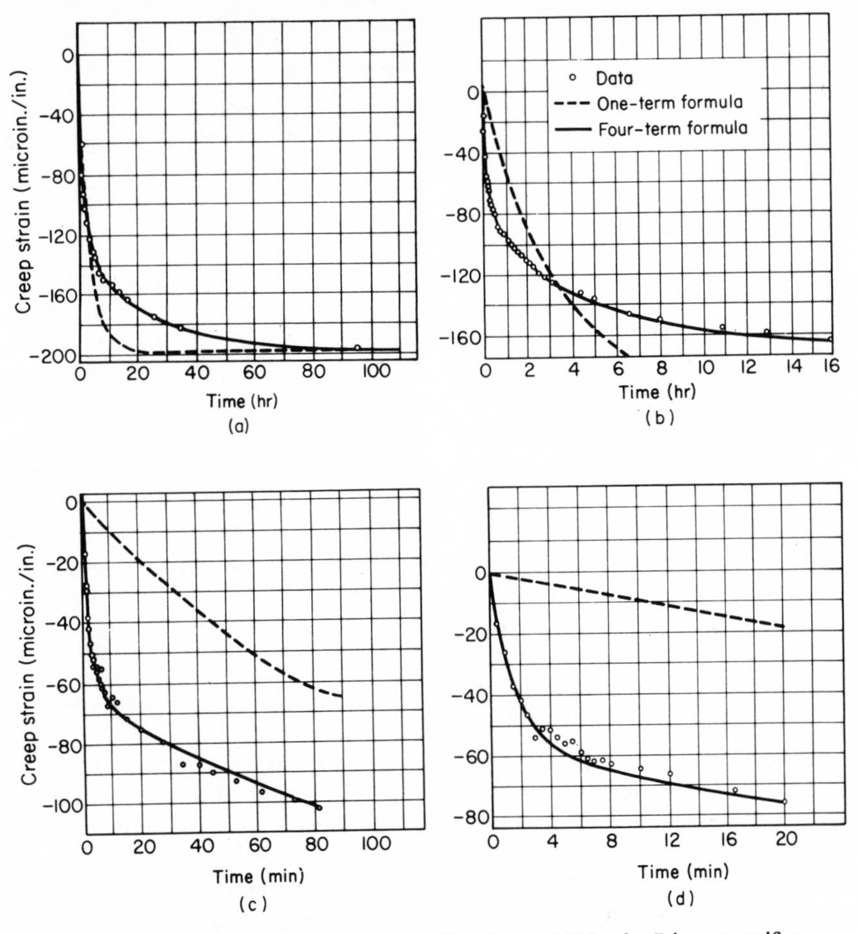

Fig. 16.14 Strain-time curve after unloading from 46,000 psi; *B* is a magnification in time of *A*; likewise *C* of *B*, and *D* of *C*. 800°F. From (R16.3).

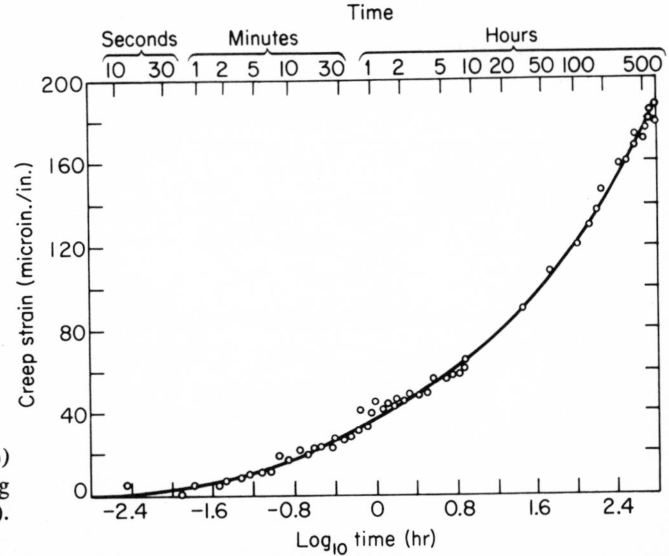

Fig. 16.15 Fit of Eq. (16.20) experimental points to using twelve terms. From (R16.3).

The creep curve in Fig. 16.15 was fitted to the data points by using twelve terms of Eq. (16.20).

STRESS RUPTURE

Not only is deformation time-dependent, but fracturing is time-dependent as well. Fracture properties are measured by means of stress-, or creep-rupture, tests. These are essentially identical to creep tests but strains are not measured during the test. Hence, the test apparatus can be simpler, and, for this reason, these tests are less costly than creep tests and are preferred, especially in studies such as material surveys.

Reduced ductility in long-time, high-temperature tests is frequently associated with voids that formed at the grain-boundaries. Grant and his associates have shown (R16.1) that fracturing is a means for relieving stress near the boundaries when slip is made difficult by restraint. In addition, many metals become embrittled by microscopic or submicroscopic structural changes at the high temperatures and stresses.

PARAMETER METHOD FOR PREDICTING CREEP-RUPTURE PROPERTIES

Prediction of rupture times at various stresses and temperatures are based on the use of time-temperature parameters. This method differs from those discussed above. In the latter the creep vs. time curve was described by an equation whose constants could be evaluated in short time tests, and the data so obtained was then used in the same equation for long times. Alternatively, predictions could be based on the obvious procedure of extending isothermal creep curves to longer times. In the parameter method an attempt is made to "buy" time with temperature. This appears reasonable since one would expect a time-temperature relationship in high temperature deformation. Unfortunately, as Dorn points out (see Fig. 16.11) this is strictly applicable only when the same reaction controls the process over the investigated temperature range. In spite of this shortcoming of the complete interchangeability of time and temperature, the parameter methods are frequently used and have been successful in many cases for predicting long time behaviors.

A temperature-compensated time relationship similar to Eq. (16.15) is the basis for two of the three common parameter methods. On taking common logarithms of (16.15), one obtains

$$\log \theta = \log t - k\Delta H/RT$$

or

$$\log t = \log \theta + \left(\frac{k}{R}\right) \Delta H \left(\frac{1}{T}\right) \tag{16.21}$$

where $k = \log e = .434$

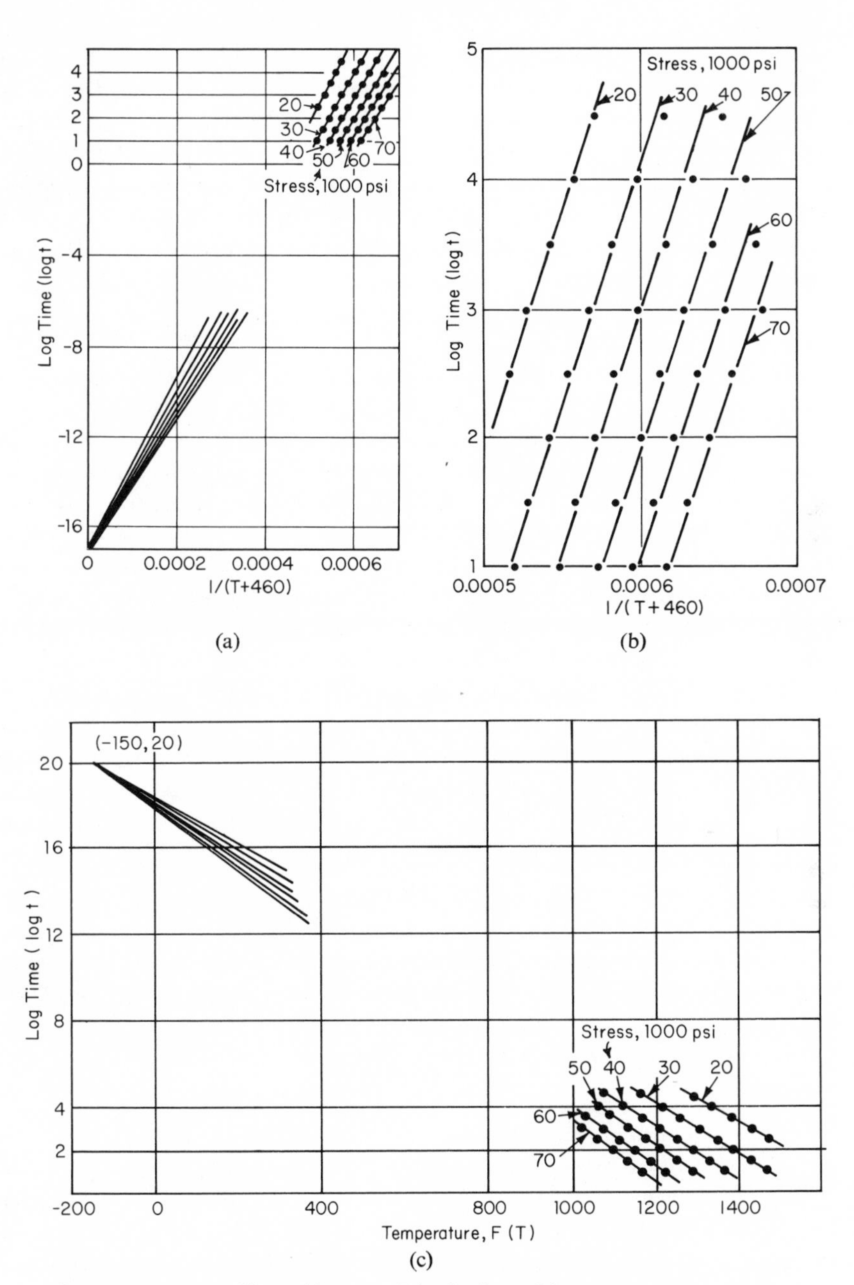

Fig. 16.16 Log t vs. $1/T$ and log t vs. T for finding: (a) Intercept, constant C for Larson-Miller. (b) Slope ΔH for Dorn. (c) Intersection point (log t_a, Ta), for Manson-Haferd parameters. From R. M. Goldhoff, *Materials in Design Engineering*, **49,** No. 4, 93 (1959).

which is linear in $\log t$ and $1/T$ at a given stress if θ and $\Delta H/R$ are functions of stress only. On this plot the ordinate intercept is $-\log \theta$ and the slope, $k\Delta H/R$. The *Larson-Miller parameter* is based on these authors' conclusion that the $\log t$ vs. $1/T$ curves have a common intercept when plotted with stress as the parameter. By the use of reaction rate theory, they found the intercept to be -20. The slope (i.e., $k\Delta H/R$), on the other hand, is a function of stress, as shown in Fig. 16.16a. With temperature in degrees Rankin, Eq. (16.21) is rewritten as

$$(T + 460)(\log t + C_1) = P_1 \tag{16.22}$$

where C_1 = intercept on $\log t$ axis on a plot of $\log t$ vs. $(1/T + 460)$
and $P_1 = f(\sigma)$ = slope = kH/R is the Larson-Miller parameter.

With (16.22) as the relation between time, temperature, and stress, the stress to cause rupture is plotted as a function of P_1 (for each combination of T and t that led to fracture). A single master curve for each σ is thus obtained (Fig. 16.17). Although Larson and Miller suggested the value of -20 for C, these originators of the parameter, as well as other investigators who have evaluated the agreement between data and the master curve, found that the intercept varied with material, so that the use of the parameter is improved by experimentally establishing C for the material of interest.

Orr, Sherby, and Dorn propose the use of θ in Eq. (16.15) as a parameter. It has been used by these authors for pure metals and nonferrous alloys, but has not been generally accepted for high temperature alloys. Again taking the logarithm of the equation, and assuming ΔH to be constant, curves of $\log t$ vs. $1/T$, with stress as the parameter, would be a family of parallel lines (Fig. 16.16b). Hence the *Orr-Sherby-Dorn parameter* would be written as

$$P_2 = te^{-\Delta H/RT_k} \tag{16.23}$$

where T_k = degrees Kelvin

A parameter based on an empirical relationship between rupture time and minimum creep rate was also proposed by Monkman and Grant. They found for a variety of alloys that

$$\log t_r + m \log \dot{\varepsilon}_c = \dot{C} \tag{16.24}$$

where t_r = rupture life,
$\dot{\varepsilon}_c$ = minimum creep rate,
m and C are material constants.

The most successful of the parameter methods was proposed by Manson and Haferd. Again the time-temperature relationship is empirical and, in this case, involves two disposable constants rather than just one as proposed by Larson and Miller. The parameter is based on the observation that constant stress plots of $\log t$ vs. T are more nearly linear than $\log t$ vs. $1/T$ curves. Because of their use of T rather than $1/T$, their procedure is referred to as the *linear parameter method*. Further, all of the $\log t - T$ curves

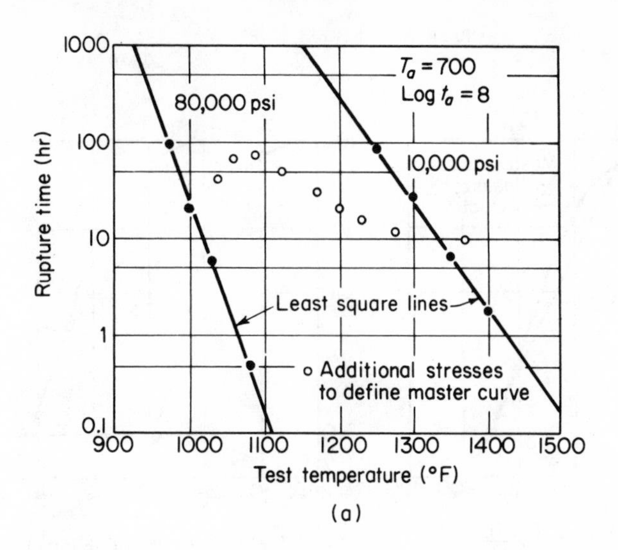

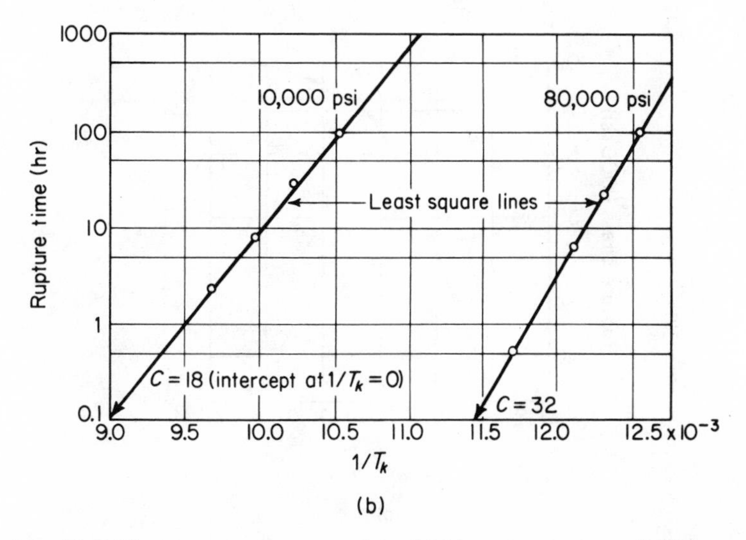

Fig. 16.17 Time–temperature curves for: (a) Linear parameter and (b) Larson-Miller and Dorn parameters. From S. S. Manson, G. Succop, and W. F. Brown, Jr., *Trans. Am. Soc. Metals*, **51**, 911 (1959).

intersect at the point $T = T_a$ and $t = t_a$ (Fig. 16.16c) leading to a family of curves,

$$T - T_a = P_3(\log t - \log t_a)$$

where $P_3 = f(\sigma)$

The *Manson-Haferd parameter* then is written as

$$P_3 = \frac{T - T_a}{\log t - \log t_a} \tag{16.25}$$

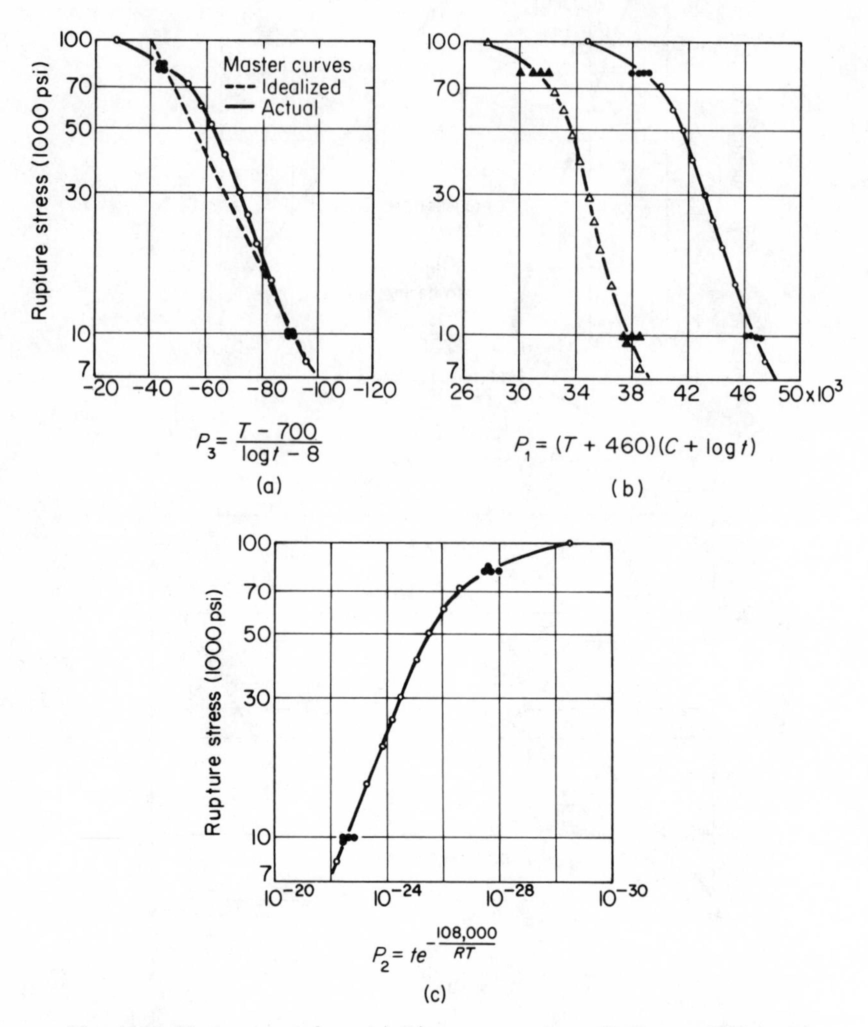

Fig. 16.18 Master curve for: (a) Linear parameter. (b) Larson-Miller. (c) Dorn. Same ref. as Fig. 16.17.

To develop a better understanding of how the parameter methods are used, the isothermal rupture stress vs. rupture time curves will be obtained using the linear, Larson-Miller and Dorn parameters. This example was taken from a paper published by Manson, Succop, and Brown in 1959 who demonstrated a detailed procedure for obtaining the stress rupture characteristics of a material with a minimum amount of total testing time.

A Cr-Mo-V steel commonly used at moderately elevated temperatures ("17-22-A" S) was selected for the study. Isostatic (constant stress) tests at two widely different stresses, which were chosen on the basis of a tensile strength vs. testing temperature curve, were first carried out to establish two rupture time vs. testing temperature curves (Fig. 16.17a). The intersecting point of these curves, found analytically, gives the values of T_a and $\log t_a$ for the linear parameter. Using these values for the intercept, and the additional data of Fig. 16.17a, rupture stress vs. P_3 was plotted (Fig. 16.18a).

The data in Fig. 16.17a are replotted as rupture time vs. $1/T$ for the Larson-Miller and Dorn parameters (Fig. 16.17b). The first of these methods requires the curves to converge to a negative point on the abscissa axis, and, to satisfy the latter, they must be parallel. Since the data followed neither of these behaviors, both an average value of 25 and the commonly accepted value of 20 were taken for C in Eq. (16.22), and an average of $\Delta H = 108{,}000$ was used in Eq. (16.23). Rupture stress vs. P_1 is plotted on Fig. 16.18b, and

Fig. 16.19 Isothermal stress-time curves based on master curve in Fig. 16.18. Same ref. as Fig. 16.17.

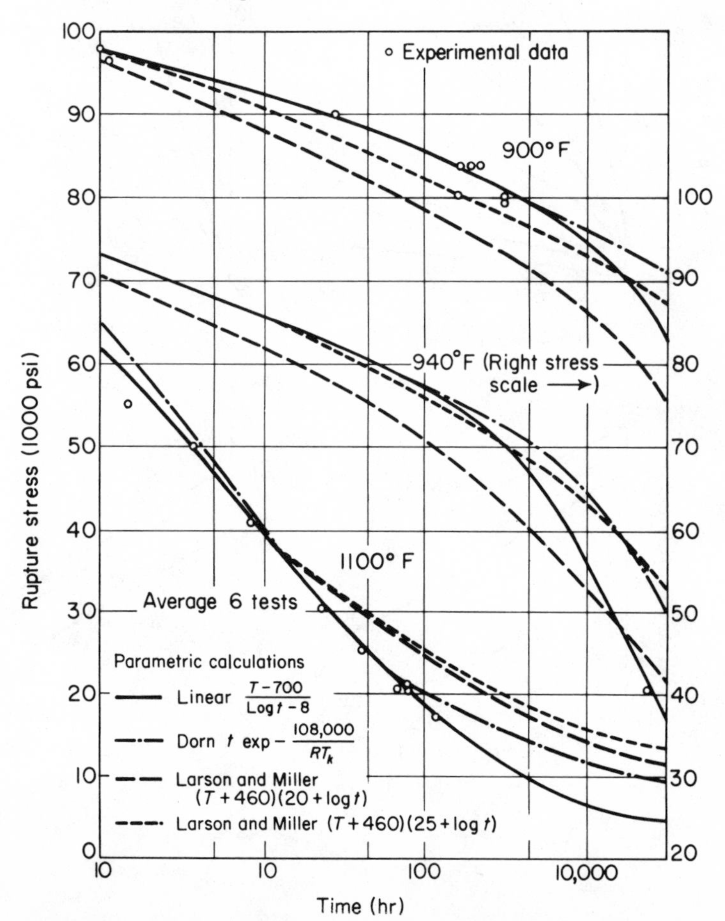

vs. P_2 in Fig. 16.18c. Additional tests, shown by the open circles in Fig. 16.17, were run to establish more accurately the rupture stress parameter curves. By using the latter, the isothermal rupture stress vs. rupture time curves, shown in Fig. 16.19, were obtained. Note in these that the linear parameter gave good agreement with the data points at times longer than 100 hours, even though the master curves were based on testing times shorter than this.

Although most investigators agree that the Manson-Haferd parameter is the most reliable, extrapolations are uncertain even with this, especially when the times are much longer than those for which data have been collected. Indeed, Lubahn compared the accuracy of predictions, using both the Larson-Miller and Manson-Haferd parameters, with simple extrapolation of isothermal curves. For his single steel and the data used, the parameter

Fig. 16.20 Different ways of plotting constant-load creep-test data. Cr–Mo steel at 525°C., stresses in tons (2,240 lb.) per in.² From I. Finnie and W. R. Heller, (R16.2).

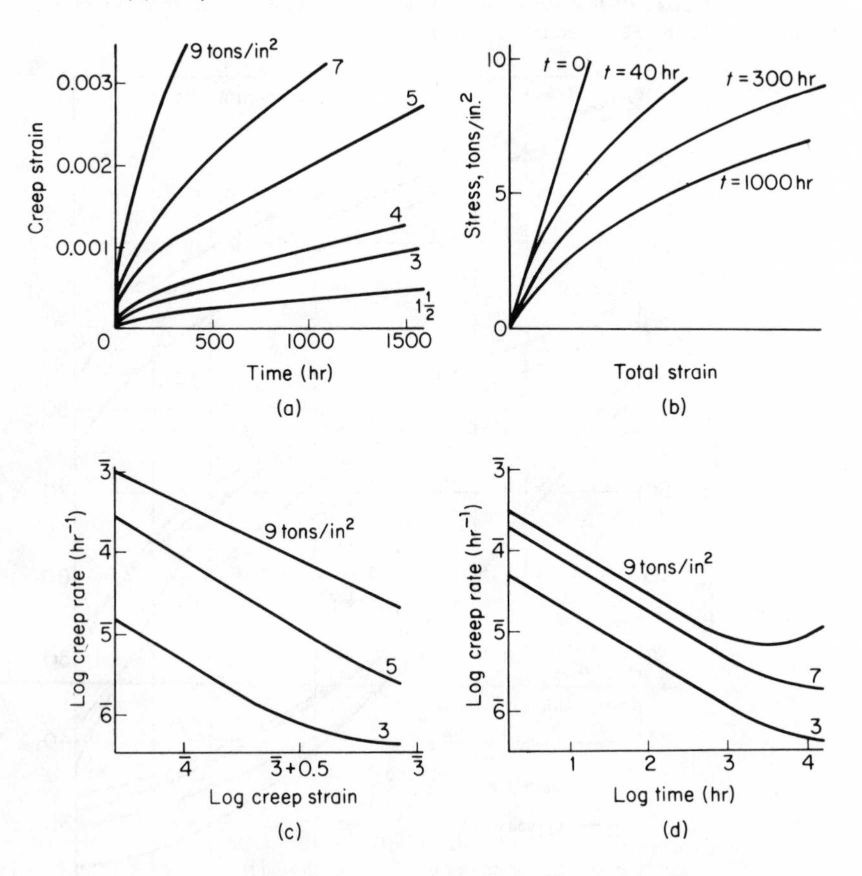

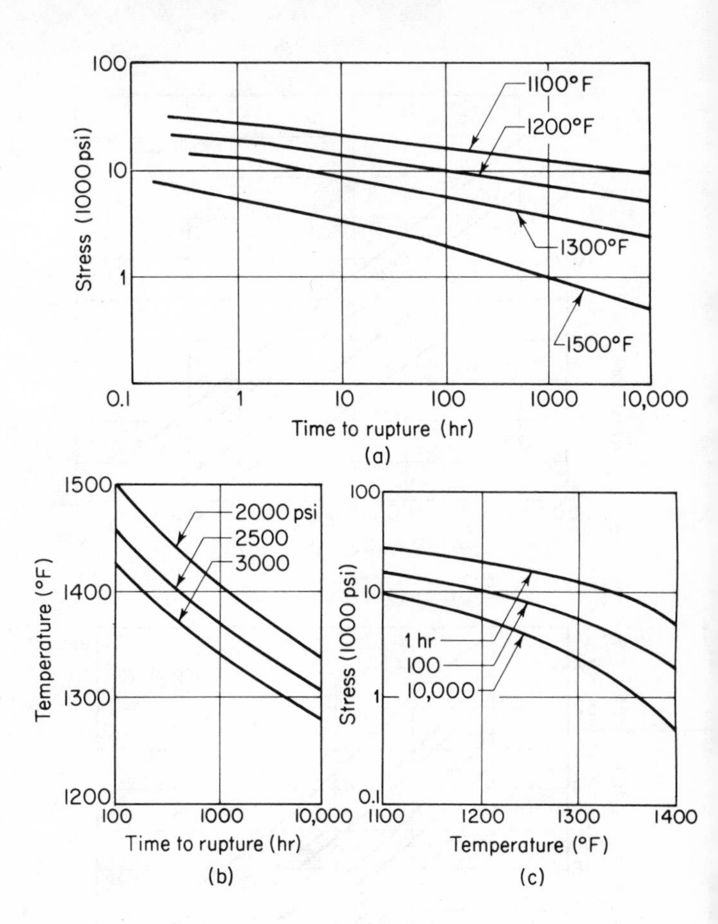

Fig. 16.21 Different ways of plotting rupture data. From I. Finnie and W. R. Heller, (R16.2).

methods were no more accurate than the isothermal extrapolation. It must be mentioned, however, that the extrapolations of an isothermal curve require a considerable amount of data at one temperature while, as shown by the example above, all the time-temperature data available on a particular material may be used in the parameter methods.

PRESENTATION OF CREEP AND STRESS-RUPTURE DATA

Depending on their intended use, creep and stress rupture data may be presented in a number of different ways. The constant temperature-creep-time data, in Fig. 16.20a, are replotted in three ways in Fig. 16.20b, c, and d. The fictitious isochronous stress–strain curves in (b) are useful because they can be described by the type of equations that are used for short time tensile tests and, hence, are helpful in problems of stress analysis. Curve (c) is also useful because of the ease with which it can be described analytically, and (d) is valuable for expressing the early stages of creep.

Stress-rupture data are also given in a variety of ways; three of these are shown in Fig. 16.21. It is sometimes helpful to combine the creep and rupture

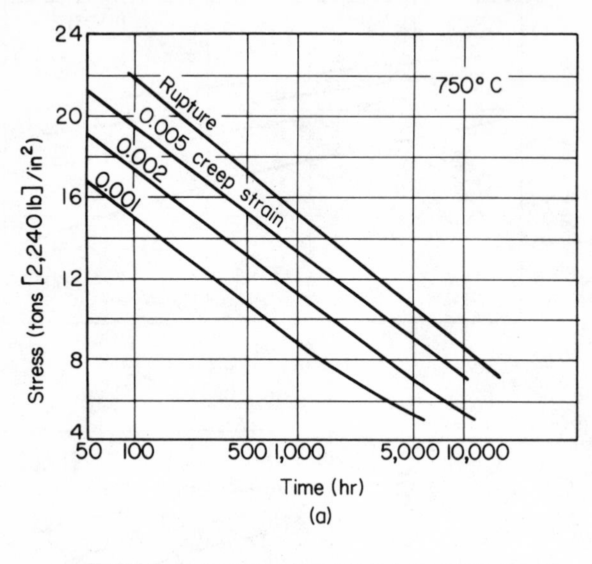

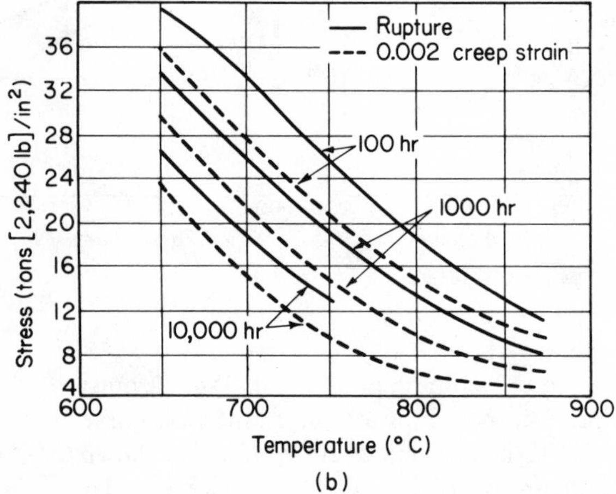

Fig. 16.22 Combined rupture and creep-strain data for a steel. From I. Finnie and W. R. Heller, (R16.2).

data (Fig. 16.22). The fact that these rupture and creep curves are parallel suggests that the parameter methods might be used for creep as well as rupture, and, indeed, this is frequently done. Rather than defining a master curve for rupture in this case, the curve refers to some specific amount of creep such as 0.02 per cent.

STRESS ANALYSIS OF CREEP DEFORMATION

Calculation of the stresses in an element subjected to creep follows procedures similar to those in elasticity problems, excepting that the relationship between stress and strain is no longer a linear one. As is the case with plastically deformed members, this nonlinearity complicates the problem, not only because the mathematics for even elementary problems is more involved but also because aids, such as the superposition principle, are no longer applicable.

Creep bending is analyzed as an example of a time-dependent deformation problem. The assumptions used in elastic bending (see Chapter 13) are also required here:

1. The beam is long, compared to its cross-sectional dimensions, so that shear deflections can be neglected.
2. Initially plane cross sections remain plane under load.
3. Bending deflections are small compared with the length of the beam.

To simplify the calculation of moment of inertia,

4. A beam, with a plane of symmetry normal to the neutral surface, is considered in this example.

To select a particular relationship between stress and strain:

5. Equation (16.9) is assumed to apply to every fiber of the beam.

Following the procedure used for elastic (13.7) or plastic (13.24) bending, the bending moment is

$$M = b \int_{-y_2}^{y_1} \sigma_x y \, dy \qquad (13.24)$$

where b = beam width
y_1 and y_2 = distance from neutral axis to outside fiber of beam
σ_x = fiber stress
y = distance from neutral axis to "test-element" fiber.

Integration of (13.24) requires an expression relating σ and y. Since the fiber strain (parallel with the beam axis) in each element, ε_x, is proportional to its distance from the neutral axis, we have

$$\varepsilon_x = \varepsilon_{x_1}\left(\frac{y}{y_1}\right) \qquad \text{(a)}$$

Differentiating with respect to time,

$$\dot{\varepsilon}_x = \dot{\varepsilon}_{x_1}\left(\frac{y}{y_1}\right) \qquad \text{(b)}$$

According to assumption (5) above,

$$\dot{\varepsilon} = B\sigma^n \qquad (16.9)$$

so that

$$\sigma_x = \sigma_{x_1}\left(\frac{y}{y_1}\right)^{1/n} \qquad \text{(c)}$$

Substituting (c) in (13.24), and integrating between the neutral axis and

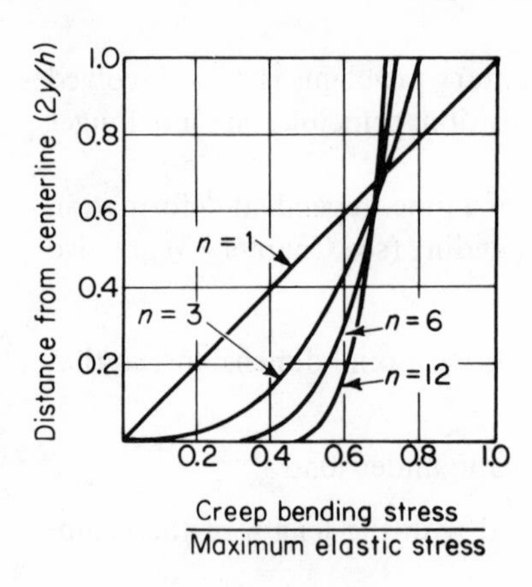

Fig. 16.23 Stress distribution in bending of a rectangular beam for different values of the exponent n. From I. Finnie and W. R. Heller, (R16.2).

the beam extremities (i.e., the limits $y_1 = h/2$ and $y_2 = -h/2$) yield for the maximum stress in the outer fiber

$$\sigma_{x_1} = \sigma_{\max} = \frac{Mc}{I}\left(\frac{2n+1}{3n}\right) \tag{16.26}$$

The fiber stress, σ_x, at any distance, y, from the neutral axis is found by solving Eq. (c) for σ_{x_1} above and substituting the latter in (16.26) to yield

$$\sigma_x = \frac{Mc}{I}\left(\frac{y}{c}\right)^{1/n}\left(\frac{2n+1}{3n}\right) \tag{16.27}$$

Equations (16.26, 16.27) are identical with the elastic solution (13.12) when $n = 1$. The manner in which the stress distribution varies with n is shown in Fig. 16.23. Note that the maximum surface stress in creep is lower than the value obtained by elastic analysis.

CREEP UNDER MULTIAXIAL STRESS

Because stress and strain or strain rates are not linearly related during creep, the method of superposition cannot be used. Hence, the stresses must be considered as a combined effect rather than considering the effect of each of them separately. This, of course, was also the case for plastic deformation, and it greatly complicates creep problems when multiaxial stresses are active. Nevertheless, procedures have been developed for treating this problem, and one of these is given here. By following this treatment, expressions similar to (5.2a) and (6.14), used for the elastic and plastic cases, respectively, will be developed. The relationship between strain rates and the three principal stresses is found by starting with the Eq. (8.3)

$$\dot{\varepsilon}_1 + \dot{\varepsilon}_2 + \dot{\varepsilon}_3 = 0$$

Equation (6.15) is then written in terms of strain rates rather than strain increments,

$$\frac{\sigma_1 - \sigma_3}{\dot{\varepsilon}_1 - \dot{\varepsilon}_3} = \frac{\sigma_2 - \sigma_1}{\dot{\varepsilon}_2 - \dot{\varepsilon}_1} = \frac{\sigma_3 - \sigma_2}{\dot{\varepsilon}_3 - \dot{\varepsilon}_2} \tag{16.28}$$

By combining (8.3) and (16.28), a generalized form, relating strain rate and stress, can be written as

$$\begin{aligned} \dot{\varepsilon}_1 &= \tfrac{2}{3}C[\sigma_1 - \tfrac{1}{2}(\sigma_2 - \sigma_1)] \\ \dot{\varepsilon}_2 &= \tfrac{2}{3}C[\sigma_2 - \tfrac{1}{2}(\sigma_3 - \sigma_1)] \\ \dot{\varepsilon}_3 &= \tfrac{2}{3}C[\sigma_3 - \tfrac{1}{2}(\sigma_1 - \sigma_2)] \end{aligned} \tag{16.29}$$

which is analogous to (6.14). The problem now becomes one of finding a value for C. To do this, it is assumed that (6.14) has the same relationship to (16.29) as (6.17) has to the sought equation. Essentially, this implies that the effective stress and effective strain have the same relationship as effective stress and *effective strain rate*, the latter being defined as the time derivative of the effective strain, $\dot{\varepsilon}_e$. Experiments have shown this to be the case.* From Eqs. (5.17c), (5.19), and (5.20) with $\nu = \frac{1}{2}$, values for the sought relationships (16.30) can be found from uniaxial tensile data since $\dot{\varepsilon}_e = \dot{\varepsilon}_1$ and $\sigma_e = \sigma_1$. Using these, one obtains

$$\begin{aligned} \dot{\varepsilon}_1 &= \frac{\dot{\varepsilon}_e}{\sigma_e}[\sigma_1 - \tfrac{1}{2}(\sigma_2 + \sigma_3)] \\ \dot{\varepsilon}_2 &= \frac{\dot{\varepsilon}_e}{\sigma_e}[\sigma_2 - \tfrac{1}{2}(\sigma_3 + \sigma_1)] \\ \dot{\varepsilon}_3 &= \frac{\dot{\varepsilon}_e}{\sigma_e}[\sigma_3 - \tfrac{1}{2}(\sigma_1 + \sigma_2)] \end{aligned} \tag{16.30}$$

PROBLEMS

1. The following data were collected on a $\frac{1}{2}$ inch diameter steel bar at 1000°F. over a 10 inch gauge length:

Time	*Total Extension*	*Time*	*Total Extension*
0	0	5	15.5×10^{-3} inches
5 sec	15.1×10^{-3} inches	10	16.0
10	15.1	30	17.1
30	15.2	1 hour	17.8
1 min	15.3	2	18.5
2	15.4	5	19.7

* A. E. Johnson, Metallurgical Reviews, *5*, 447 (1960).

Time	*Total Extension*	*Time*	*Total Extension*
10 hours	20.7	360	28.4
24	21.2	480	29.0
48	22.8	720	30.0
96	25.9	960	31.1
192	26.6	1200	32.0
240	27.2	1320	33.3
		1410	fracture

Plot strain vs. time and strain rate vs. time on both rectangular and log-log. What is the elastic strain, inelastic strain, primary, secondary, and tertiary creep strain? What is $\dot{\varepsilon}$ during secondary creep?

2. What is the value of A and a in Eq. (16.2) for primary and secondary creep?

3. Using the data shown below, plot rupture stress vs. the Dorn, Larson-Miller, and Manson-Haferd parameters. Use these as a direct extrapolation to obtain the rupture stress at 900 and 1000 F for 15,000 hours.

RUPTURE DATA FOR LOW ALLOY *Cr-Mo*-V STEEL*

Test Temp., °F.	*Stress, psi* $\times 10^{-3}$	*Rupture Time, Hours*
900	90	37
900	82	975
900	78	3,581
900	70	9,878
1000	80	7
1000	75	17
1000	68	213
1000	60	1,493
1000	56	2,491
1000	49	5,108
1000	43	7,390
1000	38	10,447
1100	70	1
1100	60.5	18
1100	50	167
1100	40	615
1100	29	2,220
1100	22	6,637
1200	40	19
1200	30	102
1200	25	125
1200	20	331

Test Temp., °F.	*Stress, psi* $\times 10^{-3}$	*Rupture Time, Hours*
1350	20	3.7
1350	15	8.9
1350	10	31.8

* From R. M. Goldhoff, and R. F. Gill, "Discussion to Manson, Succop and Brown, *Trans. Am. Soc. Metals*, 911 (1959).

4. Using the curves in Fig. 16.9, find the values of B and n in Eq. (16.9).

REFERENCES FOR FURTHER READING

(R16.1) *Creep and Recovery*. Cleveland: Am. Soc. Metals, 1957.

(R16.2) Finnie, I. and W. R. Heller, *Creep of Engineering Materials*. New York: McGraw-Hill Book Company, 1959.

(R16.3) Lubahn, J. D. and R. P. Felgar, *Plasticity and Creep of Metals*. New York: John Wiley & Sons, Inc., 1961.

(R16.4) Smith, G. V., *Properties of Metals at Elevated Temperatures*. New York: McGraw-Hill Book Company, 1950.

(R16.5) *Structural Processes in Creep*, Iron and Steel Institute, London, Spec. Report No. 70, 1961.

17

RESIDUAL STRESSES

The term *residual stresses* is used to describe the system of stresses that are "locked-in" a body even though external forces are not acting on it. These were mentioned in Chapter 5, where it was pointed out that they arise whenever the various parts within a member or structure are dimensionally mismatched. For the body to remain continuous, the mismatched portions, and the material in their vicinities, must distort so that each element has a shape that is compatible with that of its neighbors. The strains associated with this distortion give rise to stresses, and the latter are linearly related to the former, since only elastic stresses can be locked into a member. If the stresses were beyond the elastic limit of the material in some small volume, this element would deform plastically until the stresses were reduced to σ_o.

ORIGIN AND CLASSIFICATION OF RESIDUAL STRESSES

As an example of how residual stresses might arise in a body, consider the thermal-stress history to which the adhesive in a joint is subjected. Assume that the system consists of aluminum adherends (members being joined) and an epoxy adhesive assembled as in Fig. 17.1. Further, assume that the epoxy does not shrink during curing, nor creep appreciably over the range of temperatures between its curing temperature and ambient. When the liquid epoxy is poured into the shaded void in

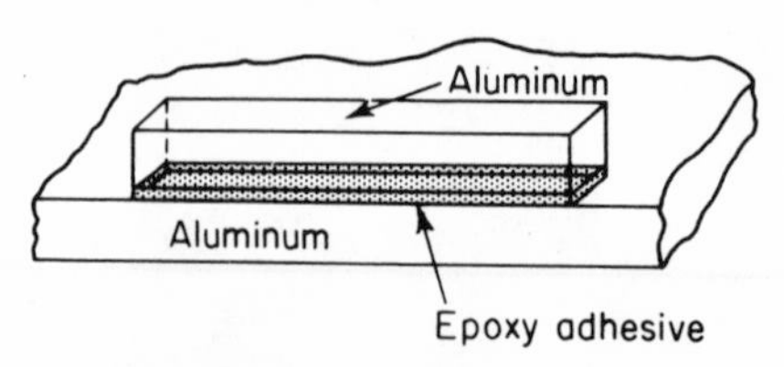

Fig. 17.1 Residual stresses are developed in heterogeneous structures such as an adhesive joint when its temperature is changed.

Fig. 17.1, it completely fills the cavity and adheres to both aluminum surfaces. After solidifying, both the aluminum and epoxy cool from the curing temperatures to ambient. The epoxy, however, has a higher coefficient of thermal expansion than the aluminum so that it tends to contract more than the latter. Adhesion at the interface prevents the epoxy from contracting as much as it would if it were free, and in so doing stresses are produced in the epoxy, while balancing opposite stresses arise in the aluminum. These residual stresses are not relieved unless the adhesive creeps so that they are added onto any stresses developed in the adhesive by the application of load during service.

In the adhesive example, residual stresses developed in a heterogeneous structure even though its temperature changed essentially isothermally. Nonuniform cooling of a homogeneous body can also cause residual stresses as pointed out in Chapter 5 and shown in Fig. 17.2. The ingot is initially stress free in (a) and at a uniform high temperature. As the outside of the ingot cools, (b), it contracts. The center of the ingot is still hot and hence extended, so that it opposes the contraction of the case, resulting in a tensile stress in the case and compressive stresses in the core. Because the core is hot, its yield strength is low, so that it readily contracts plastically, following the colder and, therefore, much harder outer case. As cooling continues, (c), the center which had flowed plastically continues to contract thermally. Eventually the entire body reaches a uniform temperature; because the core

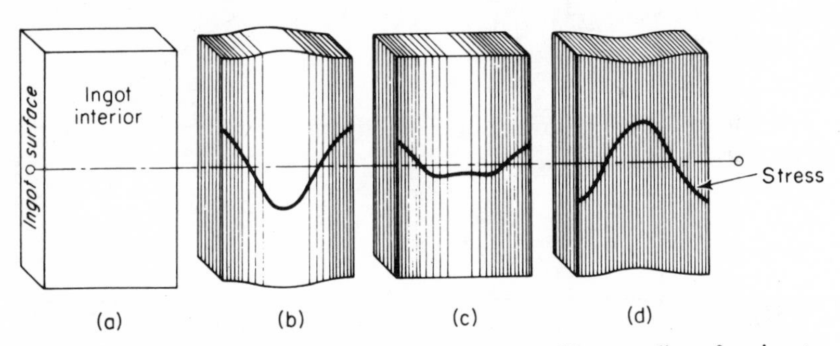

Fig. 17.2 Residual stress pattern produced by non-uniform cooling of an ingot (light areas are hot; shaded ones are cold).

had previously shrunk plastically, its total thermal plus plastic contraction exceeds that of the case. The differential contraction is opposed by the case so that, finally, a tensile stress develops in the core and a compressive one in the case, (d).

The question of whether or not a stress in an assembly is a residual stress is one of definition. There is no ambiguity in defining the stresses developed in the cooling ingot as residual. In the example of the adhesive joint, however, the assembly might be free of external loads, but its components are not. The same circumstance was found in the shrink-fit assembly discussed on p. 115 and would be encountered in any structure fastened by bolts. The stresses developed in the various components of an assembly are referred to as *reaction stresses*; they are not considered in this chapter because they are the province of structural mechanics.

Not only are residual stresses produced by the nonuniform volume changes resulting from transient temperature gradients on heating and cooling, but they occur due to the nonuniform shape changes that accompany deformation as well. This is readily seen by considering the stresses developed in a beam loaded beyond its elastic range and then unloaded. For simplicity, assume that the material does not strain-harden so that the stress distribution on bending is as shown in Fig. 17.3a (also see Fig. 13.24). All the section underwent plastic flow in varying amounts excepting near the neutral plane. Indeed, OAB is the tensile and ODE the compression stress-strain curve of the material. If the beam material has the same tensile and compressive yield strength, does not work harden, and the elastic deformation is small enough to be neglected then, by using the procedure leading to

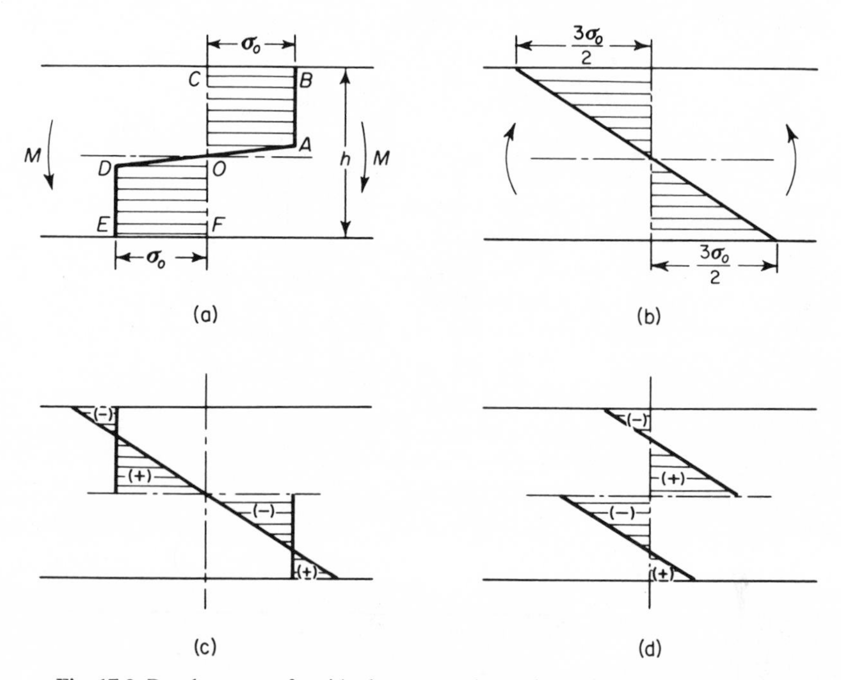

Fig. 17.3 Development of residual stresses when a beam is plastically bent and then the load released.

(13.11), the bending moment is $\sigma_0 bh^2/4$. When the beam is unloaded, it springs back and the applied bending moment vanishes by elastic deformation. The elastic-stress distribution that has the same magnitude of moment but an opposite direction from the plastic one is shown in Fig. 17.3b. On subtracting the plastic stresses that existed under load from the elastic ones released in unloading, the stress pattern shown in Fig. 17.3c is left so that the residual-stress pattern consists of two equal antisymmetrical parts on either side of the neutral surface (Fig. 17.3d).

Scale of Residual Stresses

A further classification of residual stresses is based on the relative sizes of the stressed volume and the entire body of which the former is a part. Note that the stresses developed in the cooling ingot varied rather slowly over the dimensions of the ingot. These are termed *macro-residual stresses*, or *body stresses*. In other cases, the stresses vary drastically over a small volume of the body. For example, in heat treating aluminum alloys very small particles of new phases are formed by precipitation due to solid-state reactions, and it is this that accounts for the strengthening. The precipitated particles crystallize in a system different from that of the matrix from which they formed and hence would occupy a different volume than the parent material. Obviously, this results in a dimensional mismatch, and large stress gradients develop within small volumes. These are called *microstresses*. The stress fields around dislocations are another example of the latter. It might be noted that a nonuniform distribution of microstresses can result in body stresses. If the ingot in Fig. 17.2 were an alloy that undergoes transformation on slow cooling, but not on fast, it would develop far more of the transformed phase in its interior than on its surface. This would result in a high density of microstresses in its interior which, in turn, would cause body stresses across the ingot (see Fig. 17.4).

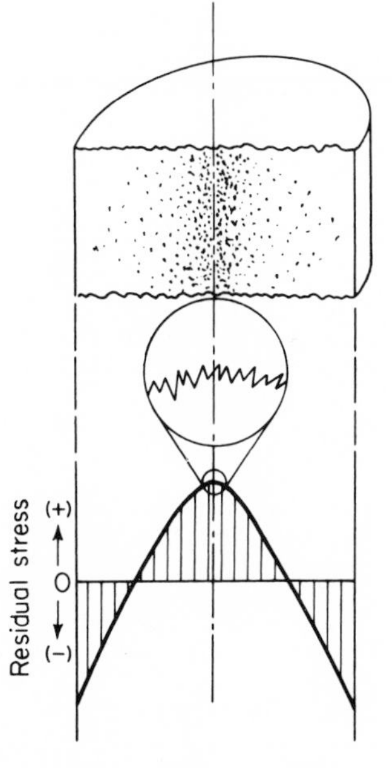

Fig. 17.4 A larger amount of precipitate in the center of the ingot than near the surface causes transverse residual stresses. The macro-stresses are caused by the wide difference in micro-stresses shown enlarged.

Residual Stresses Arising from Thermal Effects

One of the most common commercial thermal operations is the rapid cooling (quenching) or heating of metals such as in heat treating. This gives rise to thermal stresses as discussed on p. 113 to 116 while the various parts of the member are at different temperatures, and to residual stresses after the temperature is uniform. Normally materials contract on cooling, and the residual patterns that develop are similar to those that arise in the ingot discussed above. Steel, in addition, undergoes a phase change from face-centered cubic (austenite) to body-centered cubic (ferrite) (see Chapter 4) on cooling. Because the latter structure is less dense than the former, the metal expands rather than contracts in the temperature range of the phase change during cooling. If the steel has a low concentration of alloying elements, the phase change generally occurs at such a high temperature that it does not influence the residual-stress pattern. Certain alloying elements such as nickel cause this to occur at lower temperatures, and as shown in Fig. 17.5, a 17% nickel steel undergoes its transformation between

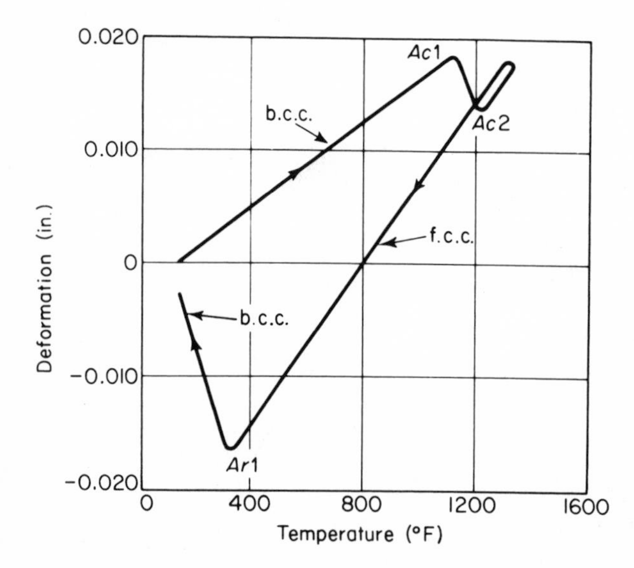

Fig. 17.5 Linear dimensional changes in a 17% nickel steel rod during heating and cooling. (On heating transformation b.c.c.→f.c.c. begins at A_{c1} and is complete at A_{c2}. On cooling f.c.c.→b.c.c. begins at A_{r1}.) From (R17.6), after Bühler and Scheil.

400°F and ambient. Hence, in cooling through this range, the metal expands rather than contracts. As a consequence, the residual stresses in a nickel-free steel and one containing 16.9% nickel are essentially opposite each other (Fig. 17.6).

The magnitude of residual stresses developed in any quenched member such as the cylinder described by Fig. 17.6a depends on two factors: one,

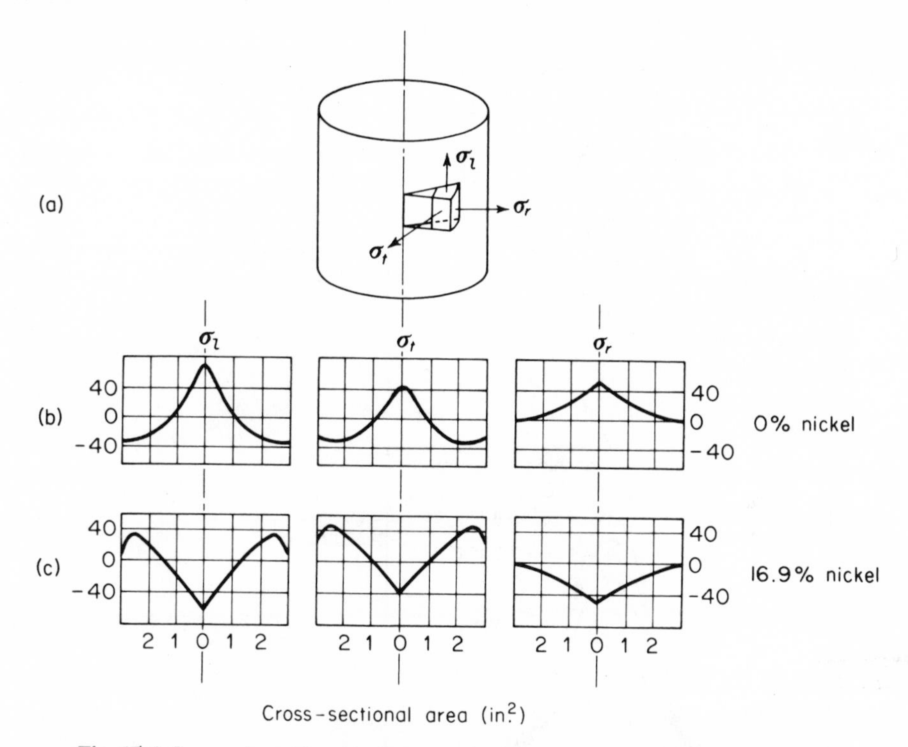

Fig. 17.6 Comparison of residual stresses in water-quenched 1.89 inch diameter bars: (a) symbols for stress direction; (b) Nickel-free steel and (c) 16.9 percent nickel-steel. From (R17.6), after Bühler and Scheil.

the stress-strain relationship for the metal, and two, the amount of elastic mismatch. Considering first a particular mismatch (or strain), the residual stresses increase with increasing elastic modulus of the material, but the maximum stress, regardless of other factors, is always limited by the yield strength of the metal at the high temperature.

The amount of mismatch is determined by a number of factors, the first two of which depend on the mechanical and physical properties of the material:

1. The shape of the yield strength-temperature curve: For any difference in case and core temperatures, σ_o at the higher temperature limits the magnitude of elastic stress that can be locked-in. As stated above, stresses in excess of σ_o are relieved by plastic flow. Further, when a member is quenched, the temperature difference between its case and core becomes less with increasing time. If, on cooling, σ_o remains low until the case and core temperatures are essentially identical, the maximum residual stress developed is also low. Materials such as Co and Ni base alloys (superalloys) and

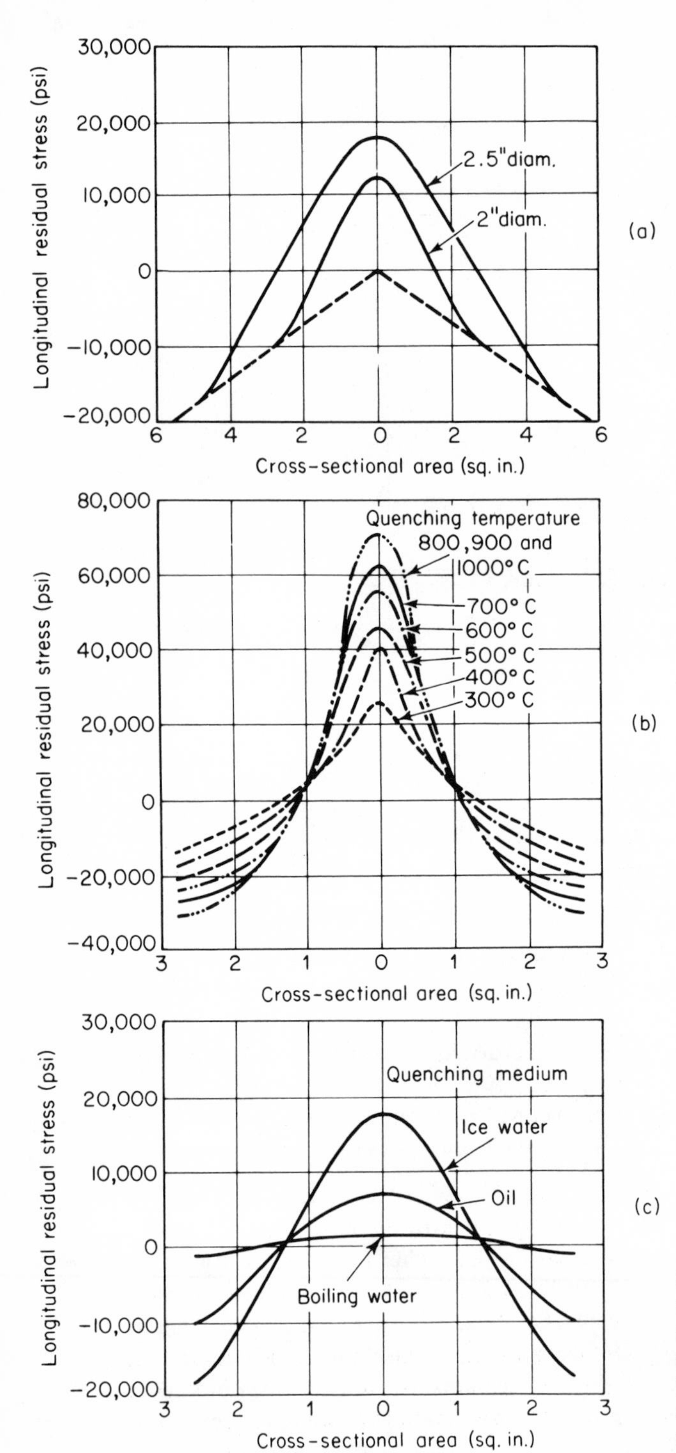

Fig. 17.7 Influence of (a) rod section size (b) temperature difference between rod and quenching medium (water in this case); and (c) quench severity on magnitude of residual stress. From (R17.1).

hot-work tool steels have high yield strengths even at high temperatures so that the residual stresses developed in these during quenching may be very high. Because these metals frequently are also brittle at ambient temperatures, the residual stresses can be extremely damaging.

2. The thermal-physical properties: Obviously, the strains that arise in cooling or heating a member are directly proportional to its thermal coefficient of expansion, κ, for any instantaneous temperature difference between case and core.

The value of the temperature differential is determined by the materials *thermal diffusivity;* the more quickly thermal energy is transferred through the body, the lower are the temperature differences that arise in it. The diffusivity $1/\lambda$ is defined as $\kappa/\rho C$, where κ is thermal conductivity, C is specific heat and ρ is density.

Another factor affecting the cooling or heating stresses is the size of the section, the larger the members the larger the stresses; also the higher the temperature differential and the more efficient the heat transfer from the cooling or heating medium, the higher the stresses. This influence of each of these three is shown in Fig. 17.7.

The above discussion is based essentially on quenching cylinders. If the shapes are more complex than this, the residual stresses that are developed may be even larger. In casting or molding metals or plastics, they frequently are allowed to cool at least partially in molds. Not only does this result in nonuniform extraction of heat, but also normal contraction may be hindered by sections of the rigid mold.

Another case in which residual stresses are developed due to nonuniform heating is in welding. As one might expect, spot welding results in a stress pattern that has rotational symmetry and, as shown in Fig. 17.8, the stress distribution is opposite to those developed in a shrink-or-press fit (Fig. 5.8). Welding along a line (such as in making a butt joint) causes a tensile residual stress across the weld due to the constraint exerted by the portion already welded and cooled down.

Fig. 17.8 Residual stress due to heating a spot; (a) hot area contracting away from surroundings when it cools; (b) Radial stress; (c) Tangential stress.

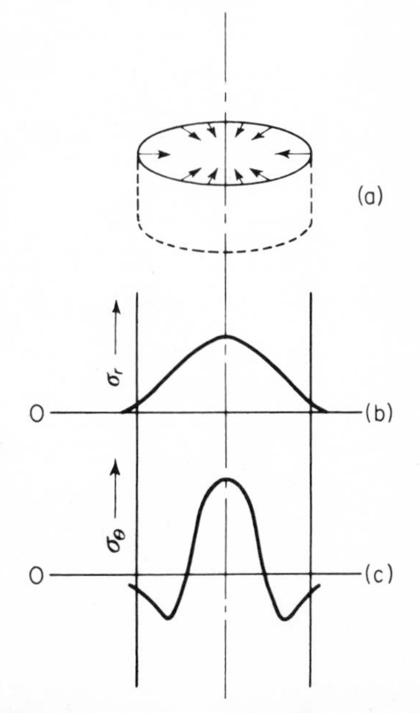

Residual Stresses Due to Nonuniform Deformation During Working

In the bending example above, it was shown that nonuniform plastic flow results in residual stresses when the applied load is removed. Essentially all cold working operations on metals (except simple stretching) cause nonuniform flow so that almost all worked members have in them residual stresses unless they have been subsequently removed by special treatments as discussed later in this chapter. Even a qualitative prediction of the stress patterns in worked shapes is difficult since there are two behaviors acting in opposition to each other, causing nonuniform flow. On the one hand, there is a tendency for the material to lag at the metal-tool interface due to friction. When this occurs, the deformation near the surface is less than in its interior. On the other hand, when the total deformation is small or when the tools are narrow (see Fig. 11.19) plastic flow does not penetrate the entire metal thickness, so that most of the deformation is concentrated near the surface. Rolling of a metal strip can serve as an example of the conditions under which these two occur. Figure 17.9 is a schematic diagram showing a strip of a particular thickness being rolled by small and by large rolls. In case (a) the penetration is shallow, and surface stresses are compressive (diagram b), because the surface metal wants to elongate more than the relatively rigid core permits. In (c), a lagging zone borders the rolls causing the center to deform more than the surface and the surface residual stresses are tensile (diagram d). The latter condition is analogous to the compression example given in Fig. 11.14. Surface working is favored by small rolls and small reductions, while the opposite conditions tend to work the center more. This same situation

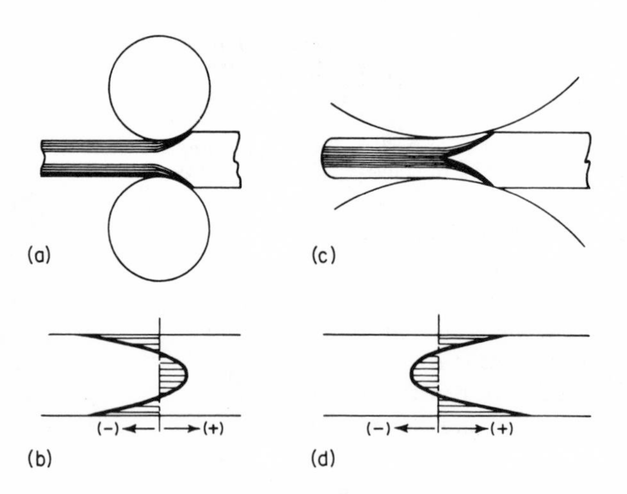

Fig. 17.9 Schematic diagram of metal flow and consequent residual stress patterns from rolling with small and large rolls. (Shaded area is most heavily deformed.)

occurs in drawing through a die with a small die angle being somewhat analogous to the large rolls. Differences of residual stress patterns in long members can be found by slitting them along their lengths. Opposite residual stresses are seen in the two drawn rods of Fig. 17.10. After being slit, the one on the left tended to open, and the one on the right to close.

In many cases, special mechanical working treatments are purposely carried out to produce a desirable residual-stress pattern. These processes are termed *prestressing* or *setting*, and in practice consist of loading the member to a larger stress than the one to which it will be subjected in service to produce some plastic flow.

Two of the most common prestressing processes are the setting of springs and torsion rods. Setting of a leaf spring is done by plastic bending (Fig. 17.3) and its advantage is demonstrated by the diagrams in Fig. 17.11. Assume that a leaf spring has the residual-stress distribution shown in Fig. 17.11a and that a typical load applied to it normally would produce the stress distribution shown in Fig. 17.11b. The applied stress is added to the residual one and, as a consequence, the maximum stresses to which the member is subjected are greatly reduced, diagram (c). Because the residual and applied stresses are additive, the former are sometimes called *initial stresses.*

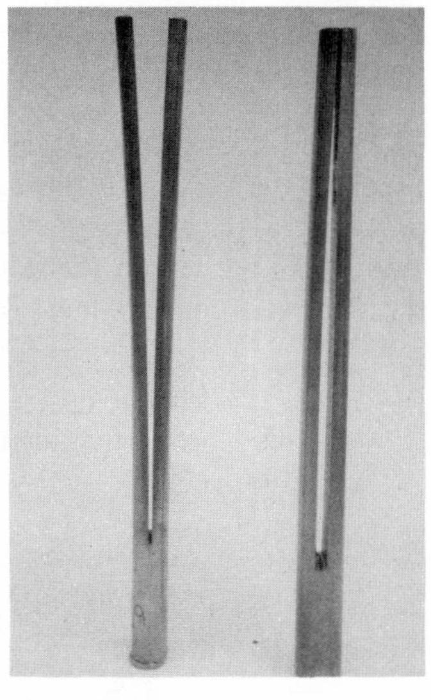

Fig. 17.10 Difference in residual stress pattern in two rods causes one to open and the other to close during slitting.

Helical springs are also set, and in the manufacture of heavy compression springs such as those used in railroad car wheel assemblies, the springs are compressed until the coils are touching before the springs go into service.

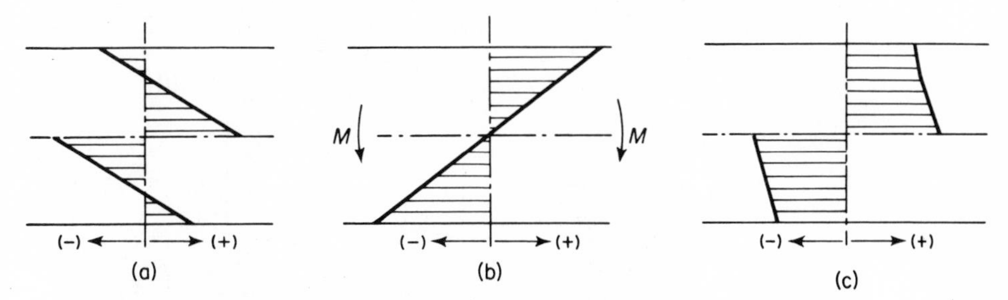

Fig. 17.11 Lowering of stress developed in a member because of initial stress: (a) Initial stress; (b) Applied stress; (c) Sum of (a) and (b).

Torsion rods are another example of a structural member that is set. If the bar is twisted enough so that essentially the entire cross section yields, the distribution of plastic stresses for a nonstrain-hardening material is as shown in Fig. 17.12b. An oppositely directed torque of the same magnitude, but with only elastic stresses, is shown in (c). Subtracting (b) from (c) as was done for bending leaves the pattern shown in (d).

The strength of thick-walled tubes can also be appreciably increased by prestressing. This process, known as *autofrettage*, is used on thick-walled gun barrels. The process consists of subjecting the tube to a large internal hydraulic pressure to develop plastic flow in the tube walls prior to finally machining it to shape. As in all the prestressing operations, the maximum benefit is derived, in the absence of strain-hardening, by applying sufficient pressure to just cause plastic flow throughout the total wall thickness. With strain-hardening, the resistance to yielding continuously increases with the amount of prestressing. The sketch of the tube is shown in Fig. 17.12e,

Fig. 17.12 Residual stress development in (a) to (d) setting torsion rods and (e) to (h) autofrettage.

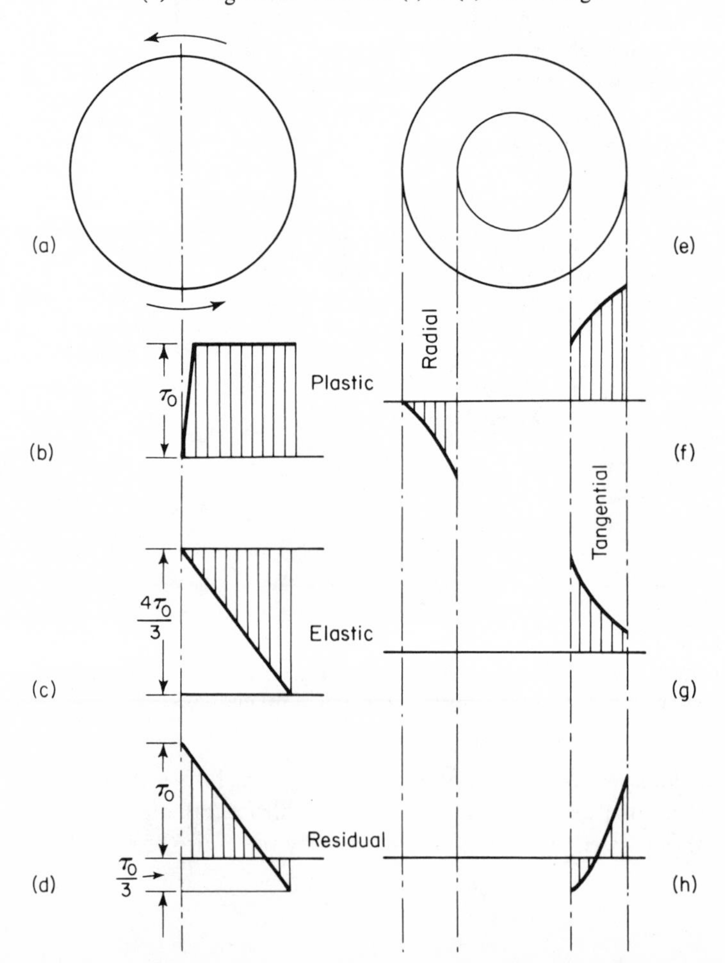

the radial and tangential stresses in the plastic range in (f), and the elastic stresses computed from Eq. (2.27) in (g). The largest stresses (and consequently those that control fracture) are tangential, so special attention is paid to these. Note that the maximum tangential stress in (f) occurs on the outside but in (g) on the inside. The reason for the former is that the plastic stresses are limited by the yield criterion and hence are dependent on the radial stress according to Eq. (6.2).

$$\sigma_\theta - \sigma_r = \sigma_o$$

Subtracting the elastic stresses from the plastic ones results in the pattern shown in (h). When the tube is subjected to an internal pressure in service, the sum of the service stress and initial stress results in a more uniform stress distribution than the purely elastic one, thus increasing the allowable service stresses in the part.

Surface Residual Stresses

Many parts are subjected to bending and torsion in use. These, of course, fail due to tensile stresses that develop on their surfaces, and their life can be extended by purposely induced surface compressive stresses. Light reductions by rolling or drawing produce surface-compressive stresses, but in many cases these operations are not practical; for example, it can not be used for complex cast or forged shapes such as propeller blades.

The most common method for inducing surface-compressive stresses without bulk distortion is by *shot peening*. In this process small cast iron or steel balls, of approximately 0.005 to $\frac{3}{16}$ in. diameter, are projected at a high velocity onto the part being treated. The energy from each impact is sufficiently high to cause plastic flow in a small volume. The accumulation of these indentations stretches the surface, producing a mismatch between it and the material beneath, causing a compressive-residual stress much as was the case in taking light roll passes. The amount of strain as well as its depth can be controlled by the peening variables: stresses between 0.5 and 0.9 σ_o and depths of penetration between 0.005 and 0.050 in. have been reported. The stress distribution obtained with a systematic change in shot size and air pressure is shown in Fig. 17.13.

Another method for mechanically working the surface to produce compressive stresses is by *surface rolling*. In this process, a hard roller with a small contact radius is pressed against the piece to be worked while both the roller and work turn on their own axes (Fig. 17.14a). By moving either the roller assembly or work piece in an axial direction, long lengths can be stressed. The method is particularly attractive for stressing fillets in shafts (Fig. 17 14b). Although analytical expressions for calculating the stresses developed in the plastically deforming surface are not available, the pressures are probably much higher than those occurring in shot peening. The depth of work can also be appreciable and values as high as 0.5 inch have been

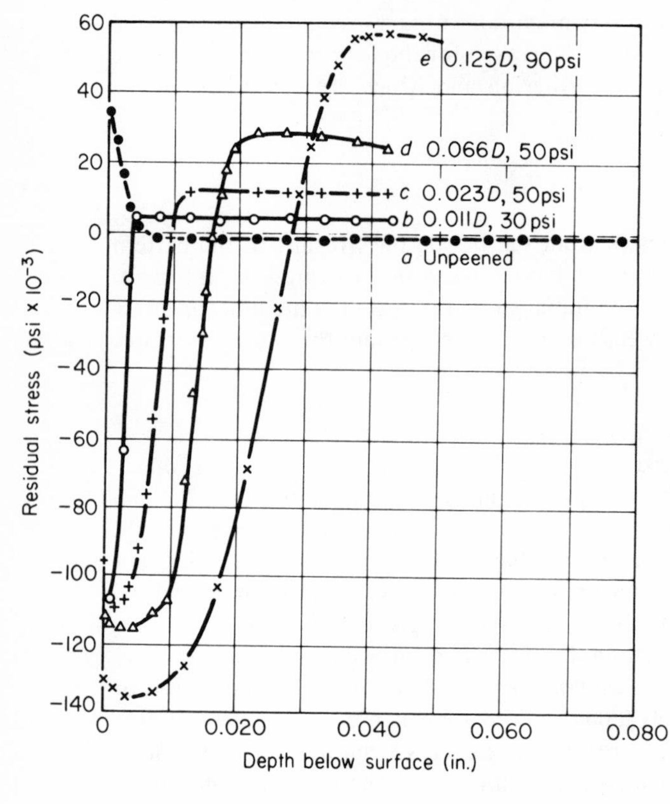

Fig. 17.13 Effect of shot diameter and air pressure on stress distribution for steel with a hardness of R_c 42. From J. M. Lessels and R. F. Brodrick, in *International Conference on Fatigue in Metals*, London: Inst. Mech. Engineers, 621, 1956.

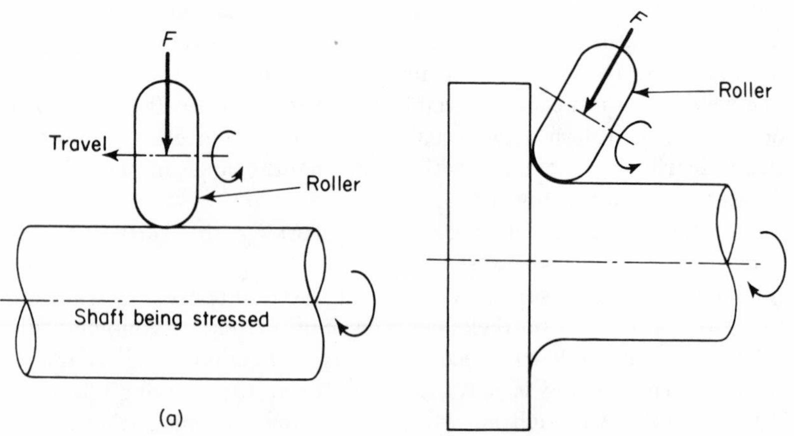

Fig. 17.14 Surface rolling: (a) Shaft; (b) Fillet.

reported. The method has the obvious limitation that it can only be used on members which have rotational symmetry.

Surface-compressive stresses are also produced by *nitriding* and *carburizing*, two common chemical surface-hardening treatments applied to steels. These are carried out by diffusing nitrogen or carbon, respectively, into the iron lattice at high temperatures, where diffusion rates are high. Absorbing either or both of these by the superficial layer causes it to expand relative to the interior, resulting in a high compressive-stress state on the surface. In carburizing, an additional hardening is achieved by quenching to form martensite.

In grinding or machining, the surface stresses are usually tensile, as illustrated in the machined piece of epoxy (Fig. 17.15). In brittle materials this can be very detrimental.

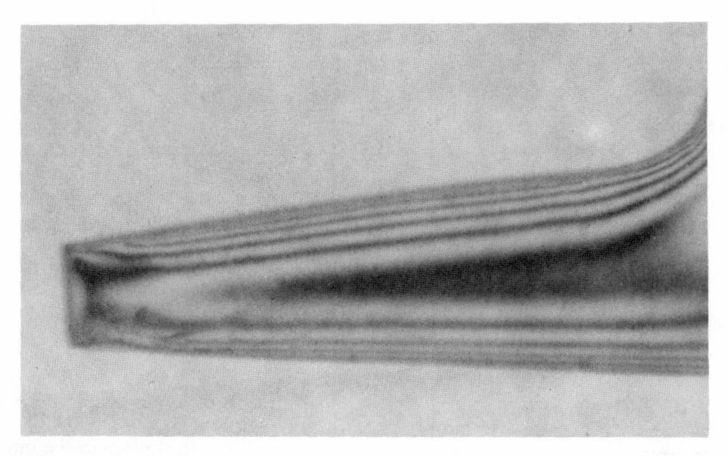

Fig. 17.15 Tensile residual stresses shown by photoelastic pattern along milled sides of epoxy pin.

MEASUREMENT OF RESIDUAL STRESS

Residual stresses can be measured either mechanically or by means of X-rays. The mechanical method is a destructive one involving the removal of layers of the body in which the residual stresses are being determined. As each layer of the body is removed, the stresses within it are relieved causing a redistribution of the remaining stresses. By measuring the dimensional changes that result from the incremental stress relief, the complete stress system in the body can be deduced.

Layer Removal Techniques

To demonstrate how this method of successive removal of layers is used, a simple residual-stress system in a cylinder containing a longitudinal tensile stress in the outer hull and compressive stress in the core will be considered. As early as 1914, Heyn suggested an analogy to this pattern consisting of a stretched spring in the outer layer of a rod and a compressed one in the center (Fig. 17.16a). Both springs are in contact with the ends of the rod so that the forces remain in equilibrium. Obviously, if the outer layers of such a composite rod are removed, the load on the center compressed spring is partially relaxed so that it can extend (Fig. 17.16b). If the initial length of the rod were l_o, and it stretched by an amount Δl, the extension $d\varepsilon_1 = \Delta l/l_o$ and, since the deformation is elastic, the amount of stress relieved,

$$\sigma = E\, d\varepsilon_1 \tag{a}$$

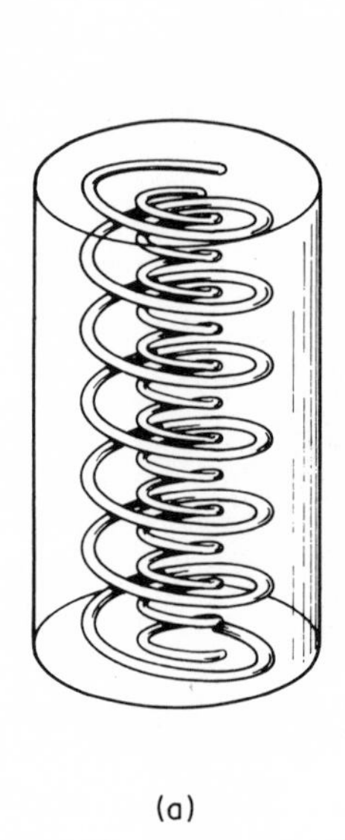

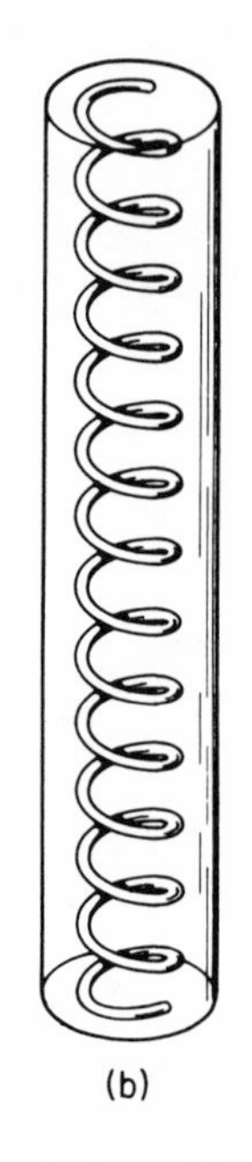

Fig. 17.16 (a) Spring analogy of compressive stresses on the surface and tensile stresses in the interior of a rod; (b) Relaxation of spring load and increase in core length when case is removed.

The load relieved by removing the outer layer must equal the change of load on the remaining core. Hence,

$$P_{\text{core}} = A_1 E d\varepsilon_1 = P_{\text{skin}} = \sigma_1 dA_1 \tag{b}$$

where A_1 = remaining area of core
dA_1 = area of skin removed

so that

$$\sigma_1 = \frac{A_1 E\, d\varepsilon_1}{dA_1} \tag{c}$$

Ordinarily, the stresses will not vary abruptly between skin and core but instead will change slowly across the bar. To determine this stress pattern over the complete radius, additional layers are successively removed and the length of the bar noted after each step. The calculation after each removal would be identical with (c) excepting that the residual stress pattern was altered by making all the preceding cuts. If after removing the second layer, the new set of measured values were substituted into (c), the apparent stress obtained at the second point would differ from its true value by the amount of stress relieved by the first cut. This error is avoided by adding (or subtracting) the previously relieved stress to each new determination. Hence after removing the second layer

$$\sigma_2 = \frac{A_2 E\, d\varepsilon_2}{dA_2} - E\, d\varepsilon_1 \tag{d}$$

where the second term on the right-hand side is the correction for the previous layer removal. As layers are successively removed, the correction is required for all previous layers so that the stress in the n-th layer is given by:

$$\sigma_n = \frac{A_n E\, d\varepsilon_n}{dA_n} - E(d\varepsilon_1 + d\varepsilon_2 + \cdots d\varepsilon_{(n-1)}) \tag{e}$$

If the layers are removed in differential thicknesses,

$$\sigma = E\left(A\,\frac{d\varepsilon}{dA} - \int d\varepsilon\right) = E\left(A\,\frac{d\varepsilon}{dA} - \varepsilon\right) \tag{17.1}$$

where ε = total strain after each removal.

Equation (17.1) is most easily applied to finding longitudinal residual stresses by the use of graphical methods. In practice, this is done as follows: thin layers are removed from the bar (most commonly by machining) and the total strain ε after each removal is plotted as a function of the area of bar remaining (Fig. 17.17). These points are connected by a smooth curve, and then A and ε can be read directly at any radius position while $d\varepsilon/dA$ is the slope of the curve at the particular radius.

Fig. 17.17 Graphical method for determining the residual stress in a rod by Eq. 17.1. From (R17.1).

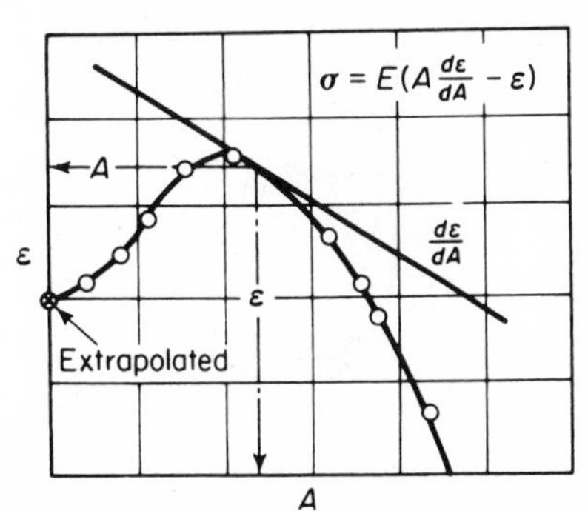

If one could be assured that a bar had only longitudinal residual stresses in it, the *Bauer-Heyn method* described above would give accurate stress values so long as thin enough layers were removed during each step to avoid averaging out stress peaks. However, the length of the bar might also change on removal of successive layers because of the relief of radial and tangential residual stresses. Hence, in a real case where stresses would be expected in all three directions, the Bauer-Heyn method would yield only approximate results. It underestimates the stresses when the longitudinal and tangential ones are of like sign, and overestimates them when the two are of opposite sign at a particular point.

This failing of the Bauer-Heyn method was recognized as early as 1919 by Mesnager, who proposed a procedure and set of equations for determining all three stresses, longitudinal, transverse, and radial in cylindrical members. Sachs (1927) developed easier equations for finding the stresses in the longitudinal, tangential, and radial direction, i.e., σ_l, σ_t, and σ_r, respectively, and was the first to carry out the experimental procedures, so that the process is now known as the *Sachs method* or *Sachs boring-out method.*

The process is applicable to rods or tubes, and it assumes that the residual stresses in these are uniform along their lengths and have rotational symmetry. Because of the manner in which cylindrical members are made and treated, both of these are realistic assumptions.

Experimentally, the process consists of measuring the surface strains in the longitudinal (ε_l) and tangential (ε_t) direction as successively larger holes are bored in the piece. Resistance wire gages are almost always used for measuring the strains (see p. 57) and contact with the gage leads are generally made through a mercury cup (as in Fig. 17.18) so that the connection can be readily made and broken for each boring step.

Sachs introduced the parameters

$$\lambda = \varepsilon_l + \nu\varepsilon_t$$

and

$$\theta = \varepsilon_t + \nu\varepsilon_l$$

so that the equations for the three stresses could be written in a form similar to that of Eq. (17.1).

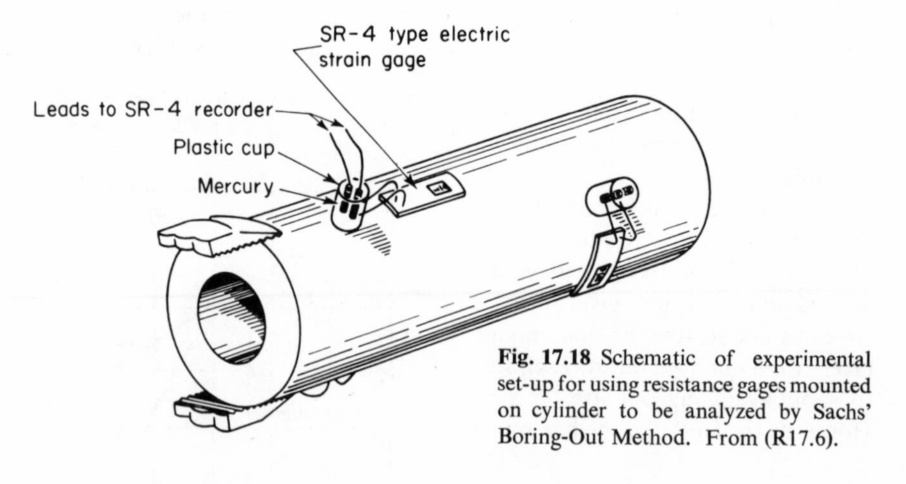

Fig. 17.18 Schematic of experimental set-up for using resistance gages mounted on cylinder to be analyzed by Sachs' Boring-Out Method. From (R17.6).

$$\left.\begin{aligned}\sigma_l &= E'\left[(A_o - A_b)\frac{d\lambda}{dA_b} - \lambda\right] \\ \sigma_\theta &= E'\left[(A_o - A_b)\frac{d\theta}{dA_b} - \frac{A_o + A_b}{2A_b}\theta\right] \\ \sigma_r &= E'\left[\frac{A_o - A_b}{2A_b}\theta\right]\end{aligned}\right\} \tag{17.2}$$

where $E' = \dfrac{E}{1 - \nu^2}$

A_o = original area of cylinder

A_b = area of bored out portion of cylinder.

In these equations the first term represents the stress relieved in a particular boring step and the second term is the sum of the stresses previously relieved. A graphical method similar to that shown in Fig. 17.17 would be used for solving (17.2).

Methods for measuring the residual stresses by stepwise removal of layers in members other than cylindrical shapes have also been developed. One of these, the *Treuting-Read method* is used for measuring the residual stresses in flat plate or strip. It is also based on the removal of thin layers of the strip, but rather than direct measurement of strains, changes in curvatures were used as the measure of stress relief as the layers are removed. (Strains

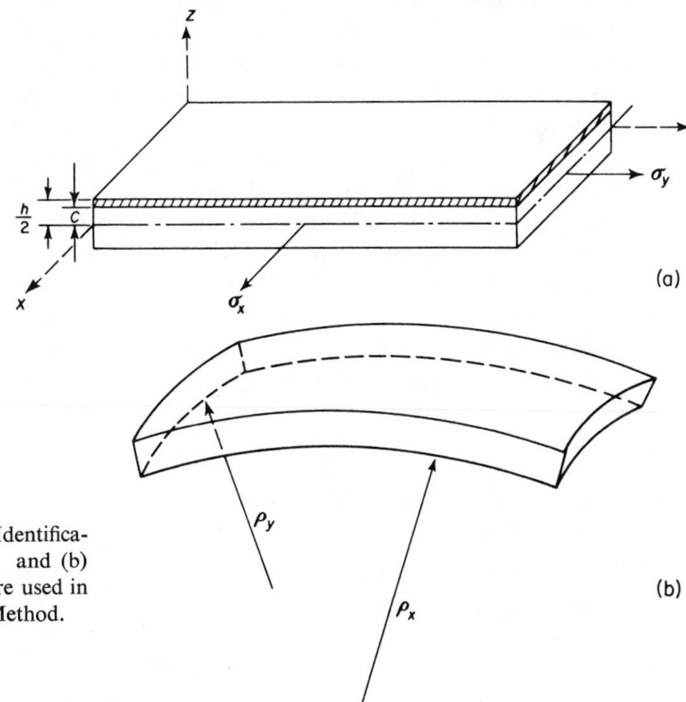

Fig. 17.19 (a) Identification of symbols; and (b) Radii of curvature used in Treuting-Read Method.

could have been used as well.) The method assumes that the stresses vary through the plate thickness, but not in the plane of the plate, and that $\sigma_z = 0$. In practice, layers are removed by first grinding and then etching to dissolve the thin layer of metal distorted by grinding. After each layer is taken off, the thickness and curvature in the two major directions are measured. If the radius of curvature in the x direction is r_x and in the y direction r_y (Fig. 17.19), the curvature ρ_x and ρ_y can be defined as:

$$\rho_x = \frac{1}{r_x} + \frac{\nu}{r_y} \quad \text{and} \quad \rho_y = \frac{1}{r_y} + \frac{\nu}{r_x} \tag{17.3}$$

and the equations for the stresses in the two directions are:

$$\sigma_x = \frac{E}{6(1-\nu^2)}\left[\left(\frac{h}{2}+c\right)^2 \frac{d\rho_x}{dc} + 4\left(\frac{h}{2}+c\right)\rho_x - 2\int_{h/2}^{c} dc\right] \tag{17.4}$$

$$\sigma_y = \frac{E}{6(1-\nu^2)}\left[\left(\frac{h}{2}+c\right)^2 \frac{d\rho_y}{dc} + 4\left(\frac{h}{2}+c\right)\rho_y - 2\int_{h/2}^{c} dc\right]$$

where h = half original plate thickness
and c = distance from original plate center plane to cut surface.

Deflection Techniques

In many cases a complete residual-stress distribution need not be determined. For example, in some cases it is only necessary to know whether the surface stresses are tension or compression. In others, for example, in comparing various forming processes, the one that causes the largest or least residual stress is sometimes all that is required. Approximate means for calculating stresses based on the beam formulas (see Chapter 13) can be used for these. Fortunately, in most forming operations the processing is continuous so that the stresses are constant over long lengths; or else, because of the rotational symmetry of the forming process, e.g., when drawing sheet metal cups, the stresses do not depend on θ (in polar coordinates).

Rods drawn through a die to reduce their diameter develop rotationally symmetrical residual-stress patterns. When the beam formula is used to calculate the surface stresses on the slit rod, however, the stresses are assumed to vary linearly from the neutral plane to the surface and to be constant on each horizontal plane parallel with the cut. Although the second assumption is particularly unrealistic in round bars, residual stresses are approximated in slit rods, using the formula:

$$\sigma_l = \frac{1.65\,\delta E r}{L^2} \tag{17.5}$$

where σ_l = surface longitudinal stress
δ = deflection
r = radius of bar
L = length of cut

The deflection method can be more reasonably used on members in which the stress varies in only one Cartesian direction. The section cut from a wide plate and slit along its center plane. (Fig. 17.20 is such a section.) As was the case with the rod, the two rectangular halves curved away from each other as the residual stresses were relaxed during cutting. According to the beam formula, which assumes that the stresses are distributed linearly over the plate thickness, the maximum surface stress is

$$\sigma = \frac{E'h\,\delta}{2L^2} \tag{17.6}$$

$E' = E/(1 - \nu^2)$ (is used since the wide thin plate deflects under a condition of plane stain)

h = plate thickness

δ = deflection

L = length of cut

The deflection method has also been applied to thin-wall tubes. The longitudinal stresses can be approximately assessed by cutting out a tongue along the length of the tube (Fig. 17.21a), and the circumferential stresses are evaluated by cutting a circumferential tongue (Fig. 17.21b) or by slitting the complete tube length (Fig. 17.21c). As one might expect, the deflection of a longitudinal tongue will depend on the width of the slit section. The maximum deflection occurs when the tongue width equals 0.1 to 0.2 of the tube diameter. The appropriate formulas

Fig. 17.21 Deflection techniques to approximate residual stresses in tubes. Shaded area is stress distribution based on Beam formula while curved might be actual distribution. From (R17.1)

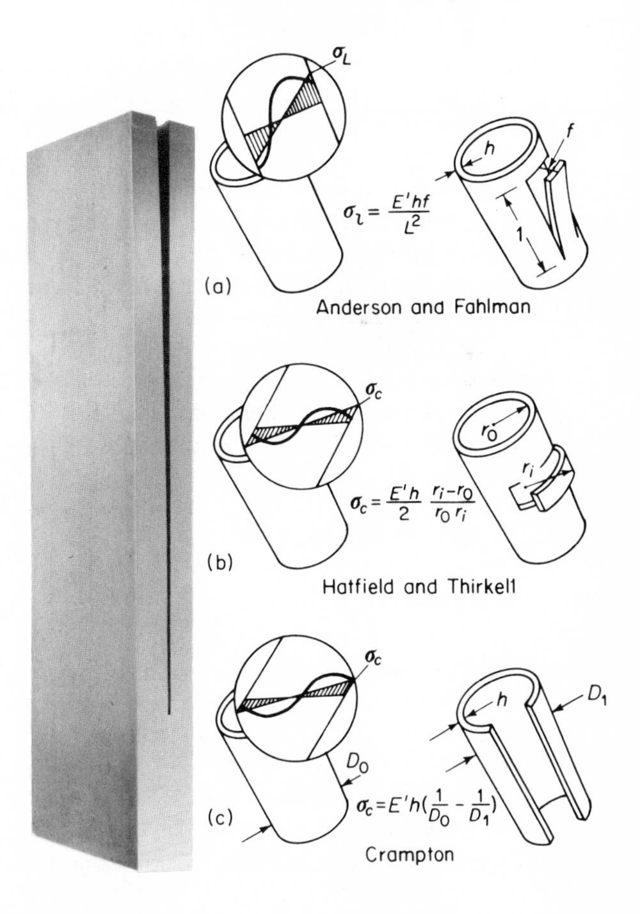

Fig. 17.20 Opening in a slit wide plate due to surface tensile stresses.

applied to the slitting procedures are shown in the figure. In each of these, it need not only be assumed that the stress varies linearly over the wall thickness, but in addition, that the forces within the tube wall are not balanced over one wall section. Only unbalanced forces between diametrically opposite walls cause bending when the tube is slit.

All these single-cut techniques must be viewed with extreme caution and should not be relied upon when exact values of the residual stress are needed. However, they are useful for comparative purposes, to assess the relative levels of locked-in stresses in variously processed but otherwise identical products.

Local Strain Techniques

All the measuring procedures described above were applicable to problems in which the stress varied over the thickness of the plate or shell. In a process such as spot welding where a very small area is heated while the bulk of the piece remains undisturbed, the residual stresses of interest vary over the plane rather than through the thickness of the piece (see Fig. 17.8). Local strain techniques have been developed for measuring elastic strains over short lengths such as are of interest in welding. These can use either mechanical or X-ray procedures. The mechanical means, like those discussed above, are based on measuring the strains produced when some of the material is removed. By their very nature, these have the disadvantage of being destructive. The X-ray methods on the other hand are nondestructive and are based on measuring the distortion of the atomic lattice, which is a direct measure of the strain between adjacent atoms produced by the elastic stress. The latter are described in detail in (R17.3,5).

Two methods are commonly used for mechanically measuring local strains in plates. One of these consists of marking gage lengths on the plate in two perpendicular directions (x and y) after which the plate is cut into strips. The stresses are then calculated by the equations:

$$\sigma_x = \frac{E}{1 - \nu^2}(\varepsilon_x - \varepsilon_y)$$

and

$$\sigma_y = \frac{E}{1 - \nu^2}(\varepsilon_y - \varepsilon_x)$$

The second mechanical method for evaluating local stresses is based on measuring the distortions that occur near a hole drilled into a large member. Because the hole can be small compared with the overall size of the item, the method is not necessarily destructive since it may be possible to plug the hole at the completion of the test. The smaller the hole, the more sensitive must the equipment be, whereas the maximum permissible hole size is determined by its weakening effect on the structure.

If the residual stress in the member is uniaxial and its direction is known, it is relatively simple to determine its magnitude. Gage marks are placed on the member in the direction of the known stress slightly farther apart than the diameter of the hole to be drilled (Fig. 17.22). If, after drilling the hole, the distance between the points increases, the relieved stress was tensile; if it decreases, the stresses were compressive. The stresses can be calculated from measured strains by Hooke's Law (Chapter 5). For more accurate determination of the stress, a calibration bar is used. This consists of a tensile (or compression) bar of the same width, thickness, and material as the tested item and containing a hole of the same size as used in the test. If wide enough pieces to duplicate the actual structure are not practical, a series of bars of various widths are used so that the test results can be extrapolated to an infinite width. The calibration bars are loaded in tension (or compression) and the extension (or contraction) across the hole is compared with the amount of extension (or contraction) that would have occurred in the absence of the hole.

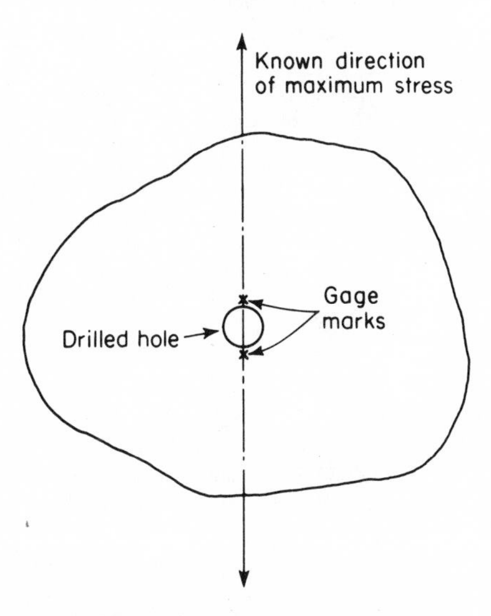

Fig. 17.22 Schematic diagram of procedure used for measuring local strains if the largest stress direction is known.

If the stresses are not uniaxial, or if the direction of the uniaxial stress is not known, a strain rosette such as shown in Fig. 3.9c can be placed on the piece, and the hole drilled in the center of these. From the technique described on p. 57 the principal strains and their directions can be calculated. By Hooke's Law these are converted to stresses.

RELIEF AND REDISTRIBUTION OF RESIDUAL STRESSES

An unfavorable residual-stress pattern can be altered either by decreasing its magnitude or superimposing on it a more favorable stress distribution. Shot peening, surface rolling, or carburizing (p. 469) are all methods for accomplishing the latter. In these processes, the piece must be stressed to a sufficient depth. Otherwise, a thin, superficial shell of material under a compressive state of stress forms on the surface, leaving large and possibly dangerous tensile stresses just below it.

Most commonly high residual stresses are reduced by a *stress relief anneal* which consists of heating the part to a high enough temperature to produce extensive stress relaxation. The yield strength of most materials decreases with increasing temperature so that they flow at the high temperature, thus reducing the stress magnitude as shown schematically in Fig. 17.23. The process is actually one of relaxation (see pp. 210 and 442) and it is a function of both temperature and time. Figure 17.24 shows how stress in a steel

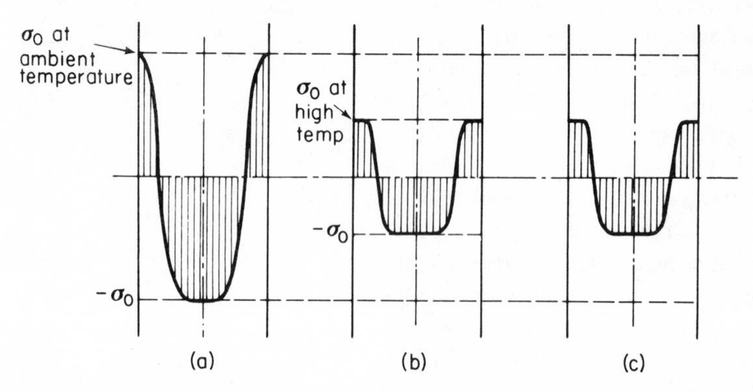

Fig. 17.23 Lowering of residual stress by stress relief annealing: (a) Initial stress pattern; (b) Pattern at high temperature; and (c) Stresses at ambient temperature after annealing.

Fig. 17.24 Relaxation of stress in a steel as influenced by time, initial stress level, and stress relief temperature. After E. A. Rominski and H. F. Taylor, Trans. Am. Foundrymen's Association, **51,** 707, (1943).

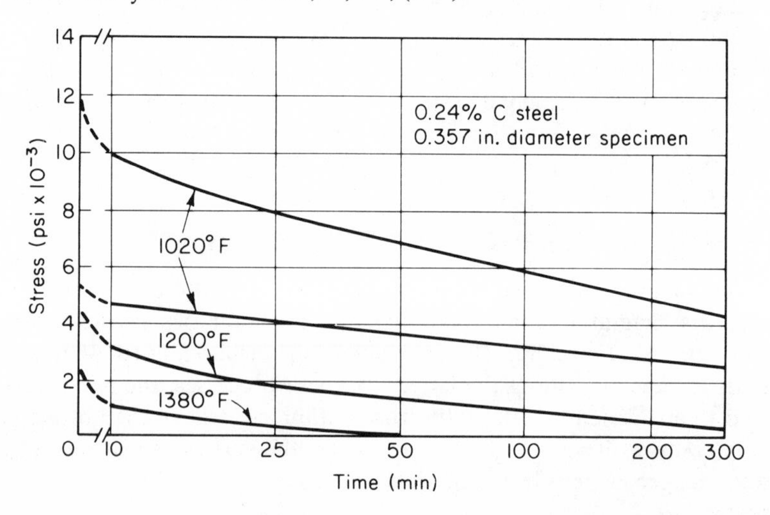

decreases with time, temperature, and initial level of stress. Stress relief temperatures are kept below the recrystallization temperature, and as a consequence the high tensile strength and hardness of a strain-hardened metal is essentially preserved.

Hardened steels are almost always tempered after quenching in order to make them tougher. Heating the steel for tempering, like stress relief annealing, reduces the residual-stress level in the metal; the higher the tempering temperature the more is the stress lowered.

Thermal treatments are also used to establish favorable residual-stress patterns in non-metallic materials. *Tempered glass* is heat treated to form a compressive stress in the skin and tensile stresses in the interior. The process consists of cooling the surface of a hot glass sheet by means of air jets causing the surface to cool quicker than the interior to form surface compressive stresses as was the case for the cooling ingot in Fig. 17.2. In this condition, tempered glass plate is three to five times more resistant to fracture in static loading and impact shock loading by blunt objects, and about four times more resistant to thermal shock. Rubber tire tread is put into a compressive stress state by *heat softening*. The stress state in this case is developed by bending the tread which was initially shaped as shown in Fig. 17.25a to the shape in 17.25b. While the tread is under a tensile stress, it is heated by steam jets. At the steam temperature, the tread flows, relieving the stress. Now when the applied bending stress is removed and the tire cools, the elastically bent inner material springs back to its original shape producing a compressive stress in the tread surface.

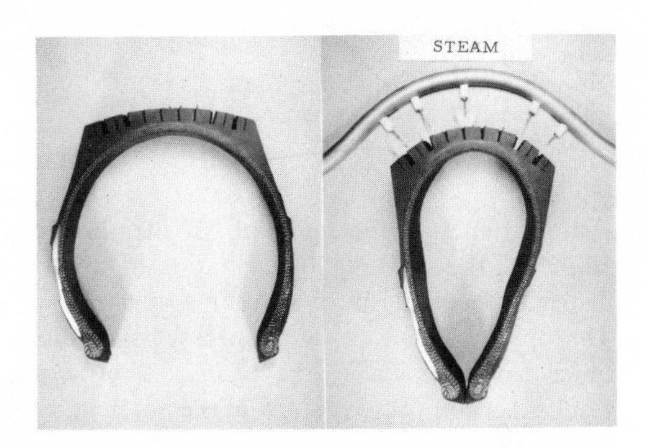

Fig. 17.25 Residual stress produced in tires by heat softening.

Just as residual stresses are established by nonuniform plastic flow, they can be relieved by uniform flow. Assume that a member has the stress pattern shown in Fig. 17.26a. If the piece is stretched, the first elements to have their yield strength exceeded are those with the highest residual tensile stress, (b). As the extension is continued, eventually the entire cross section undergoes plastic flow, (c). When the applied load is removed, (d), the initial stress has either vanished or been greatly reduced, depending on the strain hardening

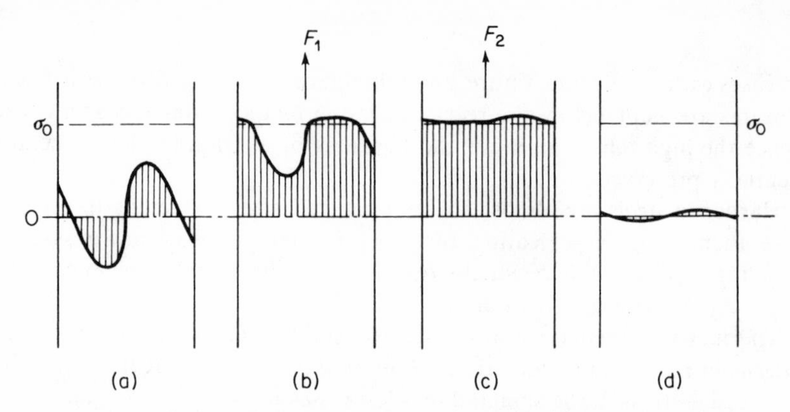

Fig. 17.26 Relief of residual stresses by stretching: (a) Initial stresses; (b) Stress distribution when plastic flow begins; (c) Distribution when entire cross-section extended plastically; (d) Stresses after release of load.

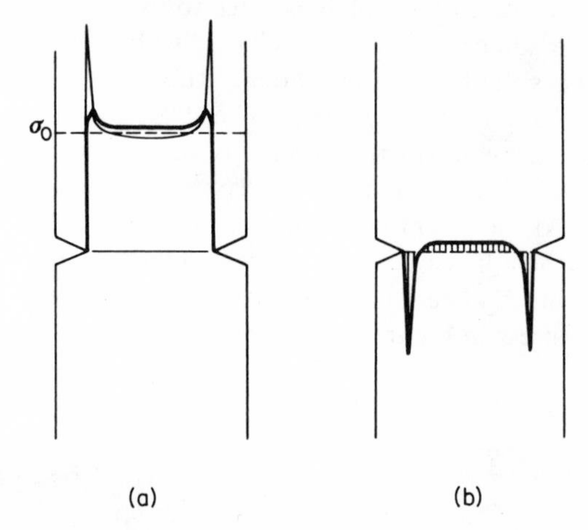

Fig. 17.27 Development of compressive residual stresses at a notch bottom by prestretching into the plastic range; (a) Heavy curve, "plastic" stress distribution; light curve, equivalent "elastic" stress; (b) Residual stresses on unloading.

characteristics of the material. If members containing holes or fillets are stretched, the resulting redistribution of stress is particularly beneficial. The distribution of plastic flow near a surface notch was discussed on p. 403. The "plastic" and "elastic" stress distribution for an identical load on a notch bar is shown in Fig. 17.27a. Subtracting the latter from the former results in a compressive stress at the notch bottom (Fig. 17.27b) on unloading.

EFFECTS OF RESIDUAL STRESS

As discussed throughout this chapter, tensile surface residual stresses promote fracturing while compressive stresses counteract it. This becomes most apparent when the member is subjected to alternating loads as discussed in Chapter 18 on Fatigue. In some cases, the residual stresses are so high that the member fails without the need for any additional stress, as is the case for the drawn rod in Fig. 17.28. This steel rod was hardened to a very high yield strength prior to drawing. It cracked spontaneously at room temperature after processing, probably due to a slow phase change adding stress to those left after drawing.

In machining operations, any residual stress can be damaging because of warpage of the finished part as shown by the slit rod in Fig. 17.10 and the

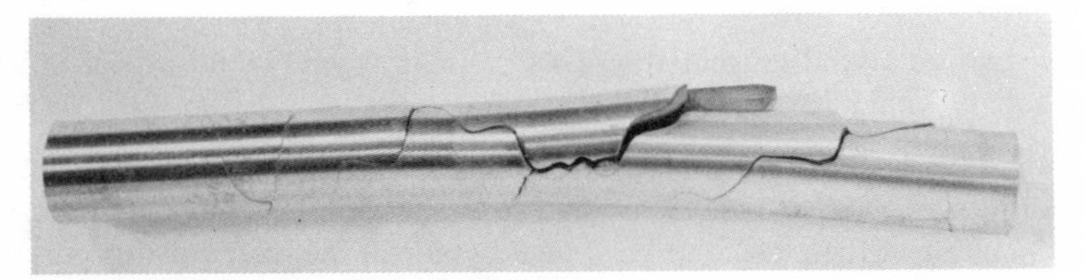

Fig. 17.28 Severe cracking of a drawn rod due to residual stresses.

plate in Fig. 17.20. In some instances the harm done by residual stresses is not readily apparent. For example, when a radial hole is drilled in a drawn rod, the residual stresses are relieved causing the hole to become elliptical. This shrinking of one diameter may cause sticking of the drill, and the rod is considered to possess poor "machinability." Residual stresses across the width of rolled strip cause it to *camber* after slitting, i.e., curve so that after slitting the stock the narrow strips show a lateral bow when laid flat (Fig. 17.29).

Resistance to tensile fracture in certain aggressive environments is greatly depressed. The attacking medium is selective and one that may be harmful to one material may have no effect on another. For example, iron and steel are sensitive to hot caustic solutions, aluminum to salt water, copper alloys to ammoniacal solutions, and polyethylene to oils or organic acids. The manner in which the stress and the corrosive medium interact to lower the fracture stress is not completely understood, but the failures are always brittle. This phenomenon of brittle cracking of an ordinarily ductile material due to an aggressive

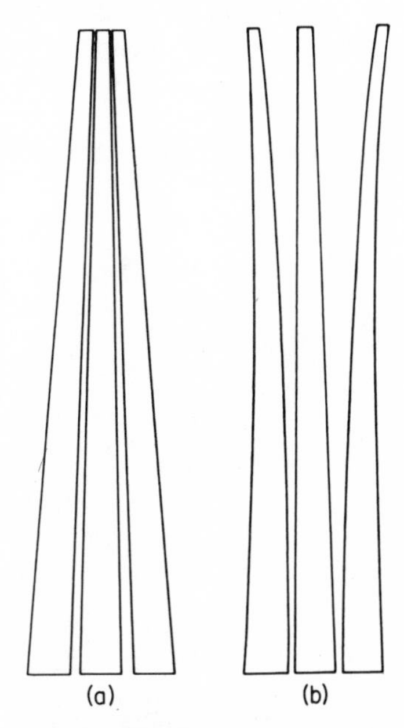

Fig. 17.29 Slit strips from a coil are: (a) Straight in the absence of residual stress; and (b) Cambered or bowed due to residual stresses across the coil width.

environment is known as *stress corrosion cracking* or *environmental stress cracking*. The effect of water vapor on the stress corrosion cracking of glass was discussed on p. 96. Because such cracking can occur at stress levels far below σ_o, residual tensile stresses are a frequent cause of this type of failure.

PROBLEMS

1. Prove Eq. (17.6).

2. Draw the radial and tangential stress distribution you would expect in wheel and axle of Fig. 5.8.

3. It should be possible to form two large rings by slitting a long steel plate along its center plane. What type of residual stress pattern does this require through the thickness of the plate, and how would you roll the plate for this purpose.?

4. The following strain increments were measured as 0.020 in. thick layers were turned from the inside surface of a tube with a 2″ O.D. and 1″ I.D. Draw the residual stress pattern.

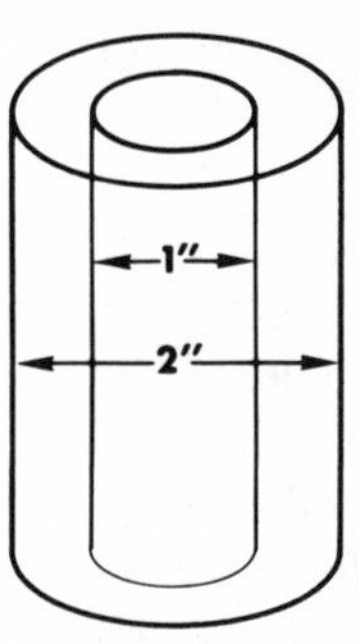

Layer Number	$\varepsilon_\theta(\%)$	$\varepsilon_l(\%)$
1	0.011	0.014
2	0.013	0.016
3	0.016	0.018
4	0.019	0.020
5	0.023	0.023
6	0.027	0.026
7	0.031	0.029
8	0.030	0.032
9	0.028	0.034
10	0.024	0.031
11	0.019	0.027
12	0.014	0.021
13	0.009	0.016
14	0.005	0.011
15	0.002	0.007
16	−0.001	0.003
17	−0.004	−0.001
18	−0.008	−0.005
19	−0.012	−0.010
20	−0.017	−0.016
21	−0.021	−0.022
22	−0.023	−0.018
23	−0.024	−0.013
24	−0.025	−0.010

5. The slit in the steel plate of Fig. 17.20 has the following dimensions:

$$L = 8 \text{ in.}$$
$$n = 1 \text{ in.}$$
$$\delta = 0.20 \text{ in.}$$

Draw the approximate residual-stress pattern across the tnickness.

6. If the actual stress distribution in the wall of a tube is as shown below, calculate δ when a narrow tongue is slit 10 in. long (use graphical integration).

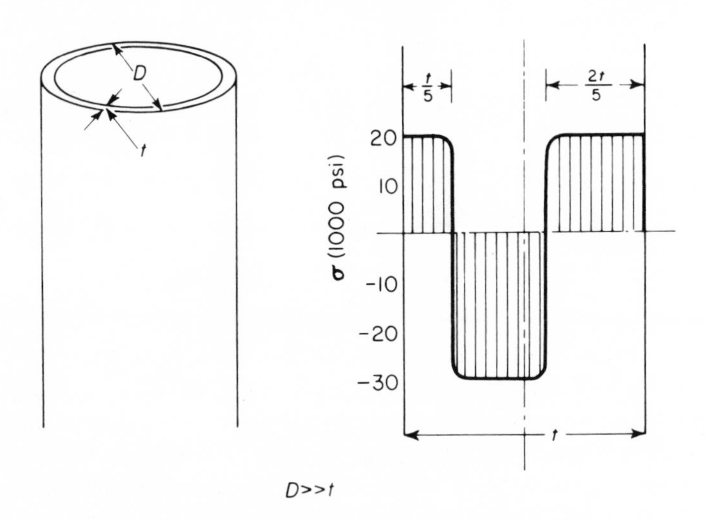

REFERENCES FOR FURTHER READING

(R17.1) Baldwin, W. M. Jr., "Residual Stresses in Metals", *Proc. ASTM*, **49**, 1949, pp. 1–45.

(R17.2) Brodrick, R. F., *New Methods of Measurement of Residual Stress*, WADC Technical Report 54-3. Wright-Patterson Air Force Base, Ohio: Wright Air Development Center, 1954.

(R17.3) Heindlhofer, K., *Evaluation of Residual Stress*. New York: McGraw-Hill Book Company, 1948.

(R17.4) Horger, O. J., "Residual Stresses", in *Handbook of Experimental Stress Analysis*, ed., M. Hetenyi. New York: John Wiley & Sons, Inc., 1950.

(R17.5) Society of Automotive Engineers, *The Measurement of Stress by X-rays*, Subcommittee on X-ray Procedures, Division IV, Residual Stresses of the Iron and Steel Technical Committee.

(R17.6) Treuting, R. G., J. J. Lynch, H. B. Wishart, and D. J. Richards, *Residual Stress Measurements*. Cleveland, Ohio: ASM, 1950.

18

CYCLIC STRESS AND FRACTURE BY FATIGUE

The fast-propagating, brittle fractures considered in Chapters 9 and 15 occur by the application of a single, sufficiently high load, usually coupled with the preexistence of a crack. A different type of fracture is caused by *fatigue*, which is a cumulative result of a large number (often many millions) of application of stresses, none of which reaches the ultimate tensile strength and, usually, not even the macroscopic yield stress. Fatigue may result from a repetition of a particular loading cycle or from an entirely random variation of stress. The former case is naturally more amenable to experimental reproduction and subsequent analysis.

The number of fatigue failures in engineering practice is very large and they probably account for more than 90% of all the mechanical fractures known. Breakage of suspension and valve springs in automobiles, propellers in ships and aircraft, steel cables in cranes and elevators, axles of railway rolling stock, and even huge rotors in steam turbines are representative examples, although the list can be extended almost indefinitely.

A fatigue crack starts, as a rule, at a point of high-stress concentration like a groove, oil hole, keyway corner, bottom of thread, or shaft fillet, and gradually works its way through the material. Often the stress concentrator is a small preexistent accidental crack or flaw, e.g., inclusion, in the part.

A fatigue crack will propagate when acted upon by tensile or shearing stresses, but probably not by compressive stresses. In operating equipment, working and rest periods alternate and the loads vary. As a result, a fatigue crack will spread intermittently, and characteristic *lines of arrest* or *beach markings* become visible on the fracture surfaces (see Figs. 18.1 through

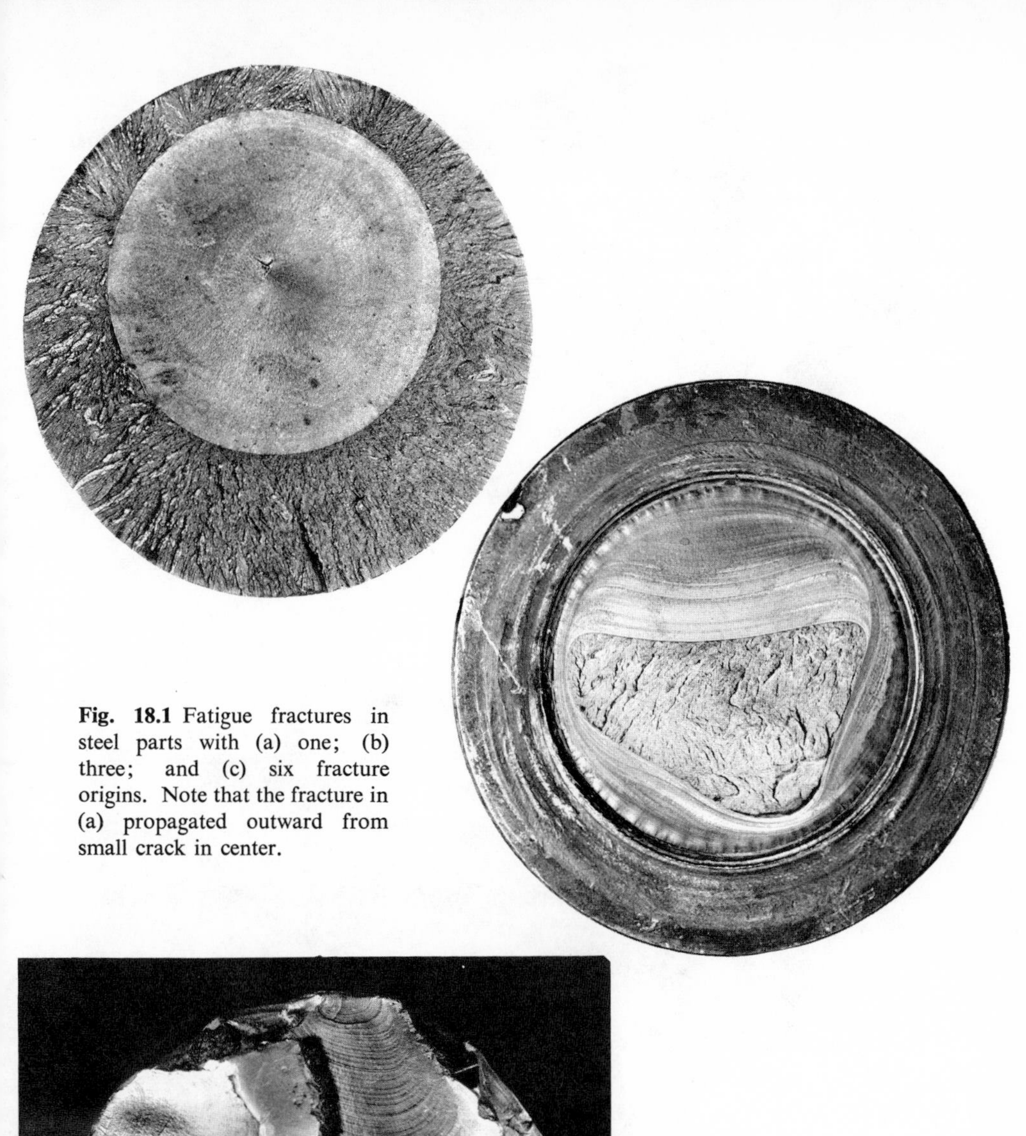

Fig. 18.1 Fatigue fractures in steel parts with (a) one; (b) three; and (c) six fracture origins. Note that the fracture in (a) propagated outward from small crack in center.

Fig. 18.2 Peeling type fatigue fracture which started from keyway corner of a 3.375 in. dia. dynamometer shaft.

Fig. 18.3 Fatigue failure of crankshaft showing numerous lines of arrest.

18.3). After a sufficiently large fraction of the original cross section is traversed by the crack, so little load-carrying area is left that the member fails in an ordinary ductile or brittle fashion.

It will be apparent from this brief statement of facts that two basic elements are involved in fatigue: (1) the initiation of a crack and (2) its propagation through the material. The discussion in the following pages summarizes the present state of understanding of the principal factors underlying these two phenomena.

STRESS AND STRAIN NOMENCLATURE USED IN CONNECTION WITH FATIGUE ENDURANCE AND FATIGUE LIMIT

A very common type of repeated loading is a cyclically *alternating* or *reversed stress* in which $\sigma_{max} = -\sigma_{min}$ (Fig. 18.4a). The *mean stress*, $\sigma_a = (\sigma_{max} + \sigma_{min})/2$, is then zero. An alternating stress can be produced by tension-compression, reversed bending, or reversed torsion cycles. In tension-compression (*direct stress*) the stress is calculated over the minimum cross section; in bending and torsion it is the maximum fiber stress.

A piston rod of a double-acting, reciprocating pump is subjected to alternating direct stress; an axle of a railroad car undergoes alternate bending, each wheel rotation producing a full-stress cycle in the outer layer. Cyclic torsion is found in various reversible mechanisms, but usually with a superimposed static load, e.g., torsion bars in automobile suspensions.

Fig. 18.4 Sinusoidally variable cyclic stresses: (a) Alternating; (b) fluctuating; and (c, d) pulsating stress.

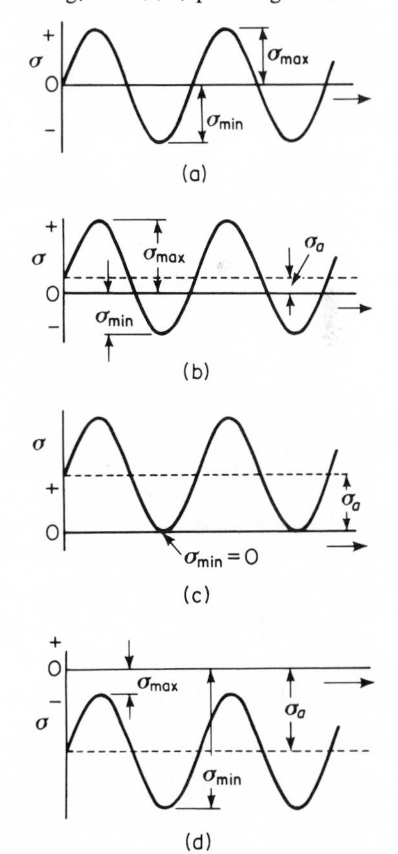

The *stress amplitude* is defined as $S = (\sigma_{max} - \sigma_{min})/2$ and the *stress range* as $2S = \sigma_{max} - \sigma_{min}$. Superposition of a static preload upon the alternating stress results in an asymmetrical stress cycle with $\sigma_a \neq 0$. This is referred to as *fluctuating stress* (Fig. 18.4b). In the particular cases when $\sigma_{min} \geq 0$ or $\sigma_{max} \leq 0$, one speaks of a *pulsating stress*, diagrams (c) and (d). This type of cycle is often described by the *stress ratio*, $R = \sigma_{max}/\sigma_{min}$, which becomes -1 for alternating stress, 0 or ∞ when either $\sigma_{max} = 0$ or $\sigma_{min} = 0$, and anything in between for various fluctuating stress patterns. Another useful stress ratio is $A = S/\sigma_a$. A cycle is defined completely by specifying σ_{max} and σ_{min}, σ_a and S, or various combinations of σ_{max}, σ_{min}, S and R or A.

The number of cycles that a material will withstand at any given cyclic-stress level before fracturing is called *fatigue life* or *endurance*, N_e; a specimen or part that fails after only several score of very high-stress reversals may last for millions of cycles under sufficiently low stresses. Certain metals exhibit a definite *fatigue limit* or *endurance limit*, σ_F, and they will withstand an infinite number of cycles when $S \leq \sigma_F$. Unless a mean stress is specifically indicated, the σ_F values quoted in the literature refer to reversed stresses, with $\sigma_a = 0$. Various carbon and alloy steels exhibit a true fatigue limit below about 500°F., and experience indicates that if such material can sustain 10 million cycles without fracture, it will "live" under these conditions forever. On the other hand, nickel, copper, titanium, and their alloys, as well as aluminum alloys used in aircraft construction, do not have a true σ_F and they may break at 20, 30, or 100 $\times$ 10^6 cycles. In these cases, the figures cited in the literature specify a safe stress for a finite fatigue life, N_e, usually for 10^8 or 5 $\times$ 10^8 cycles. *Fatigue strength* may denote either a true or a "qualified" fatigue limit, depending on the material it refers to.

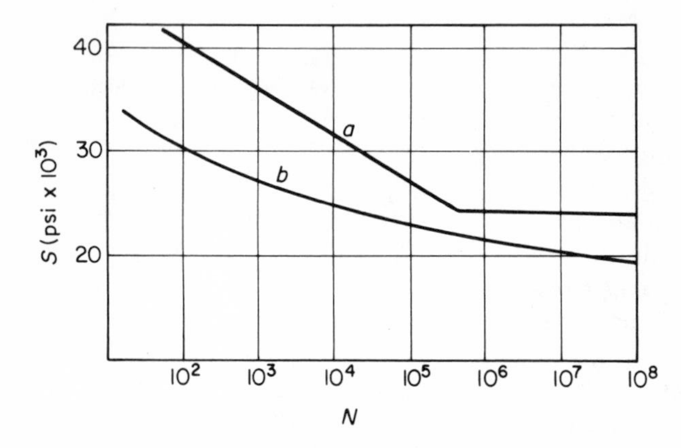

Fig. 18.5 Typical appearances of S-N curves for materials (a) with and (b) without definite fatigue limits. Pure alternating stress with zero mean stress.

Fatigue data are presented in the form of *S–N* curves, with the logarithm of the number of cycles, *N*, plotted as the abscissa, and the stress amplitude, *S*, as the ordinate. In Fig. 18.5, material *a* exhibits a true fatigue limit, $\sigma_F = 24.5 \times 10^3$ psi, while that represented by curve *b* will eventually fail even under relatively low cyclic amplitudes if *N* is high enough. Although in most cases we are interested in allowable cyclic stress for long lives, endurances as low as 10,000 cycles or less are satisfactory in certain cases, with correspondingly high design stresses. Cannon barrels, for example, belong in this category.

PRINCIPLES OF FATIGUE TESTING

As in most other testing methods, the fatigue properties are normally determined on small specimens representative of the materials which are subjected to cyclic stressing. Only in exceptional cases are full-scale tests used, e.g., on airplanes and on certain gears and shafting used in automobile power transmission trains.

Fatigue testing machines are based on a variety of principles. For example, the alternating bending cantilever tester in Fig. 18.6 has only undergone minor changes since it was introduced by Wöhler nearly a century ago. The maximum fiber stress occurs at the point where the cylindrical part

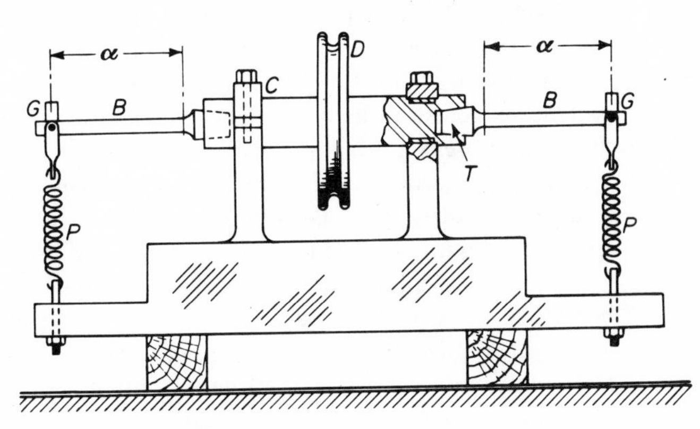

Fig. 18.6 Wöhler-type, cantilever beam rotating bending fatigue tester.

of specimen B merges with the enlarged tapered end T. The bending moment being Fa the stress is computed from (13.12). Force F is developed by adjustable loading springs P although dead weights are used in modern machines. Noting that σ_{max} depends on both the bending moment and the section modulus, it is possible, by using a suitably tapered specimen, to locate σ_{max} at some distance from the clamped end. The concern with the stress concentration due to the fillet is thus eliminated.

Tension-compression cycles can be produced by various means and one is illustrated diagrammatically in Fig. 18.7. Two pairs of unbalanced masses A and B rotate at high speed and cause housing H to move to and fro, thereby inducing alternate tension and compression in specimen S. By changing the

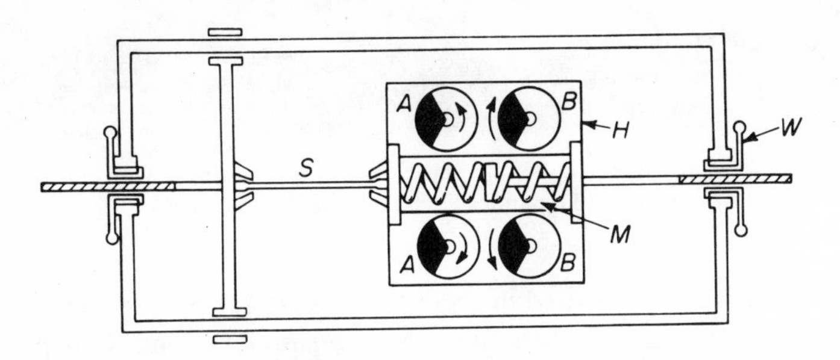

Fig. 18.7 Principle of the Schenck tension-compression maching. After R. Cazaud (R18.2).

relative angular position of pair A relative to B, the sinusoidal axial force in specimen S varies from a maximum when the masses work in unison to zero when they oppose each other. A static load can be superimposed by handwheel W through spring M. The same machine can be adapted to cyclic torsion testing by using a special specimen mount. Description of other fatigue testing machines can be found in (R18.2).

Owing to the large number of cycles needed to obtain a single point on the S–N diagram, the speeds of conventional fatigue test equipment are high, varying from about 1,500 to 30,000 cpm. Cycle counters and automatic switches which stop the machine when the crack is reasonably large, or when the sample breaks, are standard equipment.

Complete structures and subassemblies also undergo fatigue tests. For a large structure, such as an airplane, the testing setup is built around it, special rigs being used to test various wing and tail elements. To imitate the conditions prevailing in service, "whiffle-tree" arrangements, adhesively bonded to the various surfaces are used for load transmittal (Fig. 18.8). The loads are applied through loading cells attached by cables to the "whiffle-trees" and to rigid beams forming a framework within the test building.

Fig. 18.8 Testing rig for full-scale static and fatigue tests of Convair 880 jet liner. (Courtesy General Dynamics/Convair, San Diego, Calif.)

Unlike the procedures used in conventional fatigue tests, variable stresses may be applied to aircraft structures so as to simulate conditions prevailing on takeoff, landing, and flight under various atmospheric conditions. Planes are expected to fly a limited number of hours, and the existence of an absolute fatigue limit for every element is not expected or even aimed at. Similar procedures are used for testing automobile components, such as wheel bearings, in which case a typical road contour would be programmed into the test machine.

INSTABILITY OF PROPERTIES OF METALS UNDER FATIGUE CONDITIONS AND MECHANICAL HYSTERESIS

A Hookean solid could withstand an infinite number of stress cycles since the applied stress range fits between the positive and negative values of the elastic limit. Real solids are never perfectly elastic (Chapter 8) although glasses well below their softening point come very close to the ideal. In metals, true elastic limits persumably do not exist, and an application of cyclic stress must result in at least minor plastic deformations. No matter how small these deformations are, they are bound to result in some structural rearrangements and in consequent changes of stress-strain properties.

If a metal has a true fatigue limit σ_F (Fig. 18.5), it apparently does not, or eventually ceases to, suffer cumulative changes when stressed at or below σ_F. Hence, it must either behave elastically within the range $\pm\sigma_F$, or any initial plastic flow must terminate or else become completely reversible. From the practical point of view both situations are equivalent except that a full-loading cycle will be represented by a straight line of slope E in the first case, but by a closed antisymmetrical loop in the second (Fig. 18.9). Here, (*a*) may be regarded as a special case of (*b*), when the width of the loop along the ε axis is vanishingly small. The area of the loop in $\sigma \cdot \varepsilon$ units represents irreversible work of deformation per unit volume per cycle (lbs./in.^{2} = lb.-ins./in.^{3}) and is termed *hysteresis*.

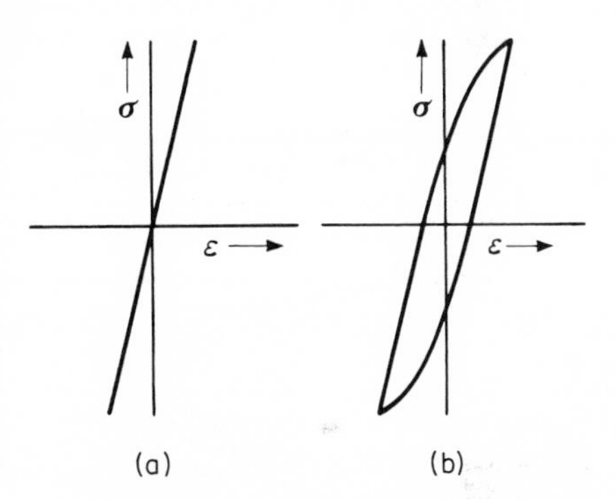

Fig. 18.9 Perfect elasticity (a) and "elastic" hysteresis loop (b) exhibited by a cyclically stressed material below the fatigue limit.

A loop of constant width can persist right from the start of the experiment, curve *a* in Fig. 18.10, or it can change as cycling progresses. In the second case, the width of the loop at any constant stress range $\pm\sigma$ may pass through a transient stage during which it decreases (curve *b*), or at first increases and then decreases to a constant level, (*c*). A gradual widening of the loop is indicative of increasing structural instability attributable to mobility of dislocations or nucleation and growth of fatigue cracks, e.g., curves *d* and *e*.

A wide hysteresis loop portrays a strong deviation from perfect elasticity and indicates that the elastic limit or yield σ_o is low compared with the applied cyclic stress. A progressive narrowing of the loop at a constant $\pm\sigma$ range means that the static stress-strain curve becomes steeper, an effect usually attributable to work-hardening.

The physical criterion defining the maximum hysteresis (or limiting stress-strain loop) that can be sustained by various metals indefinitely is unknown. But for equal stress ranges and lives, a material exhibiting a larger hysteresis loop is often considered more desirable inasmuch as it has a higher damping capacity and will develop lower peak stresses under resonant vibrations (see p. 218.)

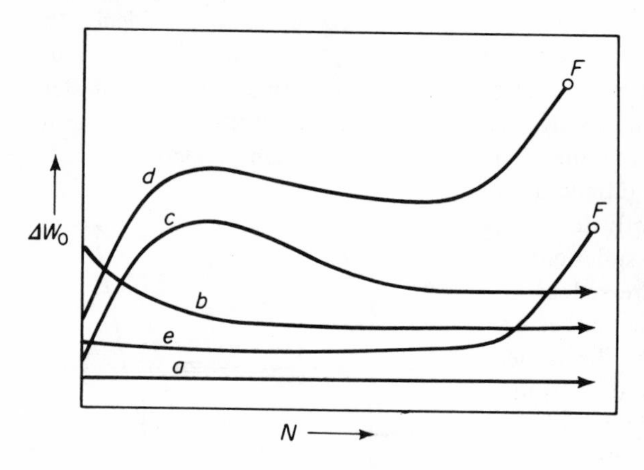

Fig. 18.10 Changes of hysteresis loop area ΔW_0 during cycling in the safe (curves *a*, *b*, *c*) and unsafe (*d*, *e*) stress ranges. *F*—fracture (schematic).

The steep increase of damping prior to final fracture *d* and *e* (Fig. 18.10) is caused by the high stresses and consequent large plastic strains near the tip of the propagating crack. The deformation work heats the metal in the reduced cross section, softens it; even more plastic flow ensues, the crack spreads, and so forth, until final separation occurs.

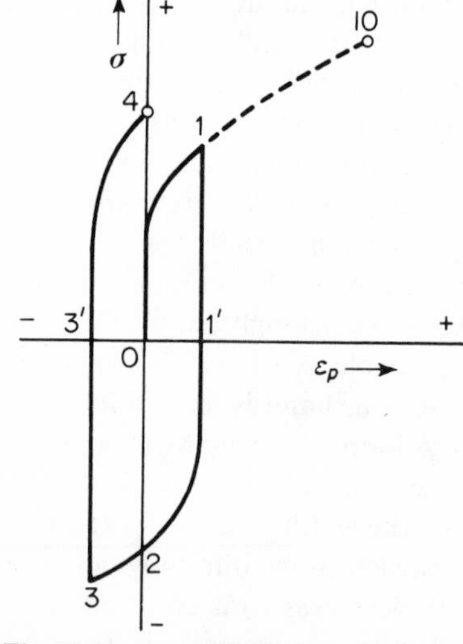

Fig. 18.11 Comparison of work-hardening in a full strain cycle with that in an equivalent unidirectional deformation.

The Bauschinger Effect and Work-Hardening Under Cyclic Strains

Let us consider the cyclic stress-strain relations, disregarding for simplicity the elastic strain component. For this purpose, let the following strain sequence be applied to an annealed, ductile metal with yield strength σ_o: (1) $\varepsilon_p = 0.2\%$, (2) $\varepsilon_p = -0.4\%$, and (3) $\varepsilon_p = 0.2\%$. The net plastic strain is zero ($0.2 - 0.4 + 0.2 = 0$), and the entire *strain cycle* is represented in Fig. 18.11 by the distorted spiral 0–1–1′–2–3–3′–4.

To facilitate an analysis of successive cycles, it is convenient to discuss them by reference to the terminal or maximum stress points reached in each "forward" or "reverse" half-cycle. Note that the stresses at the terminal points 1, 3, 4 increase in this order because of progressive work-hardening. Yet the total hardening at 4 is much less than at point 10, which corresponds to strain $4\varepsilon_p$ applied unidirectionally. This is due to a partial annihilation by reversed plastic strain of the hardening induced in the preceding half-cycle, conceivably through generation of dislocation loops of opposite sign and which wipe out some of the existing ones. The occurrence of such *work-* or *strain-softening* can be ascertained through hardness measurements during reversed deformation, the hardness then changing according to Fig. 18.12.

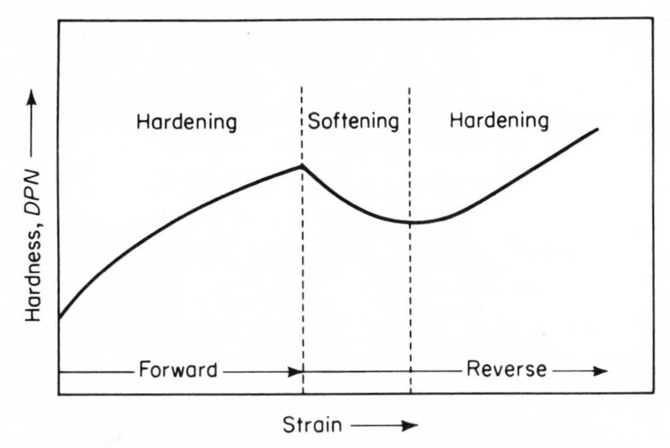

Fig. 18.12 Transient strain-softening which accompanies the Bauschinger effect during reversed deformation (schematic).

When the strain cycle is repeated over and over again between the $+\varepsilon_p$ and $-\varepsilon_p$ limits (Fig. 18.13), the difference in stress at the neighboring end points of adjacent loops like 1–3, 3–5, or 5–7 decrease rapidly and disappears after a limited number of cycles. This condition is described as *saturation-hardening* and its level increases with cyclic strain ε_p. At small alternating strains the saturation stage is approached rather slowly, but at large ones the steady-state condition is reached after only a few cycles. When saturation-hardening occurs, the hardness of a cycled metal oscillates between fixed limits which are reached quickly and do not change to the point of fracture (Fig. 18.14).*

* N. H. Polakowski, *Iron Age*, **170**, Aug. 28, 106 (1952).

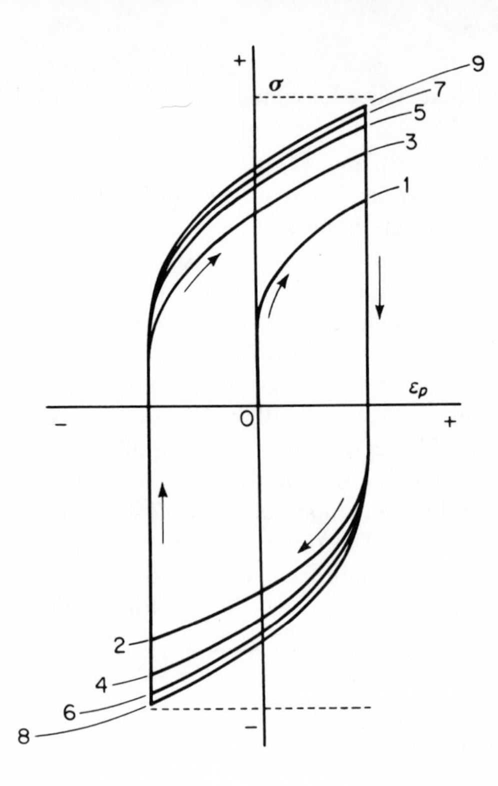

Fig. 18.13 Saturation hardening of an annealed metal cycled between fixed positive and negative strains. Dashed horizontals indicate the saturation stress range (schematic).

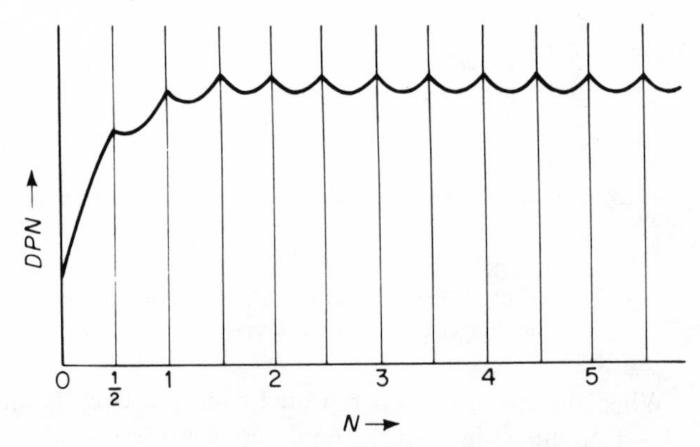

Fig. 18.14 Rapid saturation-hardening during large-strain cycling. Note alternate softening and hardening in each half-cycle.

The occurrence of saturation-hardening at levels less than the maximum found in undirectional deformation for much smaller total strains proves that cyclic strains are nonadditive in their work-hardening effect. Associated with it is a low rate at which the inherent ductility is used up by the process. Both phenomena indicate that fatigue fractures cannot be attributed to "excessive work-hardening in fatigue or a consequent *exhaustion of ductility* although beliefs to the contrary were held for many years.

Another important conclusion that can be drawn from the foregoing is that the apparent difference between the large, plastic stress-strain loops just discussed and the narrow hysteresis loops recorded at low, nearly elastic strains is only in their size but not in any characteristic features. Thus, in a particular material, a fatigue failure is certain to develop very rapidly at, say, $\varepsilon_p = \pm 0.02$ while it may be known from experience that one will never occur at $\varepsilon_p = \pm 0.0005$ ($= 0.05\%$). Yet it is not possible so far to establish from the appearance of single loops taken at various intermediate cyclic strains, where the boundary between the safe and unsafe ranges lies. It is evident, however, that the fatigue limit σ_F is, in effect, an elastic limit, in the sense that stressing at or below σ_F results in reversible, non-cumulative (and therefore non-damaging) structural changes.

Cyclic Bauschinger Effect and Work-Softening in Cold-Worked Metals

If 0–1–2 in Fig. 18.15 is the tensile stress-strain curve and the specimen is unloaded at 1 to 1′ and then reloaded in compression, a compressive curve 1′–1″ results. Inasmuch as the Bauschinger effect B_1 is due to the interaction between the applied external stress and the locked-up, microresidual stresses, or stress fields at dislocations in the metal, their intensity must control the magnitude of the *B*-effect. Since the level of internal stresses increases with prestrain, the *B*-effect must increase accordingly, as illustrated by B_2, which results from unloading at 2 and loading in reverse along 2′–2″. This is certainly true for strains below 1 to 2% which are of interest here.

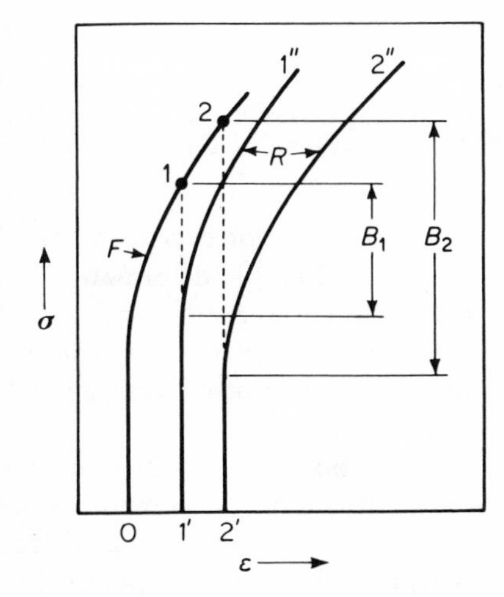

Fig. 18.15 Variation of Bauschinger effect (*B*) with prestrain. *F*—forward, *R*—reverse straining direction.

A weakening of the *B*-effect is manifested by the initial portion of the reversed flow curve becoming steeper, and, as a consequence, the reversed yield stress being less depressed. These events, illustrated in Fig. 18.16, show that wherever the maximum stress in the preceding half-cycle was reduced, the following reverse curve was steeper, and vice versa. A progressive reduction of the maximum stress on each reversal is seen to cause the difference between the forward and reversed, torsional (torgue-twist angle) curves to gradually decrease and eventually disappear. The *B*-effect, defined as the difference between the yield stresses for two opposite straining directions, can thus be eliminated by

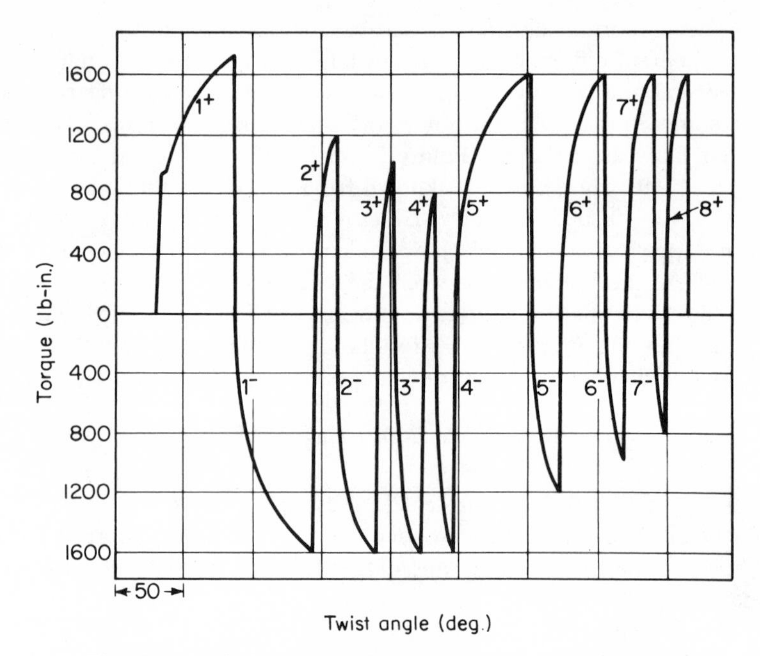

Fig. 18.16 Variations of the Bauschinger effect in annealed 1045 steel with torque applied prior to reversal ($d = 0.459$ in., $l = 1$ in.). From N. H. Polakowski, *Proc. ASTM*, **63**, 535 (1963).

cyclic straining of decreasing amplitude. When this is accomplished and the original tensile deformation is resumed (Fig. 18.17), the metal yields at a markedly lower stress, σ_0', than at the end of the first stretching phase (i.e., point X). In effect, the material is now *strain-softened*, the amount of softening apparently increasing with the intensity of the Bauschinger effect B in Fig. 18.17.

The above changes have interesting applied implications. For instance, common cold-working methods are drawing through dies (bar stock, wire) and rolling (sheet and strip). These products are often not sufficiently straight (flat) for their ultimate applications directly after processing, and additional straightening operations may be required. The latter involve the use of machines which bend the rod or sheet in alternate direction at a decreasing amplitude. Since a Bauschinger effect develops under these conditions and follows the changes just described, it is apparent that a cold-processed bar or sheet will have its yield strength depressed after straightening. If such change is objectionable, every effort must be made by the manufacturer to get the cold-worked material as straight as possible from the principal cold-working operation, to minimize the need for subsequent straightening.

Strain (or work)-softening may be accompanied by an interesting side

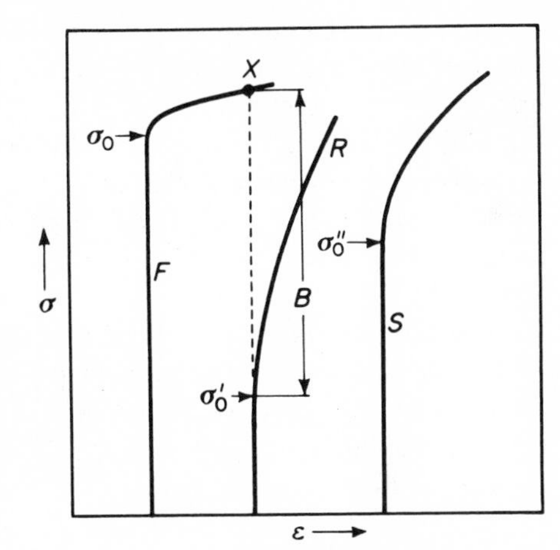

Fig. 18.17 Initial Bauschinger effect *B* and work-softened condition *S* of a statically cold-worked metal as a result of cycling at a gradually decreasing amplitude. *F*—forward, *R*—reversed straining.

effect. According to Eq. (10.7), the tensile necking strain, ε_n, increases with the strain-hardening exponent *n*. While the plastic portion of an *S*-curve in Fig. 18.17 starts at a stress $\sigma_o'' < \sigma_o$, its slope is obviously greater than that of the relatively flat upper part of the as-cold-worked curve *F*, and so is its exponent *n*. Hence, the initially hard material can be expected to exhibit, after cycling, an increased uniform elongation before fracturing, as is indeed demonstrated in Fig. 18.18. The cyclic Bauschinger effect can thus be utilized to regenerate some of the ductility exhausted by prior cold-working operations.

Work-Hardening and Softening in Fatigue

The work-softening processes, discussed in the preceding section, were observed during slow cycling, and it is important to ascertain that like effects

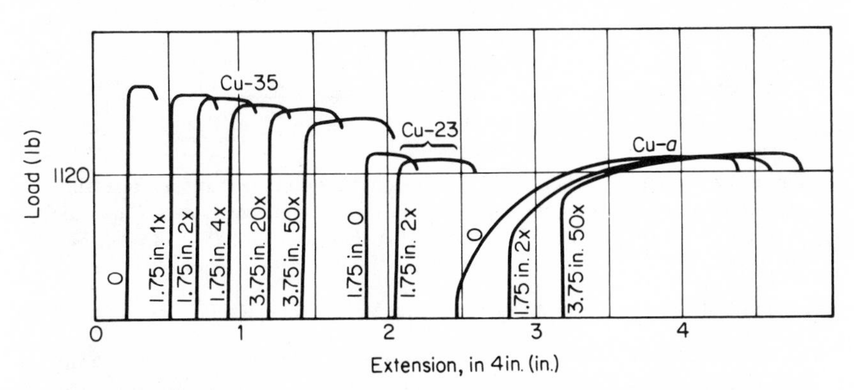

Fig. 18.18 Effect of reversed bending on tensile properties of annealed (Cu-a), and cold-rolled (Cu-23, Cu-35) copper strip. From N. H. Polakowski, *Proc. ASTM*, **52,** 1086 (1952).

can take place in high-speed cycling which is more in line with the common image of true fatigue conditions. Numerous experiments, run at frequencies up to 9,000 cycles per minute, all gave affirmative results; Fig. 18.19 displays some of the earliest, obtained on annealed and on cold-drawn copper rods. As expected by reference to Fig. 18.14, the very soft, annealed rod increasingly hardened with rising amplitude or number of cycles whereas the cold-drawn metal, with its high yield strength, progressively softened under similar conditions.

The diagrams in Fig. 18.19 were obtained from compression tests on the "fatigued" specimens, to eliminate possible effects of fatigue cracks upon static properties. Later experiments gave analogous results in tension. They also indicated that fatigue softening of cold worked metals occurs only above the fatigue limit.

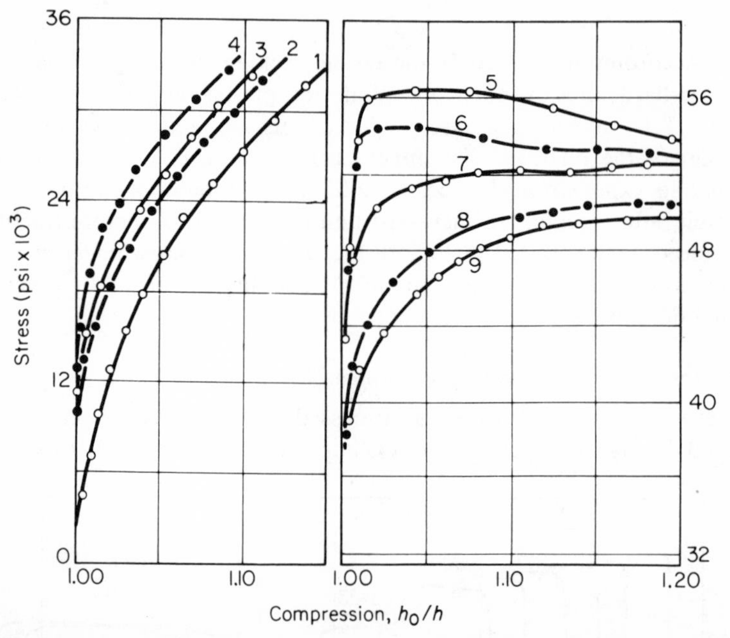

Fig. 18.19 The effect of alternating tension-compression fatigue stresses on static compression properties of annealed (Curves 1 to 4) and cold-drawn (Curves 5 to 9) copper. From N. H. Polakowski and A. Palchoudhuri, *Proc. ASTM*, **54**, 701 (1954).

STRUCTURAL CHANGES ASSOCIATED WITH FATIGUE

(a) *Slip Markings.* Fatigue is intimately connected with certain structural changes which accompany cyclic stressing. As far back as in 1903, Ewing and Humfrey found in their fatigue experiments on rectangular wrought iron specimens that slip bands formed in the highest stressed regions. Gough and

Hanson later (1923) demonstrated that a high slip band density can be developed by cyclic stress below the fatigue limit, i.e., in the *safe range* and there is no significant difference in this density when annealed metals are stressed in the safe or unsafe stress ranges. These findings were since confirmed by many others, especially on metals whose fatigue limits were higher than the static yield strength ($\sigma_F > \sigma_o$), like pure aluminum or coarse-grained low carbon steels.

One characteristic difference between the appearance of "static" and "fatigue" slip lines in f.c.c. metals was reported by Forsyth.* Whereas these lines were typically straight in the first case, they were irregular and wavy in the second. This feature is ill-understood. All these changes were observed on soft, non-work hardened metals and single-phase alloys. Prior static cold working by extension or rolling introduces a high and reasonably uniform slip band density so that subsequent changes due to fatigue are difficult to follow. Forsyth's work on aluminum alloys and copper, indicates, however, a stronger delineation of sub-boundaries during fatigue following cold rolling.

Whether and in what manner fatigue produces slip in high-strength (hardened or hardened and tempered) steels remains a mystery. There are no effective techniques for studying gross slip in these materials even during static deformation, and no attempts seem to have been made to follow up the more discrete changes to be expected as a result of fatigue stressing.

(b) *Surface Effects and Crack Nucleation.* It was long known that fatigue failures ordinarily originate at a free surface. Indeed, it was found that if a thin surface layer of a fatigued specimen is periodically etched off during the test, the life of the material can be spectacularly increased. More recent tests reveal, however, that while the surface roughening due to slip band formation can be substantially removed by electropolishing or etching, some persistent and deep-penetrating bands remain. Cracks eventually developed in these persistent bands.

Since cyclic stressing of a soft metal causes work-hardening, it was logical to expect, as was subsequently confirmed by tests, that most of such hardening occurs during the first few hundred cycles, when the yield stress has its lowest value. The persistent bands develop at the same time and possibly the cracks within them. The incipient cracks first appeared after as little as 5% of the specimen life which amounted to several million cycles. During the remainder of the test some of them propagated until one grew to the critical size. Occasionally small cracks can develop early in a fatigue test but remain stagnant thereafter. Nonpropagating annular cracks were found along the

---

* P. J. E. Forsyth, *Journal of Institute of Metals*, **83,** 173 (1954–55); C. A. Stubbington and P. J. E. Forsyth, *ibid.*, **86,** 90 (1957–58).

notch root during fatigue tests on V-grooved round specimens of low-carbon steels (<0.15% C). The cracks were up to several tenths of a millimeter deep but specimens so cracked endured up to 10^8 cycles without failing. Similar findings were later made on a Ni–Cr steel and an aluminum alloy.

(c) *Particle Coalescence and Phase Transformations. Understressing (Coaxing).* Mechanical working of metastable solid solutions or polyphase alloys accelerates transformations leading to a more stable condition. For instance, a pronounced softening was detected in high-strength aluminum-base alloys as a result of fatigue. These changes were apparently associated with a gradual precipitation and coarsening of intermetallic compounds from solid solution in the matrix. They were reminiscent of and, in their effects, comparable with overaging beyond the optimum strength conditions. The above circumstance probably accounts for the fact that despite their high strength (up to about 90,000 psi), these alloys do not develop a true fatigue limit and the cyclic amplitudes they can carry at endurances $N_e = 10^7$ to 10^8 are rather low.

On the other hand, the fatigue limits of carbon and certain alloy steels can be raised up to 25% by subjecting them to prolonged cycling, beginning from well below σ_F and then increasing the amplitude periodically in small steps. This effect is known as *coaxing* or *understressing*, and it is attributed to slow strain-aging during the test and the attendant expansion of the elastic range. Related to it is the increase of the fatigue limit of carbon steel on heating above room temperature. A maximum in σ_F is reached between 400° and 600°F, since in this range aging is instantaneous. The fatigue limit begins to fall systematically above this range in view of the thermal softening which then becomes the overriding factor.

There are indications that the "knee" on the *S–N* curve *a*, in Fig. 18.5, and even the existence of a true fatigue limit may be due to solute nitrogen and carbon atoms which promote strain-aging effects in steels and in certain other metals (e.g., titanium).

(d) *Grain Fragmentation.* The internal stresses brought about by accumulation of slip packages result in a fragmentation of grains into many subgrains separated by low-angle boundaries (Figs. 7.27, 28). Early evidence to that effect was indirect, through X-ray photographs taken at the same spot prior to and after cyclic stressing. In annealed metals with a low-yield stress, such as copper or cartridge brass, the initially sharp reflection spots became blurred, this indicating that each grain was subdivided into disoriented parts. The effect was similar to that caused by static, unidirectional cold working although not as strong. More recently, direct evidence of grain fragmentation was obtained with the aid of electron microscopy.

(e) *Extrusions and Intrusions.* An example of this unique phenomenon discovered in 1954 by Forsyth is seen in Fig. 18.20. These extruded

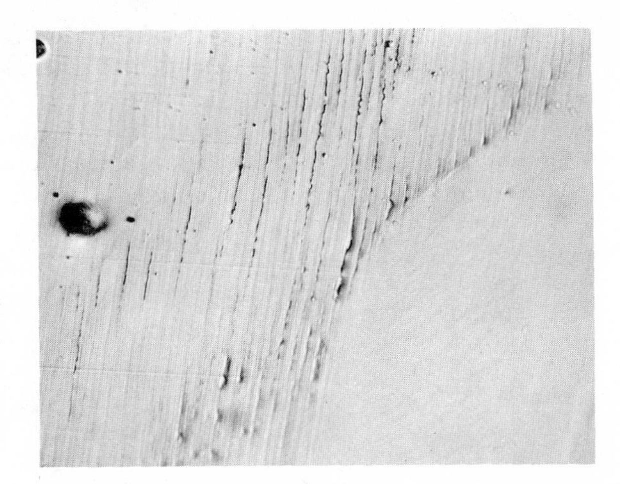

Fig. 18.20 Slip band extrusions on surface of fatigued Al-10% Mg alloy. From A. J. McEvily, R. L. Snyder, and J. B. Clark, *Trans. Met. Soc. AIME*, **227**, 452 (1963).

"tongues" occur invariably in areas thickly covered with slip bands, and they leave behind crevices in which a fatigue crack occasionally starts. Since such *slip band extrusion* was also found on statically cold-worked and subsequently fatigued metals, the implication is that it may have been caused by local softening, the soft material being squeezed out.

Cottrell and Mott suggested certain dislocation models to account for both extrusion and the inverse effect, *intrusion*. Nevertheless, the origin and significance of these effects are still an enigma, especially since a fatigue crack starts sometimes within the extrusion area and sometimes outside it.

EFFECT OF MEAN STRESS ON FATIGUE LIMIT (LIFE)

In numerous applications a cyclic stress of constant or variable amplitude is superimposed upon a static stress. For instance, oil well pumps may be several hundred feet below the surface, and the uppermost end of the piston rod must carry not only the pulsating pressure but also its own, quite respectable weight. Also, aeroplane wings in flight are exposed to all sorts of sudden stresses and vibration in addition to those caused by the weight of the plane.

A basic relation between σ_a and $2S(=\sigma_{\max}-\sigma_{\min})$ can be determined by carrying out complete series of fatigue tests at various fixed σ_a values. It is evident that a specimen will sustain only a single application of a mean stress equal to the true stress σ_u associated with the ultimate tensile strength S_u. The appropriate cyclic range will thus be zero. When the mean stress is gradually reduced, the specimen can sustain a superimposed cyclic stress as long as $\sigma_a + S < \sigma_u$. Experience shows, however, that its life N_e is very short when this inequity is barely satisfied. To extend the life, the cyclic amplitude S must be reduced gradually until a safe condition, i.e., the fatigue limit σ_F, is reached.

By applying a succession of decreasing mean stresses, one can determine for each σ_a a cyclic amplitude that would secure a specified life N_e. In this manner, a series of σ_{max} and σ_{min} values can be established for endurances of 10^4, 10^5, 10^6, etc. cycles. By joining all the points for a given N_e, the diagram shown in Fig. 18.21 is obtained and the cyclic stress S, as well as σ_{max} and σ_{min} for any value of mean stress σ_a, can be read off it directly.

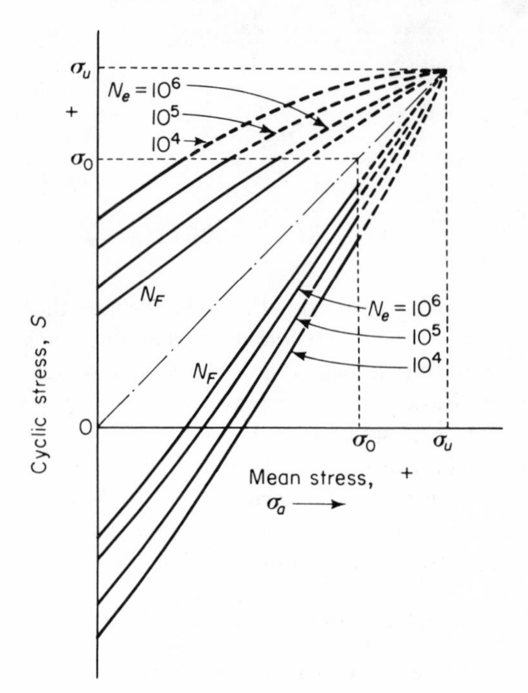

Fig. 18.21 Goodman-type diagram for finite (N_e) and infinite (N_f) fatigue lives (schematic).

The portion of the diagram where σ_{max} exceeds the yield strength σ_o is of little practical significance, for it entails a permanent deformation, which cannot normally be tolerated in a machine or structure. This is why the upper right-hand portion of the diagram is dashed, and σ_o is the upper limit for $(\sigma_a + S)$. The part of the diagram to the left of the $\sigma_a = 0$ axis is rarely of interest because compressive stresses do not promote fatigue, whereas the maximum stress may be limited by the buckling load when the member in question is relatively long and slender.

Plots of this general type were first suggested by Goodman and, although subsequently modified, they are still referred to as *Goodman diagrams*. A number of them are reproduced in various textbooks and manuals.*

* E.g., C. Lipson and R. C. Juvinall, *Handbook of Stress and Strength*, New York: The Macmillan Company, 1963.

Physically, the decrease of safe amplitude S when σ_a increases is due to the inherent ability of shear and tensile stresses to enlarge voids and propagate nascent cracks. This effect may operate even when $S + \sigma_a < \sigma_o$.

RESIDUAL STRESSES AND FATIGUE

In internally sound materials, fatigue cracks always start on a highly stressed, free surface. The ease with which they nucleate and propagate at a given cyclic stress is determined by the inherent hardness (strength) of the material and by the sign and level of the residual stresses in the surface layer. From the considerations of Chapter 17, it is apparent that residual compressive stresses should increase, while tensile stresses should decrease the fatigue limit. This conclusion is corroborated by extensive factual evidence, such as the data in Fig. 18.22.

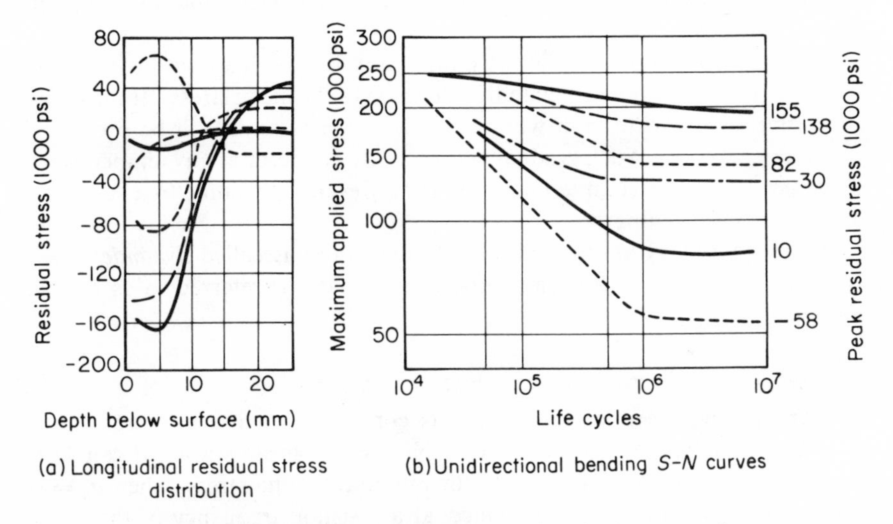

Fig. 18.22 Residual stress and fatigue data for heat-treated and peened 1045 steel. From R. Mattson and J. G. Roberts in (R18.10).

Whereas work-hardening increases the fatigue limit by raising the yield stress, its overall effect greatly depends upon the residual surface stresses it introduces. The influence of various processing methods upon these stresses was described in the preceding chapter and the conclusions reached then are directly applicable to fatigue, inasmuch as the effect of a residual stress on fatigue is essentially the same as of a superimposed mean stress. However, the beneficial effects of shot peening or surface rolling on the fatigue limit should not be hastily extrapolated into the range of high stresses, well above

σ_F; the perceptible plastic strains induced in these conditions cause the residual stresses to gradually fade away.

Various topics pertaining to this particular subject are discussed in detail in (R18.1).

COMBINED FATIGUE AND CREEP

Although in most high-temperature applications creep properties are the major limiting factor in design, fatigue may also be important. For instance, turbine blades are usually exposed to vibrations in addition to the centrifugal and working forces. These vibrations are principally due to mechanical resonance at certain speeds. Transient vibrations coupled with high-fatigue stresses can also arise in jet and rocket engines, especially during unsteady speed periods, e.g., take-off. Pipe lines operating at high temperatures may be similarly affected by vibrations transmitted from machinery, such as pumps or compressors.

If the temperature is sufficiently high, any fatigue load with a nonzero mean stress will promote creep in the direction of the mean stress. If the latter is positive, a slow extension will occur while compression will be produced under a negative σ_a. Such gradual deformation will occur at any temperature. However, below a certain temperature it may eventually come to a standstill due to work-hardening.

Creep caused by an asymmetrical fatigue stress is called *dynamic creep*, and the failure by which such a process terminates is referred to as *fatigue rupture*.

Presentation of Fatigue Rupture Data

For plotting creep-fatigue data it is convenient to use the stress ratio $A = S/\sigma_a$ i.e., the quotient of cyclic amplitude and mean stress. A can vary from zero for pure creep to infinity for pure alternating stress, when $\sigma_a = 0$ and creep is essentially absent. Since, at a constant frequency of the alternating component, the S–N diagram can also be regarded as a stress-time (S–t) plot, it can be used to present combined fatigue-creep rupture data.

It is customary to lay the maximum stress on the ordinate so that separate A curves must be given for various S values (Fig. 18.23a), t_f being the time to failure. Such set of curves is valid for a single temperature only. An alternate useful representation of combined creep-fatigue data is in the form of stress-range diagrams (Fig. 18.23b). The mean (creep) stress is the abscissa and the alternating, fatigue component, the ordinate. Straight lines emanating from the origin represent different stress ratios, A_1, A_2, A_3, etc., A_n being the tangent of the angle between the appropriate radius and the mean stress axis.

A stress-range diagram can be reproduced directly from a family of appropriate σ–t curves by transferring the intersection points of each "t"

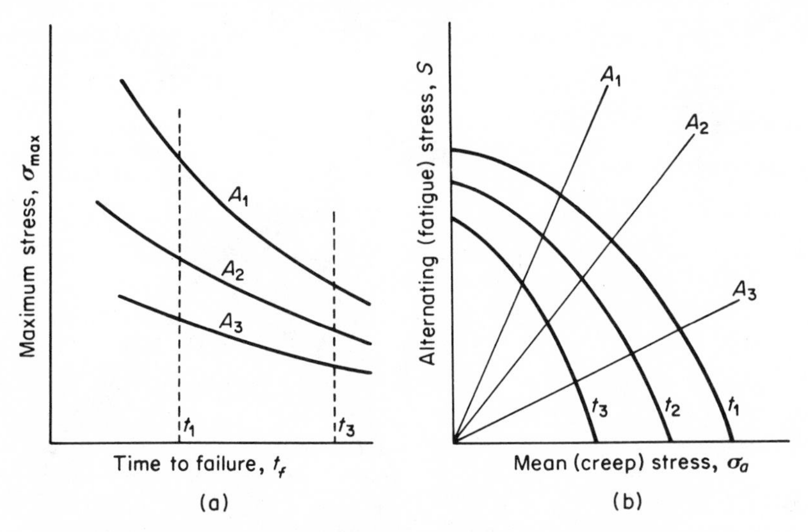

Fig. 18.23 Diagrammatic representation of fatigue rupture data. A—ratio of cyclic amplitude S to mean stress σ_a.

vertical with successive A curves in Fig. 18.23a onto the appropriate A radii in diagram b, followed by joining all the points for each t value to form a continuous curve. With the aid of these curves one can interpolate the allowable fatigue or creep values for any specified σ_a or vice versa.

The effect of temperature can be shown on this same diagram by plotting the individual curves for constant temperatures T_1, T_2, T_3, etc. and a constant rupture time t_f or, conversely, by plotting a set of constant t_f curves for each temperature. For $\sigma_a \to 0$, each such curve tends to become tangent to the ordinate axis, the point of tangency being the fatigue limit.

For design purposes, the time to rupture is usually less important than that required to produce a specific permanent strain, say, 0.5 or 1% extension. The appropriate t-curves will be similar to those in Fig. 18.23b except that they will approach the $\sigma_a = 0$ axis asymptotically.

A set of experimental curves for a 403 type, chromium stainless steel is reproduced in Fig. 18.24. The plot is similar to the one in Fig. 18.23b but it reveals an interesting feature in that, within a limited range, a somewhat higher mean stress σ_a can be used in conjunction with an alternating stress than without it. This is believed to be caused by precipitation-hardening which is, at first, intensified by the cyclic stress. Clustering and coarsening of the precipitates at still higher stresses results in overaging and a consequent loss of strength.

The frequency of the fatigue stress has little effect on fatigue life at low temperatures. At high temperatures, on the other hand, it is very prominent.

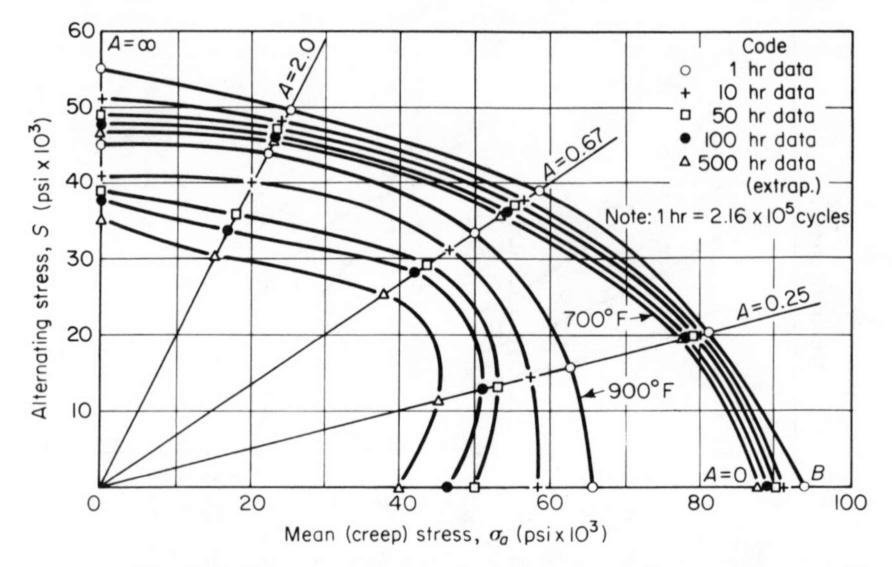

Fig. 18.24 Stress-range diagrams for S-816 alloy for 100 h life or 2.16×10^7 cycles. From F. H. Vitovec and B. J. Lazan (in) *ASTM Spec. Tech. Pub.*, No. 196.

At the lowest frequencies the effect of creep (including cyclic creep) is predominant. At high frequencies, fatigue damage is the major factor limiting the life of the parts.

The fracture appearance under creep-fatigue conditions tends to vary with *A*, i.e., to be intercrystalline for low, and intracrystalline for high *A* values.*

FATIGUE CRACK PROPAGATION

As has already been stated, the exact process of fatigue crack formation in a sound material is not yet fully clarified. However, from the practical point of view, the details of the submicroscopic mechanism that cause a crack nucleus to form "from nothing" is generally of little importance; most structural materials contain minor voids, inclusions, and internal discontinuities, or even surface defects (scratches, nicks) induced during manufacture, handling, and subsequent fabrication into assemblies and complete structures.

The propagation of an existing crack by fatigue is basically similar to the movement of a sharp notch (stress raiser) during static fracture. The large stresses developed at the crack tip cause it to propagate into previously undamaged material. The stress then increases, pushing the fracture farther, and so forth. In view of the repetitive action of a cyclic stress, a fatigue

* A. H. Meleka, *Metallurgical Reviews*, **7**, 43 (1962).

fracture exhibits the characteristic on-and-off appearance, resulting from the alternate opening and closing half-cycles, as illustrated in Fig. 18.25. This gives rise to the characteristic beach markings not only in metals (Figs. 18.1–3) but also in plastics (Figs. 18.26).

When the crack length reaches a critical value, l_c, a final, catastrophic failure takes place. According to Christensen,* the following semi-empirical formula agrees well with fatigue crack propagation tests on sheet metal:

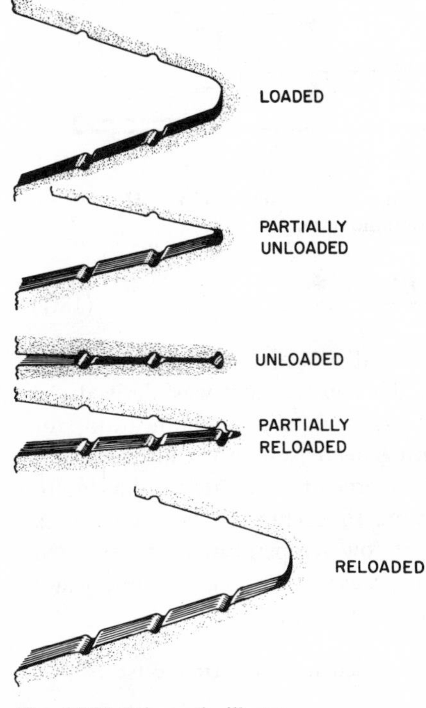

Fig. 18.25 Schematic illustration of deformation at crack tip during one loading cycle. From A. J. McEvily, Jr., R. C. Boettner, and T. L. Johnston, "On the Formation and Growth of Fatigue Cracks in Polymers", (in) *Fatigue—an Interdisciplinary Approach*, John J. Burke, Norman L. Reed, Volker Weiss, eds., Syracuse University Press, 1964.

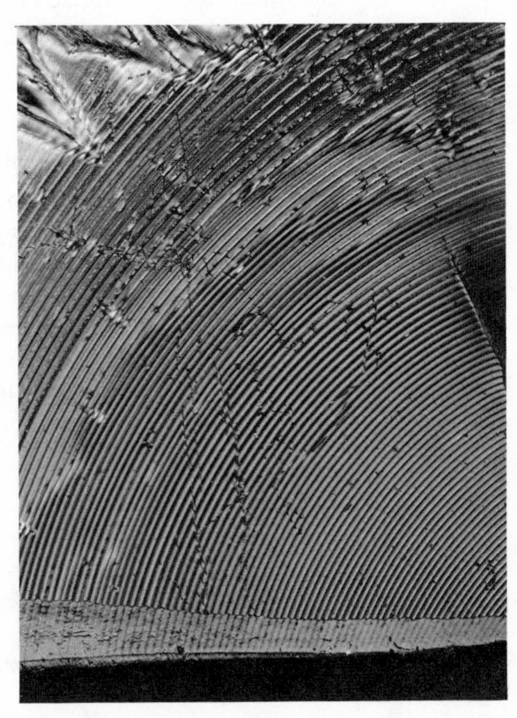

Fig. 18.26 Striations produced by high-strain fatigue on the fracture surface of a polyethylene rod (same reference as Fig. 18.25).

* R. H. Christensen, Douglas Paper No. 3000, Missile and Space Div., Santa Monica, Calif.: Douglas Aircraft Co., 1964.

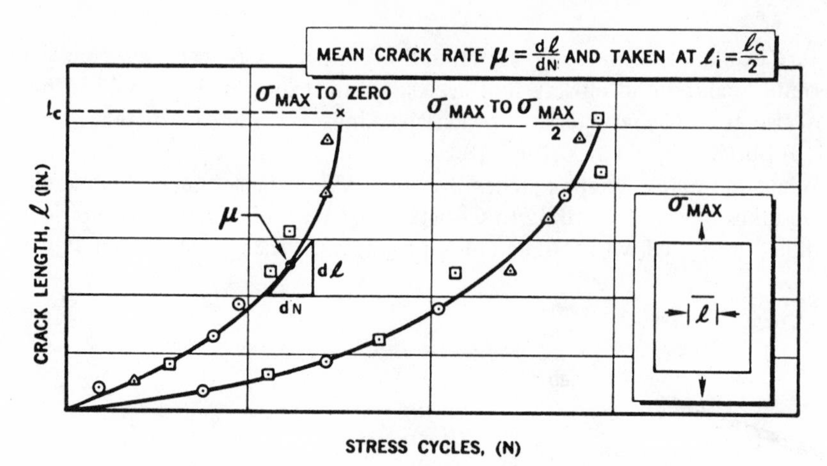

Fig. 18.27 Growth of crack as a function of stress ratio. From R. H. Christensen, from reference cited in footnote on p. 509.

$$\frac{dl}{dN} = \mu \frac{l/l_c}{1 - l/l_c} \tag{18.1}$$

where l is the instantaneous crack length and μ is a constant equal to the crack rate when $l = l_c/2$ (Fig. 18.27). This expression was derived for uniaxial loading but it is believed to hold, with slight modification, for biaxial loading as well. The critical length l_c decreases with falling temperature, but the number of cycles required for a crack to nucleate (or develop to some minimum size) increases at the same time (Fig. 18.28). The crack propagation rate, dl/dN, is also higher at low temperatures. At elevated temperatures, on the other hand, creep cracking becomes an important factor, and it intensifies the damage attributable to fatigue.

Fig. 18.28 Growth of fatigue cracks as a function of temperature (same reference as Fig. 18.27).

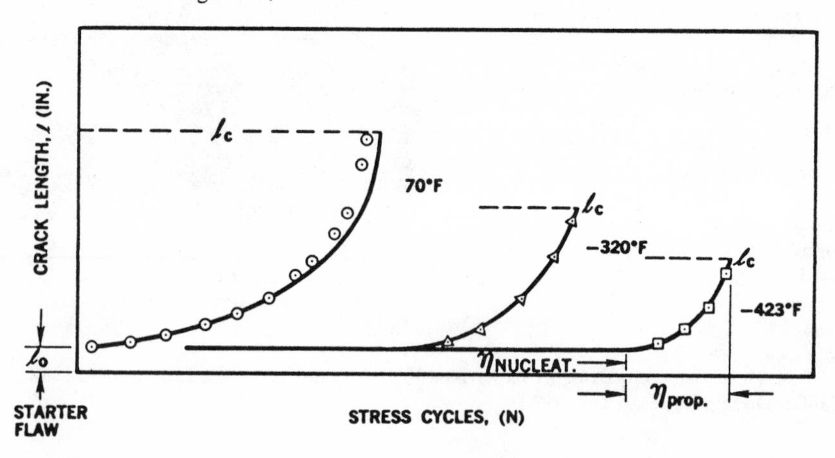

These and similar considerations play an important part in the design of supersonic aircraft and indicate the choice of certain structural materials in preference to others (e.g., titanium vs. aluminum alloys for skins).

An extensive discussion of various aspects of fatigue crack propagation, especially in connection with aircraft design, is presented in Harris' book (R18.7).

OVERSTRESSING AND CUMULATIVE DAMAGE

If a part is exposed for a certain number of cycles to a stress in excess of the fatigue limit, it suffers some permanent damage. Such *overstressing* reduces the number of cycles that this part could withstand at some lower stress. Also, the original fatigue limit may be impaired.

Equipment exposed to heavy intermittent loads is often designed for finite fatigue life, earth-moving machinery and aircraft being typical in this respect. For design purposes, one must start by assuming a typical loading sequence and establish the number of cycles excepted at each stress level during the useful life of, say, an airplane. Guiding data of the required type are often available from actual measurements on prototypes or previously built planes (Fig. 18.29).

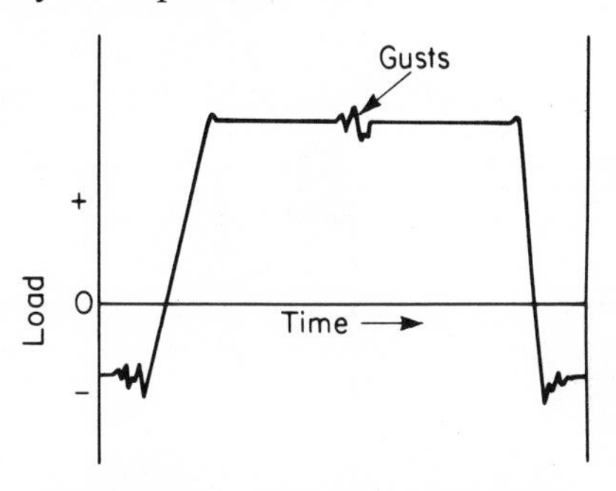

Fig. 18.29 Load on a wing of a transport plane during taxiing, take-off, flight, and landing.

To put the engineering work on a quantitative basis, the law relating the degree of damage D to the number of cycles n at a given stress must be known or reasonably assumed. Although the damage is probably slow at first but becomes rapid toward the end of the expected fatigue life, a linear relation between D and N is often assumed (*Miner's rule*). According to it, if the fatigue lives at amplitudes $S_1, S_2, \ldots S_m$ are $N_1, N_2, \ldots N_m$, respectively, a fatigue failure will occur when

$$\frac{n_1}{N_1} + \frac{n_2}{N_2} + \cdots + \frac{n_m}{N_m} = \sum \frac{n_m}{N_m} = 1 \qquad (18.2)$$

Miner's rule is generally inaccurate, but it is a simple and useful means of arriving at a sensible preliminary figure.

FATIGUE PROPERTIES OF NOTCHED ELEMENTS. THE SIZE EFFECT

Fatigue cracks are produced as a result of repeated small plastic strains, and the higher the amplitude S, the faster these cracks grow and propagate.

From the theory of elastic stress concentrations (Chapter 15) it is known that notches of increasing acuity gradually increase the stress level at the root or, more exactly, the ratio of the maximum axial to the nominal stress. (In this context, the nominal stress is computed on the reduced area neglecting the stress concentration.) In a ductile metal under stress, the yield stress is reached rather quickly along the notch-root perimeter and the ensuing plastic flow limits the true axial stress to a value significantly lower than that expected under elastic conditions. The ability of the material to flow acts thus as a sort of "safety valve." However, the narrow, annular layer at the notch bottom, once yielded, will exhibit a Bauschinger effect, and some plastic deformation will take place at quite low stresses on each cyclic reversal. This situation is clearly conducive to fatigue, and the higher the stress-concentration factor, the lower will be the fatigue limit that is conventionally based on the "nominal" stress value. In Fig. 18.30, the plain and notched fatigue properties are compared.

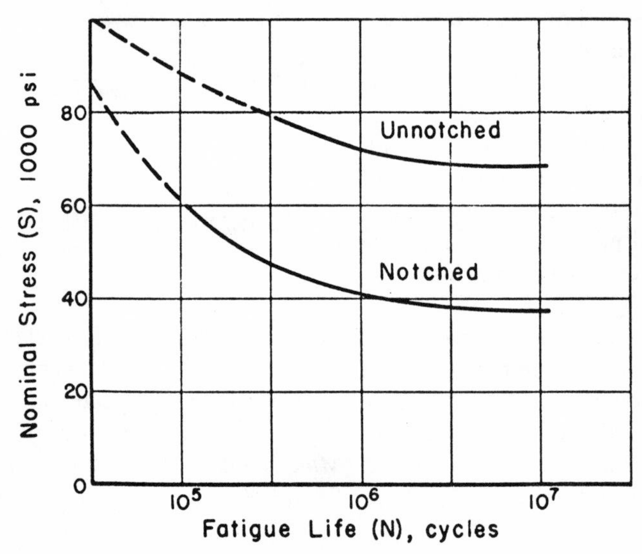

Fig. 18.30 Effect of semicircular groove ($K_t = 2$) on the rotating-bend fatigue strength of a round bar of SAE 4130 steel (R18.6).

To create a reference frame for comparison and discussion, the term *fatigue notch sensitivity*, q, defined as the ratio

$$q = \frac{K_f - 1}{K_t - 1} \tag{18.3}$$

is introduced. In it, $K_t = \sigma_N/\sigma_1$, where σ_N is the maximum stress at the notch root calculated from elastic theory, while σ_1 is the nominal stress in the same location but in the absence of a notch. K_f is the experimentally determined ratio of the unnotched fatigue limit σ_F and the true fatigue limit after notching, σ_{F_n}.

If the notch had no effect at all on σ_F, then $K_f = 1$ and $q = 0$. If, on the other hand, the notch sensitivity were determined by the theoretical stress-concentration factor alone, then $K_f = K_t$ and $q = 1$. The observed q values are between these two extremes. For instance, K_t was 3.5 in a 0.333 in. dia. bar equipped with a 45° circumferential notch, 0.036 in. deep and with a 0.010 in. tip radius. The experimental K_f values were 1.75 for low carbon steel and 2.8 for an aluminum alloy.*

The physical meaningfulness of K_t and q under fatigue conditions is questionable in view of the plastic deformation at the notch bottom. An added complication consists in the *fatigue notch-size effect*, displayed in Fig. 18.31. These diagrams refer, schematically, to a sheet of width b in which a center hole of radius r was drilled. As expected, the notch effect becomes weaker with increasing r. But while the elastic stress-concentration factor K_t is extremely high at very small notch radii, K_f passes through a maximum at some intermediate r_c and again approaches unity for $r \to 0$. One might then conclude that very sharp but shallow cracks may do little damage from the fatigue viewpoint.

Not only the notch geometry, but also the size of the cross section affects fatigue strength, even in the absence of any notches. Because of this *size effect*, fatigue data obtained on small, laboratory type specimens, are usually higher than those recorded on massive shafts. The actual difference varies, but may reach 30% over a diameter range from 0.5 to about 10 ins. The size effect is very pronounced in reversed bending. In reversed direct stressing, it is sometimes negligible and sometimes substantial. The latter is particularly true of aluminum alloy sheets, with and without notches (drilled holes). Large notches in large specimens depress the fatigue strength more than

Fig. 18.31 Schematic illustration of the fatigue notch-size effect.

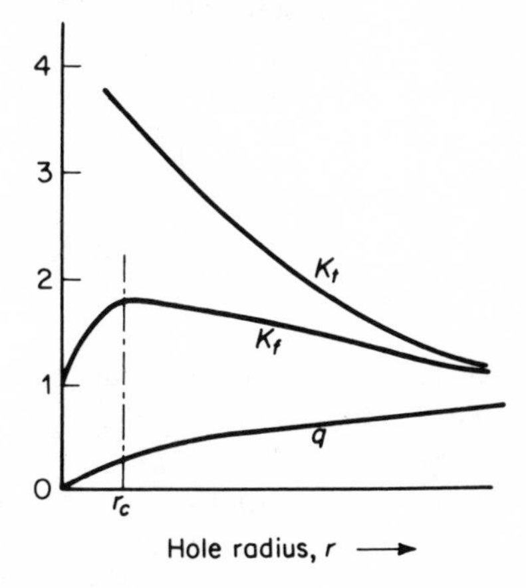

* P. G. Forrest, (in) *Proc. Interntl. Conf. on Fatigue*, London: Inst. Mech. Engrs., 1956, p. 171.

geometrically similar notches in small specimens.*

No convincing explanation of the size effects is yet available but the various concepts advanced so far are reviewed in detail in (R18.6–8).

THERMAL FATIGUE

The origin of thermal stresses is explained in Chapter 5, p. 113 and it is shown that, within the elastic range, their magnitude is determined by the temperature difference, according to Eq. (5.23). *Thermal shocks*, which accompany sudden temperature changes, such as occur in the skin of a space vehicle during re-entry, or in furnace refractory linings when the fuel is ignited, result in particularly severe stresses.

In the aforementioned situations, the thermal stresses were due to the constraint exerted by certain portions of a body on their neighbors. However, even if heating is uniform across the section, like in electric resistance heating of long bars or tubing, thermal stresses will arise when the heated element is rigidly confined so that its free expansion or contraction is prevented.

Repeated inducement of thermal stresses leads eventually to *thermal fatigue*, which manifests itself in multiple checking or "craze cracking" of the surface in massive bodies. Forging dies, metallic casting molds, and pipelines conveying alternately hot and cold liquids are afflicted with these defects. The cracks in exhaust valve stems (Fig. 18.32) are due to a combination of thermal and mechanical fatigue.

In a brittle material, a thermal shock causing surface tensile stresses may produce an instant failure. In ductile materials, capable of plastic flow, the total strain, ε_T, may consist of elastic and plastic components, ε_e and ε_p, respectively. Thus

$$k \cdot \Delta T = \varepsilon_T = \varepsilon_e + \varepsilon_p = \frac{\sigma}{E} + \varepsilon_p$$

$$\varepsilon_p = k \cdot \Delta T - \frac{\sigma}{E} \tag{18.4}$$

where ΔT is the temperature difference, k–coefficient of thermal expansion, and E–Young's modulus.

Relation (18.4) is valid only when the holding time at the higher temperature is short so that stress relaxation by creep does not occur. Even with this reservation, however, calculation of stresses and strains during repeated heating and cooling (*thermal cycling*) is difficult and somewhat uncertain in view of Bauschinger effects that are likely to develop.

---

* H. Grover, (in) *Fatigue*, Syracuse University Press, 1964.

Fig. 18.32 Cracks in an exhaust valve stem due principally to thermal fatigue.

A detailed analysis of thermal cycling was developed by Manson* while a comprehensive review of the subject of thermal fatigue was published by Glenny.**

MISCELLANEOUS TYPES OF FATIGUE DAMAGE

The several fatigue fractures in Figs. 18.1 and 18.2 were caused by tensile and shear stresses. The failure occurred in each case essentially along one continuous surface, and the material outside the fracture area probably remained sound.

The picture can be changed radically when a cyclically stressed part is exposed to chemical attack. Water, especially salty, various acids, and even certain corrosive gases (e.g., SO_2) contained in industrial atmospheres, result in *corrosion fatigue*. The latter leads to failure faster than purely "mechanical" fatigue, the difference being spectacular in many cases.

Corrosion fatigue is usually accompanied by a formation and growth of numerous cracks. Because of this, a steel that normally exhibits a true fatigue limit ceases to do so in corrosive surroundings, and its endurance

* S. S. Manson, *Machine Design*, 1958–61 (serialized articles).

** E. Glenny, *Metallurgical Reviews*, **6**, 387 (1961).

at any stress level is also greatly reduced. The same is true of other metals, but while the volume of factual information concerning this type of fatigue is tremendous, no generally valid rules can be formulated. To slow it down, corrosion resistant coatings or special alloys must be used despite the extra cost of such protective measures. It is apparent then, that corrosion fatigue must be related to stress corrosion (Chapter 17).

Pitting fatigue of gears and ball bearings (Fig. 18.33) although related to the high local pressures, is probably a form of corrosion fatigue. *Spalling*, on the other hand, appears to be of purely mechanical origin, and the fracture surfaces are substantially parallel to the affected surface (Fig. 18.34).

Fig. 18.33 Manifestations of pitting fatigue on roller bearing and gear.

Fretting is a poorly understood form of fatigue caused by minute oscillating motions between materials in rubbing contact.* It was first observed in press-fitted joints, but it can be expected to arise even under light pressures between parts of equipment designed for operation in high vacuum, e.g., in a lunar surface vehicle. The small cyclic movements under pressure result in extensive and very characteristic surface damage (Fig. 18.35) through formation of powdery debris (often composed of oxides) with cracks developing at a later stage.

* P. L. Teed, *Metallurgical Reviews*, **5**, 267 (1960).

Fig. 18.34 Spalling of gear teeth caused by fatigue.

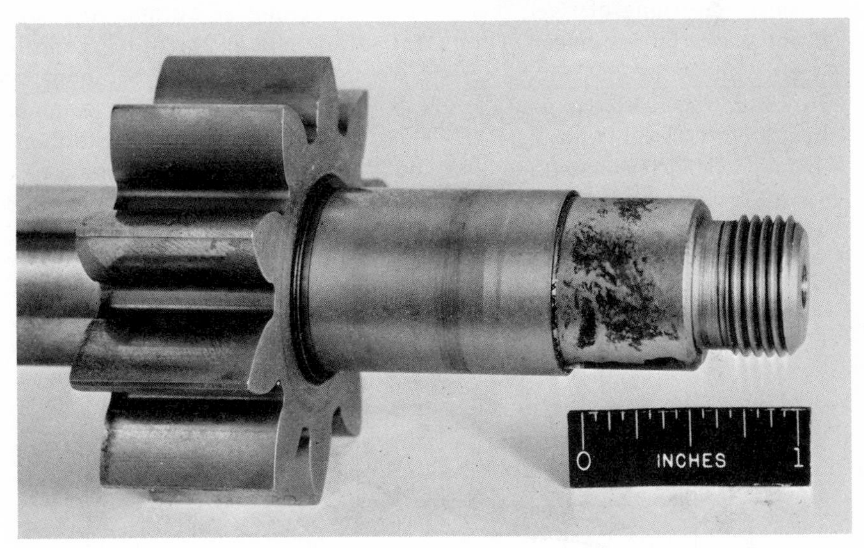

Fig. 18.35 Surface damage by fretting.

The high noise level of jet aircraft brought about the problem of *acoustical fatigue*, and many failures traceable to this factor occurred in American and British jet bombers.* The main sources of the acoustical energy are the jet stream of the engine, the turbulent boundary layer, and the wake of the flying craft. They all may generate resonant vibrations in the fuselage, and measurements made on planes indicate that the takeoff period is the most

* W. J. Trapp and D. M. Forney, Jr., in *Fatigue*, see p. 514.

critical. The problem is expected to become more acute in supersonic vehicles and it will probably result in an increasing use of high-damping, composite structural materials consisting of alternate layers of metals and viscoelastic polymers or of cellular, honeycomb-type cores sandwiched between two face plates, with a layer of adhesive between them.

RELATION BETWEEN STATIC AND FATIGUE PROPERTIES

As soon as fatigue was recognized as a major cause of failure of structures and equipment, a search started after an easy-to-determine and yet reliable relation between fatigue strength and yield, tensile strength, and hardness. Unfortunately, such relations as were found, turned out to be very approximate only, and their usefulness is limited to rough estimates at best.

In any metal or alloy, the fatigue strength increases with increasing yield or tensile strength. They all increase as the grain becomes finer or when the temperature is lowered. All can be improved by suitable heat-treatment (quench hardening of steel, precipitation hardening of Cu and Al alloys). From a very large number of tests, the mean value of the $\sigma_F : S_u$ ratio for steels with up to 200,000 psi tensile strength was found to be about 0.5. However, the scatter is considerable and $\pm 30\%$ deviations are common. In still harder steels with $S_u = 220{,}000$ to 300,000 psi, the above ratio tends to fall with increasing strength to about 0.3–0.25. The wide variations of $\sigma_F : S_u$ in copper, nickel, and aluminum alloys is illustrated in Fig. 18.36 which is based on numerous tests.

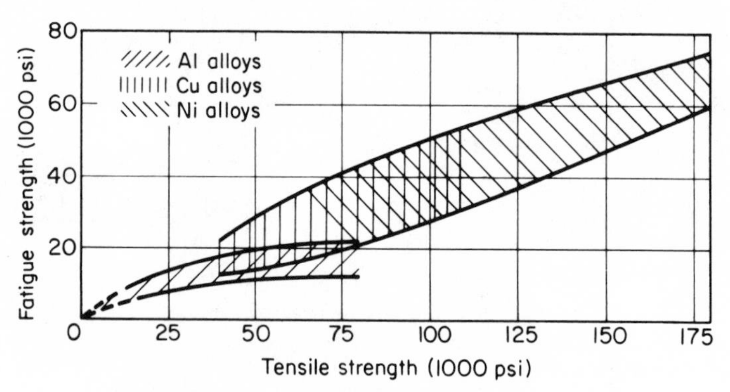

Fig. 18.36 Fatigue strengths of aluminum-base alloys ($N_F = 5 \times 10^8$) and copper and nickel alloys ($N_F = 10^8$).

FATIGUE UNDER COMBINED (MULTIAXIAL) STRESSING

In view of the innumerable combinations of stresses that can act on a specimen in various directions, fatigue tests under complex stress systems are costly and time-consuming. The most important combinations encountered in engineering practice are bending and torsion stresses, biaxial tension, or a combination of direct and shear stresses.

Such tests as were performed on steels and aluminum alloys indicated that the use of the significant stress, σ_e, together with the simple, uniaxial fatigue data leads to conservative and, therefore, relatively safe design. There is no fundamental justification for this, and the criterion $\sigma_e < \sigma_F$ must be regarded as empirical.

One must realize that most structural materials are anisotropic and that the fatigue limits, or fatigue lives for any given stress amplitude may differ significantly in various directions. The difficulty of establishing a reliable criterion for multiaxial strength thus becomes quite obvious.

THE STATISTICAL NATURE OF FATIGUE DAMAGE

The incidence of a fatigue failure in any particular specimen (element, assembly) subjected to cyclic stress is governed by highly localized properties and/or changes in the material. Whereas a composition of matter can be designed so as to provide certain average properties, their point-to-point variation is beyond control. As a result, fatigue properties, with their sensitivity to localized weaknesses, tend to inherently develop considerable scatter. For example, if one performs 20 or 50 routine fatigue tests on nominally identical specimens at a constant stress level, the number of cycles sustained by individual specimens before failing may vary over a range as wide as 20:1 or even 100:1. When all the data are plotted on a single S–N diagram for various S values, they form a wide *scatter band* (Fig. 18.37).

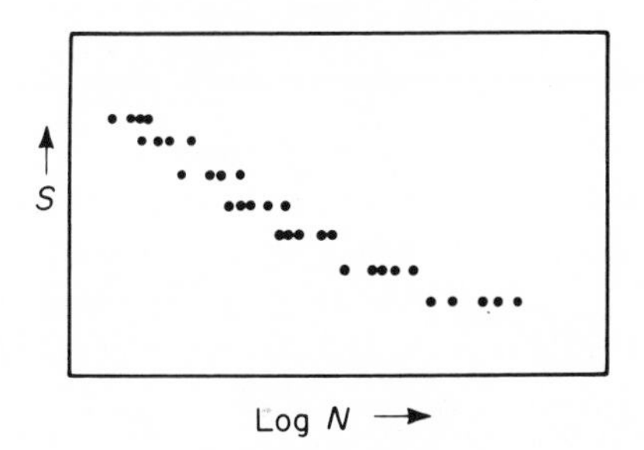

Fig. 18.37 Typical scatter of fatigue data. Dots denote failures.

In these conditions it is more meaningful to speak of the *probability of failure* at any given stress level (or *probability of survival* for a given number of cycles) than to specify an absolute, single-valued relation between stress amplitude and fatigue life.

The principle of the method of statistical analysis can be appreciated by reference to Figs. 18.38–18.40, cited from a book compiled for the U.S. Navy by Battelle Memorial Institute (R18.6).

A total of 53 welded specimens was *step-tested*, meaning that each specimen was run for at least 10^7 cycles at 30,000 psi, whereupon the stress was raised in 5,000 psi steps and the test was repeated at each step. The number of

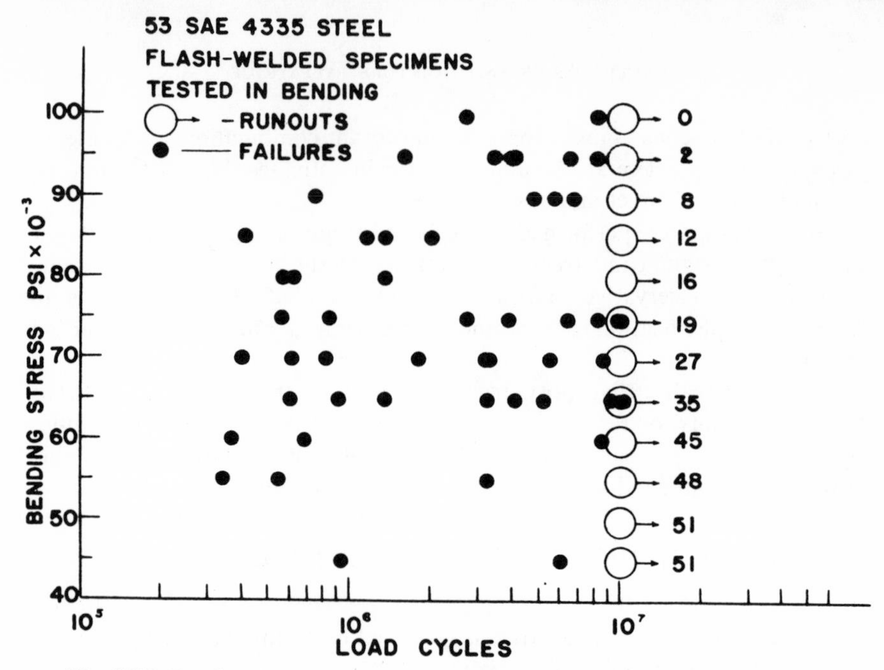

Fig. 18.38 Results of reversed bending tests of flash-welded specimens of SAE 4335 steel. After Henry, in (R18.6).

cycles that a specimen survived at the highest stress was recorded, and the entire set of results is displayed in the seemingly confusing Fig. 18.38.

The trend becomes much clearer, however, when the data are presented in a histogram (Fig. 18.39), the Gaussian, or normal distribution being indicated by the dashed curve. After calculating the cumulative percentage of survivors

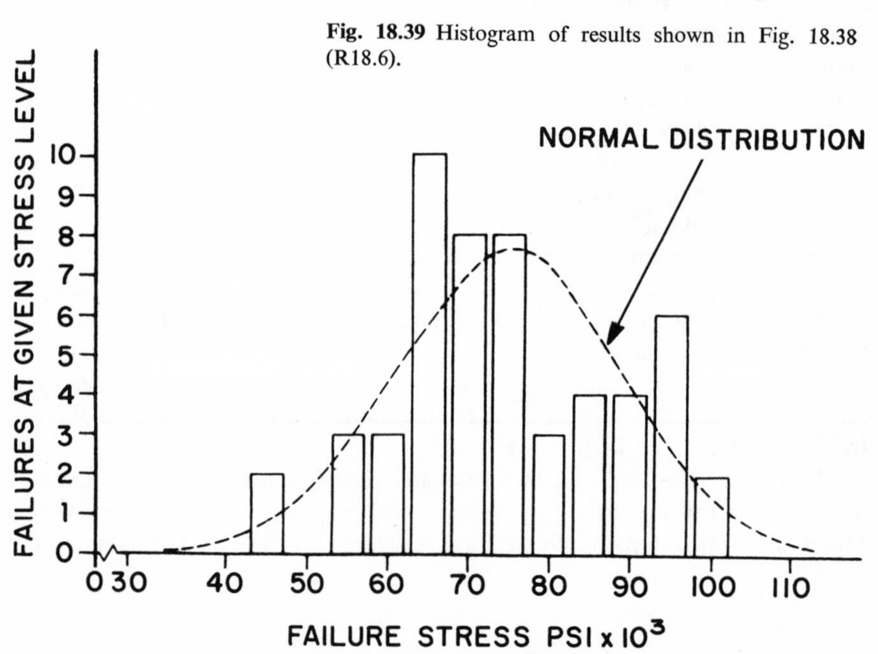

Fig. 18.39 Histogram of results shown in Fig. 18.38 (R18.6).

at each stress level from the histogram and plotting them on probability paper, an orderly picture results (Fig. 18.40). The straight line is drawn to best fit the points, assuming a normal distribution of mortalities (i.e., failures). By definition, the "*mean fatigue limit*", $\bar{\sigma}_F$ is taken as the stress for 50% failures.

Using standard methods of statistical analysis, it is possible to establish distribution curves, and hence probabilities of survival for any given N as the cyclic stress varies. One can then draw continuous *S–N* curves for any selected probability of survival (Fig. 18.41) and select by interpolation the proper design values for a particular application.

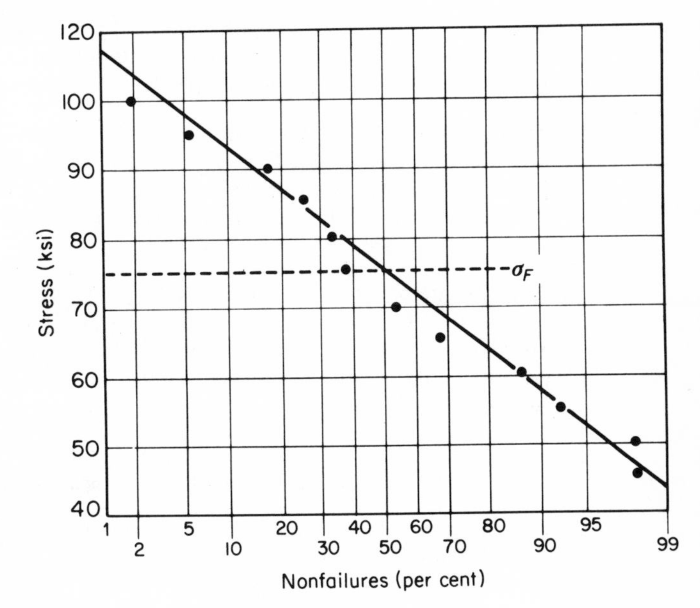

Fig. 18.40 Probability plot of data in Fig. 18.39 (R18.6).

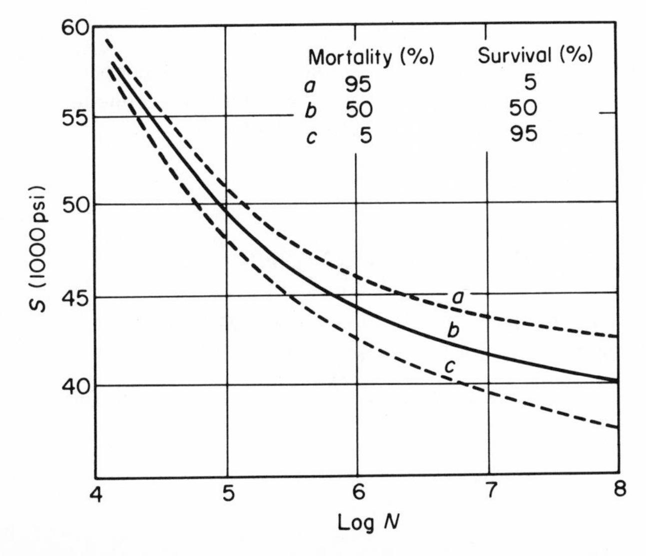

Fig. 18.41 *S–N* curves based on probability of survival (or failure) at particular stress levels (R18.6).

PROBLEMS

1. It has been observed that when a cold drawn low-carbon steel rod is cyclically stressed ($\pm S$) in the presence of a tensile mean stress σ_a such that $S + \sigma_a$ is slightly less than σ_o, the rod begins to elongate but only after several hundred (or thousands) cycles were applied. Explain the cause of this phenomenon.

2. Sketch several successive stress-strain cycles (loops) leading from condition *F*, *R* to *S* in Fig. 18.17.

3. When a soft copper wire is put under a slight tension while plastically twisted to and fro, it begins to elongate, and the elongation increases with the amplitude of the torsional strain. Speculate on the possible reason for this effect.

4. The fatigue life of a certain element is 20,000 cycles at $\pm 60{,}000$ psi, and 1,000,000 cycles at $\pm 40{,}000$ psi, the $S - \log N$ relation being a straight line between these points. Using Miner's rule, calculate the expected endurance at $\pm 45{,}000$ psi if the element in question already received 20,000 cycles at $\pm 52{,}000$ and another 30,000 at $\pm 48{,}000$ psi.

5. Attempts are being made currently to facilitate the processing (drawing, rolling) of hard-to-deform metals by superposition of ultrasonic vibrations. Draw the approximate shape of the tensile curve, indicating the change experienced when the ultrasonic transducer is first coupled to, and then uncoupled from, the system.

6. Draw a Goodman diagram for a metal with $\sigma_F = \pm 28{,}000$, $\sigma_o = 37{,}000$, $S_u = 46{,}000$ psi and determine the safe mean stress at $\sigma_F = \pm 23{,}000$ psi.

7. Describe the various fatigue conditions existing in a gear-key-shaft assembly used in the traversing mechanism of an overhead crane. Consider each component as well as the entire group.

8. Compare the situation in (7) with that in a like assembly in the hoisting mechanism.

9. Cyclic stressing above the fatigue limit of a material such as soft copper is accompanied by some work-hardening. Will periodical annealing eliminate fatigue or, at least, extend the life of the specimen and why?

REFERENCES FOR FURTHER READING

(R18.1) Almen, J. O., and P. H. Black, *Residual Stresses and Fatigue in Metals.* New York: McGraw-Hill Book Company, 1963.

(R18.2) Cazaud, R., *Fatigue of Metals.* New York: Philosophical Lib. Inc., 1953.

(R18.3) Forrest, P. G., *Fatigue of Metals.* New York: Pergamon Press, Inc., 1962.

(R18.4) Freudenthal, A. M. (ed.), *Fatigue in Aircraft Structures.* New York: Academic Press, Inc., 1956.

(R18.5) Gough, H. J., *The Fatigue of Metals.* London: Scott, Greenwood & Sons, 1924.

(R18.6) Grover, H. J., S. A. Gordon, and R. L. Jackson, *Fatigue of Metals and Structures.* Battelle Mem. Inst., Bureau of Naval Weapons, U.S. Navy, 1960.

(R18.7) Harris, W. J., *Metallic Fatigue.* New York: Pergamon Press, Inc., 1961.

(R18.8) Kennedy, A. J., *Processes of Fatigue and Creep in Metals.* Edinburgh: Oliver and Boyd, 1962.

(R18.9) Moore, H. E., and J. B. Kommers, *The Fatigue of Metals.* New York: McGraw-Hill Book Company, 1927.

(R18.10) Rassweiler, G. M., and W. L. Grube (ed.), *Internal Stresses and Fatigue in Metals.* Amsterdam: Elsevier Pub. Co., 1959.

INDEX OF NAMES

INDEX OF SUBJECTS

T